MW00447506

Research Methods for Public Administrators

Research Methods for Public Administrators contains a thorough overview of research methods and statistical applications for advanced undergraduate and graduate students, and practitioners. The material is based on established social science methods. Concepts and applications are discussed and illustrated with examples from actual research.

The book covers research design, methods of data collection, instructions on formulating research plans, measurement, sampling procedures, and statistical applications from basic statistics to more advance techniques. The basics of conducting experiments, survey research, case studies, and focus groups are discussed. Data organization, management, and analysis are also covered, as are data analysis and hypothesis testing. Descriptive and inferential statistics are discussed and illustrated with examples. The book also includes a chapter on obtaining and analyzing secondary data (data already collected for other purposes) and a chapter on reporting and presenting research results to a variety of audiences.

This is a general textbook written primarily for students of public administration and practitioners in public and not-for-profit organizations. It includes materials shown to be useful in gathering and assessing information for making decisions and implementing policies. The material is discussed at a level to be accessible and with enough detail to be useful.

New to the seventh edition:

- Additional and expanded material on qualitative research, big data, metadata, literature reviews, and causal inference
- New material on experiments and experimental research
- New examples and case studies, including those dealing with public policy
- Expanded material on using computers for data management
- Information on new NSF and NIH ethics and protection of human subjects requirements for researchers
- New datasets and PowerPoint slides for each chapter.

Gary Rassel is Emeritus Professor in the Department of Political Science and Public Administration at the University of North Carolina at Charlotte.

Suzanne Leland is Professor in the Department of Political Science and Public Administration at the University of North Carolina at Charlotte.

Zachary Mohr is Associate Professor in the Department of Political Science and Public Administration at the University of North Carolina at Charlotte.

Elizabethann O'Sullivan, Professor Emeritus in the School of Public and International Affairs at North Carolina State University, passed away in 2014.

Research Methods for Public Administrators

Seventh edition

Gary Rassel, Suzanne Leland, Zachary Mohr, and Elizabethann O'Sullivan

NEW YORK AND LONDON

Seventh edition published 2021
by Routledge
52 Vanderbilt Avenue, New York, NY 10017

and by Routledge
2 Park Square, Milton Park, Abingdon, Oxon, OX14 4RN

Routledge is an imprint of the Taylor & Francis Group, an informa business

First edition published by Longman Publishing Group 1989

Sixth edition published by Routledge 2016

Library of Congress Cataloging-in-Publication Data
Names: Rassel, Gary R. (Gary Raymond), 1944– author. | Leland, Suzanne M.,
    1971– author. | Mohr, Zachary, 1965– author. | O'Sullivan, Elizabethann, author.
Title: Research methods for public administrators / Gary Rassel, Suzanne Leland,
    Zachary Mohr, Elizabethann O'Sullivan.
Description: Seventh Edition. | New York : Routledge, 2021. | Revised edition of
    Research methods for public administrators, 2017. | Includes bibliographical
    references and index.
Identifiers: LCCN 2020032342 (print) | LCCN 2020032343 (ebook) |
    ISBN 9780367334345 (hardback) | ISBN 9780367334369 (paperback) |
    ISBN 9780429319860 (ebook)
Subjects: LCSH: Public administration—Research—Methodology.
Classification: LCC JF1338.A2 O78 2021 (print) | LCC JF1338.A2 (ebook) |
    DDC 351.072/1—dc23
LC record available at https://lccn.loc.gov/2020032342
LC ebook record available at https://lccn.loc.gov/2020032343

ISBN: 978-0-367-33434-5 (hbk)
ISBN: 978-0-367-33436-9 (pbk)
ISBN: 978-0-429-31986-0 (ebk)

Typeset in Times
by Apex CoVantage, LLC

Visit the eResources: www.routledge.com/9780367334369

Contents

15 Completing the Project and Communicating Findings

Figures

Preface to the Seventh Edition

The authors are pleased to present the seventh edition of *Research Methods for Public Administrators*. Designed for advanced undergraduate and introductory graduate courses in research methods and applied statistics, the intended audiences are those who are or soon will be administrators in public and nonprofit organizations and those who teach them. The material is not limited to this audience—students and faculty in other disciplines will find the methods covered here useful as well. We are pleased that teachers and students have found previous editions valuable. The dynamic nature of public administration as a field and of the methods and statistics used to study it continues to amaze us.

Many technical and substantive innovations in research methods and data applications have been developed since the publication of the previous edition of *Research Methods for Public Administrators*. New trends in public administration have affected practice in and study of the field. These include more emphasis on behavioral public administration and greater use of concepts from psychology and sociology. This has led to more use of the methods of psychology, as well as including experimental research. This emphasis is seen in the articles and journals of the field. Edition seven of the text addresses these methods and includes examples from this genre. We take these issues into account and continue to offer a practical introduction to the methodological tools that public administrators and policy analysts use to conduct research in our world.

As the authors completed work on the manuscript, students, faculty, and the world of higher education were facing a changed world, largely due to a global viral pandemic but for other reasons as well. The format for delivering instruction had to change along with it. The authors of this edition of the text were mindful of these changes as we worked.

New to This Edition

The seventh edition continues in the tradition of previous editions and emphasizes method, analysis, and application. We have revised the book to incorporate changes in research design methods and emphasis and in media and how they are used to gather and analyze data and report results.

- We have updated information and exercises, changing existing material where we thought it useful. Sections on newer topics such as replicability of data and metadata have been added. We have added more information and examples of experimental research in public administration.
- Software packages are easier to use and have greater capability. More statistical packages with greater capacity of managing and analyzing data are available. Geographic information systems (GISs) and the capability of GIS software to combine geographic analysis with statistical application have grown extensively. The coverage of these topics has been expanded.

- Qualitative research methods are more widely used and more generally accepted. Accordingly, we have expanded the coverage of qualitative approaches and included additional examples. Practitioners often use them in combination with or instead of quantitative methods. Today's students need to understand and be able to use both quantitative and qualitative methods.
- New examples and references have been added and earlier ones updated. We have included more visual displays to illustrate concepts. A wealth of resources, interactive sites, and exercises are now available online.
- All chapters have been revised, some more than others.

  - Chapter 3 includes more information on experiments based on stronger interest in behavioral public administration in the profession.
  - Chapter 4 on measuring variables has new examples.
  - In Chapter 5, the discussion of nonprobability sampling designs has been expanded.
  - Chapter 6 on survey research has been updated to present current information on Internet surveys, the impact of cell phones on data collection, and survey data collection by tablet computers.
  - Chapter 8 on ethics has been updated to include new National Institutes of Health and National Science Foundation requirements for researcher training. It details how institutional review boards have adapted to these changes and discusses what this means for research involving human subjects, including research using social media.
  - Chapter 9 updates the discussion of the decennial United States census and incorporates material looking forward to Census 2020 and new procedures for conducting the census and using census data. The discussion of big data has been expanded, and the chapter includes a new focus on the importance of metadata.
  - The discussion of analysis of variance (ANOVA) has been moved to Chapter 12 from Chapter 13 to better emphasize its primary use as a test of statistical significance.
  - The discussion of logistic regression in Chapter 14 has been updated and expanded with a new example.
  - Chapter 15 has an updated and expanded discussion of the importance of social media for communicating about research and disseminating research results. We expanded the discussion of the importance of saving data for research replication, transparency, and openness.

- Changes linked to the Internet and the tremendous expansion of digital technology continue to influence the research process and dissemination of results. The amount and availability of data have expanded. The Internet and additional software programs have increased the ability of researchers to communicate with each other and the public. This edition of the text includes more Internet references and more material on using the Internet for research.
- As with previous editions, study items, knowledge questions, homework exercises, and problem assignments are available for each chapter. We revised many of these and developed new ones; those found to be unclear or ineffective were omitted. Also, as with previous editions, an *Instructor's Manual* has been prepared to accompany the seventh edition. The *Instructor's Manual* includes answers and solutions to questions and problems in the text. It also includes instructor's resources and additional questions and exercises. The *Manual* can be found on the book's web site at www.routledge.com/9780367334369.
- A set of PowerPoint slides has been prepared for each chapter. These are new to edition seven.

Using This Textbook

As with edition six, an E-version of the text will be available in addition to a printed copy. Faculty teaching in master of public administration and master of public policy programs use this text in both research methods and statistics courses. They often assign exercises using computers. Although several excellent statistical packages for computers are available, we have found that the most common, easiest to use, and easiest to teach set of programs is SPSS (Statistical Package for the Social Sciences). It is also comprehensive, covering most techniques used in a first-year graduate methods and statistics course. A temporary license for SPSS is available for students at a substantial discount. Stata has also become very common in universities, and many students and faculty prefer to use it. The datasets available to users of edition seven are compatible to use with either SPSS or Stata to complete the exercises requiring a computer. Some of the exercises can be completed using Excel.

Practitioners are likely to have Excel on their work and home computers and have experience with it. Furthermore, data can be transferred easily from SPSS and other statistical programs to and from Excel to SPSS and most other statistical packages.

Several supplemental files were prepared for this edition. These include two sets of data, a copy of the questions used in a sample survey, and a report of a qualitative study using focus groups. One data file includes data from the sample survey—the General Social Survey (GSS). These data are in an SPSS file, a Stata file, and a separate Excel file. Codebooks or data dictionaries are available for the survey data. The second dataset is a file of county-level data from a large southeastern state. It includes data on 78 variables for 100 counties. This is an Excel file and can be exported to SPSS if desired. Another file contains descriptions of the variables and sources for the data for the county variables. Those files not containing data are Word files.

Instructors will find these datasets helpful and convenient for students to use in applying computer software for analysis. They can be used to illustrate concepts, techniques, and applications discussed in the chapters and allow for a variety of exercises, including statistical applications, data management, and measurement. They are also a source of data for lengthier research projects. All files are available at this link: www.routledge.com/9780367334369.

Along with revisions and new material, we have retained the existing material that forms the basis of a body of knowledge about research methods and applied statistics. Although the text includes basic as well as more advanced topics, we attempted to present information in such a way as to be useful to students and administrators with a wide range of experiences. We intend for instructors to be able to select from the material, adapt it to their needs, and have more than enough for a one-semester course. Chapters can be omitted or used in a different order depending on the instructor's needs and approach. Many instructors have used the text in a two-course sequence: the first for research methods and design, the second for statistics and statistical applications.

In writing and revising this book, one of our goals was to help administrators and researchers collaborate more effectively. We wanted administrators and students to use the book as a reference when they want to learn more about a topic, technique, or procedure. Administrators often read research reports to decide whether the findings can be used in their job. We intended to write a text for applied research that administrators would find useful and that related to the work they do. We also wanted to have a book that would teach students about topics relating realistically to work they would do in public or nonprofit organizations. To that end, we selected examples and created problems similar to those public and nonprofit administrators often face. We also strongly advise students to *read the endnotes*. Some primarily provide the sources for items of information. Many endnotes, however, contain additional information immediately relevant to the text topic. Others will lead to useful sources of information.

As authors, we drew on our experiences and those of administrators, researchers, teachers, colleagues, and students. As a field of study, research methods can be surprisingly dynamic. Keeping up with changes and new applications has been interesting and challenging.

Acknowledgments

Numerous individuals helped in the revision and production of this text, while others provided help specifically to recent editions. Unfortunately, we are unable to list them all. Numerous reviewers have commented on previous editions, helping us to improve subsequent editions. They, instructors, former students, and colleagues provided voluminous and helpful feedback and information. We are very thankful for the support, guidance, and technical help of the capable people at Taylor & Francis, especially Hannah Shakespeare, senior acquisitions editor for research methods, and her assistant, Matt Bickerton.

Professor Jacqueline Chattopadhyay of the University of North Carolina at Charlotte prepared the original diagrams for Figures 5.2, 5.3, 5.4, and 5.5 on sampling for the sixth edition. We continue to use those in this edition. Corey Correll, MPA, provided valuable assistance while a student at the University of North Carolina at Charlotte.

Those familiar with earlier editions of *Research Methods for Public Administrators* will note the introduction of two new co-authors—Suzanne Leland and Zachary Mohr. They are both experienced teachers, researchers, and administrators. Maureen Berner and Jocelyn Taliaferro, who had been co-authors of previous editions, transitioned to other projects and responsibilities. As with most, if not all, subsequent editions of a work, separating and identifying the contributions of authors can be difficult. The current authors thank the previous authors for their contributions to and support of a seventh edition. Of particular note are the contributions of Elizabethann O'Sullivan, who initially conceived of such a methods text and worked diligently to see that it was published.

As with all book authors, we ultimately take responsibility for any errors or lack of clarity in the material. We also appreciate the patience and understanding of our family, friends, and co-workers while we worked, fretted, and fumed during work on this revised manuscript.

1 Beginning a Research Project

In this chapter, you will learn:

1. Why knowledge of research methods is valuable for public administrators.
2. How to develop a research project.
3. About building and using models.
4. Strategies for presenting models.
5. Definitions of common research terms, including *variables* and *hypotheses*.
6. About selecting a research topic and stating a research question.

Public administrators often ask questions that begin "how many," "how much," "how efficient," "how effective," "how adequate," and "why." They may want to learn something about a group of people, how much a program will cost, or what it can accomplish for each dollar spent. They need to decide how serious a problem is, whether a policy or administrative action solved a problem, what distinguishes more effective programs from less effective ones, and whether clients are satisfied with program performance. They are accountable to politicians, parents, citizens, recipients of program services, and the courts for providing public services. Public and nonprofit organization employees may also be accountable to funding agencies.

Administrators rely on data to make better decisions, to monitor results, and to examine effects. *Data* is another word for information. Understanding research methods is key to gathering, using, and evaluating information appropriately. As a current or future public administrator, you know that adequate information is essential to making effective decisions.

In the role of administrator, you may need to collect and summarize data and act on your findings or supervise others who do so. You may conduct studies or contract with others to perform studies for you to answer questions about programs under your jurisdiction. You may receive regular reports to monitor the performance of your organization and employees. You may read research and get ideas you wish to implement. Even if you never initiate a study, your knowledge of the research process should leave you better able to determine the adequacy of data, interpret reports, question results, and judge the value of published research.[1]

This text will provide you with the skills to produce information using a variety of research tools. More importantly, we hope it will provide you with the tools to make empirically informed judgments as you *sort* and *use* information produced by others to make decisions in your role as a public administrator.[2]

In this text, we discuss both research design and research methods and why each is important. Research design is the overall plan for a research project and includes components to ensure that when the research is completed, the purpose of the project will have been met. Research methods deal with specific research design issues and will include criteria regarding the amount of control the researcher has over the research context, assignment of research subjects, and extent of comparison among subjects and over time. Types of specific designs include experimental, time series, and cross-sectional, to name just a few. Different designs will be used depending on the purposes of the studies.

Research methods are the activities that implement the plan and refer to the various components of selecting study subjects, such as sampling, and types of data collection. The various types of samples and procedures for collecting data with surveys and questionnaires are typical topics of methods texts.

Starting the Project

Research should begin with careful planning. Even though determining the purposes of a study and whether it will produce the desired information is tedious, experience has convinced us that time used to clearly define the purpose of a study and carefully critique the research plan is time well spent. Administrators who make this preliminary investment will have a better understanding of the problem being studied and will avoid conducting poorly conceived research studies and collecting useless data. In this chapter, we present guidelines for better research planning. No matter who actually conducts a study, administrators who spend time defining the problem, planning the study, debating it with others, and reviewing related research markedly improve their work and experience fewer disappointments and wasted efforts.

An excellent starting point for any administrator involved with research is to understand *if* a decision needs to be made, *when* it needs to be made, the *nature* of that decision, and *what* information would be helpful to the decision makers. Focusing on the decision can help identify the true purpose of a study. In terms of timing, if it is not possible to influence the decision, then one must consider whether the study will really be of value. When considering the nature of the decision, what level of importance does the decision warrant? Trying to measure the impact of a multimillion-dollar program affecting thousands of citizens may be more important than evaluating an internal office recycling program. Are lives, jobs, or public safety at stake? Or is it a pet project of the mayor or council? Last, identifying the information decision makers need in order to make a decision is fundamental and will provide a starting point for developing research questions.

Establishing a study's purpose goes beyond stating exactly why it is being done. An investigator must know who wants the study done, how and when the person plans to use its findings, what resources exist to support the study, and what research has previously been done on the issue. After an investigator answers these questions, she can list the research questions and decide what evidence will provide adequate answers. She also avoids planning a study that exceeds the available resources or yields information only after it is needed. It is seldom wise to plan a large study when a smaller one will obtain the necessary information. A study's purpose evolves to become more focused and better understood as investigators and decision makers begin their work together.

Students and practitioners who have never conducted formal research are often unsure of how to proceed and unaware of the various activities involved. Courses and books on research are intended to help by teaching the activities involved and skills needed. Research methods textbooks typically present a format for conducting research projects, and course instructors include similar guidelines in course material. These formats are helpful instructions for carrying out the research. When told of the early steps in the research project, beginners often ask about the reasons for tasks and how they relate to the overall project. An overview of the entire process can help show how parts of the project fit together and why they are necessary. Familiarity with this information can also help managers understand how researchers approach their work and help them work together. Illustration 1.1 outlines the typical steps in conducting a research project. We advise the reader to review this outline as a prelude to the topics discussed in this book.

Experienced and beginning researchers should be familiar with the following procedures. They will usually need to understand and complete each step before the research project is complete. Beginning researchers should keep in mind, however, that although all steps in such formats must usually be completed, and many in the order presented, in doing their work, researchers often go back and forth among these steps. They may work on more than one step at the same time.

Formulating the Research Problem

Researchers and clients should clarify why the research is being done. They should address the following questions at the beginning of a study. In most projects, answers to many of these questions will be included in the early part of the final research report.

What is the research question to be answered? What are the purposes and objectives of the study? To what cases are the results to apply? These issues are usually addressed in a section titled "Problem Statement."

Literature Review

Review what others have written about the selected topic. From the review, the researcher should learn about theories and hypotheses used by others, about data collection techniques, and about ways of measuring variables.[3]

Research Objectives and Hypotheses

Researchers usually develop hypotheses or state research objectives. They may do both. The researcher should explain why he or she expects each to be supported. Variables are also defined and described in this section and identified as independent or dependent. Possible control variables should also be discussed.

Research Design and Data Collection Method

This includes the plan for how the research will be conducted: what data will be necessary, how they will be collected or what data source(s) will be used, how variables will be measured (operationalized), and sampling procedure. (Some of this information may appear in the previous section, especially with regard to variable definitions.) The cases or units of analysis for the study must be identified—these are the subjects about which data will be collected.

The method of collecting data must be described. From whom will the data be collected and how often? Will this be an experiment, a survey, or document research? The specific technique should be discussed in detail.

Describe the research population, and if a sample will be selected, describe the sampling procedure in detail.

Data Analysis and Interpretation

After collecting data, the researcher analyzes it to test the hypotheses and describes the analysis and the results in this section. The nature and strength of the relationships between variables should be described, and the researcher must clearly explain whether the hypotheses are supported. If the study included research objectives, the researcher should explain if they were met.

Implications

In addition to analyzing the data and testing the hypotheses, the researcher should explain what the results mean. How do they contribute to a body of literature, analysis of a policy, or the solving of a problem?

Reporting

The researcher prepares a report. In doing so, the researcher must keep in mind the audience for the research. In addition to or instead of written reports, researchers may also present the research at conferences and other professional meetings.

Illustration 1.1. A Format for Conducting Research Projects

Developing a Research Question

Once the researcher and administrator understand what a study can provide—and what it cannot—the researcher should begin by stating the research question. The research question, or questions, when answered, should provide information necessary to accomplish the purpose of the research. Empirical, that is, observable, information is required to answer it. By definition, research involves the study of observable information, so without it, no research can take place. This text stresses numerical or quantitative information; however, qualitative, that is, non-numerical, information is also empirical and important and may be used to answer research questions. The authors of a well-known source on research design state that principles of good research apply to both qualitative as well as quantitative research and that both can be systematic and scientific.[4] They also note that much research cannot be easily categorized as either qualitative or quantitative—that projects often use a combination of the two approaches. They cite excellent examples of this.[5]

Consider the research question: "To what extent does employee telecommuting improve organizational productivity?" This question has more than one possible answer, and obtaining an answer requires empirical information. Still, this simple question can mask the amount of work that lies ahead. What is meant by "telecommuting"? How would you define and measure "productivity"? Which employees? All employees or just people holding certain positions? Does it involve the productivity of the entire organization or just certain parts of it? How much improvement is expected? To help limit the study's focus, the administrator indicates the purpose of a study. A study to determine whether to adopt a change in employment procedures will be different from a study to evaluate an existing procedure or policy.

Using Models to Organize the Research Study

After stating the research question and the study's purpose, the researcher should build a preliminary research model. Investigators use models to simplify reality by identifying important items and eliminating irrelevant details.

Research models consist of *variables* and *relationships*. A single variable does not constitute a model; it must be linked with another to be part of a model. Variables that are not related to at least one other variable should be eliminated from a model. Variables that are only weakly linked to others also may be eliminated. Explicit models are the words, schematics, or equations that represent the variables and their relationships. The strength of explicit models is that they give others access to the researcher's proposals and allow them to critique, replicate, or improve them.

Thinking About Models

One can think about a model visually by sketching out the variables and the expected relationships or links among them. For example, Example 1.1 presents a model developed as part of a study to identify ways to slow the rate of increase in Medicaid costs.[6] A research question this model might be intended to answer is: "What factors are related to the increase in Medicaid costs?"[7]

Example 1.1 Applying a Model

Research question: How may the rapid increase in Medicaid costs be reduced?
Purpose: To recommend change(s) in the Medicaid program that will contain the costs of the program.

Procedure: An initial model is sketched out with the major variables representing general strategies to reduce costs: reducing these variables will reduce costs. The research will identify and evaluate the basic types of strategies available.

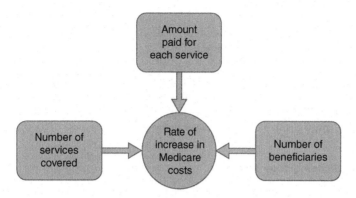

Discussion: The sketch places each variable in a separate "box." The variables, which also may be identified as "inputs," are placed in boxes. Lines indicate the links between variables, and arrows indicate the direction of the links. Here, each strategy was linked to the outcome of reducing the rate of increased costs. The arrows suggest that each strategy should reduce the rate of increase in Medicaid costs. A sketch keeps track of the model's variables and relationships, helps communicate the model to others, and facilitates discussion about the model and the research plan.

With the variables identified, the investigators may:

1. Identify feasible strategies associated with each variable. Should the number of beneficiaries be reduced by changing income requirements or age limits? What services should be reduced? How might less be paid for services?
2. Incorporate the specific strategies into the model.
3. Develop a research plan: What variables and relationships will be studied first? How will they be studied?

Knowing Which Variables to Include and How to Link Them

In deciding what to include, researchers draw on their own ideas and experiences and those of colleagues. Researchers will usually review previous research and analyze data already collected on the topic and will often conduct a preliminary study. In the Medicaid example, a researcher may consider the number of procedures covered and the rate of increase in the number of people eligible.

Before we study *model building* in depth, let's look further at the preliminary model designed to guide the Medicaid research project. Investigators identified the variables representing the general strategies available to reduce costs. The central variable was the rate of increase of Medicaid program costs. The other variables were the number of beneficiaries, the number of services covered, and the payment for each service; these variables were linked to program costs in this model. If these were reduced, the rate of increase in Medicaid costs would also decrease.

Example 1.1 presents the model as a schematic or diagram. The arrows link the variables and show those thought to act on other variables. A model also may be presented as a statement.

Another way to present the model would be with a sentence: "If the number of beneficiaries, the number of covered services, or the amount paid for each service were reduced, Medicaid costs would increase more slowly." A model also may be presented as an equation, as shown in the following:

> The rate of increase in Medicaid costs is a function of number of services covered + amount paid for each service + number of beneficiaries.

The major relationships are between reducing the rate of increase in program costs and the other variables; however, the other variables may be related to each other as well. For example, reducing the number of services covered may very well reduce the number of beneficiaries. The model could then show a link between these two variables.

Organizing and Refining Your Thoughts

Building a model requires researchers to organize their thoughts and to communicate them effectively to others. The model in Example 1.1 should center policy makers' attention on possible solutions and facilitate communication among themselves and others concerned with Medicaid costs. With the preliminary model sketched, investigators can move on to defining precisely what they mean by each variable. They may also decide to include additional variables and explore the linkages among them. They can decide which linkages to investigate further and with what priority. A model may be altered and refined often during the early stages of designing a study.

How detailed should a model be? It depends. Investigators may brainstorm to identify all possible variables, and they may build a very detailed model. To confirm points of agreement, they may develop a simplified version with just a few variables included. The model in Example 1.1 was limited to the basic components of a study. Other models may elaborate on the basic model. For example, a model can spell out how specific strategies for reducing the number of beneficiaries would affect Medicaid rates. Another model can focus on ways of reducing services and the effects of alternative strategies on Medicaid rates. Regardless of the purpose or level of detail, a research model should be based on *theory*. That is, we should have a well-developed explanation for expecting the linkages between the variables that our model posits and be able to explain it to others. These linkages represent hypotheses about the relationships between the variables. The hypotheses should be based on the theory.

In addition, you should think about the assumptions underlying the theory. For example, if you are planning to research the relationship between automatic seat belts in cars and vehicle accident deaths, theory would suggest that the greater the percentage of cars with automatic seat belts, the lower the number of vehicle accident deaths. However, this expectation assumes that automatic seat belts are actually used and not disabled or turned off.[8]

Maps provide a simple illustration of how models vary according to their purpose. For a mountain hike, you need a topographical map with details about the terrain. Such a detailed map, if you could find one and fit it into a car, would be virtually useless on a drive from New York to California. On the other hand, imagine using an ordinary road map to hike through the Grand Canyon! Variables important for one purpose may not be useful for another. And just as developing a research model is based on good theory, choosing a map is based on understanding where you want to go and how you think you might get there.

Consider the Model Preliminary

The model should be considered preliminary. It may change during the course of a study. Model building starts the researcher on an iterative process to collect, analyze, and present data consistent with the study's purpose. Models allow us to go beyond simply collecting and looking at data. They help identify what relationships are expected and why. Precise-looking data can be wrong, and apparent relationships among variables may be due to statistical errors. By the end of a study, models enable users to organize their information and reach reasonable conclusions about the importance of variables and their relationships to one another.

Even though models rarely stay the same throughout a study, a researcher should not begin collecting data without an explicit model and should be reasonably satisfied that this model includes all relevant variables. The included variables and their relationships become important components of decisions about which data to collect and how to analyze them. Data collection and data analysis often require costly amounts of time, expertise, and funding. Without a proper road map, researchers can wander in the wilderness of data—gathering numbers and other pieces of information without a clear direction. Without a proper map, a researcher risks collecting data that are not needed and not collecting data that are necessary.

Try to avoid "falling in love" with your models. This can happen when a person has labored over a research plan and becomes trapped by his own inflexibility. Instead of considering the model preliminary and adjusting or rejecting it as appropriate, the researcher goes through all sorts of data manipulations to demonstrate that the preliminary model is correct. The model should not be the focus of your work. The focus should be having the best information for the right people in time for them to make an informed, considered decision.

Building the Model

Research presentations, both written and oral, typically follow a standard pattern to facilitate effective communication. An audience can more easily follow a presentation that proceeds from a statement of the problem to a description of the model, its variables, and their relation to each other. The researcher then describes the data collection process, analysis, and results. Normally, a researcher does not report on false starts, mistakes, and backtracking when presenting the results of a study.

The process of developing a research project, however, is seldom as systematic and logical as textbooks, research reports, or public presentations make it appear.[9] Designing a study requires creativity, insight, and willingness to be flexible. Nor does each investigator follow the same strategy for identifying variables and their relationships. In general, however, to identify the variables, defend their relevance to the model, and postulate the nature of their relationships, investigators integrate their own ideas and research experiences, specific knowledge of the topic, the observations of others, and the existing research literature. Furthermore, researchers must invest time in understanding the specific problem at hand in order to engage in good model-building.

Ideas

Ideas refer to the knowledge, beliefs, or impressions one has about a research question. We rarely approach a situation with no knowledge or insight. Most of us retain a wealth of information to help us solve problems. Do not downplay the value or importance of this information. If you do not make use of your experience, knowledge, or opinions when faced with problems, you will not be efficient as an administrator or as a researcher.

Building a model requires you to formalize your existing ideas by making them explicit and communicable to other people. Through words, drawings, or equations, you identify the variables you consider important and explain how you believe they are related to each other. Thus, you clarify fuzzy ideas and expose them to critical examination. You may find that some of your beliefs will not make sense. Some of your ideas suddenly may seem naive or incomplete. Some will be based on assumptions with which others disagree. Nevertheless, unless you are willing to risk having your ideas challenged, you may miss variables and relationships critical to problem solving.

Peer Interaction

Peer interaction refers to the discussions and debates among colleagues about the research question and possible answers. To understand the role of peer interaction in model building better, we recommend reading about how scientists go about their work.[10] Doing so will quickly dispel the myth of the isolated scientist working alone. A major lesson is the value of criticism and debate. Without it, errors may go uncorrected and oversights may abound. Few people find criticism easy to accept or arguments easy to verbalize, but the process of peer discussion about a research project uncovers assumptions and blind spots.

A particularly important reason for peer interaction in public administration is that our professional and educational backgrounds tend to shape our ideas in ways that we may not recognize. For example, our disciplines shape the questions we ask, how we approach problems, the data we use, and how we use them. Imagine, as a city administrator, you are interested in studying ways to address homelessness in your city. As a first step, you decide to gather data on the causes of homelessness in your area. An analyst with an economics background may want to focus on quantitative data from city employment and support programs. Another analyst with a background in social work may turn first to interviews with clients of local shelters to learn about their experiences. An analyst with a public safety background may wish to test a possible relationship between homelessness and parolees from prisons in the area. If they had to work together, we suspect they might experience frustration and tension. Yet, ideally, the conflict of ideas and the resulting compromises should yield a research plan superior to that developed by any one of the individuals working alone. One name for the concept of drawing from other disciplines is "borrowing strength."

Review of Existing Knowledge

Research often is an iterative process in which investigators build on the work of others. After identifying the purpose of a study, investigators search for information on similar studies. Some people assume that quantitative researchers have little need to do this. Applied researchers may not see their studies as part of a larger body of knowledge. Nevertheless, a few hours reviewing academic journals and reference documents should convince you of the value of published materials whether you are conducting quantitative or qualitative, academic or applied research.[11]

Too often, researchers studying an unfamiliar problem start by "reinventing the wheel." In other words, they may begin to build a model that has already been developed and refined many times. The recommended approach is first to conduct a *literature review*, locating and examining what has already been written on the topic. The objective is not to ensure the originality of ideas. Rather, it is to identify information and ideas that we may incorporate into our own research. As we conduct the literature review, we think about the planned study and better understand its purpose and what we can expect to accomplish. An examination of previous research identifies the following:

- Models used by other investigators
- Variables included in studies with a similar purpose
- Definitions currently used for the variables
- Techniques for measuring the variables
- Sources of data
- Strategies for collecting data
- Theories used to link the variables
- The strength of the relationships between variables

- Results and conclusions
- Suggestions for further research

With this information, investigators avoid wasting time. They may find that their research question has already been answered. They may learn that others failed to confirm relationships that at first seemed important or obvious.

"State-of-the-art" articles that summarize the nature of existing research are especially helpful. They cite the major research work in the field and the major research themes. You will find that such articles quickly bring you up to date. They should have covered the relevant literature, easing your burden. They may also help you avoid the temptation to go too far afield in your investigation.

You will find that a reference librarian is an invaluable resource in locating appropriate literature. Most article abstracts are now available electronically and accessible online. A university reference librarian can introduce you to these and other useful online resources and suggest efficient search strategies for locating relevant research. He or she can show you how to use appropriate search engines, some of which focus on particular scholarly fields. Conducting online research can be both valuable and frustrating. Search engines can uncover a wealth of information or a pile of useless or, worse yet, erroneous information. One must be careful about depending on poorly sourced material. The most useful literature is published in peer-reviewed academic journals and books. The peer-review process ensures that experts have reviewed the articles and judged that the research employed and correctly carried out appropriate techniques of data collection and analysis. Other sources, such as trade journals, may also be helpful. However, keep in mind that the information in them typically does not undergo the same peer-review process.

Bringing Information Together to Build a Model

Example 1.2 suggests how an investigator can bring together ideas, peer interaction, and the existing literature to build a model. The example applies model building to the process of conducting citizen surveys. If given the opportunity to build an original survey, you should develop the model that you plan to use *before* developing the survey in order to ensure that the survey will obtain the needed information. Like many investigators, you may be inclined to design a questionnaire. Or you may plan to adapt one that has been used before. If done before explicitly building a model, however, you may find that the questionnaire or questions are not appropriate for your study. You may not be able to examine explanatory research questions using the resulting data. Try to avoid the temptation to skip the model-building step.

Example 1.2 Building a Model

Research questions: What do residents think about their town? Are they satisfied with available services? What changes do they want?

Purpose: To consider citizen perceptions in preparing the town's comprehensive plan.

Procedure: The planning director planned to survey residents to learn how they judge the town and its services.

To identify variables of interest, she:

1. solicited existing surveys from local planners (peer interaction)
2. examined handbooks published by the American Planning Association and the International City/County Management Association to find sample surveys (literature review)

3. drew on what she learned during her professional training, from her career experience (ideas), and from shop talk (peer interaction)

Several variables seem linked to perceptions of public services. She selected the following variables as relevant:

1. where residents live; if they are homeowners or renters (based on the town council's interest)
2. the gender and age of residents; whether they work in town or in the nearest city (based on the planning staff's interest)

She met with the Planning Advisory Board to consider what services to include. They discussed the value of linking the demographic variables to satisfaction with services.

Summary of model: The survey will gather data on citizen satisfaction with public services, including trash collection, medical facilities, and housing. To see if the town government meets the needs of all residents, the planning staff will examine whether residents with different characteristics (gender, age, location of home or job, whether they rent or own) rate services differently.

To organize her presentations, she sketched the model:

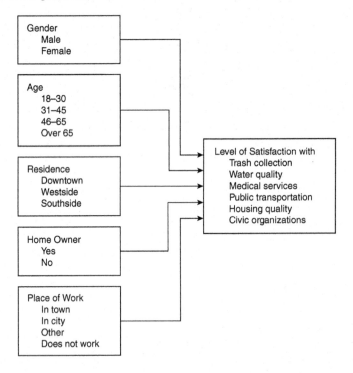

Discussion: To develop the survey, the director had to decide which variables were relevant for the town's planning. To do this, she used her own ideas, peer interaction, and the literature. She met with her staff, the manager's staff, an advisory board, and the town council to review and refine the survey's content. The director read the literature to identify possible survey questions and to get ideas on how to construct the sample, how to achieve a higher response rate, and how to analyze the data. Next, she will scrutinize the specific questions, pretest the questionnaire, and pilot the study. As she conducts these steps, she may make some modifications to the preliminary model.

Examining a Dataset

Investigators with access to existing datasets, that is, data from previous studies, may examine and analyze some of that data as they focus and develop their ideas. In this way, investigators may get a better idea of the importance of variables and the strength of their relationships.[12] If a dataset does not contain the specific variables desired, the investigator may study similar variables, which can be considered stand-ins or proxies.

For example, a neighborhood or postal ZIP code may act as a proxy measure for racial or ethnic groups, social class, or even family composition. Or the researcher may decide that a dataset does not offer an acceptable way to capture one or more variables that are part of the planned model. The number of datasets available online and the variety of topics covered is immense. Many are easy to access. (Chapter 9 has a lengthier discussion of accessing and using secondary data, data already collected.)

Investigators analyzing an existing set of data may rethink their study's purpose and their model's adequacy. Through the process, they may gain greater insight into their own study and revise their original ideas.

Pilot Testing the Model

Before collecting data on a large scale, the research plan should be rehearsed. A small study, called a pilot study, is launched to test the adequacy of the proposed data-collection strategy. During the pilot study, investigators discover the feasibility of their research plans and how much time and effort will be needed to collect, compile, and analyze the data. In the pilot study, investigators should carry out the entire plan, including the data analysis and interpretation of results. Often investigators fail to analyze and interpret the pilot test data. This is unfortunate, as they then miss opportunities to test the model's appropriateness, identify important variables not included in the original model, or find that some variables in the model are not needed.

Collecting data without a preliminary model is not a sound research practice. Too often, research projects begin prematurely. After identifying the research question, some investigators begin collecting data eagerly. They cut off further planning with comments such as, "Let's see what the data show," or, "I'll decide after I look at the data." Questionnaires are constructed, subjects are questioned, and data are analyzed with too little attention given to the investigators' objectives. Unfortunately, the investigators may then not recognize problems until the end of the research project.

Types of Models

At this point, you may be wondering: "How do investigators present a model's variables and relationships?" While many possibilities exist, ranging from physical models, such as mock-ups of buildings, to highly abstract verbal presentations, administrative researchers usually work with two types of models: schematic models and symbolic models. Schematic models refer to models that use pictures, lines, points, and similar paper-and-pencil–type products to designate the variables and illustrate their relationship to each other. The sketches in Examples 1.1 and 1.2 present the model as a schematic, as do blueprints, flowcharts, and maps. Symbolic models refer to models that use words or equations to represent the elements and describe their relationships.

The type of model chosen depends on the model's purpose, the audience, and the investigator. More than one type of model may be used to present a set of variables and relationships. Consider getting directions to a person's house. Some people distribute maps—a schematic model. Others give verbal directions, listing roads, landmarks, and distances—a symbolic model. Others may give both a map and verbal directions. The choice depends largely on a host's preferences and perceptions of what will be easiest for his guests.

Many investigators build both a symbolic and a schematic model, as is done in Example 1.2, "Building a Model." The example's symbolic model summarizes the planning director's actual verbal model, which establishes the importance of linking each specific demographic characteristic to the residents' perception of community services. The schematic model sacrifices the detail of the symbolic model, but it effectively identifies the model's essential features.

Schematic Models

Schematic models work well in summarizing the thinking behind the model and in drawing attention to its major features. Schematic models with a few variables can be understood quickly and focus people's attention. Consequently, schematics help an investigator think through the model and explain it to others. A schematic also may serve as an elaborate checklist. For example, an investigator can record what data are needed to measure a variable and when those data are gathered or when a relationship has been examined. Schematic models are less effective if they include too much detail. Only so much information can be included about a variable and its relationship to other variables before a schematic becomes cluttered and confusing.

Most administrators and students of public and nonprofit administration are familiar with schematic models such as flow charts and logic models. Logic models illustrate the components in program development and operation. Logic models use a standard format, and components are fitted to this format for specific programs.[13]

Figure 1.1 shows the standard form of a logic model with an illustration of a housing program.

Symbolic Models

Symbolic models include verbal and mathematical models. Verbal models use words to describe the variables and define their relations to each other. Verbal models are often found in the introductory or theoretical sections of research articles. A newspaper account of a research study and its findings may also be considered a verbal model.

This type of model has distinct strengths. The investigators have all the potential offered by language to describe the model, allowing a full and detailed explanation of complex relationships. The researchers are not constrained by requirements to reduce the model to equations. A fuller range of users can understand and interpret a verbal model than can work with mathematical models. Verbal and mathematical models can complement each other in describing relationships. The verbal model provides the rich details, and the mathematical model looks at the precise nature of the relationships.

Mathematical models can condense a great deal of information by using equations to specify relationships. The mathematical models indicate which relationships exist or are expected, indicate their directions, and measure their strengths. Mathematical models allow administrators to predict needs, to estimate the impact of policy decisions, and to allocate resources more efficiently. In some areas of interest to administrators, such as operations research or management science, mathematical models guide the model-building process.

Input→	Activities ->	Outputs->	----------- Outcomes -----------		
Wages	Staff training	# of clients counseled	# of clients with good credit	# with new housing	# homeless
Supplies	Record keeping				

Figure 1.1 A Logic Model Applied to a Housing Program

Source: Adapted from Figure 4.2 in O'Sullivan, Rassel, Taliaferro, *Practical Research Methods for Nonprofit and Public Administration*, Longman-Pearson, 2011.

Complicated mathematical models often are handled best with computers. In such instances, a software program can quickly manipulate the variables and depict their relationships statistically and graphically. Unlike an analyst, the computer can keep track of a large number of variables and the relationships among them. A computer model differs from other models primarily in the amount of information it can manipulate, its speed, and its accuracy. It is less likely to make recording or computational errors. As with other models, however, the accuracy of a computer model is a function of human judgment that creates the rules for selecting and manipulating variables.

Limitations of Models and Model Building for Specific Users

No matter how many variables are included, models still represent a simplification of reality. They represent one view of which variables are relevant to the problem at hand and which variables can be ignored. Differing views may be entertained. The investigators who build a model and the people who critique it are subject to the full range of human weaknesses. Model builders contend with limited time, money, and knowledge. Their viewpoints may be colored by biases that lead them to ignore others' comments and criticisms.

Users may be seduced by a model's clarity and its apparent usefulness. They may assume erroneously that it is accurate and adequate. Users who read a study or sit in an audience tend to focus on the details of the verbal description, a diagram, or an equation. They may find it useful to step back and reflect on what the model leaves out.

Researchers who initiate a study at the request of a policy maker or administrator may have a similar concern. The public administrator or policy maker who requests the study should discuss its audience and use with the research team and should also ask the research team to explain the logic of its model in clear terms. For full-time researchers, models serve as a logical mechanism to organize their thoughts and studies, but this is not necessarily the case for public administrators as "clients." If a client must adapt to the investigator's methods, the study may be logical and well organized, but it may be ill suited to the client's needs. The research team should not force the administrator or policy maker who requested the study to conform to its own way of approaching a problem.

Understanding Users' Goals

In conducting research for specific users, researchers should take the time to understand why users want the study done and what they plan to do with the findings. The more that researchers know about the topic they study and the circumstances surrounding it, the more the user—the client—will trust the researchers and the study results. Conversations[14] between administrators and the investigators may seem far removed from identifying the model's variables and relationships. But a researcher also has to deal with the reality of trying to satisfy client demands. Nevertheless, the research study's clients have valuable perspectives on potential models. For example, to evaluate strategies encouraging parents to have toddlers vaccinated, the analyst may ask the client to speculate on which factors motivate parents to do so and which factors lead them to delay or reject vaccinations. The analyst can then organize the information into a symbolic or schematic model.

Researchers should also keep in mind that administrators tend to judge information and data differently and make decisions differently than researchers, a point made earlier in this chapter.[15] Administrators and academic researchers may also have different time frames, with administrators expecting results more quickly than academics.

The researcher may leave many of the research details in the background. A client may receive a list of research questions with a strategy for answering them or a description of an experiment. A user might review graphs and tables similar to those planned for the final report. This strategy mirrors the model-building process. Beginning with the clients' perspective avoids the mistake of ignoring the observations and concerns of users who are not used to articulating their ideas within the context of an explicit model. Focusing on the clients' concerns and needs ensures that a study will be planned to address a specific problem. Otherwise, researchers or analysts may plan a study around their own analytical preferences. Involving the client in the model-building process can help the client identify and clarify the true purpose of the study.

In conducting research to answer theoretical questions, such as the nature of organizational leadership, investigators pursue model building as described earlier. They include peer interaction, a literature review, and pilot-test information in building and refining their models. If they report their findings in a scholarly journal, they will use the shorthand afforded by research jargon. In theoretical research, a study evolves from the research that precedes it. Theoretical studies develop a discipline's body of knowledge and set the stage for further empirical research. Consequently, the researchers must provide a detailed discussion of their methodology and conduct extensive statistical analysis of quantitative data.

Applied studies generally are done with intended users in mind. In these, the literature review may be less thorough and systematic and more narrowly focused. Specific users combine research findings with other information to guide their thinking about a problem or to make a decision. Investigators avoid research jargon in communicating with most administrators and policy makers. The jargon serves as a checklist for the researcher, so he does not overlook the numerous details that can weaken a study, but he avoids using jargon otherwise. If the terms are unfamiliar to audiences, their attention may shift to understanding the words, or they may lose interest and ignore the important points that the investigator wants to emphasize. This point is important as you read this chapter's next section. We introduce a wealth of terminology, but if you read a report done for an administrator, you may never see these specific terms. Nevertheless, in any report, you still should find ample evidence that the researchers built models and applied standard research procedures based on this terminology in designing their study.

The Components of Models: Variables and Hypotheses

The primary components of models are variables. Variables are observable characteristics that can have more than one value, that is, characteristics that *vary*. Some examples of variables and their values are shown in Table 1.1.

If the characteristic has only one value in a study, it does not vary and is called a *constant*. Researchers often use as a study population those cases with only one value of a characteristic. For example, in a study of the relationship between employee age and preference for team projects, the researcher included only men. In this study, gender was a constant. Another constant was type of industry—only one manufacturing industry was included.

Table 1.1 Variables and Variable Values

Variable	Values
Gender	Male, Female, Non-binary
Job satisfaction	Very satisfied, Satisfied, Dissatisfied, Very dissatisfied
Salary	Actual dollar amount of salary
Age	Number of years since birth

Hypotheses

To test a model, researchers examine the relationship between variables linked in the model. The investigators express the relationship between two variables in a simple model called a *hypothesis*. A hypothesis is a statement that specifies or describes the expected relationship between two variables in such a way that the relationship can be tested empirically. Hypotheses form the foundation of a research effort. A clearly written hypothesis helps researchers to decide what data to collect and how to analyze them.

Consider the following four examples of hypotheses. Can you identify the variables in each hypothesis?

H_1: Persons jailed for burglary are more likely to be rearrested once released than are those people jailed for assault.

In this hypothesis, the variables are the type of crime for which someone is jailed and the probability of being arrested again.

H_2: City governments keep more supplies in storage than do county governments.

The variables in the second hypothesis are the type of government and the amount of supplies kept in storage.

Note that hypotheses H_1 and H_2 are reasonably specific. We have a good idea of what data to collect to measure the type of crime, whether a person arrested has been arrested before, type of government, and the amount of supplies kept in storage.

H_3: Training programs improve the skills of the chronically unemployed.

Now in this hypothesis, imagine the frustration of the researcher trying to test it. And yet it is a reasonable beginning point for a research project. It is just not a useful hypothesis, since it is not easily testable as stated here. To what type of training program does the hypothesis refer, and what does its author mean by "skills"? We may even debate who qualifies as chronically unemployed. The vagueness of the hypothesis leads us to suspect that its author has a poorly developed model.

H_4: The city manager form of government is more common in the Southern states.

In hypothesis H_4, a value is given for the independent and dependent variables; however, the variables themselves are only implied. They are "region of the country" for the independent variable and "form of city government" for the dependent variable.

The decision on how specific a hypothesis should be is relative. Some variables, such as skills and who is chronically unemployed, can be defined as the researcher decides how to measure a variable and from whom to collect data. Even apparently specific variables may need further definition. Consider the crimes of burglary and assault. Each of these has several categories, from less to more serious. What types of supplies should be included in collecting data to test the second hypothesis?

Two Kinds of Variables

The most useful hypotheses typically include two variables, an independent and a dependent variable, and imply that a change of value in one variable is accompanied by a change of value in the other variable.

- The *dependent variable* is the characteristic whose changes the researcher wants to explain. The *independent variable* is considered the explanatory or causal variable.
- The *independent variable* is used to explain variation in the characteristic or event of interest. It is sometimes referred to as an "input" or "causal variable." The *dependent variable* represents or measures the characteristic or event being explained. It is also referred to as an "outcome" or an "effect."

One may visually identify the independent and dependent variables with a schematic model; the arrow leads from the independent variable to the dependent variable. Another way to think about dependent and independent variables is to ask, "Which variable depends on the other?" The goal of research often is to identify the causes of problems or social conditions we wish to change. To confidently claim that one variable is really the cause of another, however, requires four types of evidence, discussed later in the text. Good hypotheses usually state a relationship between or among variables, and researchers test for those relationships. That two variables are related does not mean that one causes the other.

Some people find it helpful to rephrase a hypothesis as an "if-then" statement. For example, "If the age for Social Security eligibility is raised, then the rate of increase in Social Security costs will be slowed." The "if" statement contains the independent variable, that is, age for Social Security eligibility. The "then" statement contains the dependent variable, that is, rate of increase in costs. In our example hypotheses, the independent variables were category of crime, type of local government, presence or absence of training programs, and region of the country. The respective dependent variables were the probability of being arrested again, the amount of supplies in storage, quality of life, and form of city government.

We have defined hypotheses as consisting of independent and dependent variables. Consider, however, a hypothesis such as the following with two independent variables: "Women and older adults use public libraries more often than men and younger adults." Is it necessary that both women and older adults use public libraries more for the hypothesis to be supported? What if women and younger adults more often use public libraries? Or does the independent variable have four values: older women, younger women, older men, and younger men? If this is the case, the hypothesis should make the values explicit. Otherwise, to avoid the ambiguity of having one part of a hypothesis supported and the other part unsupported, it is conventional to have one hypothesis for each independent variable. In our example, the new hypotheses would be:

1. Women use public libraries more often than men.
2. Older adults use public libraries more often than younger adults.

Characteristics of Good Hypotheses

Although the most useful hypotheses state a relationship between two or more variables, we can also state and test hypotheses containing only one variable. For example, a state demographer may hypothesize that "the population of the state's capital city is over 250,000." Or the director of economic development for a resort community may claim that "summer concerts bring $2 million of economic activity to the community." The only variables in these hypotheses are "size of population" for the first and "number of dollars of economic activity" for the second. Testing such hypotheses can provide important information. However, testing hypotheses relating two variables helps us understand why the dependent variable varies. Explanatory research questions always involve testing hypotheses that have both a dependent and an independent variable.

Good hypotheses have the following characteristics:

- They include one independent and one dependent variable.
- Each variable is clearly stated.

- From the wording, it is clear how each variable varies.
- The expected relationship between the variables is clearly stated.

Units of Analysis

When testing hypotheses, the units of analysis used in the research should be at the same level of aggregation as those in the hypothesis. *Units of analysis* are the cases or entities for which we measure variables. For example, the units of analysis in the survey of town residents illustrated in Example 1.1 are individuals who respond to the survey. The units of analysis in the County Data file, included on the web site for this textbook, are counties. Many of the county variables are summaries of items in each county. The percent of people with a college degree, for example, is a county-level variable. It is based on counting the number of people in the county with a college degree. For individuals, we would record whether each has a college degree. The County variables, therefore, are at a higher level of aggregation than the survey variables.

Consider the following example of an error that could occur if the units of analysis for the data are not at the same level as those in the hypothesis. Assume that a researcher is interested in factors related to student Scholastic Aptitude Test (SAT) scores. He states the following hypothesis:

"The better the student's school attendance record, the higher the student will score on the SAT exam." (Note that the units of analysis are individual students. School attendance is the percent of total school days attended.)

The researcher does not have data on individual students but does have data from a large number of school districts. He tests the hypothesis and finds that the average SAT scores of students from school districts with higher average student attendance are higher than average SAT scores of students from districts with lower average attendance. This researcher cannot then accurately conclude that individuals with better attendance records scored higher on the SAT exam. They may have; however, since the data used to test the hypothesis were based on school districts, the researcher cannot assume that the results also apply to individual students.[16]

The Nature of Relationships

An important characteristic of a hypothesis is the pattern of the relationship it postulates. We refer to this patterned relationship between an independent and a dependent variable as *covariation*. Covariation between two variables commonly takes one of three forms: (1) *positive*—also called *direct*, (2) *negative*—also called *inverse*, or (3) *nonlinear*.

To describe the patterns of covariation, consider the relationship between the amount of job training received and salary on the first job after training. The relationship may be *positive*; as the number of training hours received increases, the amount of salary increases. The relationship may be *inverse*, or *negative*; as the number of training hours increases, the amount of salary decreases.

The relationship may be *nonlinear* in that a distinctive but nonlinear pattern occurs. For example, two such patterns might emerge from a study of the relationship between training and salary levels. In the first, salary increases as the number of training hours increases but only to a point, beyond which it begins to level off or decrease as training increases further. In the second, as the amount of training increases, salary increases to a point, after which the amount of salary received stays constant. Figure 1.2 shows examples of positive, negative, and nonlinear relationships.

If the independent variable has no discernible effect on the dependent variable, we say that the two variables do not vary together. If two variables do not vary together, their relationship is described as *random* or *null*.

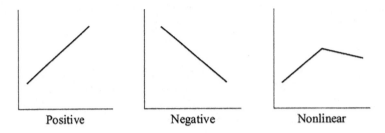

Positive   Negative   Nonlinear

Figure 1.2 Examples of Relationships—Positive, Negative, and Nonlinear

The Role of Control Variables

In stating and testing a hypothesis, we may wonder about the effects of other variables on a hypothesized relationship. Researchers may suspect a variable not included in the hypothesis is related to the independent and dependent variables and affects the relationship between them. To account for this, they add that variable to the analysis to see if it alters the relationship between the independent and dependent variables. Such a variable is called a *control variable* because including it in the analysis allows the researcher to "control for" any effect it has on the relationship between the independent and dependent variables. Many researchers use the term *independent variable* for both the core independent variable of interest in the hypothesis and the control variables. As you build a model and work on a literature review, keep in mind a distinction between the independent variable centrally involved in your hypothesis and the independent variables less central to your study that are included as control variables. For this reason, some researchers call the central independent variable the *explanatory variable* and all other independent variables *control variables*.

A control variable may show that the original relationship was in error; for example, the relationship was false or spurious. The following examples illustrate spurious relationships.

First, consider the hypothesis "arrests for assaults increase with ice cream sales." In other words, these two variables are positively related. If subsequent research shows a statistical relationship, should you conclude that ice cream causes physical violence? Of course not; average daily temperatures (a control variable) may reveal that hot weather, not ice cream sales (the independent variable), is associated with more assault arrests (the dependent variable). Ice cream sales typically are higher during hot weather than during cold weather, so is the number of assaults because of more outside activity, frustration with the heat, and so on?

Similarly, hospital administrators often challenge reports that their hospitals have a higher-than-expected death rate. They contend that their hospitals treat sicker patients. Thus, they argue that patient prognosis at admission (a control variable), not the hospital itself (the independent variable), causes the higher death rate (the dependent variable).

Moderating Variables

In addition to helping identify a spurious relationship, control variables may alter the hypothesized relationship radically. Such variables are called moderating variables—they moderate or change the relationship between the original two variables. Consider how noise affects productivity. If a task requires concentration, noise tends to diminish productivity, whereas if a task is monotonous, noise tends to increase productivity because it jolts workers out of their daydreams. In Example 1.3, we show how a control variable is used to interpret the effectiveness of three job-training programs. When the relationship between the training program and student employment is considered, one program appears to be markedly more successful than the other two. When the educational level of trainees is also considered, the difference becomes less striking. For students without a high school diploma or with education after high school, two training programs have nearly identical success in placing trainees.

Example 1.3 Illustrating Variables and Hypotheses

Problem: Identify successful job-training programs for hard-to-place individuals.

Hypothesis: On-the-job training programs will be more successful than other training programs in placing participants in permanent positions.

Independent variable: Type of training program.

Dependent variable: Placement in a permanent position.

Control variable: Educational level.

Findings: The data reported in the following list supported the hypothesis that on-the-job training programs had the highest placement success. These data also showed that work-skills training had the lowest placement success.

Placement Success by Job-Training Program

Vocational education program (attendance in a school-based training program): 31 percent of all participants placed.

On-the-job training: 41 percent of all participants placed.

Work-skills training (program to teach basic reading and mathematical skills, attitudes, and behavior needed for permanent, skilled employment): 23 percent of all participants placed.

Table 1.2 Placement Success by Program, Controlling for Years of Schooling (H.S. = High School)

Education	Vocational Training	On-the-Job Training	Work-Skills Training
<12 years	33%	35%	23%
H.S. graduate	28%	43%	25%
>12 years	38%	39%	25%

Discussion: The examination of the relationship between the independent and dependent variables for all participants shows that on-the-job training is markedly more successful in placing trainees (41 percent compared to 31 percent and 23 percent). If education level is controlled and held constant, vocational education and on-the-job training do about equally well in placing participants who did not graduate from high school and those who continued beyond high school. Based on these findings, a counselor would refer trainees who only completed high school to on-the-job training. Other participants may do equally well in either on-the-job training or vocational education. If you funded these programs, what decisions would you make? Would you want to look at other control variables? Which ones? Why?

Before you leave this example, make sure you can interpret the figures in the table with the control data. For example, 33 percent means that 33 percent of the people in vocational education who had not completed high school had been placed. And 43 percent of those in on-the-job training who had graduated high school had been placed. Researchers have different ways of presenting control data, so expect to take time to determine exactly what the reported data mean.

In the example of the relationship between noise and productivity, the relationship is not in the same direction for each value of the control variable. In the example of the relationship between job-training programs and student employment, the relationship is notably stronger for one value of the control variable than for the others.

What can you deduce from the introduction of a control variable?

1. The hypothesized relationship is spurious—it only appears to exist because the control variable is related to both independent and dependent variables, or
2. The relationship between the independent and dependent variables is stronger for some values of the control variable than for other values, does not have the same direction for each value of the control variable, or is supported for some values of the control variable and unsupported for other values, or
3. The control variable has little or no impact on the relationship.

For studies of human behavior or attitudes, common control variables are age, sex, income, education, and race. For research on agencies, common control variables include agency size, mission or purpose, size of budget, and region of the country or state. As you read the literature about a particular policy or program, you may locate other control variables that are used regularly to understand the relationship between the original independent and dependent variables.

Selecting a Topic and Formulating a Research Question

Students may be overwhelmed by seemingly limitless possibilities of topics when required to write a research paper. Analysts face a similar situation as they decide how to study a policy and its impact or even which policies to study. Administrators have to motivate themselves to do research if it is not required in their jobs. The results of research studies, however, often provide important information. Students, analysts, and administrators all want to study relevant issues, that is, issues that engage public, professional, or agency attention. Recent relevant topics include public pension reform, teacher tenure, income inequality, the implications of changing demographics, immigration, and the consequences of rising sea levels. Other relevant issues can be gleaned from professional journals, professional meetings, and the national media. Researchers often choose topics by identifying problems discussed in these sources. A typical format for writing a research paper is to begin by briefly discussing the problem the research will address and why it is a problem.

Administrators who must decide on adopting or changing a program or policy may recognize the need for information that requires a research project. They may have the resources and time to conduct it themselves or assign it to staff members but more likely will employ researchers from outside of the organization. In either case, the administrator requesting the study should be involved in selecting or approving the research questions that will drive the project.

Once a topic is selected, the researcher must focus his efforts in order to move on to the next steps in the process. Typically he does this by stating what he wants to find out as a research question. The research question can be framed by considering the questions we listed at the beginning of the chapter—questions that begin "how many," "how much," "how efficient," "how effective," "why," and "what." The research question should lead the researcher to develop hypotheses that will answer the question if tested.

Students' Selection of a Topic

Students have the most freedom in selecting a topic. A required paper serves as an opportunity to study an unfamiliar topic or to develop new skills. Public administration students may seek projects that introduce them to agency officials or add to their résumés. In selecting a topic and framing the research question, students should do the following:

1. be interested in the topic
2. have, or be able to develop, the required knowledge and skills to conduct the study
3. have sufficient time and resources to conduct the study

A student may be encouraged to select a topic simply because it is "hot." Unfortunately, a relevant topic has little value if the student cannot produce a credible study. Students may be hindered by a lack of interest in the topic or insufficient knowledge. For example, a review of the environmental policy literature may require knowledge of economics, statistics, or the physical or biological sciences—expertise that students rarely can develop within a semester. Students may underestimate the time and cost involved in empirical research—data collection and analysis take resources. Beginning and experienced researchers should keep in mind that a small, well-designed study is better than a large study that falls apart because of insufficient time or money.

Analysts' Selection of a Topic

Analysts' choices of research topics are more limited; topics must be relevant to the decision at hand. Similar to students, they should have requisite knowledge, sufficient time, and adequate resources; however, co-workers or contractors can supplement an analyst's knowledge or skills. In selecting a topic, analysts consider its relevance to the agency and their careers. A relevant topic is one that is important to the agency's mission, consumes substantial resources or affects the public's quality of life, and is amenable to change. Employees with analytical skills are often assigned to study issues important to agency managers. Studies on inconsequential topics or on policy areas resistant to change are likely to be ignored. A record of producing unused studies is unlikely to lead to individual career or organizational success.[17]

Administrators' Selection of a Topic

Administrators rely on research findings to identify problems, evaluate solutions, and make decisions. They use research findings to keep existing resources and to justify additional ones, to monitor programs, and to improve employee performance. An effective strategy for administrators in university towns or with strong alumni ties is to keep a list of research questions. The list is ready whenever a faculty member calls looking for a class project. Students get to work on a relevant project, and the agency benefits from a low-cost study. Even the most modest study can provide administrators information about a policy and its implementation.[18]

Administrators may often rely on the knowledge and research skills of others. Nevertheless, as stressed earlier, they should be sufficiently engaged to oversee the project and to question the research strategy and the findings.[19] In our experience, working with researchers to define a research question, build a model, and critique the method and the findings contributes to administrators' knowledge of their agency and their programs. In the end, the topic should matter to someone. Otherwise, there is no point in doing the research.

Summary

Effective quantitative research requires that investigators articulate the purpose of a study. With the purpose in mind, the investigators can select the variables of interest and postulate their relationships to each other. These elements and relationships constitute a model. Investigators may find that a model aids them in three ways. First, it helps them to explain their ideas to others and to solicit reactions and criticisms. Second, it helps investigators to understand their ideas better. Third, it provides a useful guide to the research. It is particularly valuable in ensuring that all specified variables are measured and analyzed.

In building a model, investigators consider their own ideas and the ideas of colleagues and review existing research. Models may also emerge as investigators work with existing data or examine the results of a pilot test. The specific way that investigators use these information sources varies from person to person and requires some creativity. Nevertheless, one can expect to rethink his or her ideas several times as information from different sources is brought to bear.

In our opinion, the review of the literature is the most important source of information. A thorough review of the literature greatly reduces the likelihood of wasting resources, such as time and effort. It provides valuable information that can be used in the later stages of research planning. Frequently, it helps investigators to identify strategies they can copy or adapt. Investigators will want to make use of the resources of reference and document libraries, particularly abstracts, statistical indices, citation indices, and the wealth of other electronic sources.

The preliminary model may be modified several times during the course of planning a study. Once a plan is developed, investigators should conduct a pilot test. It should include data collection, data analysis, and interpretation of the findings. Pilot-test information can be reviewed by investigators and major study users to ensure that the final study will fulfill its purpose.

Administrative researchers primarily use schematic and symbolic models. Schematic models help investigators to illustrate their proposals and explain them to others, but much of the detail that is needed to understand and test the model may be missing from the schematic. Symbolic models use words or equations to represent the variables and their relationships. Verbal and mathematical models often are complementary. The verbal model presents the model in all its detail; the mathematical model gives precise descriptions of the relationships between variables.

The components of research models are variables, that is, characteristics that vary. An independent variable is an explanatory variable—it is often thought of as a cause. A dependent variable is the characteristic whose variation the researcher wants to explain—it is often thought of as the result. A hypothesis is a simple model that states a testable relationship between an independent and a dependent variable. Frequently, an investigator will add additional variables, called *control variables*, to see whether they alter the hypothesized relationship. Common control variables in studies of human behavior include race, gender, and age. In studies of organizations, number of employees, size of budget, and mission are often used as control variables.

Once the preliminary model has been outlined, the investigators must decide when and how often to collect data, how much control to exert over the study, what data to collect, and from whom to collect the data. To make these decisions, they consider the purpose of a study, how it affects the timing and frequency of data collected, and the investigators' degree of control over the research situation. A variety of approaches are available and used by administrators and researchers as sources of information. These approaches include cross-sectional studies, time-series analysis, and case studies, among others.

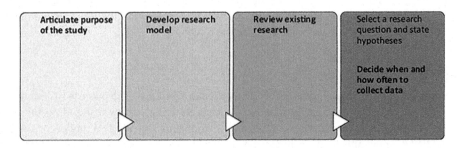

Figure 1.3 Beginning a Research Project

Notes

1. See G. King, R. Keohane, and S. Verba, "The Importance of Research Design," in *Rethinking Social Inquiry: Diverse Tools, Shared Standards*, 2nd ed., eds. E. Brady and D. Collier (Lanham, MD: Rowman and Littlefield, 2010), 181–192.
2. Susan Gooden and Rajade M. Berry-James, *Why Research Methods Matter: Essential Skills for Decision Making* (Irvine, CA: Melvin and Leigh Publishers, 2018), 10, comment in similar fashion regarding the importance of research methods to administrators.
3. We discuss literature reviews further in Chapter 15.
4. Gary King, Robert Keohone, and Sidney Verba, *Designing Social Inquiry: Scientific Inference in Qualitative Research King* (Princeton, NJ: Princeton University Press, 1994), 5.
5. King, Keohane, and Verba, "The Importance of Research Design," 6.
6. John Holahan, Diane Rowland, Judith Feder, and David Heslam, "Why Is Medicaid Spending Increasing?" *Health Affairs* 12, no. 3 (1993), available at www.healthaffairs.org>doc>full?hlthaff 12.3.177.
7. Medicaid is a joint federal and state health care program that provides services primarily to low-income individuals and those with disabilities. States are required to offer certain services but have discretion regarding others. Hence, the Medicaid program can differ from state to state.
8. James Jaccard and Jacob Jacoby, *Theory Construction and Model Building Skills: A Practical Guide for Social Scientists* (New York: The Guilford Press, 2010).
9. A. Kaplan, *The Conduct of Inquiry: Methodology for Behavioral Science*, 2nd ed. (New Brunswick, NJ: Transaction Publishers, 1998). Kaplan refers to this distinction as "logic in use"—how research is actually done—and "reconstructed logic"—how the textbooks and researchers tell us that research is done.
10. Extensive electronic resources on the scientific method are available. For an interesting look at science as a culture, see Bruno Latour, *Science in Action: How to Follow Scientists and Engineers Through Society* (Cambridge, MA: Harvard University Press, 1987). Another valuable text in this area is Paul Diesing, *How Does Social Science Work? Reflections on Practice* (Pittsburgh, PA: University of Pittsburgh Press, 1992). A continually evolving example of scientific method and practice is the research on DNA.
11. For information on writing a literature review, see: www.lib.ncsu.edu/tutorials/litreview; Jose Melissa Galvan, *Writing Literature Reviews: A Guide for Students of the Behavioral and Social Sciences*, 7th ed. (New York: Routledge, 2017) An Internet search will lead to additional guides. We discuss conducting literature reviews in more detail in Chapter 15.
12. For a discussion on empirical analysis as part of model building, see J. W. Tukey and M. B. Wilk, "Data Analysis and Statistics: Techniques and Applications," in *The Quantitative Analysis of Social Problems*, ed. E. R. Tufte (Reading, MA: Addison-Wesley, 1970), 370–390. Although an older reference, this article is still valuable. The authors warn against exploring data with no model in mind and advocate that researchers not take their models too seriously or be unwilling to change them.
13. Chapter 3 in Wholey, Newcomer, and Hatry. See specifically John A. McLaughlin and Gretchen B. Jordan, "Using Logic Models," in *Handbook of Practical Program Evaluation*, 4th ed., eds. J. Wholey, K. Newcomer, and H. Hatry (San Francisco: John Wiley & Sons, Inc., 2010), Chapter 3, 55–80.
14. Gooden and Berry-James, *Why Research Methods Matter*.
15. Ralph Hummel, "Stories Managers Tell: Why They Are as Important as Science," *Public Administration Review* 51, no. 1 (1991): 31–41. Gooden and Berry-James also cite Hummel.
16. King, Keohane, and Verba, "The Importance of Research Design," 30. The error involved in applying results from higher-level cases to individuals is sometimes called an "ecological fallacy" or "ecological inference fallacy". See S. Lieberson, *Making It Count: The Improvement of Social Research and Theory* (Berkeley: University of California Press, 1985); W. S. Robinson, "Ecological Correlations and the Behavior of Individuals," *American Sociological Review* 15 (1950): 351–357.
17. See A. J. Meltsner's, "Problem Selection," in *Policy Analysis in the Bureaucracy* (Berkeley: University of California Press, 1976), 81–113, for a provocative discussion on problem selection.
18. Graduate programs in public administration and public policy often have capstone courses in which students conduct research projects for local governments and not-for-profit organizations. For example, students in a Master of Public Administration (MPA) class at UNC Charlotte have recently completed projects on fire station location, community center feasibility, and citizen satisfaction with specific services.
19. Gooden and Barry-James, *Why Research Methods Matter*.

Terms for Review

models
model building

literature review
pilot study
schematic model
symbolic model
variables
constants
hypothesis
values of a variable
independent variable
dependent variable
unit of analysis
covariation
positive relationship
inverse or negative relationship
nonlinear relationship
random or null relationship
spurious relationship
control variable

Questions for Review

The following questions should indicate whether you have a basic competency in this chapter's material.

1. Why should an investigator state a purpose for a model before building it?
2. Why should an investigator build an explicit, preliminary model before beginning a quantitative study?
3. Identify the steps involved in model building and comment on their importance. Select one of the steps and consider the effect on the model if it were eliminated.
4. In general, would a model built by one person be superior to a model developed by a group? Justify your position.
5. a. In the following hypotheses, identify the independent and dependent variables, two or three possible values for each variable, and the direction of the relationship. Then suggest two control variables for each hypothesis:

    i. Death rates in automobile accidents are higher in less densely populated areas.
    ii. The higher the average driving speed on a highway, the higher the automobile death rate on that highway.
    iii. Defendants with records of alcohol abuse are more likely to miss scheduled court appearances.
    iv. Parents of elementary and high school students are more satisfied with their children's schools than are parents of junior high students.

   b. Revise hypotheses i and iii to improve them.

6. Develop three hypotheses, each with an independent and dependent variable on a topic of interest to you. Then suggest a control variable for each hypothesis. (Create a schematic model including the control variable for each hypothesis.)
7. Consider safety violations as an independent variable. Think of three outcomes that may be linked to safety violations. Write three hypotheses, each of which includes one independent variable and one outcome. Indicate the direction of each hypothesis.

8. Using criteria discussed in Chapter 1, evaluate the following proposed "hypotheses":

    1. H_1: School boards should not be appointed.
    2. H_2: There are more African Americans than Hispanics on U.S. school boards.
    3. H_3: Appointed school boards are more likely to have African American members than elected school boards.
    4. H_4: African American school board representation is measured by the percent of school board members who are African American.

9. Identify the independent variables and their values in the hypotheses:

    1. H_1: Older mothers with a college education are most likely to participate in parks and recreation programs; younger mothers with only a high school education are least likely to participate.
    2. H_2: Older mothers and more educated mothers are most likely to participate in parks and recreation programs.
    3. What, if any, changes would you make to H_1 or H_2?
    4. Imagine that a researcher wants to compare the percent of adults in U.S. states enrolled in the Medicaid program. Using the general information about the Medicaid program given in the chapter, what is the researcher's unit of analysis?

Problems for Homework and Discussion

1. Process models or flowcharts are often developed to describe and guide a process. One may use a process model to describe ways to build a model. Use a schematic to illustrate what you consider a "good" way to build a model. Show all linkages and their direction. You should have described either a systematic step-by-step process or a dynamic process in which the steps are repeated and the model is revised several times. Write a one-page defense of the process you described.
2. Link the purposes of a literature review with the stages of model building. At what stage(s) of the model would you conduct a review, and what would be the purpose of the review at each stage?
3. Find a newspaper article that presents a verbal model.

    a. State the apparent purpose of the model.
    b. Draw a schematic to illustrate the model.
    c. State a hypothesis included in or implied by the model.
    d. Name and identify the independent and dependent variables in your hypothesis.

4. Because of your model-building skills, you have been asked to head a committee studying the job-training needs of a town's labor force. The committee's purpose is to identify ways to improve the quality of the town's labor force to meet the needs of existing employers and to attract new employers to the area.

    The committee's first meeting is next week. A committee member who has no particular political power wants to survey citizens and community leaders; she has offered to prepare a survey for the meeting.

    a. Would you encourage her to bring a draft survey? Justify your decision.
    b. Outline your agenda for the meeting.

5. Consider the problem of productivity in the United States. Identify recent publications that address the problem. Look at three of these publications. Find a definition for *productivity*. Identify the variables the authors use to study productivity. Create a tentative model to

study productivity. Then state the purpose of your model. Sketch a schematic model that incorporates the variables you think are most important.

6.   Many communities must take water conservation measures in the summer. List strategies to get citizens to decrease their water usage. Develop a water conservation model that includes water conservation strategies and strategies for getting citizen compliance.

In class, work with three to five classmates. Compare your models and jointly develop and sketch a model that you think should be tested.

7.   Consider automobile accidents in the United States. Your instructor will divide the class into groups of three or four. Each member will locate and summarize a recent publication that studies the problem. Using this information, the group will complete the following:

a.   Identify three to five variables linked to automobile injuries or fatalities.
b.   Create a preliminary model to arrive at a policy to reduce injuries and fatalities resulting from automobile accidents. State the research question and a purpose for the model. Sketch a schematic model that incorporates the variables that you consider most important.
c.   State hypotheses to relate each independent variable in the model to the dependent variable.
d.   Prepare a presentation explaining and justifying your model. Your instructor may provide alternative topics.

8.   Select one of the following topics or a topic assigned by your instructor:

1.   Teenage pregnancy (incidence, policies, or programs)
2.   Juvenile crime (incidence, policies, or programs)
3.   Quality of drinking water
4.   Health insurance reform (policies, effectiveness)
5.   Municipal finance (innovations)
6.   Lobbying by not-for-profit organizations (policies, activities)

a.   Use a web search engine to locate and explore a site relevant to the topic chosen. What information did you find at the site that could help you design or carry out a study?
b.   Using an electronic literature database, search for research articles on the topic. How many entries did you receive? How many seem worth consulting?
c.   With a group of classmates, develop a guide to effective online searching.

Working With Data

Two data files, two computer files, and three Word files accompany this text. The data files are: (1) an Excel spreadsheet, County Data File, containing several variables for all 100 counties in a state and several variables from the 2018 version of the General Social Science nationwide survey. The data in the County Data File are from 2017 unless otherwise noted. The data were obtained from a central dataset of North Carolina data maintained by the state government and from a ranking of counties assembled by the Robert Wood Johnson Foundation. The sources of the data and the variables are described in a Word file, County Data File Notes, accompanying the data file. The items on the survey are included in a Word file; the data are in an SPSS (Statistical Package for the Social Sciences) file and in a Stata file. Stata is a more recently available, popular statistical analysis program. Another file, Focus Group Report, is a report of a focus group study. (Focus groups are discussed in Chapters 2 and 6.) While it is tempting to go immediately to a dataset and begin analysis, researchers should first state the questions they are trying to answer.

1. a. Access the County Data File and variable descriptions and review the list of variables. State three questions a state or county administrator could answer by analyzing the data. These questions should relate to policies that would normally concern elected officials and administrators.

    b. Select two variables and state (1) one hypothesis with a positive relationship and (2) one with an inverse, that is, negative relationship.

    c. For each hypothesis, identify a third variable from the dataset that could serve as a control variable.

2. Access the General Social Survey file and review the questions and variables. State two research questions that would be of interest to someone doing research on citizen satisfaction with government services.

    State two hypotheses: (1) one with a positive relationship and (2) one with a negative or inverse relationship. Ideally, your hypotheses, when tested, should provide information to help answer the research questions.

Recommended for Further Reading

Information technology is changing rapidly. Many texts exist to help Internet users with research, but most become quickly outdated. Reference librarians are an excellent source for learning about new technologies and how to use them effectively. See Hartman, Karen, and Ernest Ackermann, *Searching and Researching on the Internet and the World Wide Web*, 5th ed. (Wilsonville, OR: Franklin, Beedle and Associates, 2010).

The following volume contains excellent discussions of model and theory building. The authors include exercises for students of different disciplines. See Jaccard, James, and Jacob Jacoby, *Theory Construction and Model Building Skills: A Practical Guide for Social Scientists* (New York: The Guilford Press, 2010).

For information on writing a literature review, see: www.lib.ncsu.edu/tutorials/litreview; J. M. Galvan, *Writing Literature Reviews: A Guide for Students of the Behavioral and Social Sciences*, 7th ed. (New York: Routledge, 2017). An Internet search will lead to additional guides. The NCSU site has a good video narrated by an instructor. Galvan's text has examples of literature reviews.

2   Research Designs for Description

In this chapter, you will learn:

1. About cross-sectional, time-series, longitudinal, and case study designs, and the criteria for selecting each.
2. How to interpret findings from cross-sectional and time-series designs.
3. In-depth case studies and how to judge their quality.
4. Meta-analysis and how it is used to combine results of several studies.
5. The differences between quantitative and qualitative research.
6. Focus groups as study designs.
7. Observational data and its limitations.

After formulating a tentative model to guide the research, investigators need to select a design. In it, they outline how they intend to collect data on each variable and how they plan to analyze the relationships among the variables. When you study tables, graphs, or other quantitative presentations, you may never think of the decisions and actions required to gather and organize the data they illustrate. Yet these decisions and actions determine the value of the research, and investigators should conduct each step carefully.

Research Methodology

Research methodology is a structured set of steps and procedures for completing a research project.[1] The quality of a set of data is determined by the research methodology, even though discussion of the methodology might be limited in a final report. For this reason, it is altogether too easy for readers or listeners to underestimate the importance of research techniques and to take for granted the accuracy of reported findings.

The data collection and analysis of research methodology include the following steps:

1. deciding when and how often to collect data
2. developing or selecting measures to "operationalize" each variable
3. identifying a sample or test population
4. choosing a strategy for contacting subjects
5. planning the data analysis
6. presenting the findings

Research designs are plans that guide decisions about when and how often to collect data, what data to gather, from whom and how to collect data, and how to analyze data. The term *research design* has a general and a specific meaning.

- *General meaning:* The general meaning of research design refers to the plan for the study's methodology. The design should indicate the purpose of the study and demonstrate that the plan will answer the research question(s) and is consistent with the study's purpose. Frequently, research designs are described as blueprints for the final research product.
- *Specific meaning:* The specific meaning of research design refers to the type of study. Common types of studies are cross-sectional studies, time-series studies, case studies, and experiments. These types of studies or designs dictate when and how often to collect the data and how much control an investigator will exert over the research environment.

Designs to Find Relationships and Show Trends

Cross-sectional studies, time-series studies, and case studies place researchers in an environment where they have little, if any, control over the events. In general, all these studies use what is referred to as "observational data."[2] In contrast, experimental studies allow the researcher to control the research environment. The experimental setting and subjects are selected carefully, and the investigator decides who will be exposed to a treatment or intervention, at what intensity, and for how long. Only with experimental data can we clearly test whether one variable *causes* a change in another. Accordingly, we label the experimental designs as designs for explanation and will discuss them later in the chapter. Yet there are many phenomena that we might think of as important explanatory variables, for which random assignment is impossible. One example is the unemployment rate across different cities. We cannot randomly assign some cities to one level of unemployment and other cities to a different level. Another example is the occurrence of a particular gene across people. A researcher cannot randomly assign a gene associated with cancer to some participants in a study. Therefore, when studying such phenomena as explanatory variables, we are limited to observational data and accordingly limited in our ability to make causal claims. With observational data, we can at best test whether one variable *relates to* another in a manner that may *suggest* causality.

Thus, we label and categorize cross-sectional studies, time-series studies, and case studies as designs for description. These three designs for description may be used separately, or they may be combined. Cross-sectional studies are often combined with time-series studies, and case studies may incorporate features of cross-sectional or time-series studies. We refer to this as triangulation when multimethods are employed.[3]

Descriptive designs provide a wealth of information that is easy to understand and interpret. They may identify problems and suggest solutions. Such studies can be undertaken to answer questions such as: How many? How much? How efficient? How effective? How adequate? The designs are used frequently to produce the data needed for planning, monitoring, benchmarking and evaluating. For instance, administrators can combine findings from descriptive studies with other information and decide what, if any, action to take. Consider once again the problem of prison suicides. With stricter enforcement of drunk-driving laws, reports of prison suicides have increased.[4] These reports led investigators to collect data on how many prison suicides were committed each year and who committed suicide. For each reported suicide, investigators noted whether the prisoner had been incarcerated for a violent crime, a nonviolent crime, or being intoxicated or was being held in isolation. Upon analysis, investigators found that suicides tended to occur among males incarcerated for drunkenness or kept in isolation.

The analysis did not tell prison officials *why* one male prisoner chose to commit suicide and another did not. But the data did tell them how widespread the problem of prison suicide was and what characterized the inmates who were more likely to commit suicide.

What could prison officials do with the information? They could identify how many inmates in their institution were at risk of committing suicide. In other words, how many males did they have incarcerated for drunkenness or placed in isolation each day? Was the number of such incarcerations steadily increasing? Did the number go up on weekends or during certain times of the year?

Officials could propose and evaluate solutions to prevent those prisoners at risk of committing suicide from doing so and determine the feasibility of implementing these solutions. If a solution involves separate facilities or increased staffing, how large a facility is needed to take care of anticipated peak needs? What changes, if any, need to be made in the current staff work schedules? If a solution is tried, its effectiveness can be monitored. Did the rate of suicides drop? If a suicide occurred, had the inmate been identified as being at risk? If identified, had he received the available treatment for a potential suicide?

To answer these and many other questions, data are required. Other examples would lead to different questions requiring different data. Administrators require data to carry out their tasks. Administrators who understand how data were collected, whether they were involved in the collection, will be better prepared to use the data effectively and efficiently.

The administrator who wants to know what *caused* a particular event or outcome will find descriptive designs somewhat limited. Nevertheless, all three descriptive designs discussed in this chapter can eliminate untenable explanations and furnish valuable leads. As King, Keohane, and Verba state in their discussion of this topic: "Good description is better than bad explanation."[5] And, if the designs are planned and analyzed carefully, they may produce reasonable estimates of a treatment's effect on a dependent variable. Thus, in some specific circumstances, descriptive designs may suggest causality.[6]

Cross-Sectional Designs

A cross-sectional design collects data on all relevant variables at one time. A researcher may collect data from various sources—such as surveys, experiments, forms, or a database—and create a unique database. For example, to study factors associated with traffic fatalities, a researcher may collect data from several sources for each state on the number of fatalities, road conditions, traffic density, arrests and penalties for various traffic offenses, and so on.

Two analogies are often used to describe cross-sectional designs. In one analogy, the design is viewed as a physical "cross section" of the population of interest. In the other analogy, the design is seen as a "snapshot." Both analogies underscore the static, time-bound nature of the design. The cross-sectional design depicts what exists at one point in time. Clearly, events may change markedly at a later time, even in the next time interval.

Cross-sectional designs should be used to

- answer questions of how many? How much? Who? How efficient? How effective? How adequate?
- gather information on people's knowledge, attitudes, and behavior
- stimulate exploratory research and identify hypotheses for further research

Cross-sectional designs are particularly suited for studies that involve collecting data

- on many variables
- from a large group of subjects
- from subjects who are dispersed geographically

Using Surveys With Cross-Sectional Studies

We normally think of cross-sectional studies in conjunction with surveys, whereby an individual or a representative of an organization answers a questionnaire or fills out a form. Analysts then compile the data and analyze the variables. Imagine a survey to answer the research question, "How do the careers of executives working in the public and nonprofit spheres compare?" The model may have included variables such as gender, age, professional education, and professional experience. An advantage of well-designed, well-documented, and carefully implemented cross-sectional designs is that researchers with different interests and models often can work with data from a single cross-sectional study. This particular set of survey data could then be analyzed by other researchers whose primary interest is how gender, age, or academic preparation is related to career outcomes. The use of data by researchers who did not participate in their collection is called *secondary data analysis.*

The media, foundations like Pew Charitable Trusts, or state or local government themselves often sponsor surveys on current issues or citizen satisfaction. For example, live interviewers or an online survey might ask respondents about their knowledge, opinions, or behavior. Then the particular media outlet reports the percent of respondents in the various categories: how many are familiar with an issue, how many agree or disagree with a stated opinion, how many act in a particular fashion. The reports may indicate how responses vary according to demographic characteristics: Do women feel differently from men? Do people who live in urban areas and rural areas behave differently from citizens who live in other areas of the United States? And so on. Undoubtedly, as you read this, you also will begin to note the many studies conducted and reported with the goal of better understanding ourselves and our society.[7]

Over 100 U.S. federal agencies use cross-sectional designs to collect data. The U.S. Census Bureau, a major source of demographic and economic data, regularly surveys individuals, governments, and businesses. These data, which constitute official statistics, may provide a snapshot of the state of the nation as they report the status of its people and of the natural and economic environments. The data guide the decisions of public agencies, nonprofit organizations, and businesses. For example, data on age distribution help estimate demands for services. With knowledge of how many young children are in a community, planners can predict how many children will enter public schools each year. Childcare providers can develop business plans based on the number of children eligible for care, where they live, and their characteristics. Nonprofits may use the data to support the need for a proposed service. Some online surveys are conducted on a regular basis from different cross-samples in each year, such as YouGov, which is the Cooperative Congressional Election Study (CCES survey), and the General Social Survey (G.S.S). It uses a matched random sample methodology to solicit respondents from an online opt-in survey pool that match a sampling frame of adult citizens. The sampling frame used the American Community Survey and weights create a sample that is representative of the demographics of the nation.[8]

Assembling Data in Cross-Sectional Studies

A researcher can create a cross-sectional database by combining data from different sources, as previously mentioned. A specific strategy is to collect data from records. Employee files, volunteer applications, client records, budgets, and grant applications are all potential data sources. Employee files can yield detailed information on employee characteristics, which can be linked to performance evaluations, patterns of absenteeism, or turnover data. Applications for volunteer positions provide similar information. In both cases, the information may identify gaps in employee skills or volunteer interests that can help in designing training or recruiting strategies. The possibilities are almost endless and depend on the needs of an organization and the imagination of a researcher.

Cross-sectional studies may provide data on just one variable (e.g., the educational level of the population), but their greatest value is in describing the relationships among several variables.

When the data are analyzed, cases are divided into different groups based on values of the independent variables. Basing a cross-sectional study on a carefully developed model helps to ensure that appropriate information is collected and provides a guide for data analysis. Although such studies can provide useful information and be effective in testing hypotheses, they may suffer from less-than-careful research techniques. For example, if models are not thought out or surveys are not adequately tested in advance, important questions may be omitted or misunderstood by those completing the survey.

Limitations of Surveys

Although surveys can be very useful for collecting data, they unfortunately are often poorly designed and implemented. An investigator may attempt to obtain information from too many people, be unable to make return calls or send follow-up mailings, or be plagued with a low response rate. Surveys using telephone interviewers are increasingly difficult to conduct because the heavy use of mobile phones has led to more sophisticated ways of screening calls from unknown persons. This has led to the reliance on online surveying techniques that attempt to match random sampling. Questionnaires may also be constructed without enough care given to purpose, design, and question wording. Investigators should conduct a pretest, that is, a small study to test the adequacy of a data collection instrument or procedure. If they fail to do so, the survey instruments might also ignore relevant variables, or people may just not understand the survey. Respondents, especially those contacted by persons wanting data on public and nonprofit agencies and their administrators, may suffer from "respondent fatigue." A 20-minute survey, which may seem short to its designer, can be burdensome to an administrator who receives several such requests, has limited interest in the topic, or has to take additional time to think about an answer or gather data. The result may be a low response rate. If the investigator has no way of contacting nonrespondents or knowing their important characteristics for comparing samples, this problem is compounded. In some instances, surveys may produce the wrong data or have little or no usable data. They may have excluded critical variables. For example, a survey of state government employees received 33,000 responses and took 550 hours of staff time to process. All that could be done with the data was to report the answers to each question. The survey included only one independent variable, and that variable turned out to be unusable. The question asking respondents to indicate where they worked was phrased in such a way that the analysts could not tell whether a person who answered "Personnel" was referring to the state office of personnel or another agency's personnel office. Such problems are preventable. Administrators should not begin collecting data without critically evaluating the survey items and data collection process to see whether the survey will be able to answer the research questions. An important safeguard is to conduct a pretest of the survey well in advance. Also, before developing questionnaires, steps should be taken to understand how respondents conceptualize the problem. This can be done by conducting intensive interviews or holding focus groups to make sure the researcher is asking the correct questions.

Analyzing Data in Cross-Sectional Studies

The data produced from cross-sectional designs can be analyzed in any number of ways using a variety of statistical tools. In this chapter, we will use only frequencies and percentage to illustrate how cross-sectional data can be analyzed descriptively. Example 1.3 in the previous chapter, drawn from a cross-sectional design, was implemented to learn the placement record of state-funded job-training programs.[9] Prior to the study, administrators had no data on the success

Table 2.1 Age and Employment Status of Participants in On-the-Job Training Programs

Age	Employed After Training
16–21 years	40 (51%)
22–24 years	37 (74%)
25–54 years	84 (70%)
>54 years	3 (75%)

of training programs in placing participants. The data on job-training program participants were collected from state government files. The dependent variable was employment status at the end of the training program. The independent variables were type of training program, age, gender, race, and educational attainment.

For each type of training program, analysts examined placement outcomes by age, gender, race, and educational attainment. They found that participants in on-the-job training programs were most likely to be employed. Males were more likely to be employed than females; employment status varied with age but did not vary with race or educational attainment. Tables similar to Table 2.1 were produced. Administrators could study the detailed tables to learn whom the programs served and how well a program did in placing its various client groups.

Example 2.1 comes from a cross-sectional study conducted for the U.S. Internal Revenue Service to

1. determine satisfaction with online or electronic filing (E-filing) as well as ideas for product improvement
2. identify reasons for not filing online
3. compare perceptions of online filers and other taxpayers
4. determine ease of communicating with the IRS

The study, typical of cross-sectional studies, gathered data on many variables. The example here, "Application of a Cross-Sectional Design," is from one part of the analysis. We chose an example that illustrates how cross-sectional designs can guide marketing decisions. We assume that the research team generated ideas (variables) to explain why some taxpayers choose to use E-file and others do not. We included the IRS table, which considered both how users varied in their perception of E-filing and what attributes of filing taxes were important to them.

Example 2.1 Application of a Cross-Sectional Design

Problem: The Internal Revenue Service (IRS), the U.S. tax collection agency, compares taxpayer satisfaction with its products and services. One study compares online filers (users of E-file), lapsed users (those who discontinued using E-file), and nonusers. Comparison variables include demographic characteristics, ease of communicating with the IRS, perceptions of E-file features, and perceptions of the IRS. Are attitudes about E-file attributes associated with product use?

Design: A telephone survey of 1,000 randomly sampled employed taxpayers between the ages of 18 and 74.

Findings: From 977 people who provided information.

Table 2.2 IRS Survey Results

Agreed That	Type of User		
	Current Users (599)	Lapsed Users (106)	Nonusers (272)
Having accurate return			
is really important	95%	92%	91%
is true of E-file	69	53	39
Being assured return is private/secure			
is really important	92	92	87
is true of E-file	55	34	26
Easy to use/little hassle			
is really important	78	75	73
is true of E-file	65	30	32
Being inexpensive			
is really important	71	70	60
is true of E-file	53	45	40
Getting return to IRS			
is really important	71	58	46
is true of E-file	82	67	61
Getting refund faster			
is really important	61	44	42
is true of E-file	78	62	54
E-file is a better way to file your federal income taxes	63	26	27

The analysts concluded that attitudes were related to usage. They noted that while accuracy, security, and ease of use are important to all taxpayers, nonusers and lapsed users "have NOT gotten the message of E-file's benefits in three areas they actually care a lot about—Accuracy, Privacy/Security, and Ease of Use."

Source: Adapted from "Findings from the 2005 Taxpayer Satisfaction Study," U.S. Department of Treasury, Internal Revenue Service, July 2005. Publication 4241, Catalog 37303Q. Available at https://web.archive.org/web/20151207231454/www.unclefed.com/IRS-Forms/2005/p4241.pdf.

Limits of Cross-Sectional Designs

Cross-sectional studies are generally inappropriate if investigators want to learn *why* something happened. As noted previously, one cannot demonstrate with cross-sectional, observational data that a treatment, intervention, or other independent variable *caused* a given outcome. Except in the case of "experimental research designs," investigators cannot exert the needed amount of control over the intervention or over the environment surrounding the study. Investigators may be unable to rule out alternative explanations as to why something happened. For instance, in the preceding job-training program example, investigators cannot eliminate the possibility that participants in on-the-job training programs were more likely to be employed than nonparticipants because of *observable*—and *unobservable*—factors associated with self-selecting into job-training programs.[10] This kind of concern is the major peril of using observational data, and cross-sectional observational data in particular. Nevertheless, a major use of cross-sectional designs is to uncover relationships that can be studied further in experimental studies.

Studies on the effects of estrogen are a good example of how cross-sectional research can inform and motivate experimental studies. In 1985, a noted medical journal published two cross-sectional studies on estrogen.[11] One study examined data on 1,234 postmenopausal women living in a Boston suburb. The researchers found that estrogen users in this sample were twice as likely as nonusers to experience heart disease. In another study, researchers surveyed 121,964 postmenopausal nurses and found that estrogen users were one-third as likely to experience heart disease. Design features may have led to the different findings. The studies had different sample populations. One study used a mailed questionnaire to collect data; the other relied on personal interviews and physical examinations. One study collected data over a four-year period; the other over an eight-year period. The results of both studies were plausible theoretically.[12]

Statistical association is insufficient evidence that a treatment or program causes an observed outcome. Experiments are designed to produce evidence of causality and the effect that a treatment has on a dependent variable. To see whether hormones were actually beneficial, the Women's Health Initiative recruited and randomly assigned 27,347 women to take either hormones or a placebo. The pills looked the same, and the participants did not learn whether they had received hormones until the end of the study. The study was discontinued after 11 years when women who had been taking hormones were found to be at greater risk for heart disease.[13]

Time-Series Designs

Time-series studies collect and present data on a single unit or set of subjects. The data are collected on the same variable(s) at frequent, closely spaced regular intervals over a relatively long period. The data can depict both short-term changes and long-term trends in a variable. Most of us are familiar with time series that regularly report indicators of some aspect of the nation's economic or social climate. Such time series include consumer price indexes, the unemployment rate, and crime rates.

Time series are suited for situations in which an administrator wants to do the following:

1. establish a baseline data measure
2. describe changes over time
3. keep track of trends
4. forecast future trends
5. evaluate the impact of a program or policy

Time-series data are neither hard to gather nor hard to arrange for analysis. The data may be gathered by investigators or taken from existing databases. The data come from one or more units—such as a state, county, or office—and may be collected on the first workday of every month, the fourth Thursday of every November, or any other interval appropriate to the study. An accompanying account of events should be kept and consulted to help explain unexpected patterns: For example, a drop in swimming pool usage may be associated with an unusually cold or rainy summer.

Administrators find time-series data presented in graphs or tables easy to interpret and to combine with other information. They use time series to monitor programs and activities under their jurisdiction. For example, administrators in a social services department may decide to reallocate resources within the department based on the number of clients served, the number of services offered, and the length of service time.

A time series may or may not include an explicit independent variable. Time, however, is an implied independent variable. The data are frequently presented in a graph with time along the horizontal axis and the dependent variable along the vertical axis. The reader's attention focuses on the dependent variable and its changes or variations over time. Of course, the passage of time itself is not a variable. Rather, events and actions that take place during the passage of time may correlate to the variable graphed in the time series. However, using time allows us to understand the trend, up or down, in the variable of interest. Then investigators can ask questions about what is happening over time to cause the trend.

Looking at a Time-Series Study

Let us consider an example. Figure 2.1 shows two time-series lines—one represents the unemployment rate and the other the property crime rate. Property crimes are burglary, larceny-theft, motor vehicle theft, and arson. The data help illustrate several features of time series and show how to examine the relationship between two time series.[14]

Note that this example reports U.S. data for the entire country; the employment data are collected from a sample of U.S. households, and the property crime statistics are from individual law enforcement agencies. You could have done the same analysis on a single state, county, or city.

Now, what do you look for in this time series? Your eye should scan each line to identify the overall patterns and noticeable dips and rises in the line. You look for four types of variations within a time series that reports yearly or monthly data.[15] These variations and their definitions are included in Table 2.3.

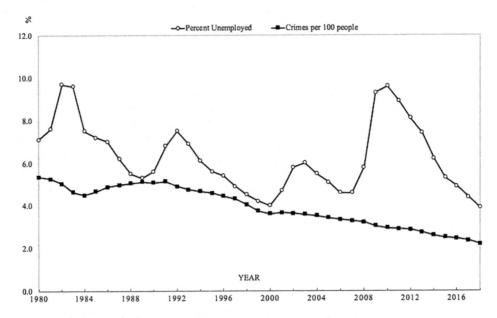

Figure 2.1 U.S. Property Crime Rate and Unemployment rate 1980–2016

Source: Estimated property crime rate from FBI Uniform Crime Reports, prepared by the National Archive of Criminal Justice Data, located April 13th, 2020, at www.ucrdatatool.gov/. Raw data are the number of reported offenses per 100,000 population. These were adjusted to the scale shown. Rate for 2018 is based on FBI estimate located April 13th, 2020. The unemployment rate is the seasonally unadjusted rate, located April 14, 2020, at http://data.bls.gov/timeseries/LNU04000000?years_option=all_years&periods_option=specific_periods&periods=Annual+Data.

Table 2.3 Variations Within a Time Series

Variation	Definition
Long-term trends	General movement of a variable, either upward or downward, over a number of years
Cyclical variations	Changes in a variable that occur within a long-term trend; cycles recur in one- to five-year intervals
Seasonal variations	Fluctuations traceable to seasonally related phenomena; for example, holidays or weather
Irregular (or random) fluctuations	Changes that cannot be attributed to long-term trends, cyclical variations, or seasonal variations

LONG-TERM TRENDS IN THE DATA

Look first at the changes in the property crime rate:

* 1980–1984: Property crime rate decreases.
* 1985–1991: Property crime rate increases.
* 1992–2000: Property crime rate decreases.
* 2001–2002: Property crime rate increases.
* 2003–2018: Property crime rate decreases but more slowly.

Variables neither increase nor decrease indefinitely. Changes in social, economic, and environmental conditions may lead to a change in direction. Since 1991, the U.S. crime rate, including property crimes, has steadily dropped, except for a small increase in 2001. Explanations for the decline have included the state of the U.S. economy, the increasing age of the U.S. population, and changes in the criminal justice environment (e.g., more police and more people in prison).[16] The time series for the property crime rate from 1984 through 2004 shows two distinct long-term trends. From 1985 through 1991, the property crime rate increased. Since 1991, it has decreased. The only deviation was a 1 percent increase in 2001. With a longer data period, we can see that 2001 and 2002 were slight deviation in a larger time trend.[17] The 2003–2018 data indicate that the increase was an irregular fluctuation during 2001–2002.[18]

CYCLICAL VARIATIONS IN THE DATA

Cyclical variations are regularly occurring fluctuations within a long-term trend that last for more than a year; frequently, cyclical patterns recur in one- to five-year intervals. In some cases, cycles may be very regular. For example, the percentage of Americans voting reaches a peak every four years during presidential elections. A complete cycle is from "peak to peak" or "valley to valley." The unemployment data from 1980 to 2004 have two cycles of approximately 10 years each. The data show long stretches when the unemployment rate kept decreasing. Between 1980 and 2012, despite increasing steeply at some points, periods of increasing unemployment rarely lasted more than three years. We can assume that political interventions brought about the change in direction. The cyclical variations, as illustrated by the unemployment rate, are more common than evenly spaced cycles. That is, up-and-down movements reoccur throughout a time series but not in a regular pattern of equal time periods.

SEASONAL VARIATIONS IN THE DATA

Seasonal variations describe changes that occur within the course of a year. Data must represent time intervals that recur within a year, such as days, weeks, months, or quarters. Seasonal variations include fluctuations traceable to weather, holidays, or similar seasonally related phenomena. The observed fluctuations occur within a single year and recur year after year. If we had included monthly data in Figure 2.1, you would have seen several fluctuations in any given year. Such fluctuations are considered seasonal variations only if a similar pattern is seen year after year. An administrator may use seasonal information to decide how to staff public facilities—for example, how many staff to hire for city parks and for how long to hire them (for 10, 12, or 14 weeks). A jail administrator finding that the number of people in jail is highest on Monday mornings can plan ahead for the increased demand. Ignorance of seasonal demands may result in erroneous conclusions. Imagine the disastrous consequences if a merchant were to assume that his December sales of toys marked a business upswing that would carry through to January and February. Think of the problems of the jail administrator who orders food for the week based on the number of inmates on Monday morning.

IRREGULAR FLUCTUATIONS IN THE DATA

Irregular fluctuations are variations not associated with long-term trends, cyclical variations, or seasonal variations. Irregular fluctuations may be the result of nonrandom or random movements. A *nonrandom movement* is brought about by a condition or set of conditions that can be identified and explain the variation. The conditions may be inferred from records of concurrent events. For example, negative publicity about how police handle crime reports may explain a drop in reported crimes, or a community may be struck by a natural disaster, which impacts employment. Another example is the increase in the unemployment rate from 4.9 percent in the fourth quarter of 2007 to 10.1 percent in the fourth quarter of 2009. If such factors are identified—for example, the United States experiencing a major recession in 2008–2009—we are likely to conclude that the variation was nonrandom. Otherwise, we may assume the variation is random. *Random movements* are unexplained variations that most often are relatively minor. However, many analysts would consider the large increase in the unemployment rate during that time more than minor.

The 1 percent increase in the property crime rate in 2001 was an irregular fluctuation that seems relatively inconsequential and random. A Google search suggested that the increase was largely ignored. Yet one item illustrated the hazards of prematurely assuming that a one-year shift marks the beginning of a longer trend. A research firm summarized the 2001 crime statistics and claimed confirmation for its prediction that beginning in 2000 the crime rate would increase. In fairness to the firm, it looked at more than the one data source. The researchers assumed that a weakening U.S. economy, increased spending on fighting terrorism, and the anticipated release of inmates reaching the end of their sentences would cause an increase in crime.[19]

For most administrative purposes, one simply needs to know the types of fluctuations and to recognize evidence of their occurrence in a table or a graph. Otherwise, you may misinterpret the ups and downs in a graphed time series and erroneously attribute changes in a time series to a specific event or administrative action.

It is important to draw a graph of the time series and analyze it visually. We will often see graphs with more than one variable, as is the case with Figure 2.1. However, for administrative purposes, the clarity of time series diminishes if several variables are examined together. In Figure 2.1, the graph included two variables: property crime rate and unemployment rate. You could scan the graph visually and note where the two varied together and where they went in different directions. Between 1980 and 1988, the two rates seemed to diverge. From 1992 through 2006, they both appear to be declining. After 2006, they again appear to diverge. If more variables were placed on the graph, it would become increasingly difficult to follow and interpret the data.

Forecasting With Time-Series Data

Time-series data are used frequently in forecasting or evaluating the effectiveness of a policy. In both cases, statistical techniques take into account the fluctuations of a variable. For the most part, these techniques are beyond the scope of this text. Nevertheless, here are some helpful comments about forecasting:

1. *Quantitative methods of forecasting that depend on time-series data work best for short-term forecasts* (e.g., forecasting for up to two years). A parks director or jail administrator may use quantitative techniques to forecast park usage or jail occupancy. This information may help in making staffing decisions or ordering supplies. But forecasts made for longer periods are less likely to be accurate. A major problem is correctly forecasting a change in direction of the long-term trend.

2. *Assuming that existing long-term trends may continue unchanged can work reasonably well, but the forecaster must be aware that rates of change eventually will vary and long-term trends will change direction.* For example, a town whose population has grown by

10 percent each decade will at some time experience slower growth. An unanticipated slowdown has serious implications if public facilities are planned around the assumption of a much larger population.

3. *Identifying long-term trends, cycles, and any seasonal variation is important in assessing the impact of a program or policy.* Without knowing how the variable has been changing over time, one may mistakenly conclude that a change in a time series is the result of the program or policy.

4. *Qualitative techniques are important for long-range forecasting.* An administrator who wants to identify the types of demands that her agency will face in 10 years may wish to use focus groups or interviews to obtain opinions from citizens and experts.

Other Longitudinal Designs

Longitudinal designs collect information on the same cases or comparable cases for two or more distinct time periods. Time series are also longitudinal designs. However, we treated time series in a preceding section because of their specific characteristics and common usage.

Panel designs refer to longitudinal designs that gather data on the same individuals each time.[20] Typically the cases are individuals, but cases can consist of families, organizations, or some other unit of analysis. The design combines some characteristics of the cross-sectional design and the time-series design. Because these studies can reveal which individual cases change between data collections, the researcher is able to determine the changes that take place within the group and establish a time order for variables. For example, a training director can organize unemployment data as a time series and note changes in the unemployment rate. However, unless a panel design is used, she cannot tell how individuals in this population have changed. That is, some people employed at one time may be unemployed at a later time; some unemployed at the earlier time may be employed later. Since a panel design follows the same individual cases over time, shifts by individuals from one condition (employed) to another (unemployed) can be measured.

Until the late 1960s, poverty was considered a relatively permanent characteristic of individuals. Repeated cross-sectional surveys showed the same proportion of Americans living in poverty, and so it was assumed that the same people were living in poverty each time. These studies showed almost no change from one year to the next in the distribution of income. However, a study of American families, the *Panel Survey of Income Dynamics*, found that roughly one-third of those who were poor one year were not poor the next and had not been poor the year before. The number of poor was stable because the number coming into poverty and the number leaving just about equaled each other.[21]

The way that information about the number of elderly receiving institutional care was obtained provides another example of the importance of panel studies. Federal, state, and local administrators have faced an increasing burden of providing care for elderly Americans. Cross-sectional samples of older persons taken at various times had consistently shown that about 5 percent of people age 65 and older were in nursing homes, hospitals, or other long-term care facilities. This often was interpreted to mean that any person 65 and older has only a 5 percent chance of being institutionalized. But researchers tracked a panel of individuals over several years and found that many people enter and leave long-term care facilities several times in their later years. At any moment, only 5 percent may be in nursing homes, hospitals, or long-term care facilities, but over a period of years, at least 20 percent of those age 65 and older will spend some time in one or more of these places.[22]

Framingham, Massachusetts, Heart Study

Another example of a large-scale panel study is the Framingham, Massachusetts, Heart Study.[23] In this study, 5,209 residents of the town of Framingham between the ages of 30 and 62 were recruited to submit to a detailed medical examination and lab tests every two years. The information from this study is being used to issue advice about diet, exercise, and other factors related to health conditions. In 1971, 5,124 children and spouses of the original participants were recruited for the offspring study. In 2001, researchers began the third-generation study, consisting of 3,900 of the original participants' grandchildren. These follow-up studies have allowed further investigations of risk factors for health problems and identification of genes that contribute to various diseases. The Framingham Heart Study is continuing—new exams of participants and offspring took place in 2014, collaborative studies are continuing, and medical history updates are being collected on a regular schedule.[24]

A major problem of panel studies is in obtaining an initial representative sample of respondents willing to be interviewed at set intervals over an extended period of time. Panel members also may drop out of the study for one reason or another. This is called subject attrition or experimental mortality. Another difficulty is that when people are the cases, the repeated interviews and observations may influence their behavior. When interviewed repeatedly over a period of months or years, people may change the way they answer questions in order to be consistent from one time to the next.[25]

Cohorts

When studies follow groups of cases, we often refer to these groups as *cohorts*. A cohort consists of cases having experienced the same significant event at a specific time period. They may be individuals, organizations, or some other unit of analysis.[26] The term *cohort* may refer to birth cohort: those cases born in a specific year or period. Cohorts may also be defined by the year they entered the study or by the occurrence of or exposure to a particular event, such as the year of having graduated from college or having fought in a particular war. In a panel study, investigators obtain information on the same individuals each time. But a study using cohorts may be conducted in any number of ways. Note that not all studies involving cohorts are longitudinal studies. A different sample may be taken from the cohort group each time data are gathered, or a researcher may survey members of a cohort, such as residents of New Orleans, and compare the responses of those who relocated after two hurricanes with those who returned.[27]

Meta-Analysis

Meta-analysis is a systematic technique to locate, retrieve, review, summarize, and analyze a set of existing quantitative studies.[28] It is also defined as "the statistical synthesis of data from separate but similar (that is, comparable) studies, leading to a quantitative summary of the pooled results."[29] Researchers conduct meta-analysis to draw general conclusions from several empirical studies on a given program or policy; develop support for hypotheses that merit further testing; and identify characteristics of a program, its environment, or its clients that are associated with effectiveness.

Imagine you want to learn whether early childhood education programs for disadvantaged children, such as Head Start, are successful. You probably will begin by reviewing the literature. Assume you find 10 recent research articles examining the effectiveness of different programs. You identify similar and dissimilar findings among these studies. As you classify the findings, you may compare program clients, note how cases were chosen for analysis, note how program success is defined and measured, and record any unique features in the program or the research. If the findings are dissimilar, you may look for an explanation as to why they differ. Differences in program features or in the conduct of research may account for dissimilar findings. If the findings are similar, you may ask whether the program would seem as effective if the research methodology were different. Or you may question whether the program would be as effective with other groups or in other locations.

Proponents of meta-analysis argue that simple reviews of literature such as that described here are ineffective for arriving at conclusions about research results, such as the effectiveness of Head Start. A researcher cannot summarize a large number of studies effectively, much less understand and interpret how the studies are integrated. If a subset of studies is selected for closer analysis, the reviewer's selection of studies may be colored by his own biases. In reviewing even 10 studies, the researcher may be unable to avoid inconsistent judgments. He may focus on sampling problems in one study and errors in data analysis in another.[30]

Meta-analysis, on the other hand, enables investigators to review a large body of literature and to integrate its findings. Meta-analysis requires the researchers to focus on the same specific components of every study analyzed. If you read a description of a meta-analysis, you should be able to repeat the researchers' steps and to come to the same conclusions. Unlike literature reviews, meta-analysis uses quantitative procedures in synthesizing the results of several studies.[31]

Conducting a Meta-Analysis

In conducting a meta-analysis, investigators record the same information from each study. Their goal is to identify hypotheses that are supported in study after study. In doing this, they have to eliminate alternative explanations, such as chance, which may account for the observed relationships. The analysts record the statistical information reported on the dependent variables of interest and the statistical relationships between independent and dependent variables. They also record information on the study itself, such as the dates of data collection and of publication, who the research subjects were, how the subjects were selected, the research design, evidence that research involved sound measures, and the type of publication (e.g., book, academic journal, or unpublished paper).

Once the data are collected, the investigators integrate the findings to create one database.

Integrating study findings involves quantifying specific dependent variables. Analysts also identify relevant independent variables and incorporate them into a common database. To illustrate the meaning of "integrating the findings," we consider two examples:

1. In one meta-analysis, analysts examined 261 citizen surveys. They identified survey questions that asked people to rate specific urban services such as trash collection and categorized the questions by the type of service being rated. Neither the question wording nor the possible responses were identical. The analysts ignored variations in question wording and created a common scale so that responses to similar questions from different cities could be combined. After the data were combined, the investigators observed that citizens in cities throughout the United States gave the highest ratings to arts programs and public safety and the lowest ratings to planning.[32]

2. The second meta-analysis examined studies on survey response rates. Analysts identified 115 studies to evaluate strategies to improve the response rate for mail surveys published between 1940 and 1987, with response rate as the dependent variable. Across the studies, the analysts identified 17 different independent variables. Each of these variables had been tested in at least three studies. After detailed analysis, the investigators identified two factors that increased response rate to mail surveys: including a cover letter with an appeal to respond and keeping the questionnaire short (fewer than five pages).[33]

Meta-analysis allows one to "see" a body of research rather than focusing on the quality of individual studies. A researcher may be skeptical if a hypothesis is supported only in poor-quality studies. On the other hand, findings may have even greater credibility if a hypothesis is supported no matter the quality of a study's methodology. Information on the dates of data gathering and publication help identify trends and changes in them. For example, the investigators in the response-rate study found that beginning in the mid-1970s, preliminary notification of subjects improved response rate.

Limitations of Meta-Analysis

A major problem in meta-analysis is locating a set of studies. At a minimum, the topic must have been of interest long enough for a research history to develop. Little or no research may exist on a current "hot" research topic. The researcher conducting a meta-analysis wants to include all appropriate studies or a representative sample of such studies in his analysis. Consequently, he must do a relatively exhaustive literature search. If he selects articles from a few journals or from a short time period, he runs the risk of working with a biased sample.

Each article selected does not have to include exactly the same dependent variables, nor must it study the same relationships. Of the 261 cities whose citizen surveys were analyzed, 70 percent rated their police. Less than 25 percent rated animal control or street lighting. The independent variables used in the response-rate study varied from study to study.

Critics of meta-analysis point to the problems introduced by biased article selection. Let's go back to our early childhood education example in which 10 publications were studied. First, we mentioned the problem of bias introduced if the studies analyzed do not represent the larger body of studies conducted on early childhood education. A critic may raise the related question of the "file drawer." The argument goes as follows: If many studies of a subject are done, chance alone will cause some hypotheses to be supported and the research reporting them to be published. Analysts who include only published studies may have a biased sample because many other unpublished studies on the same topic may be stashed in researchers' file drawers. Researchers put studies aside when preliminary data fail to support the hypotheses and the studies seem to be going nowhere. Social scientists in recent years have discussed the reasons researchers should publish null findings, but doing so is still arguably rare.[34] A statistical solution to this problem requires the analyst to calculate how many unpublished papers have to exist for the findings to be contradicted.[35]

Another criticism of meta-analysis has been termed the "apples and oranges" problem. The critics argue against combining dissimilar studies. Consider the citizen-survey data in which findings based on different question wording and different response patterns were grouped together to create a variable. Gene V. Glass, a major proponent of meta-analysis, argues that these critics want replication of research, not comparison of similar findings. Meta-analysis is intended to mine information from related, but not identical, studies. Furthermore, he argues that combining different studies is not much different from combining responses from different subjects.[36]

Meta-analysis may seem deceptively simple. Yet it is time consuming to identify the appropriate studies and to conduct the data analysis. Understanding the quantitative procedures used to synthesize results requires statistical knowledge. Readers who wish to try their hand at meta-analysis should be familiar with the statistics discussed in this text. Then they should review one or more of the books included in the "Recommended for Further Reading" section at the end of this chapter. They should read several meta-analysis studies for examples. Given the amount of research literature available and the difficulty and expense of researching some topics, a good meta-analysis may be a very efficient alternative to large-scale original research and increasingly feasible to conduct given advances in technology and data access.

Qualitative Research and Designs

Quantitative studies typically involve many cases and many variables that can be measured in a predetermined, specific way. The data are numeric and can be summarized numerically. Since an important goal of quantitative studies is to compare cases on several variables, factors unique to individual cases are not included and information about context is often ignored. For example, investigators use the designs described thus far to obtain information on a standard set of items for a large number of cases, as with cross-sectional studies, or information on as few as a single case for many time periods, as with time series.

While using quantitative research designs is important, these designs are not suitable for obtaining detailed information about the context in which events or behaviors occur, nor do they allow flexibility in the type of data obtained from case to case in the same study. To obtain this detailed information, investigators use more qualitative and fewer quantitative approaches. Qualitative research methods have long been important in basic disciplines as well as in applied areas such as administration. Studies using a qualitative research approach typically obtain more in-depth, detailed information on fewer cases than do studies using more quantitative designs. Two important qualitative designs, described later, are the case study and the focus group.

Qualitative Research

Qualitative research methods have received increased attention in recent years from both practitioners and researchers.[37] Qualitative research produces information or data difficult or impossible to convert into numbers without losing valuable context. The qualitative study is defined by its extensive use of such information, its preference for developing full information on cases typically below the number needed for quantitative analysis, and its consideration of the unique features of each case. Researchers may draw on both quantitative and qualitative methods, often known as mixed methods, in conducting any one study.

Qualitative studies may include information on the unique features and the environment of each case. They describe specific features of each individual, organization, jurisdiction, or program. Qualitative studies may involve extensive fieldwork; the researcher goes to where the cases are located and obtains information on them in their natural setting.[38] In this way, the researcher does not attempt to manipulate any aspect of the situation being studied but reports it as it is. Nevertheless, the qualitative researcher's background and personality influence data collection and interpretation. The researchers use their experiences and insights to design a study and to interpret the findings. A researcher's interactions with subjects affect what he is told and what information he is given.

Researchers using qualitative techniques need different skills than those required when using quantitative designs. An interviewer in a quantitative study receives a list of questions that she asks every respondent. All other interviewers would use the same set of questions and ask them in the same way. In a qualitative study, the interviewer may have a suggested set of questions but asks them as the situation dictates. Based on the response to one question, the interviewer asks another question; she needs to ask the question, listen, interpret, and phrase a proper follow-up question.[39]

The researcher using qualitative methods must be able to record information accurately, write clearly, differentiate trivial from important details, and draw appropriate conclusions from the information. Since data from qualitative studies tend to be descriptions, observations, and responses to interview questions, a large amount of information is obtained. To make sense out of it may take time and effort. Rather than doing statistical analysis of numerical data as in quantitative studies, the researcher looks for words, themes, and concepts in the analysis of qualitative data. These become the data from which the researcher draws conclusions, answers research questions, and develops additional hypotheses.[40]

Case Studies

Case studies examine in some depth persons, decisions, programs, or other entities that have a unique characteristic of interest. For example, a case study may be designed to study women in nontraditional jobs, a new approach to budgeting, or a high school health clinic. Authors of one text on the design of research state that case studies are essential for description and thus fundamental for social science. They also maintain that without good description, explanation is pointless.[41] Although case studies may be either qualitative or quantitative, here we only present the case study as a qualitative design.[42]

Except for studies that involve a single case within an administrator's jurisdiction, administrators seldom initiate or participate in the design of case studies. Even the single case study may be a by-product of someone else's research needs. For example, some state and local administrators seek out university students to conduct case studies. The studies give the students a "real-world" experience, and the administrators learn more about their agency and its programs.

Case studies are the preferred research strategy if an investigator wants to learn the details about how something happened and why it may have happened. Administrators may conduct a case study to investigate the following:

- a program or a policy that has had remarkable success (or one that has failed)
- programs or policies that have unique or ambiguous outcomes
- situations in which actors' behavior is discretionary

Although historical events can be the focus of a case study, the case should be contemporary so that the investigator has access to information. Ideally, the investigator will have direct access to the people, program, and practice details involved. Whereas analysts implementing cross-sectional or time-series designs may never contact program administrators, employees, or clients, an investigator conducting a case study cannot be so detached.

Information Derived From Multiple Data Sources

One of the hallmarks of a case study is the combination of several different sources of information. The sources of information used in case studies include documents, archival information, interviews, direct observation, participant observation, and physical artifacts, which may even include quantitative documents such as public budgets.

The inclusion of information from multiple sources is a major strength of case studies. First, each data collection strategy affects the types of questions a researcher can answer. For example, from direct observation, an investigator learns how people behave. From interviews, he hears their explanations of their behavior. The two sources of information give a more complete picture than either piece alone.

Second, the investigator can corroborate information gained from one source with information gathered from another source. For example, staff in a crisis intervention center may report a high level of community support. The claim will have greater credibility if an independent review of agency records confirms extensive community support.

Requiring information from multiple sources also is a drawback of case studies. Typically, different information sources are studied using different research techniques. For example, interviewing skills are needed for face-to-face meetings with subjects, questionnaire design skills are needed to conduct a mail or electronic survey, and content analysis skills are needed for archival research. Most of us are skilled in only one or two research techniques. Thus, a case study may require a research team or else suffer from unevenness. Administrators may find that they, and their staff, lack both the training and the time needed to do effective case studies.

Each research technique and data collection strategy takes time to design, pretest, and carry out. Incorporating multiple data sources and employing different techniques and possibly multiple researchers demands time, expertise, and energy. Consequently, researchers who conduct case studies often find that studying multiple cases is impractical because of the effort and resources required.

Focusing on the Components of a Case

The case study may focus on the case as a whole (all of its components) or just on certain components. For example, to study a state pretrial release program, the investigator may look at the program in its entirety: why it was developed, how it was initially organized, how its organization was changed since it began, why those changes were made, how defendants are chosen for pretrial release, and how much discretion staff and judges have in making pretrial release recommendations. Alternatively, the investigator may focus on the implementation of the program in one location or in a few locations. In the first instance, the program is the case. In the second, each specific location's program could be a separate case. Cases can also be nested, and the researcher may find more variation and depth to understanding why something is occurring by looking at smaller units within a larger case by studying counterfactuals within a particular unit of analysis. For example, a researcher wishes to learn about why polio has not been eradicated in Nigeria.[43] The researcher realizes that there are differences across Nigeria in the number of outbreaks based on the quantitative data. Reading academic articles and newspaper accounts leads to an understanding that in specific regions of Nigeria, people have different belief systems about the purpose of vaccines, and in a few specific communities public officials tell people to boycott polio vaccines as opposed to receiving them. Comparing different regions within a particular country may yield more insight into polio outbreaks than simply studying the country as one homogenous unit.

Most administrative case studies seem to focus on components. A case study may focus on one provision of a program rather than evaluating the entire program. For example, in a study of a pretrial release program, "successful" defendants who showed up for scheduled court appearances were compared with "unsuccessful" defendants who failed to appear. This is a limited case study. Limiting the study to the analysis of a single component does not take advantage of the strength of the case study design. A more valuable analysis would be to examine the circumstances of what promoted success, perhaps by exploring work, education, family support systems, and other variables in an in-depth manner.

Case studies may be conducted on a single case or on a set of similar cases. Each approach (using a single case study of a program or comparing two or more programs through case studies of each) would be valuable in determining the effectiveness of the program. An individual case study of a phenomenon can provide research depth. Multiple cases of a phenomenon provide research breadth. Ideally, researchers would like to have both depth and breadth. In reality, limited resources mean researchers often have to choose between the two, striking a balance that is appropriate for the task at hand. In some cases, this means either sacrificing some contextual information to include more cases or sacrificing having some more examples in order to explore just one or two in much greater detail.

Because of the potential value of case studies and the rich details they provide, you may want more information about them. If you read a case study, you may gain greater insight into how to approach or solve a problem. Practically speaking, administrators' direct experience with case studies may be limited to their professional reading. Case studies may alert you to new management techniques, programs to solve chronic problems, or strategies to improve the quality of agency or community life. If the case study is well done, you should have a good idea of whether you could implement a similar solution. Even if the case study documents a failure or otherwise does not apply, it may spur you to think more about your work environment and responsibilities. A creative administrator may read a case study, contact the researcher or program administrator to learn more details, and then be prompted to implement a new program or strategy.

What Makes a Quality Case Study?

Robert Yin carefully distinguishes between the case study as a type of design on the one hand and the data collection typically conducted in a case study on the other.[44] He also implies that most people incorrectly assume that any qualitative study is a case study. Although case studies tend to be qualitative, Yin recommends an approach for case studies that follows the scientific method. This approach entails the researcher stating a problem; formulating a research question, objective, or hypothesis; identifying the case to be studied; planning the data collection; collecting the data; analyzing the data; and writing a report. This also means the researcher should be able to assess the overall internal and external validity of the data collected.

In certain instances, particular case studies are intended to be only *preliminary* or *exploratory*. These are used for descriptive inference only. An investigator conducting an *exploratory case study* will have a research question but appropriately may decide to forgo formulating initial hypotheses. Instead, they will develop hypotheses from the case study once it is completed and draw generalizations from later studies. The exploratory case study serves as the basis for establishing new research questions, new hypotheses, and a continuing research agenda. Then, subsequent case studies will serve to thoroughly "test" that hypothesis.

Researchers may even conduct qualitative case analysis (QCA) of multiple cases paired with quantitative data to examine different ways in which different factors interact to yield a particular outcome.[45] This improves generalizations (but still not outside the sample) and allows for more in-depth insight because it integrates case studies with variable approaches, usually in samples ranging from 8 to 200. It also relies on Boolean algebra that requires each case to be reduced to a series of variables where context of the case and mechanisms produce an outcome.[46]

For example, if a public administrator wants to learn about whether neighborhoods are using a new application developed by the police department to report crime across the city, QCA may be a good tool to use. QCA could be used to find out if the neighborhoods behave in a similar fashion to what previous studies have found about the demographics, of users of other social media outlets such as Twitter or Facebook. If the neighborhoods are purposefully selected on key control variables, they do not need to sample all of them, just a purposeful sample based on demographics and key differences in the neighborhoods.

Case-study researchers need to be particularly careful to follow sound research practice.

- Be purposeful in multicase studies. Think carefully about case selection. Random samples with small numbers do not work well. Make sure the cases are intentionally selected based on the variation in the independent variable and not the outcome. The strength of the case study is being able to understand the context, which is often time consuming. Consider that you have more depth than breadth.
- Try to minimize and acknowledge biases, errors, and exogenous factors by thinking about how generalizable the relationships are across persons, settings, and times. Consider the interaction of selection, setting, and history with treatment. Provide an assessment of uncertainty.
- Make sure you are aware of the ethical and political issues involved in this type of research. Gaining access must be possible, and understand that specific data may not be accessible because of privacy or permission concerns. In particular, individual records may need to be de-identified and presented such that a person is not easily identified.

As part of the design, researchers must decide what constitutes a case and if this is the same as the unit of analysis. Doing so may not be easy.[47] Deciding on what the case is to be studied, for example, may be a problem if the case involves a program. Consider an agency program to aid abused children. The agency may have a program for abused children specifically, but it may also have services for abused mothers and for families with a potential for abuse. Should these services be included in the case study? The program may have evolved from an earlier program. At what point should the case start? Should the case study include the board of directors, clients, or others in the community? The investigator must answer these questions in order to limit, or set boundaries around, the case study. An ill-defined case can distort data, just as can a case without predetermined criteria for judging whether a hypothesis is supported. A poorly defined case can also allow the researcher to wander from her original research question.

As an administrator, you may want to evaluate the quality of the design so you can determine whether you are reading more than just a good story.

1. Look for evidence that the investigator had a specific, answerable research question.
2. Assess whether the investigator had a model before starting the study.
3. Determine whether the investigator decided the specific boundaries of the case.
4. Be sure that the case study's procedures from design through implementation are thoroughly documented, including any potential biases on behalf of the researcher.

Case studies may involve the investigator intensely in the case and require that the researcher interpret qualitative information. Consequently, case studies are hard to replicate, and great care must be taken to document what was done, how, and why. Transparency of a case study is important, as in any form of research, so that the strengths and limitations of the work are evident.

Focus Groups

Focus-group methods use group interviews to obtain qualitative data. Researchers have long used group interviews to save time and money by getting a number of people together to provide information.[48] Focus-group procedures have evolved in recent years and include a set of characteristics that distinguishes them from other group techniques. Although focus-group research is qualitative, it can be triangulated with more quantitative procedures. Survey researchers use focus groups to generate and test the items to be used on a questionnaire or survey instrument. Focus groups also are used to elaborate on data collected in surveys.

Focus-Group Discussions

Focus groups are semi-structured discussions by small groups of participants about a common topic or experience.[49] They are useful in obtaining information that is difficult to obtain with other methods such as individual interviews or survey data. Investigators use a focus-group interview to get in-depth information about and reactions to a relatively small number of topics or questions— typically fewer than 10 questions and often around five or six.

A study using focus groups usually includes several group interviews. Each group is relatively homogeneous with respect to its participants' background characteristics. This homogeneity helps to ensure that individuals will not be afraid to express their feelings about the issue at hand.[50] Homogeneity also helps guarantee that the group has a common experience to discuss, whether it is opinions about marketing material, crime victims' perceptions of police response, or need for a new service. What distinguishes focus groups is the presence of group interaction in response to researchers' questions.

The focus-group discussion is led by a moderator, who should be an experienced interviewer skilled at group facilitation. The moderator also needs to be familiar with the questions and the purpose of the study. The moderator asks the questions and guides the discussion to ensure the participation of all members of the group. The hallmark of focus groups is the explicit use of group interaction to produce data and insights that would be less accessible otherwise.

Focus groups constitute an accepted research method.[51] Not every group using discussion techniques is a focus group, however. Focus groups have specific defining characteristics; they are created by research teams for well-defined purposes. Focus groups rely on the strengths of qualitative methods, including exploration and discovery, understanding things in depth and in context, and interpreting why things are the way they are and how they got that way. To serve these purposes, focus groups, like other qualitative methods, require a great deal of openness and flexibility.[52]

Although the overall process of focus-group research usually is highly structured, the interviewing of the group must be flexible itself. The information obtained is qualitative and usually voluminous.

The researcher must:

- identify themes
- find answers to questions
- summarize the discussion of the group members

Having well-designed questions based on a clearly defined purpose will facilitate the analysis and use of the data. Additionally, the researcher must do the following:

- carefully plan the session prior to assembling the group(s) and the interviewing
- clarify the purpose of the study and discuss it with colleagues
- be clear about what information is needed, how it will be gathered (audio or video recordings, notes, transcription) and why, who will use it and how, and its confidentiality or anonymity status
- make available a written plan for the entire project, including a schedule and budget, which should be developed in advance

Focus Groups in Public Administration

Focus groups are used in the public sector in many ways. A manager may find focus groups useful for needs assessments—to identify the services clients need, where they are best delivered, and which are not appropriate—and to develop a better understanding of concerns as perceived by relevant participants. A public organization might use focus groups to address such questions as: What do various citizen groups perceive as the most important problems of the jurisdiction? Why?[53] Focus groups can be used for citizen participation and engagement in policy analysis. One common criticism of policy analysis is it replaces public participation with expert analysis. Using focus groups allows for a more inclusive model of citizen input in the public policy process.[54]

Focus groups aid in program design and planning when the objective is to develop ways to deal with a problem or situation. What are the possibilities for dealing with a problem? What options are feasible? How would the client group receive them? Focus groups can be used to evaluate existing programs and to answer such questions as: How well is a program working? Are clients dissatisfied? Why? What changes will be acceptable? How well do clients think that they would respond to a change if one were instituted? As these situations suggest, the manager would probably want to get more explanation and discussion concerning responses to questions than generally would be possible in a large-scale survey.

Focus groups have been used with:

- citizens to assess the quality of city services and to identify priorities for future activities
- groups of clients of rehabilitation programs to assess current services and explore which types of programs are most effective
- top-level city administrators to obtain input to a planned survey of citizens
- employees to find out about and improve their working environment[55]

There is widening interest in participatory focus groups, wherein participants are involved in data analysis and interpretation.[56] Although focus groups often are used in exploratory studies and as adjuncts to other forms of research, they also can be used as the major procedure without being supplemented by quantitative methods.[57]

A health needs assessment in a rural county of North Carolina is an example of the use of focus groups in combination with other methods. Administrators and researchers chose focus groups as a nonthreatening way to obtain perceptions of health, health-related behaviors, and service delivery in the county. The researchers were advised specifically not to use surveys: "These people have been surveyed to death. They're tired of being asked if they are poor."[58]

Case Study: County Focus Group

The county was fragmented by its geography, numerous neighborhoods, and variety of agencies and schools. Researchers conducted 40 focus groups ranging in size from 3 to 37 individuals, greatly exceeding the optimal size of 8 usually recommended.[59] The sessions were not recorded. A two-person team of facilitators conducted the sessions, with one team member moderating and the other taking notes. The focus groups provided important data, involved residents in the project, and helped legitimize the research and its resulting interventions. Members of the groups were asked six questions addressing health problems and potential solutions. Participants were also given a copy of the questions to provide an opportunity for a written, private response. This proved to be useful. The focus-group data were combined with statistics on disease rates; mortality; causes of death and injuries; and demographic, economic, and social data supplied by state and federal agencies to plan new services and to improve existing services and facilities. Cultural differences among residents and differences in values between the more traditional, longtime residents and the younger health care professionals were identified in the process.[60]

Summary

In this chapter, we have discussed designs that guide studies describing the occurrence of a variable or the relationship between variables. These research designs help a researcher to decide when to make observations and how many observations to make. The design selected may depend on the nature of the dataset or the problem-solving skills of the administrator.

Cross-sectional and time-series designs are particularly effective and efficient. Either singly or in combination, they provide valuable information to administrators, legislators, and the public. The data derived from both designs may be organized to communicate information quickly through graphs or tables. Cross-sectional designs show relationships among variables of interest at one point in time. Cross-sectional designs often call for the collection of many pieces of data. Innumerable investigators may access, manipulate, and analyze the resulting database according to their individual interests.

Time-series designs demonstrate long-term, cyclical, and seasonal trends and identify irregular fluctuations in the occurrence of a variable. A time-series design requires an investigator to collect data on a measure at regular intervals. To distinguish between random and nonrandom irregular fluctuations, a historical record of events that can affect the occurrence of the variable is needed. Time-series designs help a researcher to describe a variable over time. They may forecast changes in a variable and assist in making operational decisions. For example, knowing the seasonal variations in arrests allows a court administrator to make staffing assignments.

With longitudinal designs, investigators follow individual cases and obtain information on them for several time periods. These designs allow investigators to measure the changes taking place within a group as well as to measure the change in a group characteristic over time.

Meta-analysis allows researchers to assemble a set of similar studies; use their data to form a single dataset; and determine what, if any, general hypotheses have been supported consistently. The major difficulty in performing meta-analysis is to identify a representative set of studies. It behooves the researcher to search through many sources to identify appropriate published and unpublished research. The conclusions reached through meta-analysis require thorough statistical analysis in order to provide evidence either supporting a hypothesis or arguing that it could have occurred by chance.

Several qualitative approaches provide useful information for administrators. Case studies provide detail that shows how something happened and why it happened. Case studies usually include information on the natural surroundings of events. One of the strengths of case studies is that they can involve multiple sources of data. Because of this requirement for multiple data sources, we suspect that most administrators and their staff have neither the time nor the resources to conduct case studies.

Nevertheless, administrators may be interested in case-study findings and how to use them should a particular case study turn out to be more than just a good story. To determine the quality of a case study, the administrator looks for evidence that the investigator had a research question and model before collecting data, the case was clearly defined—which can at times be difficult—and the case-study procedures were documented thoroughly.

Focus-group interviewing is used to obtain detailed information from a small group of individuals. A moderator asks a well-developed set of questions and leads the discussion in the focus group. The responses of the participants to the questions and to each other's comments provide data difficult to obtain with other methods. Focus groups often are used to supplement more quantitative studies, such as those using cross-sectional designs, and have many uses in the public sector.

Figure 2.2 classifies the various designs discussed in this chapter. Three of the major types—cross-sectional designs, longitudinal designs, and qualitative research—in turn include more than one subtype. Those are also illustrated in the figure.

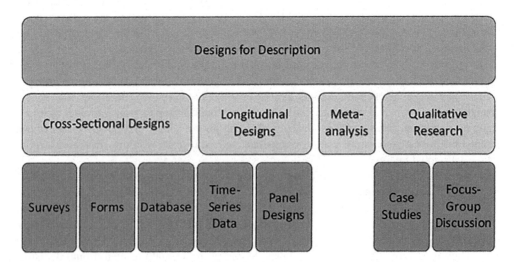

Figure 2.2 Designs for Description

Notes

1. Sometimes the term *research methodology* refers to the theoretical study of research methods.
2. See the following for a more extensive discussion of observational data and designs producing observational data: Paul Rosenbaum, *Design of Observational Studies* (New York: Springer, 2009); William Cochran, *Planning and Analysis of Observational Studies*, eds. Lincoln Moses and Frederick Mosteller (New York: John Wiley & Sons, Inc., 1983).
3. See J. Kirk, M. L. Miller, and M. L. Miller, *Reliability and Validity in Qualitative Research*, vol. 1 (Thousand Oaks, CA: Sage Publications, 1986). for a detailed explanation of the usefulness of triangulation.
4. M. Specter, "Suicide Rate of Jail Inmates Rising Sharply," *Washington Post*, February 18, 1985, sec. A1, 18. For recent data, see Bureau of Justice Statistics at www.bjs.gov.
5. Gary King, Robert Keohane, and Sidney Verba, *Designing Social Inquiry: Scientific Inference in Qualitative Research* (Princeton, NJ: Princeton University Press, 1994), 44.
6. We discuss the issue of causality in cross-sectional and time-series designs in Chapter 3. Hellevik discusses causal analysis with survey data in *Introduction to Causal Analysis: Exploring Survey Data by Crosstabulation*, Contemporary Social Research Series, no. 9 (London: Allen & Unwin, 1984). Robert K. Yin discusses using case studies to infer causality in *Applications of Case Study Research*, 3rd ed. (Thousand Oaks, CA: Sage Publications, 2011) and in *Case Study Research: Design and Methods*, 5th ed. (Thousand Oaks, CA: Sage Publications, 2013).
7. For example, see ICPSR: Inter-University Consortium for Political and Social Research at www.icpsr. umich.edu. The data in the General Social Survey File, also accompanying the text, are cross-sectional.
8. For an example of the use of CCES survey data (publicly available), see C. Maestas, J. Chattopadhyay, S. Leland, and J. Piatak, "Fearing Food: The Influence of Risk Perceptions on Public Preferences for Uniform and Centralized Risk Regulation," *Policy Studies Journal* 48 (2018).
9. B. Braddy, *An Evaluation of CETA Adult Training Programs in North Carolina Division of Employment and Training* (Raleigh, NC: Department of Political Science and Public Administration, May 1983). Unpublished manuscript.
10. See this article reviewing numerous evaluations of job-training programs: Robert J. LaLonde, "Employment and Training Programs," available at www.nber.org/chapters/c10261.pdf. Several of these evaluations compare participants with job training to those receiving no training.
11. See J. C. Bailar III, "When Research Results Are in Conflict," *New England Journal of Medicine* (October 24, 1985): 1080–1081. Summary of research results reported in "Studies Reach Opposite Conclusions About How Estrogen Pills Affect Heart," *Raleigh News & Observer* (October 24, 1985): 1A.
12. Certain biochemical chemical research suggested that the estrogen improves cholesterol levels, decreasing the probability of heart disease. Other biochemical research also suggested that estrogen increases blood clotting, increasing the probability of heart disease.
13. "WHI Study of Younger Postmenopausal Women Links Estrogen Therapy to Less Plaque in Arteries," available at www.nhlbi.nih.gov/news/press-releases/2007/whi-study-of-younger-postmenopausal-women-links-estrogen-therapy-to-less-plaque-in-arteries.
14. A similar graph for the years 1960 through 1980 accompanied Nicholas D. Kristol's article, "Scholars Disagree on Connection Between Crime and the Jobless," *Washington Post*, August 7, 1982, which discusses further the relationship between the two variables.
15. Time-series analysis may involve much smaller intervals (e.g., hours and days). These intervals will show analogous patterns. For the sake of clarity, we focused on longer time intervals; we assume that readers who are dealing with other intervals will be able to interpret their data.
16. Steven D. Levitt, "Understanding Why Crime Fell in the 1990s: Four Factors That Explain the Decline Plus Six That Do Not," *Journal of Economic Perspectives* 18 (Winter 2004): 163–190.
17. https://archives.fbi.gov/archives/news/pressrel/press-releases/uniform-crime-reporting-program-releases-crime-statistics-for-2002.
18. See www.ucrdatatool.gov/Search/Crime/State/RunCrimeStatebyState.cfm.
19. Rosemary J. Erickson and Kristi M. Balzar, *Summary and Interpretation of Crime in the United States, 2001 Uniform Crime Report, Federal Bureau of Investigation*, Released October 28, 2002 (San Diego: Athena Research Corp., 2003).
20. Scott Menard, *Longitudinal Research*, 2nd ed. (Thousand Oaks, CA: Sage Publications, 2002), 2. In practice, panel designs may be referred to as "longitudinal" designs. Someone using the data will want to check the documentation. Those interested in health and social aspects of aging will find valuable information in the following: Nathalie Huguet, Shayna D. Cunningham, and Jason T. Newsom, "Existing Longitudinal Data Sets for the Study of Health and Social Aspects of Aging," in *Longitudinal Data Analysis: A Practical Guide for Researchers in Aging, Health, and the Social Sciences*, eds. Jason T.

Newsom, Richard N. Jones, and Scott M. Hofer (New York: Routledge, 2012), Chapter 1, 1–42. The authors provide extensive detail on 12 major datasets.

21. Julian Simon and Paul Burstein, *Basic Research Methods in Social Science*, 3rd ed. (New York: Random House, 1985), 161–162. See the original study by Mary Jo Bane and David T. Ellwood, *Slipping Into and Out of Poverty: The Dynamics of Spells* (Cambridge, MA: Harvard University Press, 1983); revised and reissued as Working Paper no. 1199, National Bureau of Economic Research, Cambridge, September 1983. See a follow-up study: June A. O'Neill, Laurie J. Bassi, and Douglas A. Wolf, "The Duration of Welfare Spells," *The Review of Economics and Statistics* 69, no. 2 (May 1987): 241–248. See also David C. Ribar, Marilyn J. Edelhoch, and Qiduan Liu, "Food Stamp Participation Among Adult-Only Households," *Southern Economic Journal* 77, no. 2 (October 2010): 244–270.

22. Morton Hunt, *Profiles of Social Research* (New York: Russell Sage Foundation, 1985), 209.

23. Up-to-date information on the study and its design is at www.clinicaltrials. gov/ct/show/NCT00005121. Currently, the web site for this study is at www.framinghamheartstudy.org. To locate articles reporting analysis of the Framingham data, consult the Science Citation Index and the Social Science Index.

24. See www.framinghamheartstudy.org/participants/fhs-news.php.

25. Robert F. Boruch and Robert W. Pearson discuss in detail the advantages and disadvantages of panel designs. See "Assessing the Quality of Longitudinal Surveys," *Evaluation Review* 12 (1988): 3–18; Menard's, *Longitudinal Research,* Chapter 4, also covers the issues in panel designs. See www.Philanthropy.iupui.edu for extensive information on the Philanthropy Panel Study (PPS). This study obtained data on the same households every two years beginning in 2001.

26. Norvelle Glenn, *Cohort Analysis*, 2nd ed. (Thousand Oaks, CA: Sage Publications, 2005).

27. See Ibid. for information on analyzing panel data.

28. David Cordray and Robert Fischer, "Synthesizing Evaluation Findings," in *Handbook of Practical Program Evaluation*, eds. Joseph S. Wholey, Harry P. Hatry, and Kathryn E. Newcomer (San Francisco: Jossey-Bass, 1994), 202.

29. Robert F. Boruch and Anthony Petrosino, "Meta-Analyses, Systematic Reviews, and Evaluation Syntheses," in *Handbook of Practical Program Evaluation*, 3rd ed., eds. Joseph S. Wholey, Harry P. Hatry, and Kathryn E. Newcomer (San Francisco: John Wiley & Sons, Inc., 2010), Chapter 22, 531–553.

30. For more details on the limitations of traditional reviews of the literature, see Frederic M. Wolf, *Meta-Analysis: Quantitative Methods for Research Synthesis*, Quantitative Applications in the Social Sciences, no. 59 (Beverly Hills, CA: Sage Publications, 1986), 10–11; John E. Hunter, Frank L. Schmidt, and Gregg B. Jackson, *Meta-Analysis: Cumulating Research Findings Across Studies,* Studying Organizations, no. 4 (Beverly Hills, CA: Sage Publications, 1981), 129–130; Geoff Cumming, *Understanding the New Statistics: Effect Sizes, Confidence Intervals, and Meta-Analysis* (New York: Routledge, 2012), includes three chapters on meta-analysis. He discusses history, approaches, and tools of meta-analysis. See pages 181–262.

31. Cordray and Fischer, "Synthesizing Evaluation Findings," 200–206. See also Julia H. Littell, Jacqueline Corcoran, and Vijayan Pillai, *Systematic Reviews and Meta-Analysis* (New York: Oxford University Press, 2008).

32. Thomas I. Miller and Michelle A. Miller, "Standards of Excellence: U.S. Residents' Evaluations of Local Government Services," *Public Administration Review* (November–December 1991): 503–514.

33. Francis J. Yammarino, Steven J. Skinner, and Terry L. Childers, "A Meta-Analysis of Mail Surveys," *Public Opinion Quarterly* (Winter 1991): 613–639.

34. Sharon Paynter and Maureen Berner, "Organizational Capacity of Social Service Organizations," *Journal of Health and Human Services Administration* 37, no. 1 (Summer 2014): 111–145. The authors searched the literature for factors reported to relate to the success of large nonprofit organizations. They then applied these to small volunteer nonprofit organizations and found that the results did not hold.

35. Robert Rosenthal, *Judgment Studies: Design, Analysis, and Meta-Analysis* (New York: Cambridge University Press, 1987), 223–225.

36. Gene V. Glass, Barry McGaw, and Mary Lee Smith, *Meta-Analysis in Social Research* (Beverly Hills, CA: Sage Publications, 1981), 220.

37. For example, see Michael Q. Patton, *How to Use Qualitative Methods in Evaluation* (Newbury Park, CA: Sage Publications, 1987); Peter J. Haas and J. Fred Springer, *Applied Policy Research: Concepts and Cases* (New York: Garland Publishing, 1996).

38. John W. Creswell, *Research Design: Qualitative, Quantitative, and Mixed Methods Approaches*, 4th ed. (Thousand Oaks, CA: Sage Publications, 2014).

39. Sharon L. Caudle, "Using Qualitative Approaches," in *Handbook of Practical Program Evaluation*, eds. Joseph S. Wholey, Harry P. Hatry, and Kathryn E. Newcomer (San Francisco: Jossey-Bass, 1994), 69–95. Also see Herbert J. Rubin and Irene S. Rubin, *Qualitative Interviewing: The Art of Hearing Data*, 3rd ed. (Thousand Oaks, CA: Sage Publications, 2012).

40. To learn more, see Matthew B. Miles, A. Michael Huberman, and Johnny Saldaña, *Qualitative Data Analysis: A Methods Sourcebook*, 3rd ed. (Thousand Oaks, CA: Sage Publications, 2013).
41. King, Keohane, and Verba, *Designing Social Inquiry*.
42. Robert Wood Johnson Foundation, Qualitative Research Guidelines Project, *Using Qualitative Methods in Health Care Research*, available at www.qualres.org; Robert E. Stake, *The Art of Case Study Research* (Thousand Oaks, CA: Sage Publications, 1995).
43. For more information on vaccine boycotts, see M. Yahya, "Polio Vaccines—No Thank You! Barriers to Polio Eradication in Northern Nigeria," *African Affairs* 106, no. 423 (2007): 185–204.
44. Yin, *Applications of Case Study Research*, Chapter 1. Also see Yin, *Case Study Research*, which is especially recommended for gaining information on how to do case studies.
45. For more information on using QCA, see C. Ragin, *The Comparative Method: Moving Beyond Qualitative and Quantitative Methods* (Berkeley: University of California Press, 1987). For an example used in public administration please see L. S. Johnson and J. B. Carr, "Making the Case for (and Against) City-County Consolidation," in *City-County Consolidation and Its Alternatives: Reshaping the Local Government Landscape: Reshaping the Local Government Landscape* (New York: M. E. Sharp, 2006), 246.
46. B. Rihoux, "Qualitative Comparative Analysis (QCA) and Related Systematic Comparative Methods: Recent Advances and Remaining Challenges for Social Science Research," *International Sociology* 21, no. 5 (2006): 679–706.
47. See Charles C. Ragin and Howard S. Becker, *What Is a Case? Exploring the Foundations of Social Inquiry* (Cambridge: Cambridge University Press, 1992). Also see Yin, *Case Study Research*, Chapter 2.
48. Robert K. Merton, "The Focused Interview and Focus Groups," *Public Opinion Quarterly* 51 (1987): 550–566; Robert K. Merton, Marjorie Fiske, and Patricia Kendall, *The Focused Interview*, 2nd ed. (Glencoe, IL: Free Press, 1990); Richard A. Krueger and Mary Anne Casey, *Focus Groups: A Practical Guide for Applied Research*, 5th ed. (Thousand Oaks, CA: Sage Publications, 2015), 7–15.
49. Ralph S. Hambrick Jr. and James H. McMillan, "Using Focus Groups in the Public Sector," *Journal of Management Science and Policy Analysis* 6 (Summer 1989): 44.
50. Ibid., 48.
51. David Morgan, *The Focus Group Guidebook* (Thousand Oaks, CA: Sage Publications, 1997), 29.
52. Ibid., 31.
53. Hambrick and McMillan, "Using Focus Groups in the Public Sector," 44–45.
54. See L. C. Walters, J. Aydelotte, and J. Miller, "Putting More Public in Policy Analysis," *Public Administration Review* 60, no. 4 (2000): 349–359.
55. These and other examples are cited in Hambrick and McMillan, "Using Focus Groups in the Public Sector," 46–47. Also see Christopher McKenna, "Using Focus Groups to Study Library Utilization," *Journal of Management Science and Policy Analysis* 7 (Summer 1990): 316–329.
56. Krueger and Casey, *Focus Groups*.
57. David Morgan and Richard Krueger, "When to Use Focus Groups and Why," in *Successful Focus Groups: Advancing the State of the Art*, ed. David Morgan (Newbury Park, CA: Sage Publications, 1993), 3–19. Also see Debra L. Dean, "How to Use Focus Groups," in *Handbook of Practical Program Evaluation*, eds. Joseph S. Wholey, Harry P. Hatry, and Kathryn E. Newcomer (San Francisco: Jossey-Bass, 1994), 341.
58. Thomas Plaut, Suzanne Landis, and June Trevor, "Focus Groups and Community Mobilization: A Case Study from Rural North Carolina," in *Successful Focus Groups: Advancing the State of the Art*, ed. David Morgan (Newbury Park, CA: Sage Publications, 1993), 205.
59. Ibid., 206.
60. See "Focus Group Report," an example of a focus group study accompanying this text.
61. Data from Institute for Social Research, University of Michigan, Monitoring the Future Project.

Terms for Review

research methodology
research designs
cross-sectional design
longitudinal design
time-series designs

long-term trends
cyclical variations
seasonal variations
random variations
nonrandom variations
quantitative studies
qualitative studies
panel design
triangulation
cohort
case study
focus group
meta-analysis
observational data
experimental data
observable variables
unobservable variables

Questions for Review

The following questions should indicate whether you have a basic competency in this chapter's material.

1. What is the value of a research design?
2. a.   List the advantages and disadvantages of cross-sectional, time series, longitudinal studies, and case studies.
   b.   When should an investigator use a panel design instead of a time-series design?
   c.   Under what conditions would a focus group be the best method to use?

3. Select one of the following topics: automobile accidents, water quality, drug abuse, homeless persons, single-parent families, management information systems, personnel training, strategic management, or outsourcing. For the selected topic, pose a research question appropriate to each of the following: cross-sectional design, a panel design, a time-series design, a case study, a focus group.
4. Data have been collected annually on air quality in Smokey Mountains for the past 15 years.
       Explain why the data can be analyzed using a time-series design. What types of trends or variations should a researcher look for? How might a researcher distinguish random variations from nonrandom variations?
5. The Metro Hospital collected data on nurses every three years from 1996 to 2007. Beginning with 2007, the data were collected every year. What limitations would an analyst encounter in studying nursing trends from 1996 to the present?
6. Explain why public agencies are unlikely to conduct case studies.
7. Distinguish between units and components in conducting a case study.
8. For one of the topics in Question 3, generate a list of questions that might be asked of members of a focus group.
9. Why might government agencies use focus groups more now than in the past? What objections might managers have to the use of focus groups?
10. Compare literature reviews to meta-analyses, and discuss why researchers would invest time in conducting a meta-analysis instead of original research.

Problems for Homework and Discussion

1. For each of the following studies, identify the variables, state the implied hypothesis(es), identify the research design, and briefly evaluate its appropriateness. (Note: A study may modify a common design or combine features from more than one design.)

   a. A random audit of Unemployment Insurance (UI) sampled eight UI payments per week. If a payment error was found, auditors determined the dollar amount and classified the error by type, source, and cause. Errors were categorized as overpayment with fraud, overpayment without fraud, or underpayment. Sources of errors were the claimant, employer, or agency. Causes were identified by law or regulation violation.

   b. To evaluate the effectiveness of Head Start, an early education program, children who had been in Head Start and who were in the first through third grades were given cognitive tests.

   c. Before the adoption of the Magnuson-Moss Warranty Act, the Federal Trade Commission collected data from 4,300 respondents who had purchased a major durable good the previous year. Each respondent rated the performance and servicing of products purchased during the year. The 4,300 members of the sample were randomly selected from a national consumer mail panel. To evaluate the act, the FTC later asked 8,000 respondents drawn from the same national consumer mail panel the same questions.

   d. To develop a statistical base on private foundations, data were gathered from tax records on selected foundations' resources and expenses in 1987, 1994, 1998, and annually beginning in 2001.

   e. To assess training needs and how the government could work best with a private agency to meet them, the private agency invited two groups of its clients to participate in discussions led by a moderator from the nearby university.

2. Follow the instructions that appear after Table 2.4, which shows the percentage of seniors in high school who regularly smoke marijuana over time.

Table 2.4 Percent of High School Seniors Who Regularly Smoke Marijuana Over Time

Year	Percent	Year	Percent
1991	2.0	2006	5.0
1992	1.9	2007	5.1
1993	2.4	2008	5.4
1994	3.6	2009	5.2
1995	4.6	2010	6.1
1996	4.9	2011	6.6
1997	5.8	2012	6.5
1998	5.6	2013	6.5
1999	6.0	2014	5.8
2000	6.0	2015	6.0
2001	5.8	2016	6.0
2002	6.0	2017	5.9
2003	6.0	2018	5.8
2004	5.6	2019	6.4
2005	5.0		

   Draw a graph to illustrate the trend shown in Table 2.4 over time. Comment on the variations found in this dataset. What policy recommendations can be made based on these data?[61]
   An administrator for a state employment commission wants to study seasonal variations in the unemployment rate so she can schedule staff vacations and conferences at times when

the demand for services (as measured by the unemployment rate) is lowest. The data are shown in Table 2.5.

Table 2.5 Unemployment Rate

	Oct.	Nov.	Dec.	Jan.	Feb.	Mar.	Apr.	May	Jun.	Jul.	Aug.	Sep.
Year 1	6.8%	5.5%	5.0%	6.4%	6.3%	6.4%	5.4%	6.7%	6.4%	6.5%	6.7%	8.1%
Year 2	7.9	7.5	7.3	7.7	7.1	8.6	7.9	7.3	7.7	7.5	7.8	8.2
Year 3	9.2	9.5	9.5	10.4	9.5	8.9	8.4	8.7	8.8	8.2	8.2	8.1
Year 4	8.6	8.1	7.5	7.4	6.1	5.4	5.7	6.1	6.6	6.8	6.2	6.2

   a.   Graph the data.

   b.   Comment on any trends you notice in these data.

   c.   What times of year would you prefer for staff vacations and conferences?

4.   For each of the following problems, suggest a research design and justify your choice:

   a.   Identify revenues generated by a county sales tax first adopted in 1980 and increased periodically since then.

   b.   Learn whether change in state penalties for drunk driving was associated with fewer drinking-related traffic fatalities.

   c.   Identify what computer hardware and software are used by local governments and how the governments use them.

   d.   See whether consolidating purchasing by Middletown decreased costs of buying supplies.

   e.   See whether agency managers who attended a decision-making seminar used the skills taught.

   f.   Determine whether police personnel involved in a wellness program took fewer sick days and had fewer claims against the department's health insurance plan after enrolling in the wellness program.

5.   A state has 10 high school health clinics. These clinics have been opened within the last four years to serve the physical and mental health needs of high school students. Some of the clinics are located within a high school, and others are within a block of the school. The state department of education has decided to do a case study of these clinics to evaluate their performance and to see whether similar clinics should be established throughout the state. (In answering the following questions, you may make your own assumptions to fill in specific details about the clinics.)

   a.   Identify the units and components that could be the subjects of the case study. What information would you want on the units? On the components?

   b.   List possible data sources you would use. Indicate the type of information you would want from each data source.

   c.   Write a memorandum discussing why a case study would be a valuable research strategy.

6.   Look at Example 1.3 in Chapter 1. Assume that you are a management analyst in the department administering the job-training program. Write a memorandum to the program manager outlining what actions you would recommend based on these data. Remember that conducting further research can be a recommended action.

7.   Locate a study or article that you would classify as a meta-analysis, and answer the following questions concerning it. What was the topic? How many studies were reviewed? Was a specific hypothesis investigated in the meta-analysis? What did the authors of the meta-analysis conclude from the study?

Working With Data

1. Access the County Data File accompanying this textbook. Analyze the data to see how median household income (2015), population density (2017), and the number of active primary care physicians (2017) varies by region of the state. For each region, find the low, high, and average for each variable for each region. Report your findings in a table.

2. Cross-sectional data can be reported in tables showing differences in percentages or means. Access the County Data File accompanying this textbook. From the median income variable, create three categories of counties, for example, high income, medium income, and low income.

    a. For each county income category, determine the mean values for the number of individuals served in drug and alcohol facilities (2016) and the crime index variables. Do the data show that these items vary across low-, medium-, and high-income counties? What evidence supports your observations? Write a paragraph reporting your findings.

    b. Categorize the counties into two groups, a high and a low category according to the variables (1) number of individuals served in drug and alcohol facilities and (2) the crime index. For each income category, determine what percent of the counties fall into the high category for each variable and what percent fall into the low category. (This can be set up as a table with two rows and three columns, with one column for each income level.) In this analysis, do the data show that the items vary by income? Do these findings support the findings you had in 2(a)? Write a paragraph reporting your findings.

    c. Of the two options, that used in 2(a) comparing the means and the one used in 2(b) comparing percentages, which way of presenting this information would you recommend using in a report on this issue? Justify your choice.

3. Access the County Data File accompanying this textbook. Test the hypotheses that you formulated as part of the Chapter 1 exercise. Do the data support your hypotheses? Cite evidence to support your answer.

4. Access the County Data File accompanying this textbook. Draw a scatterplot of the crime index as the dependent variable against population density. In a short paragraph, describe the relationship. What can you conclude?

5. Access the document Focus Group Report accompanying this textbook. Create one additional question that you would have asked if you were involved as a researcher. Explain your choice of question.

Recommended for Further Reading

A good place to find information on time series, especially as a forecasting tool, is in management science texts. One widely available text is Anderson, D. R., D. J. Sweeney, and T. A. Williams, *Quantitative Methods for Business*, 11th ed. (Mason, OH: Thomson South-Western, 2013), Chapter 6 "Time Series Analysis and Forecasting." Also, Williams, D. W., "Forecasting Methods for Serial Data," in *Handbook of Research Methods in Public Administration*, eds. G. L. Miller and M. L. Whicker (New York: Marcel Dekker, 1999), 301–352.

For a discussion of various types of longitudinal designs, see Menard, Scott, *Longitudinal Research*, 2nd ed. (Thousand Oaks, CA: Sage Publications, 2002).

For information on how to triangulate qualitative and quantitative work, see Kirk, J., M. L. Miller, and M. L. Miller, *Reliability and Validity in Qualitative Research*, vol. 1. (Thousand Oaks, CA: Sage Publications, 1986), for a detailed explanation of the usefulness of triangulation.

Yin, Robert K., *Case Study Research: Design and Methods*, 5th ed. (Thousand Oaks, CA: Sage Publications, 2014) is an excellent starting point for information on case studies.

This chapter's section on case studies used this and earlier versions of Yin's book. Another book, *Applications of Case Study Research and Applications* (Newbury Park, CA: Sage Publications, 2017), by the same author, provides detailed examples of case studies.

Hancock, Dawson, and Bob Algozzine, *Doing Case Study Research: A Practical Guide for Beginning Researchers* (New York: Teachers College Press, 2017) is a short, readable text covering all steps in the process of conducting a case study.

For more information on focus groups, see the references listed at the end of Chapter 7, especially Krueger, Richard, and Mary Anne Casey, *Focus Groups: A Practical Guide for Applied Research*, 5th ed. (Thousand Oaks, CA: Sage Publications, 2015). For a more advanced and analytical presentation, see Fern, Edward E., *Advanced Focus Group Research* (Thousand Oaks, CA: Sage Publications, 2001).

The following book uses journal articles to illustrate some qualitative methods. It is brief and inexpensive: Luton, Larry S., *Qualitative Research Approaches for Public Administration* (Philadelphia: Routledge, 2015).

3 Research Designs for Explanation

In this chapter, you will learn:

1. About internal and external validity.
2. The evidence necessary to establish that two variables are causally related.
3. The questions used to evaluate evidence that a program or intervention caused an outcome.
4. The questions used to determine whether study results apply to cases other than the research subjects.
5. The common experimental and quasi-experimental designs used to infer causality.
6. About nonexperimental designs and their value.

If an administrator wants to go beyond describing the values of a dependent variable and explain why it changes over time or differs from case to case, he or she needs to show that an independent variable is causally related to the dependent variable. Often the administrator wants to show that some independent variable, such as a program activity, affected a dependent variable, such as nutrition, health care, level of absenteeism, and so forth. This is typically a program output or outcome. For example, the manager of the Women, Infants, and Children (WIC) program may want to determine if this program improved infant nutrition. Simply demonstrating that the percent of malnourished infants decreased is insufficient evidence that the WIC program is responsible. Instead, the manager needs a research design that unambiguously isolates the WIC program as the cause of decreased malnutrition.

The administrator may be able to demonstrate with descriptive designs that two variables are linked to each other. That is, he may be able to show that they are associated. However, the administrator may want to go further and demonstrate that one variable caused another. When we say that X has caused a change in Y, we mean that a change in the value of the dependent variable, Y, comes about as a result of a change in the independent variable, X, and not as a result of something else.[1] If one variable causes another, an investigator expects to find a statistical relationship between them. For example, the manager of a city's light rail transit program may want to demonstrate that reducing fares increased ridership. Simply demonstrating that the number of riders per day increased following a fare reduction is insufficient evidence that the fare reduction was responsible. He would also need to show that the number of riders would have been less or stayed the same without the fare decrease.

Establishing Causal Relationships

To establish that a causal relationship exists requires more than identifying a relationship between two variables. One of the real difficulties in determining whether public programs have had their intended impact is to separate the changes that may be due to other factors from changes that are due to the program. For example, an increase in the price of gasoline or construction on a major highway may encourage more people to use public transit. Or, to return to the earlier example of the WIC program, if the number of undernourished infants decreased after the WIC program started, the reduction may have been due to something else, such as a general improvement in the state's economy and more jobs in the community. To say that it reduced infant malnutrition, an investigator would need to demonstrate that without WIC, infant malnutrition would have been stable or greater than it was with the program.

Causality

To claim *causality*—that one variable causes another to exist or change in value—requires the following:

1. A statistical association of the two variables: The variables covary with each other, that is, a change in one variable is accompanied by a change in the other.
2. The sequential order of the variables: The independent variable—the presumed cause—must have occurred before the dependent variable—the presumed result or effect.
3. The elimination of rival independent variables as causes of the dependent variable: Variables other than the independent variable of interest must be ruled out as causes of the dependent variable.
4. A theoretical link between the independent and dependent variable: The analyst should have some logical argument for assuming that two variables covary and that one causes the other.

Descriptive Versus Experimental Designs

Social science research has two essential goals that depend upon drawing inferences from data: description and explanation. Descriptive inference describes and collects facts. We use observations from the real world so we can understand unobserved facts. Descriptive designs help translate administrative concerns and program dynamics into quantitative information. They yield findings that alone or in combination with other information support action or help avoid unnecessary further study. Cross-sectional, time-series, and case-study designs produce data that identify variables worth further study. Cross-sectional and time-series designs identify relationships among variables and describe changes in variables over time. Analysts may use them to build complex mathematical models that provide a suggestion of causal links between variables. Still, policy makers and administrators may want more conclusive evidence that an independent variable causes a value of a dependent variable, especially if they need to determine whether a program has achieved its intended objectives. To gain such evidence, they prefer to have a design that incorporates as many components of an experimental design as possible.

An *experimental design* is one in which the researcher can assign subjects to different treatments. The researcher can control who is exposed to the treatment or treatments, which are the independent variables in an experiment. The researcher can also control when they are exposed to it and the conditions under which the experiment takes place. Because most experiments have a control group, there is almost always a counterfactual, the situation as it would be absent the treatment. Next, the researcher compares the treatment group or groups to the control group to see if there is a difference in the outcome, or dependent variable.

The emerging field of behavioral public administration draws on psychology to improve our understanding of individual behavior and attitudes. Box 3.1 is an example of a 2×2 vignette experiment used in public administration to understand public blame attribution when service delivery fails under two different varying conditions: (1) in house versus contracting out for trash pick-up and (2) a budgetary condition. The experiment is called 2×2 (or two by two) because four different combinations are given to the subjects and the sample is divided into different groups that are randomly assigned. Note that in this experiment, the first treatment—independent variable—is whether the service is in house or contracted out. The second treatment or variable is whether there was a budget shortfall that may alter to whom the respondents assign blame. This allows researchers to compare and contrast a control group to three different variations of the vignette. Vignettes are scenarios that simulate the real world and, in this case, emulate a news story.

Experiments also offer the potential for practitioners to bring psychological insights into practice.[2] An example of a field experiment is how making simple changes to job ads can help a city recruit more police officers of color.[3] Researcher Elizabeth Linos (2017) conducted a field experiment demonstrating how simple changes to advertising positions for police officers could help diversify organizations. In this case, 10,000 participants were selected at random between the ages of 18 and 40. From this sample, they randomly selected five different groups. Eighty percent of the respondents were sent one of four different sets of postcards to entice people to apply. The control group (the fifth group) did not receive the postcard (20 percent of respondents). Three of the postcards (the experimental groups) varied from the traditional message, calling people to serve as a challenge, hinting at job security, or emphasizing impact. The fourth group received a postcard with the typical traditional service message aimed at motivating people to apply for the position. Interestingly, they found that the fourth group was no more motivated than those who received no postcard at all. But people of color who saw the challenge message were four times more likely to apply to the police than the other two groups (those that emphasized job security and community impact) (Linos, 2017; 2018).

Researchers should note that this kind of control is not always possible in administrative work, so administrative studies often rely on *quasi-experimental designs*. Although these designs have some of the characteristics of experiments, they lack others and some of the control afforded with them. In quasi-experiments, the researcher may rely on a naturally occurring independent variable or base her comparisons on naturally occurring groups over which she has no control. She cannot use random assignment here. Therefore, both quasi-experimental and nonexperimental designs limit the researcher's ability to judge the impact of an independent variable. Researchers often distinguish between true experiments and these other designs.[4] A true experiment is one with the characteristics noted previously and provides a means of obtaining evidence necessary to demonstrate causality. When a researcher cannot implement an experimental design, a quasi-experimental design may be used.[5]

Example 3.1 The Use of Experiments in Behavioral Public Administration

Background: Following advances in psychology, economics, and political science on the use of experiments, public administration has begun to form its own group of scholars and researchers conducting research with experiments. Traditional journals like *Public Administration Review, Journal of Public Administration Research and Theory,* and *Public Administration* have all done symposia recently on experiments in public administration. The new, open-access *Journal of Behavioral Public Administration (JBPA)* focuses exclusively on research that is conducted using experiments (www.journal-bpa. org/index.php/jbpa). To address concerns about reproducibility and to encourage replication, JBPA strongly encourages that the data and analysis be published online, and it can be found here (https://dataverse.harvard.edu/dataverse/JBPA).

Problem: In one experiment, researchers wanted to learn about the problems of contracting in a fiscally constrained environment. Governments often contract out service delivery because they are having financial difficulties or they cannot afford the large, up-front capital cost that often accompany services like trash and recycling pick-up. By contracting out, the government may push some of the cost and some of the blame if anything goes wrong onto the contractor.

Research Design & Treatments (independent variables): The study was a 2 × 2 vignette experiment. A vignette experiment is an experiment that provides a brief story where details of the story are changed depending on the treatment. The vignette in this case was a story designed to look like a local news story that describes a sanitation truck illegally dumping garbage in a creek. The experiment is designed to test whom citizens blame if a city's service has a problem and how much they are likely to blame the affected parties. The first treatment was whether the service was contracted out or provided in house. The second treatment was if there was a financial problem such as a budget shortfall that would mitigate the amount of punishment for the crime.

Subjects: 293 undergraduate students. Students were randomly assigned to one of the four groups (1) contract, no fiscal stress; (2) contract, fiscal stress; (3) in-house provision, no fiscal stress, and (4) in-house provision, fiscal stress.

Outcomes (dependent variables): There were two variables of interest. The first was whom people blamed. The baseline category is the accused employees, the second category is either the sanitation department or the contractor, and the third category was the city that legally provided the trash service. The second outcome (variable) was whether the employees should be fired or terminated from their job.

Findings: The results of the experiment showed that contracting out the service significantly changed whom people blamed in the scenario. Budget stress did not change whom people blamed. Budget stress did cause people to significantly choose other forms of punishment for the employees besides firing them. Whether the employees were contract employees or government employees did not influence whether people believed that they should be fired.

Source: Piatak, Jaclyn, Zachary Mohr, and Suzanne Leland, "Bureaucratic accountability in third-party governance: Experimental evidence of blame attribution during times of budgetary crisis." *Public Administration* 95, no. 4 (2017): 976–989.

Internal and External Validity

When evaluating the results of a program, a manager looks for a relationship between the program implementation—the independent variable—and some hoped-for results—the dependent variable. Typically, the manager wants to know if the program caused a change in some condition. As we have already suggested, a change in the target condition or population accompanied by a program change does not mean that the program change was the cause. Thus, additional evidence is necessary to determine whether lower fares by the transit system contributed to an increase in the number of riders per week or whether the efforts of WIC reduced infant malnutrition.

Internal Validity

Internal validity refers to the evidence that a specific independent variable, such as a program, policy, or action, caused a change in an observed dependent variable. If an investigator is confident that the independent variable of interest and not something else caused the observed outcomes, the investigator can say that the study was internally valid. Internal validity involves taking specific steps to give the researcher the confidence that X caused Y.

An evaluation of a national 55-miles-per-hour speed-limit law adopted in 1973 found that after the legal speed limit was reduced from 65, the number of traffic deaths also went down.[6] The conclusion that the change in speed limit caused a change in fatalities is internally valid only if the investigators were able to show that no other plausible factor caused the reduction in traffic deaths. Other factors alone or in combination with the lower speed limit could have caused the reduction. For example, the implementation of toll roads may cause people to drive less on a highway because it costs more than staying home, walking, or taking local roads, and therefore traffic on the highway is reduced.

External Validity

The external validity of a study speaks to the issue of generalizing the findings of a study beyond the specific cases involved. Typically, not all people involved in or affected by a program are studied. Only a portion is included in a study, and the investigator wants to know about the entire group. Let us say that we had established for a sample of WIC program participants that the program had reduced undernourishment in children. Would this result be true of all WIC participants? Would it be true of others eligible for but not enrolled in WIC? Would programs similar but not identical to WIC have the same results? Assume that we determine from a study conducted in North Dakota that the 55-miles-per-hour speed limit reduced fatal traffic accidents.[7] Would these results be true of other states or of the nation as a whole? These examples and questions all concern external validity.

In considering external validity, researchers are usually concerned about generalizing in one or more of three different ways:

1. generalizing from a sample to a larger population
2. generalizing from one research situation or study to another
3. generalizing from a research study to a real-world situation[8]

This last form of generalizability is often the one of interest when debating whether the programs of government or a nonprofit agency should be adopted by other organizations. A specific program in a specific location is evaluated to find out if it works. However, administrators may want to know that it would work in other locations or under different conditions.

Investigators are also concerned with validity when measuring variables. We refer to this type of validity as *operational validity*. Operational validity asks if variables have been appropriately and accurately measured. Methodologists have identified other types of validity, and some use different names for those we discuss. The interested reader is encouraged to consult the sources cited in the endnote for a discussion of these.[9]

Threats to Internal Validity

One cannot conclude that a change in one variable is responsible for changes in another unless one has ruled out the possibility that the observed change in the dependent variable was caused by some factor other than the independent variable. These other factors are called *threats to internal validity*.[10] Each threat identifies a type of plausible alternative explanation for the observed relationship between the independent and dependent variables. They may be categorized and defined as follows:

> *History:* Events other than the independent variable occurring during the same time as the study that could have affected the dependent variable.

Selection: The way that cases are selected for groups or conditions for a program or a study could affect the way they react to the independent variable. If the cases in the group receiving the intervention are different from those in other groups, then any difference in the dependent variable may be due to differences in the cases and not to the intervention.

Maturation: Natural changes taking place in the units being studied.

Statistical regression: If cases for study or action are picked because of an extreme position, such as a low performance score, the instability of extreme scores may cause the observed change. This is a special case of selection.

Experimental mortality: Cases, particularly people, who begin a study or program may drop out before the study or the program is completed. This phenomenon is also known as attrition. If those who drop out are systematically different from those who remain, the results will not be internally valid.

Testing effects: A situation in which the initial measure or test influences the subjects, which then affects the outcome of the posttest. Employees who score low on a pretest may change their behavior because of this and not because of some program.

Instrumentation: If the measuring instrument used to collect data changes between the beginning of a study and its conclusion, the results may not be valid.

Design contamination: A condition occurring when participants know that they are in a study and act differently because of it, or if subjects being compared interact. If study subjects have an incentive to behave in a specific way to make the program succeed or fail, the design has been contaminated.[11]

Threats may occur because the environments of groups being compared are different, either to start with or over time, or because the cases in the groups are systematically different. For example, if younger people, ages 18 to 22 years old, are compared to those who are 25 to 30, any difference between these groups could be due to differences in ages and not due to the intervention.

History

The threat to internal validity known as *history* arises when events or policies other than the independent variable cannot be ruled out as a source of the changes in the dependent variable. Such events are those occurring outside of the study and during the same time as the independent variable. In 1973, an oil embargo brought about a gasoline shortage; gasoline prices soared, long lines were found at gas stations, and Americans drove less. To reduce gas consumption, a national 55-miles-per-hour speed limit was established. Auto fatalities decreased, and the national 55-miles-per-hour speed limit continued for more than 20 years. The observed reduction in highway fatalities may have been the result not of reduced speed but of less driving. Improvements in automobile manufacturing may also have increased the safety of automobiles, thus reducing automobile fatalities.

Following the terrorist attacks of September 11, 2001, commentators compared changes in unemployment, consumer confidence, and confidence in government before and after September 11, 2001. Prior to the attacks, Americans were receiving tax rebate checks intended to stimulate spending. Policy observers debated if the recipients would spend the rebate money or save it.[12] In the latter case, the economy would continue to weaken. Nevertheless, the debate became moot and the effects of the attacks could not be separated from other factors affecting the economy.

Selection

Selection is a problem when the basis on which cases are chosen for a study or program condition is nonrandom. In such instances, the group of cases in the independent variable condition may be systematically different from the cases to which they are compared. If so, the difference between the participants in the groups—rather than the influence of the independent variable—may account for any observed change in the dependent variable. Consider that a professor wishes to study how well students learn in an online class versus a traditional in-person class. Students who prefer to enroll in the in-person class are older and less likely to learn a new technology, whereas students who select the online class are younger and learn to use a new technology more quickly. When the semester concludes, both are graded on an assignment that requires extensive work using a statistical software program to run regression analyses. The instructor finds big differences in how well the students did on some of the assignments. The online class did remarkably better than the students in the in-person class on running a regression analysis with the new software. The differences in results could be attributed to the fact that those who signed up for the online class were able to adapt to the new software faster because of their age and therefore did not struggle as much with the assignment than if the instruction occurred online or in person.

Maturation

Sometimes changes that occur in dependent variables during the course of the study are due to natural processes. The resulting threat to internal validity is called *maturation*. Individuals, groups, and other units of observation change over time. These changes occur naturally rather than as the result of specific, identifiable events or interventions. One may assume erroneously that an independent variable caused a change in a dependent variable when the change involved would have occurred with the passage of time, regardless of whether the independent variable was present. Most juvenile delinquents, for example, decrease their antisocial behavior as they grow older. Thus, the apparent success of delinquency programs may result from the aging of the clientele rather than program activities. If one records the behavior of one group of children in the morning and another group in the afternoon, differences between the groups may not be due to a particular independent variable but to the changes in children's behavior during the course of a day.

Regression to the Mean

A fourth threat is *statistical regression*. Consider an experience most of us have had during school. Early in our academic career, most of us learn that, on average over time, we receive roughly the same grade in courses or on exams. But on occasion, we may be pleasantly surprised by our score on a test. And we may find that on the next test, our score is noticeably lower than on the first. Or we may have a classmate who scores very low, much lower than expected, on an exam. On the second exam, she does better than on the first. Over time, the unusually high or low test scores will approach the students' average performance. In short, any time that we observe a test score that is exceptionally high or exceptionally low for a given student, we usually can predict that that student's next score will be closer to her average. Sometimes statistical regression is referred to as "regression to the mean." This occurs if a random variable is extreme the first time it is measured but will be closer to the average on its second measurement.

Social problems show analogous patterns. If a program is created in reaction to an extreme situation, then any subsequent change in the desired direction may occur because of statistical regression. A city with an unusually high number of traffic deaths one month may experience a significant reduction the next. A program installed in response to a rash of fatal accidents may appear to have been effective, when in fact the number of traffic deaths had simply regressed toward the average. The result may have been due to statistical regression, not the program. Statistical regression is a particular threat whenever cases for treatment or study are chosen because of an extreme value on some measure. If the variable is measured again, the cases will be less extreme; they will have regressed toward the average. When cases for a study are chosen because of their extreme values on the dependent variable, statistical regression represents a special case of the threat of selection.

Experimental Mortality

The threat of *experimental mortality* arises when people begin a program and later drop out before the study is completed. The difficulty with this is that dropouts may be different from those who complete the program, and the difference may affect the outcome. If the results are taken as evidence of the success of the program, they may not be valid. It may be that those who are successful stay with the program and those who are not leave. Consider the example of a successful delinquency program whose goal was to keep participants in school. Despite the program's apparent success, it may have been that the youths who dropped out of the program were those who had little motivation to remain in school. Those who continued in school might have done so even if the program had not existed.

Testing Effects

Testing effects as a threat to internal validity occur when an initial measurement changes the values of the dependent variable. In some studies, an observation is taken before and after the introduction of an independent variable. The observation taken before exposure to the independent variable is called a *pretest*. The observation taken after exposure is the *posttest*. The risk in such a procedure is that the pretest and not the independent variable may have caused an observed change in the dependent variable. For example, a person exposed to a training program to increase knowledge may show increased knowledge from the pretest to the posttest because the pretest stimulated his interest and he looked up the answers. The pretest, not the program, caused this increased knowledge. Oftentimes people do less well with an unfamiliar task; a pretest serves as practice and may improve later performances on the same or similar test. We observed this when the Graduate Record Exam (GRE) was first given on computers. Students who had not practiced and took the GRE more than once improved their scores between the first and second testing; most likely their familiarity with the new format led to the improved performance. The effect of testing also is known as the *reactivity of measurement*; people react to the measuring process and change their behavior because of it.

Instrumentation

Instrumentation is a threat to internal validity when the method of measurement changes between the pretest and the posttest. The instrument used may have changed, or the definition of the variable may have changed. Suppose that the legal definition of a particular crime changed between two measurements of the crime rate. The change in definition alone may make it appear that the crime rate had changed. If a program to reduce the crime rate were initiated during the period between the two observations, we could not be sure whether any recorded change in the crime rate was due to the way crime was measured, to the program, or to both. Investigators using agency records need to be alert as to how variables are defined and records are kept. Changes in these, if unrecognized, can cause problems of instrumentation. Instrumentation is also cited as a threat when different methods are used to measure the dependent variable in the groups being compared.

Design Contamination

The final threat to internal validity is known as *design contamination*. Several types of contamination are possible. Subjects exposed to a treatment may talk to those not exposed, and eventually both groups experience the same program or treatment. Consider the director of human resources (HR) who wishes to evaluate a program designed to prepare people for retirement. The program has been offered to some employees on an experimental basis. The HR director wants to know whether employees who participated in the program have fewer economic problems after they retire than those who did not participate. However, if participants shared the information obtained from the program with their colleagues, both groups may have made similar economic plans, effectively eliminating any difference between participants and nonparticipants. As a result, the program would appear to have been unsuccessful. As this example illustrates, one can sometimes falsely conclude that an independent and a dependent variable are not related because of a threat to internal validity.

A second type of design contamination occurs when participants guess the purpose of the research and alter their behavior so as to achieve an outcome in their best interest. Suppose a researcher wants to learn whether reducing the size of work crews on sanitation trucks from four persons to two persons will increase productivity. Fearing the breakup of their work units and increased job pressures, the workers might alter their behavior, with the smaller crews slowing down their work rate and the larger crews increasing theirs.

Yet another type of contamination occurs if a group receiving no treatment changes its behavior in the hypothesized direction because of anger or low morale. Imagine a study to determine if assigning mentors to new employees improves their productivity. Employees not selected to participate may become aware of the program and wonder why they were not included. They may assume they are less valued employees and react by decreasing their own productivity. The researcher mistakenly could assume that the mentoring program explains the difference in productivity, whereas the difference actually was caused by the poor morale of the control group.

Whenever you, as an administrator, examine empirical research, you will want to raise questions to identify any potential weaknesses in the study design and, more specifically, threats to the internal validity of the findings. A list of such questions follows. The purpose of these questions is to help you decide whether you should take action based on such findings. If a study concludes that a causal relationship exists between two variables, then you must be reasonably confident that none of the threats to internal validity account for the apparent relationship. Plausibility should be the key word in determining how much weight to put on the failure of a research design to control for any one threat to internal validity.[13] You should ask whether it is plausible that *no* factor other than the independent variable is the cause. You also should ask whether it is plausible that *some* factor other than the independent variable was the cause of the dependent variable.

Evaluating Threats to Internal Validity

Questions to help evaluate possible threats to internal validity when examining research reporting a causal relationship between two variables:

1. **Based on history**: Could an event other than the independent variable occurring during the time of the study have caused this relationship?
2. **Based on selection**: Were the people exposed to the independent variable systematically different from those not exposed? If subjects were volunteers who could choose which group to be in, could this have affected the findings?
3. **Based on maturation**: Would a similar change in the dependent variable have taken place with the same passage of time without the occurrence of the independent variable? Could the changes in the subjects have resulted from natural changes such as maturity, aging, or tiring?
4. **Based on statistical regression**: Were subjects chosen for the study because of an unusually high or low value of the dependent variable?

5. **Based on experimental mortality**: Were some people initially exposed to the treatment later assigned to the nontreatment group? Did some people who started the study drop out before finishing?

6. **Based on testing**: Was a pretest administered that could have affected responses to the posttest? Did the pretest provide practice at performing certain tasks and thus affect the posttest scores?

7. **Based on instrumentation**: Did the instrument used to measure variables or collect data change between the pretest and the posttest?

8. **Based on design contamination**: Did the program participants mingle with nonparticipants? Did program participants, nonparticipants, or both have an incentive to change their behavior and make the program succeed or fail?

Threats to External Validity

External validity refers to the appropriateness of extending or generalizing research findings to a group beyond that involved in the study. Only the findings of a study that has external validity may be generalized to other cases. Sampling strategy affects the external validity of a design. If research subjects are randomly drawn from a defined group, then the findings should be generalizable to that group, but this by itself is not enough to ensure external validity. External validity also refers to the ability to generalize the results of a study to other places, times, and programs.

An investigator may do a formal study of a program to determine whether the program has had a beneficial or a harmful effect. She first wants to know if the program works. That is the question of internal validity. But she also wants to know if the program will work for others who have not been studied. That is the problem of external validity. The study may be very informal; a manager may simply observe that his program works as he thinks it should. If he or others wish to conclude that the program will work for other groups or in other locations, they must be aware of several external validity problems. Threats to external validity come about because the conditions of the study are not duplicated for cases not in the study. Unique features of the study subjects, the study setting or conditions, and the implementation of the study program itself, as well as the fact that people may know that they are being studied, contribute to problems with external validity.

Threats to external validity include the following:

1. unique program features
2. effects of selection
3. effects of setting
4. effects of history
5. effects of testing
6. reactive effects of experimental arrangements[14]

Unique Features of Study Subjects

Sometimes in constructing an experimental study, an entire program, event, or intervention is treated as a single independent variable. However, programs typically consist of clusters of several variables. Thus, any program may have *unique features* that affect its results. The program may be unsuccessful in the absence of one or more of the unique features. Creators of experimental programs in a controlled environment may be unable to duplicate the results in a natural, non-experimental setting where they would be most needed. Consider a demonstration project to enable elderly persons to live independently. The project may have staff who are younger, less experienced, more energetic, and in other ways different from staff usually found in ongoing programs housed in social services agencies. The results of the demonstration project may be atypical and not apply to other programs.

Program managers and investigators need to be aware that programs that appear to be similar may be different enough in the way they are implemented to affect how they work. A program for reducing drug use among adolescents may appear to work. A program analyst may then infer that all such programs will work. But external validity must be considered. If the programs were only similar, not exactly the same, the analyst may not be able to generalize from one to the others. Differences in staff, client interactions, activities, or community dynamics may affect program outcomes.

Effects of Selection

The *effects of selection* arise as a threat to external validity when the subjects in the study are not representative of others to which we want the results to apply. The group being studied may have responded differently to the program treatment than others who would be likely to participate. The resulting findings are therefore not representative beyond the group being studied. As an example, consider a training program for teachers tested on a group of senior career teachers. Even if the results of this experiment are favorable, the program may not work with teachers who are much younger and less experienced.

Effects of Setting

Something about the project's *location or setting* can threaten external validity. A project based in a hospital may not work as well as one located in a senior citizens' club, simply because the hospital setting may increase feelings of dependency and work against program goals. Or a program in a New England community may have more success than one in a Southwestern city, where community support systems may be weaker.

Effects of History

History can affect external validity in a number of ways. History as a threat to external validity is somewhat different from history as a threat to internal validity. In internal validity, history is something that takes place at the same time as the independent variable of the study and creates a threat. In external validity, history as a threat is something that took place or existed in the environment prior to the study. An evaluation of a particular program may be undertaken at a time when clients and the community are unusually receptive to its services. For example, a local scandal about conditions in nursing homes may motivate aged persons and their families to seek alternatives to institutional care. A recent fire in a community may motivate homeowners to respond favorably to an education program presented by the fire department. Other communities whose historical circumstances differ may not respond in the same way. An event that happened prior to the study or during the study may make the community in which the study takes place unique. If that is the case, the ability to generalize the findings to other communities will be limited. You may think of other conditions that make a program work at one time or in one place but not at other times or in other places.

Effects of Testing

If a pretest given for study purposes affects the subject's receptivity to the program, *testing* poses a threat to external validity. Consider a parent education program. Giving the parents a pretest that asks how to handle a child's behavior may motivate the parents to become involved in the program and achieve its objectives. The pretest may then be a unique feature, and the program will not work in its absence. If a pretest affects the behavior of the study subjects, the results will not necessarily apply to clients who were not pretested. Testing here differs from the threat it poses to internal validity in that, as a threat to external validity, studied subjects are pretested and other likely participants are not. Testing is a threat to internal validity if studied subjects change their behavior as a result of the pretest, not as a result of the program.

Reactive Effects

Reactive effects of experimental arrangements refers to the fact that study situations often are necessarily artificial, and thus the study setting itself may affect the outcome. If participants know that they are being studied, they may alter their behavior accordingly. The results of such a study could not be transferred to other participants who were not being studied. As online gambling has slowly become legalized at the state level, researchers are interested in studying this as a form of addictive behavior. Assume that researchers at a university recruited a group of students to participate in an experiment that simulates gambling online. The purpose was to see how certain features of the web site encourage or discourage the amount of money they spend and how likely they would be to repeat the game even in the face of a direct loss. One issue likely to emerge is that participants will know the money they are gambling is not real and does not affect their pocketbook; therefore, they will likely be more willing to take risks and revisit sites that they would normally ignore. They also may suspect that the researchers are looking for them to continue repeating the game, especially if they feel they are being well compensated for their time. The reactive effects in this particular study therefore may misrepresent real online gambling behavior and risk taking and may also not be a good representation of how real online gaming encourages or discourages starting a new game.

Combined Threats to External Validity

The reader should note that these threats to external validity can occur in combination with each other. A problem of unrepresentative selection also may involve a problem of setting. If an investigator tested a new program by selecting a specialized group, such as Native American children, and conducted the demonstration program in a town in Tennessee, she would have both of these problems. If the program was meant to apply to all minorities nationwide, the selection of only Native Americans may bias the results. The geographical setting also may introduce a bias. Something about the particular town may contribute to the success or failure of the program. When an intervention is successful, but only in the presence of one or more of the external factors discussed, we refer to this as an interaction of the intervention with the other factor. The absence of the other factor may severely limit the extent to which results can be generalized to other cases and settings.

Increasing External Validity

External validity can be increased by replicating a program in different settings with different personnel. If a demonstration program seems to have worked in one small town in the state, and a state administrator wants to make sure that it will work throughout the state, his best strategy is to try the same program in other locations and with different-sized towns. If the program is successful in dissimilar towns, the administrator should feel more confident that it will work statewide.

A program conducted initially with a representative sample of the program's target group should increase external validity. However, replicating a program with a number of smaller, haphazardly chosen samples may offer more external validity than conducting a single study with a carefully drawn sample.[15] This argument rests on the likelihood that some program participants will quit; thus, a study starting with a representative sample may not end up with one. Experimental mortality could render the outcome of a single study externally invalid.

One approach is to select subjects who are markedly different and infer that if the program works for a wide number of different types of test subjects, then it should work for everybody. This strategy, *deliberately sampling for heterogeneity*, could result in demonstration projects targeted at the elderly in an Eastern city, a rural Southern town, and an urban Native American community. In any one study, problems with internal and external validity are unavoidable. Replication of findings under varying experimental conditions offers reasonably sound evidence of a program's transferability.

Descriptive designs such as sample surveys often have high external validity, although they offer only observational data and thus results with low internal validity. Experimental designs, on the other hand, have high internal validity but often low external validity. Administrators want to be sure that a program or policy achieves its desired ends, a matter of internal validity, before identifying the various conditions under which it will work, a matter of external validity. As with internal validity, a series of questions should be asked to help evaluate the external validity of a study. A list of such questions follows.

Evaluating Threats to External Validity

Questions to determine if a relationship between two variables can be generalized beyond the research subjects:

1. What group did the research subjects represent? Were they selected in a way to make them representative of that group? (As we will see in later sections, the best way to guarantee accurate representation is to randomly select the research subjects from the larger group they are to represent.)
2. Did the program or treatment involve demands that may have affected the representativeness of the subjects, such as excessive commitments of time?
3. Has the program been replicated among different types of subjects in different settings and locations?
4. Exactly what was the program or treatment? Were unplanned or unexamined features, such as personnel or setting, critical to program success?
5. Was the program as offered to the research group different from what will be offered to others?

Experimental Designs

Experimental designs provide the best means of obtaining the evidence necessary to infer the existence of a causal relationship between two well-defined variables. In an experimental design, the researcher can assign subjects to different groups, manipulate the independent variable, and control most environmental influences. The researcher assigns some subjects to the group that is exposed to the independent variable and other subjects to groups that are not. Using random assignment, subjects are assigned so that there is no systematic difference between the groups. If subjects are randomly assigned to groups, any change in the dependent variable should not be due to differences between the study groups. In evaluating studies claiming to have used an experimental design, you should see whether this random assignment was used to help control threats to internal validity. If subjects were randomly selected from a larger population, threats to external validity may have also been controlled.

We discuss two experimental designs in this section, the classical experimental design and the randomized posttest-only design. If you understand them, you should have little trouble understanding more complicated variations. The designs we discuss involve one or two treatment groups. A treatment group is a set of research subjects who are exposed to the independent variable and are then compared on the dependent variable to another group, the control group, whose members have not been exposed to the independent variable.

Classical Experimental Design

The *classical experimental design* has long served as the model for experimental research. It allows us to control the time order of exposure to the variables under study, to determine which subjects are exposed to the independent variable, to identify statistical association among variables, and to control for other possible causal factors. Properly utilized, it can provide the strongest evidence of a causal relationship. It is also an excellent model for demonstrating the logic of explanatory designs. Its characteristics include the following:

1. Subjects are randomly assigned to an *experimental* or a *control group* so that no systematic difference exists between the two groups. Random assignment provides that each subject has the same chance as any other subject of being in either the experimental group or the control group.
2. A pretest, measuring or providing a value for the dependent variable, is administered to both groups.
3. The experimental group is exposed to the treatment; the control group is not. Both groups should experience the same conditions, except for exposure to the independent variable. The researcher controls exposure to the independent variable, determining which group receives it and which group does not.
4. A posttest, which again measures the dependent variable, is administered to both groups following exposure to the independent variable.
5. The amount of change in the dependent variable between the pretest and posttest is determined for each group. The difference between the two groups is attributed to the independent variable.

Note that this design yields the three types of evidence needed to demonstrate that one variable causes another. Like other designs, it can provide evidence of a statistical relationship between variables. It allows us to determine whether any changes in the values of the dependent variable take place after exposure to the independent variable; with this information and by controlling exposure to the independent variable, the investigator can rule out the possibility that the values of the dependent variable changed before the independent variable occurred. Random assignment of subjects to experimental or control groups eliminates most rival independent variables, although problems associated with experimental mortality and design contamination may persist. This design has very high internal validity. It may not, however, have high external validity.

The classical experimental design is represented symbolically as follows:

R $O_1 X$ O_2 (experimental group)
R O_1 O_2 (control group)

where:
R = randomly assigned subjects
O = observation or measurement of dependent variable
X = independent variable occurs

Establishing Equivalent Experimental and Control Groups

With experimental designs, it is important to establish equivalent experimental and control groups and maintain their similarities throughout the experiment. The preferred method of creating equivalent groups is to randomly assign half of a pool of subjects to the experimental group and half to the control group. The theory underlying random assignment is that if enough subjects are used, differences among groups of subjects tend to disappear as long as no bias enters in the way that subjects are assigned to one group or the other. With random assignment, each subject has the same probability of being in either group. Random assignment allows researchers to begin their study by controlling for the threats to validity associated with nonequivalent research groups. Specifically, random assignment controls for selection bias, maturation, testing effects, and statistical regression; however, it only ensures comparability at the initial stage of the research.

An alternative method of assigning subjects is to *match* each individual or unit in the experimental group with an individual or unit in the control group. The researcher identifies characteristics that may affect the dependent variable, gathers data on them in the sample pool, and pairs units with similar characteristics. Subjects that cannot be paired are eliminated from the study. The researcher then randomly assigns one member of the pair to the experimental group and the other to the control group. For example, if she believes that education, sex, and age affect program outcomes, she identifies each subject's education, sex, and age. She then pairs the available subjects; that is, two young (18–29) females who attended college could be a pair, two older males (30–45) with high school degrees could be another pair, and so on. The assignment of the members of the pair to experimental or control groups is random.[16]

Researchers, particularly in the field of medicine, may use matching retrospectively to create comparison groups. For example, researchers match subjects who have a disease with similar persons who do not have the disease. The researchers then try to determine why one group developed the disease and the other did not.

Matching can reduce the number of subjects required because it reduces the variation within the treatment and control groups. This improves the researcher's ability to detect small changes over time or differences between groups in the dependent variable. Nevertheless, methodologists usually prefer random assignment and seldom recommend matching.[17] With more than four relevant characteristics, matching becomes difficult. It also increases the potential for bias, even if efforts are made to avoid it. Matching may reduce internal validity if it focuses on variables actually unrelated to the dependent variable and ignores variables related to it. Matching assumes that the researcher knows which other variables are related to the dependent variable and hence knows on which to match research subjects. The difficulties that stem from this last assumption are virtually eliminated by random assignment, which assumes that the two groups are equivalent in all respects except for exposure to the independent variable.

Controlling for Other Threats to Internal Validity

Experimental designs generally control for other threats to internal validity. To illustrate this characteristic, let's consider history as a threat to internal validity. Outside events (history) occurring between the pretest and the posttest may not affect internal validity if both experimental and control groups experience the same event. The event is assumed to have a similar effect on each group. If both groups experience the same conditions and events other than exposure to the independent variable, the independent variable and other factors associated with the threats to internal validity, such as history or maturation, may cause changes in the dependent variable for the experimental group. Changes in the control group dependent variable may be attributed to a threat such as history, maturation, or another threat. The posttest differences between the experimental and control groups may then be attributed to the independent variable.

Using a Classical Experimental Design

Example 3.2 summarizes an experiment structured along the lines of a classical experimental design to determine whether a training program increased the empathy of first-year medical students. Students assigned to the control group received the same small-group training immediately after the experiment was over. The strategy of randomly assigning subjects to a group that either receives the treatment first (experimental group) or second (control group) helps resolve the political and ethical problems of deciding whether a person receives treatment solely on the basis of random assignment.

Example 3.2 Using a Classical Experimental Design in the Field

Problem: Determine if small-group discussions increase medical students' ability to respond to the emotional concerns of patients.

Subjects: 134 first-year medical students registered for Medical Interviewing and History Taking. Students were randomly assigned to a small group (65 students) or a control group (69 students).

Independent variable (treatment): Small groups of approximately 16 students and a professional staff member met four times for a total of 12 hours. The students interviewed simulated patients; the groups discussed the interviews and interviewing techniques. The control group did not participate in the small group discussions.

Dependent variable: Number- out of three written scenarios where a student used an emotional term to describe a patient's concern.

Design diagram:

$$RO_1X\,O_2$$
$$RO_1\ \ O_2$$

A pretest was given to all students at the first class meeting of Medical Interviewing and History Taking. An experimental-group posttest was given at the last small-group session; a control-group posttest was given at beginning of the first small-group session (after the experiment was over).

> *Threats to internal validity:* None noted in research report. Random assignment allows us to rule out selection and statistical regression. Use of the same paper-and-pencil test for pretest and posttest allows us to rule out instrumentation and testing. Most likely, the threat to internal validity was design contamination, especially if experimental-group members shared information with control-group members. Sharing should decrease differences between the experimental and control groups.

> *Threats to external validity:* None noted; however, possible threats can be suspected. For example, the professional staff who facilitated the groups may have been enthusiastic and committed because of their involvement in implementing a new program. Also, a medical school may have its own unique culture, which affects its students' receptivity to classes and other training activities. Best evidence of external validity would be replication of this training in other medical schools.

> *Analysis and findings:* Table 3.1 shows the average number of scenarios where subjects used an emotional word in their response. It appears that the difference between the pretest average and posttest average for the experimental group is larger than that for the control group.

Table 3.1 The Average Number of Scenarios Where Subjects Used an Emotional Word

	Pretest	*Posttest*
Experimental Group	0.68	2.02
Control Group	0.89	1.13

Random assignment does not necessarily result in two identical groups; statistics help researchers decide if differences between groups could have occurred by chance. In the classical experimental design, researchers focus on the change between the pretest and posttest. The researchers compare the change in the experimental group with the change in the control group. Here, the experimental group showed an increase of 1.34; that is, it more than doubled its average use of emotional responses; the control group showed an increase of only 0.24.

Source: Based on Wolf, F. M. et al., "A Controlled Experiment in Teaching Students to Respond to Patients' Emotional Concerns," *Journal of Medical Education* 62 (1987): 25–34.

Assessing Threats to Validity

In Example 3.2, "Using a Classical Experimental Design in the Field," medical students in both the experimental and control groups showed more empathy on the posttest. Some of the change may have been caused by the lecture on patient interviewing (history), which both groups attended after taking the pretest. Since both groups heard the lecture, it is not a threat to internal validity. History would not be controlled as a threat if only the experimental group heard the lecture, since either the experiment or the lecture may have caused the change. To determine if an experiment controls for history, maturation, or another threat to internal validity requires detailed knowledge of how the experiment was conducted.

Typically, to determine the amount of change, researchers use the same procedure to measure the dependent variable at both pretest and posttest. This is not necessary to ensure internal validity. An experiment may still be internally valid even if the pretest and posttest measuring procedures are not the same, or if the pretest causes a change in the dependent variable. The equivalence of the control and experimental groups neutralizes instrumentation and testing as threats to internal validity. Both groups should respond in the same way to the pretest. Nevertheless, the pretest may sensitize the experimental group to the independent variable and affect the external validity of the design, in which case the pretest becomes part of the treatment or program. We would say that the pretest interacted with the independent variable to affect the dependent variable.

Experimental designs assume the control of the laboratory experiment, a situation that rarely applies to administrative and policy research. Field researchers, that is, researchers who work outside the confines of the laboratory, have markedly less control over subjects and events.[18] Still, researchers conduct field experiments frequently and draw reasonable conclusions from them.[19] Some notable field experiments testing public policy initiatives had many of the characteristics of a true experiment, with the exception of controlling the environment and its influences.[20]

An Adaptation of the Classical Experimental Design

Example 3.3 illustrates an adaptation of the classical experimental design. In this experiment, a university housing office created a contest among college residences with cash incentives to try to decrease natural-gas consumption. The independent variable, cash awards, was introduced several times. In a sense, this amounted to doing the experiment six times with the same groups of subjects. This example also shows how researchers combine techniques. Unfortunately, the researchers' cost-benefit analysis found that the cash awards alone cost more than the fuel savings.

Example 3.3 An Adaptation of the Classical Experimental Design

Problem: Determine effective ways to reduce energy consumption by residents in master-meter housing, where individual residents do not pay utility bills.

Subjects: Five university-owned apartment complexes. Analysts paired one set of complexes of comparable size, family mix, and energy source. One complex was randomly assigned to the experimental group, the other to the control group. The remaining three complexes could not be paired.

Independent variable (treatment): Contest with cash awards held six times two weeks apart. Indicated by X.

Dependent variable: Savings in natural-gas consumption = (predicted consumption indicated by O of the control group minus actual consumption indicated by O in the experimental group).

Design diagram:

$$R\ O_1X\ O_2X\ O_3X\ O_4X\ O_5X\ O_6\ X\ O_7$$
$$R\ O_1\ \ O_2\ \ O_3\ \ O_4\ \ O_5\ \ O_6\ \ \ O_7$$

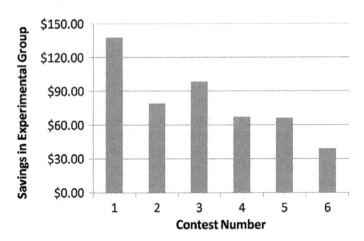

Figure 3.1 Reduction in Natural Gas Consumption

Threats to internal validity:
> History: possible problem because experimental complex also received a 15-page saving guide.
> Maturation, statistical regression, selection, testing instrumentation, controlled by the design.
> *Experimental mortality:* none.
> Contamination: unlikely; control group members expressed little interest in contest.

Threats to external validity:

Nothing indicated to suggest findings could not be generalized to other university-owned complexes.

Conclusion: The chart shows that the contest reduced natural-gas consumption substantially at first but then at a declining rate. (See Figure 3.1.) However, other data showed that the savings were not enough to offset contest costs. Questions of history and contamination as threats to internal validity were moot. Findings replicated those found in a study of an apartment complex; the approach worked better in a university that had contests among dormitories.

Source: Adapted from McClelland, L., and S. W. Cook, "Promoting Energy Conservation in Master-Metered Apartments through Group Financial Incentives," *Journal of Applied Social Psychology* 10 (1980): 20–31.

Note that in the example, the energy use declined from previous levels and leveled off. Although the data were incomplete, one could imagine that as the weeks passed, the students would revert to their previous energy-consumption habits.[21]

Randomized Posttest Designs

The *randomized posttest design* avoids the possible sensitizing effect of a pretest by eliminating it, which results in a design identical to that of the classical experiment except for the absence of a pretest. The underlying assumption is that a pretest is expendable because random assignment will result in equivalent experimental and control groups. This assumption is most likely to be valid for large study populations. The design is diagrammed as:

$$R \; X \; O_1$$
$$R \quad \; O_1$$

The randomized posttest design has a number of advantages. Eliminating the pretest reduces costs. The administration of a pretest is often difficult or even impossible. It may not be necessary. The randomized posttest design relieves the researcher of identifying a reasonable pretest. Consider a prison work-release program tested to determine whether it reduced the recidivism rate of prisoners. Qualified prisoners were randomly assigned either to a work-release program or to serve out their full terms. Upon release, all prisoners were followed to determine whether they were arrested for additional crimes. Recidivism rates for the work-release program subjects and full-term subjects were compared.[22] Consider what might happen if the researchers had to obtain pretest data. The data may not exist; the prison may not have reliable records on whether an inmate had been in prison before, how many times, or for how long. Even if sound data exist, confidentiality requirements may prohibit researchers from having access to them. And we may reasonably assume that inmates with a history of recidivism would not be among the "qualified" prisoners eligible for work release.

If experimental groups are large enough for the investigator to be confident that they are similar, the randomized posttest design is efficient and effective. If a study has relatively few subjects, a researcher needs pretest data to ensure the comparability of the experimental and control groups. Pretest data also are necessary if an investigator needs to determine the extent of change attributable to the independent variable.

Applying the Randomized Posttest Design

Example 3.4 illustrates the application of a randomized posttest design to determine whether coaching improves performance in leaderless groups. In this study, measuring performance in leaderless groups prior to introducing the independent variable would have increased the study's costs. Furthermore, the additional time required may have led to greater problems with experimental mortality and increased the opportunities for design contamination, particularly if members of the experimental and control groups had contact with each other.

Example 3.4 An Application of the Randomized Posttest Design

Problem: A university placement office wants to learn whether coaching improves performance in assigned-role leaderless groups (a technique used by major employers to select professional employees).

Subjects: 36 female students from introductory psychology classes randomly assigned to one of three groups.

Independent variable: Type of preparation for leaderless group. Two treatments were used, so two experimental groups were set up:

X_1 = Subjects attended two coaching sessions and received feedback on their performance in practice leaderless groups.

X_2 = Subjects saw a short tape of someone giving a biased account about her experience in a leaderless group.

Control group: Subjects did not attend coaching sessions, view tape, or have any other preparation.

Dependent variable: Overall performance in an assigned-role leaderless group. Scores on leadership, behavioral flexibility, oral presentation, initiative, and persuasion were averaged for each subject.

Design diagram:
$$R\ X_1 O_1$$
$$R\ X_2 O_1$$
$$R\ O_1$$

Threats to internal validity:

History: controlled by design and carefully implemented procedures for conducting the experiment.

Maturation, statistical regression, selection: controlled by the design.

Testing, instrumentation: not applicable with this design.

Experimental mortality: none.

Contamination: none documented.

Threats to external validity:

Possible interaction of selection and treatment: These were all college women who might respond to the coaching program differently from the general population of professionals seeking employment.

Possible interaction of setting and treatment: These students had no immediate prospects of using the skills to obtain a job. Thus, the experimental setting did not duplicate the natural setting where subjects would be competing for a position.

Findings:

Table 3.2 Average Performance of Three Groups

	Group		
	Coached	*Viewed Tape*	*Control*
Average performance*	5.08	2.83	2.58

*Individual mean scores on separate performance indicators. The higher the score, the better the performance.

Conclusion: Hypothesis that coaching can improve performance in an assigned-role leaderless group was supported. Tape was meant to duplicate "grapevine" advice; it may not have represented the effect of informal advice. Researchers did not determine whether coaching only makes participants test wise or whether they later apply learned behaviors to work settings. Also not known was whether coaching would work as well with persons who had management experience.

Source: Kurecka, P. M. et al., "Full and Errant Coaching Effects on Assigned Role Leaderless Group Discussion Performance," *Personnel Psychology* (1982): 805–812.

Uses and Limitations of Experimental Designs

Administrators are likely to see experimental designs in conjunction with program evaluation, although they are used in other areas as well. The purpose of program evaluation is to determine whether a planned intervention (treatment) or program has produced its intended effects. As mentioned earlier, a major problem encountered when attempting to use experimental designs for program evaluation is the loss of researcher control. This lack of control makes experimental designs impractical for evaluating established programs. Such programs are sufficiently complex and have well-established procedures. Researchers cannot expect to spell out who will receive services, exactly what services they will receive, and exactly how the services will be delivered due to ethical and regulatory issues.

Thomas Cook and Donald Campbell have identified other situations in which experimental designs will not work.[23] True experimental research in administrative settings rarely can be done quickly and requires careful planning and random assignment, which may not be practical. Important variables such as age, race, or the occurrence of a disaster cannot be manipulated or randomly assigned. The decision to evaluate most programs is made independently of the decision to implement the program and often after the program has begun, making true experimentation impossible because it is too late to manipulate the intervention (the independent variable), to conduct a pretest, or to assign subjects to groups. True experimental designs also are expensive to conduct in realistic social settings. Before implementing an experimental design, researchers should conduct preliminary studies to assess the feasibility of an experiment.

Experimental designs seem most relevant to the study of new or controversial programs at a time of scarce resources.[24] Programs that are specific, clearly described, and implemented all at once rather than phased in gradually lend themselves to experimental evaluation. The effects of modest treatments, such as changing office accounting procedures, may be more easily and reasonably studied experimentally than ambitious programs to achieve broad social goals. A researcher may have difficulty ensuring that the latter programs are carried out as designed or intended; pressures to participate in the program may make random assignment nearly impossible, and subject attrition and design contamination may be inevitable. One objection to the use of experimental designs in the study of social programs has revolved around the question of whether the control group can be offered a treatment. Although traditional experimental design calls for identifying what would happen without "treatment," the resulting practical and ethical problems have made offering alternative treatment to the control group an acceptable practice.[25] Such problems have been particularly prominent in medical research, where researchers, policy makers, and ethicists have debated the morality of withholding a new medical treatment from a group so that its effects can be verified experimentally.

Quasi-Experimental Designs

The classic experimental design relies on the researcher's ability to control the research setting. The researcher exercises control from the initiation of the experiment to its conclusion, selecting and assigning subjects and exposing the experimental group to the independent variable. As much as possible, she ensures that groups equivalent at the beginning undergo the same conditions during the study, except for exposure to the independent variable. The tradition of referring to the independent variable as a "treatment" emphasizes the objective of manipulating and controlling exposure to the independent variable.

In research conducted outside a laboratory setting, the amount of control required by the classical experimental model may be unattainable. True experiments require the following:

1. the manipulation of at least one intervention (independent variable)
2. the random assignment of subjects to groups
3. the random assignment of the independent variable to groups
4. the exposure of the experimental group or groups to the treatment in isolation from other factors

If one of these conditions cannot be met, the appropriate research design is a *quasi-experimental* one. Donald Campbell and Julian Stanley refer to "many natural settings in which the [researcher] can introduce something like experimental design into his scheduling of data collection procedures even though he lacks the full control over the scheduling of experimental stimuli which makes a true experiment possible."[26] In other words, the researcher makes use of as many features of the classical experimental design as possible and adopts other measures to control threats to internal validity. Unlike the groups in true experimental designs, members of the treatment group and the control group in quasi-experimental designs are not randomly assigned; therefore, the analyst cannot assume that the groups are equivalent at the beginning.

We discuss three of the more commonly used quasi-experimental designs: the comparison group pretest/posttest design, the interrupted time-series design, and the interrupted time-series design with comparison group.

Comparison Group Pretest/Posttest

On the surface, the *comparison group pretest/posttest design* may look like the classic experimental design, but don't be fooled. Subjects that are not selected by random assignment but instead by the researcher identifying a group of subjects that seems comparable to the group involved in the test is not the same. However, we can still learn something from this type of design. It is still a useful strategy because, frequently, programs or policies develop in a way that precludes random assignment. Suppose a police department wanted to study the effect of a compressed workweek of four 10-hour days. The study may have problems if officers object to being randomly assigned to the experimental schedule. Hence, researchers may prefer to limit the testing of the schedule to units where everyone agrees to participate. The resulting research design would require finding platoons willing to submit to the new schedule and then designating a comparison group from among the other platoons. One way of constructing a comparison group is to match each platoon in the experimental group on the basis of selected characteristics, but this is not the same as random assignment where people have an equal chance of being selected for either group.[27] Note, however, that unlike true experimental designs, members of the groups here are not randomly assigned, so the analyst cannot assume that the groups are equivalent. Nor is the "experimental" treatment randomly assigned to one of the groups.

Comparison groups also are constructed for contrast with intact experimental groups. This provided a useful strategy for one group of researchers who wanted to study how organic farming affected crop production. Since the researchers were not in a position to convince farmers to switch to organic farming, they identified known organic farms and matched each with a farm of similar size and soil composition. Similarly, in a study of the effects of day care on children, kindergarten students who had been in day care for five years formed the experimental group. Kindergarten children in the same school system who had not been in day care, and whose parents agreed to their participation, formed the comparison group. After the comparison group was selected, researchers studied the characteristics of both groups to identify systematic differences between them.

The comparison group pretest/posttest design is diagrammed as:

$$O_1X\ O_2$$
$$O_1\ \ \ O_2$$

Sometimes the comparison group design is called "the nonequivalent control-group design," but we prefer to restrict the term "control group" to groups whose members are randomly assigned. The major limitation of the comparison group design is the inability to control for biases introduced by selection. The greater the similarity between the experimental and comparison groups, the more confidence a researcher may have in making inferences from his findings. The design can control for history, maturation, pretesting, and instrumentation, but this must be confirmed. Unless a researcher can keep conditions between the experimental and comparison group the same from pretest to posttest, one group may have an experience that affects its posttest data. For example, if the comparison and experimental groups consisted of police precincts, and if the precincts used different criteria for data collection, then instrumentation would be a possible threat to validity, as the measure of the dependent variable would be different for the two groups. Experimental mortality could be checked by examining records. The analysts could determine whether any platoons discontinued the experimental schedule prematurely, essentially dropping out of the study prior to its conclusion.

Threats to Validity

Statistical regression constitutes a threat to validity in this design. This threat is of particular concern if the experimental group and the comparison group differ systematically in some dimension. Imagine that the pretest turnover rates were higher than usual in the experimental platoons and lower than usual in the comparison platoons. If each group regressed toward its average, a posttest might show a lower turnover rate for the experimental group and a higher rate for the comparison group—both due to a statistical effect, not the difference in scheduling. This is not an unlikely scenario. You would expect administrators to pick the platoon with the highest turnover rate for an experimental program to reduce turnover.

The comparison group pretest/posttest design may have high external validity if it replicates a program's effect in a variety of settings. Thus, if the change in scheduling reduced turnover in all platoons where it was tried, and the platoons included ones with high morale and low morale, ones in good neighborhoods and ones in tough neighborhoods, ones with easygoing leadership and ones with authoritarian leadership, the administrators would feel confident that shifting the entire police force to compressed scheduling would prove successful.

Evaluation of Home Weatherization Program

Another application of a comparison group pretest/posttest design is illustrated in Example 3.5. At various times when home energy costs were high, power companies and university researchers studied ways to help homeowners and renters reduce heating costs. A study conducted during one of these time periods used a comparison group pretest/posttest design to determine if a home weatherization program for low-income families reduced energy consumption.[28] Fifty-nine weatherized homes and 37 homes not yet weatherized in Minnesota in the late 1970s were compared on the amount of energy used, the dependent variable, during the same months. The independent variable was participation in the weatherization program that provided home insulation, storm doors and windows, and weather stripping. On average, weatherized homes saved 13 percent in fuel consumption.[29]

The comparison group design controls for the threats to internal and external validity in the following fashion:

> *History, maturation:* May be controlled by design. A possible threat of local history exists; that is, an event affects one group and not another. Similarly, selection and maturation may interact, that is, independently of the experiment, one group may change faster than another.
> *Experimental mortality:* Not well controlled by design but can be detected.
> *Selection, statistical regression, contamination:* Not controlled by design.
> *Testing effects:* Controlled by design.
> *External validity:* May be reduced, since neither experimental nor control group was randomly selected. However, external validity may be improved greatly by replication in different settings.

Example 3.5 An Application of a Comparison Group Pretest/Posttest Design

> *Problem:* Determine whether a home weatherization program for low-income families reduced energy consumption.
> *Subjects:* 59 weatherized homes and 37 homes waiting for weatherization in Minnesota.
> *Independent variable:* Participation in the weatherization program that provided as needed: insulation, weather-stripping, storm doors and windows, glass replacement, repairs.
> *Dependent variables:*
>
> Percentage British thermal units (BTUs) saved.
> BTUs saved per degree day per square foot of living space. $(O_2 - O_1)$
> Dollars saved per degree day. $(O_2 - O_1)$
> Design diagram:
>
> $O_1 X O_2 O_1$
> $O_1 \quad O_2$

Threats to internal validity:

History: no apparent threat noted.

Maturation: controlled by design.

Selection, statistical regression: no apparent bias, although comparison group had applied later than experimental group for weatherization; both groups had similar amount of floor space, number of occupants, percent owner occupied.

Experimental mortality: various subjects eliminated because of moves, being away from home, or inability to get accurate fuel records.

Testing, instrumentation: controlled by design.

Contamination: none noted.

Threats to external validity: Interaction between history and treatment; interaction between setting and treatment possible. Study took place in Minnesota in the late 1970s, a period of high energy costs.

Findings:

Table 3.3 Comparing Energy Use for Weatherized and Nonweatherized Homes

	Weatherized Home (Average)	Nonweatherized Home (Average)
Percent BTUs saved	10.95	−2.48
BTUs saved per degree day per square foot	1.68	−0.295
Dollars saved per degree day	$.006	$.001

Discussion: On average, weatherized homes saved 13 percent in fuel consumption; the cost of weatherization was paid back in 3.5 years. Data and design support benefits of weatherization; more evidence needed to support generalizability of findings.

Source: Hirst, E., and R. Talwar, "Reducing Energy Consumption in Low-Income Homes," *Evaluation Review* (October 1981): 671–685. (Copyright © 1981 by Sage Publications. Reprinted by permission of Sage Publications).

In the comparison group pretest/posttest design in the home weatherization study, the experimental group consisted of people who first signed up to have their homes weatherized. The comparison group consisted of people who signed up later and were on a waiting list. The researchers gathered data to make certain that the two groups were similar with respect to the size of homes, the number of occupants, and the ratio of homeowners to renters. Other unexamined differences between the groups may have existed. The initial group may have been more motivated to save energy and may have used other methods in addition to the weatherizing improvements. To decide whether the evidence suggests that a weatherization program is effective, decision makers must use judgment and experience in deciding whether plausible uncontrolled threats to internal validity may have changed the findings.

Interrupted Time-Series Design

The *interrupted time-series design* incorporates an independent variable other than time into the time-series design. The design improves greatly upon the one-group before/after design. As contrasted with the simple time series, the interrupted time-series design introduces an independent variable and can be used to trace the effects of that variable upon other variables. The simple time-series design, on the other hand, is used to describe the values of one variable over time and not to explain changes in those values.

The interrupted time-series design calls for several observations before the introduction of the independent variable. These observations help to demonstrate that exposure to the independent variable resulted in a change in the dependent variable that cannot be attributed to long-term trends, cycles, or seasonal events. A series of observations made after the introduction of the independent variable provides evidence specifying the effect of an independent variable. The independent variable may have resulted in four general possibilities.

Figure 3.2 illustrates three of these possibilities:

1. An abrupt permanent change in the dependent variable.
2. An abrupt temporary change, which lessens and eventually returns to the baseline level.
3. A gradual permanent change in which the initial change gradually increases or decreases to a point where it starts to level off.[30]
4. The fourth possibility, not shown here, is that the independent variable resulted in no change at all in the current pattern of the dependent variable.
5. The fifth possibility, not shown here, is that the independent variable resulted in a decrease in the current dependent variable.

In interrupted time-series designs, the independent variable may be introduced by the researcher. However, it is more likely that the independent variable is something that occurs naturally or is introduced by someone else. In many cases, exposure to the independent variable has already occurred when the analyst is in a position to study its possible effects.

The design may be represented as follows:

$$O_1 O_2 O_3 O_4 \, X \, O_5 O_6 O_7 O_8$$

The number of Os and their placement in relation to X depend on the number of observations made before and after the independent variable was introduced. The design controls best for maturation as a threat to internal validity because the observations capture how the dependent variable changes with the passage of time. The design also rules out the effects of statistical

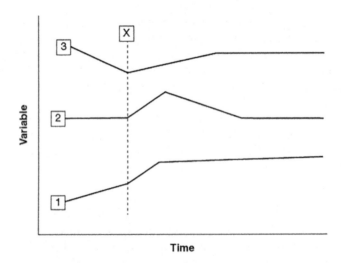

Time

Figure 3.2 Example of Impacts of Interrupted Time Series

regression. Other threats to internal validity may be eliminated after a researcher checks into the probability that each occurred. Other events, for example, may have occurred at the same time as the independent variable. A careful investigation of current circumstances can identify these. Fluctuations in a time series may be due to long-term trends, cyclical variations, seasonal trends, and random fluctuations. If enough time is considered before and after the intervention, trend patterns can be identified. The interrupted time-series design can eliminate these alternative explanations of the changes occurring after an intervention.

A researcher should be especially alert for an interaction between selection and treatment; that is, he should be confident that the preintervention data represent a population similar to the postintervention data.[31] For example, to evaluate the effectiveness of a year-round school requires a researcher to determine if the school's student body changed after the year-round school program began. Students who disliked school and dreaded the prospect of losing their long summer vacation may have transferred to other schools, and more competitive students may have transferred into the year-round school. Changes in student performance may then be due to changes in the student body and not to the longer school year. A review of the records may suggest other events that could have caused the dependent variable. A researcher should examine program records to determine the extent of experimental mortality. He should make sure that the way student performance was measured did not change during the time that the observations were made. For example, if a survey was used, were the questions reworded or different ones used? Was a different measure used, such as changing from the American College Testing (ACT) test to the SAT?

Threats to Validity

Interaction between selection and treatment in particular is a potential problem when it comes to external validity, and the design does not control against such threats. The following list summarizes how the interrupted time-series design protects against threats to internal and external validity.

Threats to internal validity:

> History, instrumentation: not protected against by the design. Researcher should review records to determine whether other events occurred at the same time as the intervention or whether any change in measurement procedure occurred.
> Maturation, statistical regression, testing: protected against by the design. Changes associated with long-term trends, cycles, seasonal variations, and random fluctuations can be ruled out if the design has a sufficient number of data points.
> Selection: not ruled out by design. Researcher needs to make sure that intervention does not coincide with a major change in the population being measured.
> Experimental mortality: not ruled out by design. Researcher should check records to determine whether subjects dropped out after the occurrence of the independent variable.

Threats to external validity:

> The design does not control against threats to external validity. Interaction between selection and treatment in particular is a potential problem.

The interrupted time-series design works best if the independent variable is expected to have a marked and immediate effect. Otherwise, its effect mistakenly may be attributed to long-term trends, cycles, or seasonal effects unless these are removed through statistical procedures or otherwise evaluated. Statistical procedures can help a researcher confirm his impressions that an effect is markedly greater than expected fluctuations.[32] This design also works best if the independent variable can be introduced all at once at a clearly identified time. If it is phased in gradually, its effects may be more difficult to identify. The researcher should also attempt to understand when the effect is likely to occur, that is, whether it will be almost immediate or will take a long time.

Interrupted Time-Series Examples

The interrupted time-series design provides journalists, politicians, and researchers a convenient, easily implemented way to track the impact of a public policy. In 1994, the police chief of New York City, newly appointed by then-Mayor Rudy Giuliani, instituted reforms that emphasized community policing targeted at crime "hot spots." The mayor's biography cites the program's success: "Under Mayor Giuliani's leadership, New York City has experienced an unprecedented 38 percent reduction in overall crime and 48 percent reduction in murder since 1993."[33] Figure 3.3 supports the mayor's boast.

Although the city's crime rate began dropping in 1989, the crime rates in 1994 and 1995 were the lowest since 1980. The crime rate decreased 11 percent in 1994 and 16 percent in 1995, the greatest drops over the 15-year period. Although it appears that Mayor Giuliani's program was successful, researchers would want to inquire about the various threats to internal validity, that is, other factors that could account for the drop in crime, and how they could be assessed. We continue the example later with a discussion of these issues. As tracking and reporting statistical data become common ways to monitor and reward performance, a serious type of design contamination may occur. That is, data collectors, administrators, or analysts will be tempted to manipulate the data to their advantage. For example, one can imagine a police department reconsidering how it classifies some deaths in order to avoid showing an increased homicide rate.

Time-series data eliminate maturation or statistical regression as threats. Yet they do not confirm the success of community policing or Giuliani's leadership. Other cities also experienced lower crime during the same time period. Other reasons cited for the drop included improvements in the economy, less use of crack cocaine, and relatively fewer teenagers in the population. Furthermore, the reported crime rate does not measure the actual incidence of crime. Cynical citizens may not call the police; busy police officers may not file crime reports. To demonstrate the effectiveness of police strategies requires more analysis. Investigators should confirm that reporting patterns were consistent. They should see if changes in the economy, in drug use, or in the population's age distribution coincide with changes in the crime rate. They should compare other cities' data with New York City's data.

Several recent studies have investigated the effect of medical marijuana laws passed by state legislature on crime rates using some form of time-series design. Alford compared time series of several crimes for states with medical marijuana laws to states without those laws. Because

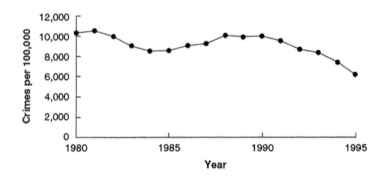

Figure 3.3 New York City Crime Rate 1980–1995

states adopted the laws at different times and to control for other variables, she used sophisticated statistical procedures for her analysis. She found that some crime rates were lower in states with medical marijuana laws, while rates for some crimes were higher.[34]

To illustrate an interrupted time-series design with more complete analysis than was done in the New York City data, we examined a study on the effect of the Illinois rape shield law. A common legal reform enacted by legislatures in the 1970s and 1980s, rape shield laws restricted the ability of defense attorneys to introduce evidence about the victim's past sexual history into a rape trial. Advocates of rape shield laws argued that a change in existing laws would encourage women to report rapes, press charges, and increase conviction rates. Illinois implemented a rape shield law in April 1978. Figure 3.4, based on data drawn from a more detailed study,[35] shows the yearly data on the conviction rates for rape or rape-related charges in Chicago.

The conviction rate went up in 1978, but in subsequent years the conviction rates dropped and did not look markedly different from those of the years before the reform. Recall that to implement a time-series design, a researcher should also collect qualitative information. The qualitative information allows the researcher to identify other factors that could have caused an observed change. In this study, the researchers interviewed judges, state attorneys, and public defenders. From the interviews and extensive statistical evidence, the researchers concluded that the law "had no impact on the indictment rate, conviction rate or incarceration rate."[36] The researchers learned that, prior to 1978, case law had started to restrict the use of sexual history as evidence;

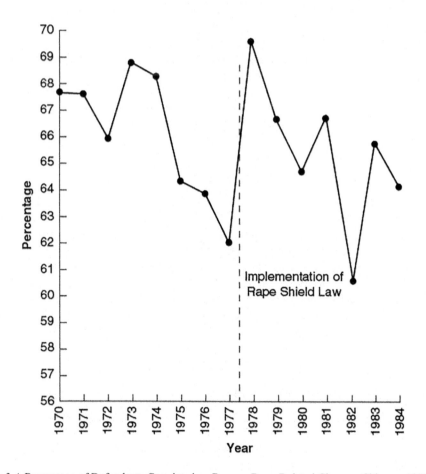

Figure 3.4 Percentage of Defendants Convicted on Rape or Rape-Related Charges, Chicago, 1970–1984

thus, the reform had been more gradual than it appeared. The researchers also observed that the law could have affected only cases a jury heard and in which a claim of consent was part of the defense. Since fewer than 10 percent of rape cases were heard by a jury, the potential of the law to affect overall conviction rates was less than its advocates expected.

This example suggests the complexity of interpreting an interrupted time-series design. The graphical information is helpful and easily understood. The graphs, however, are often only a first step; they are usually followed by detailed statistical analysis.[37] Tracking a variable and viewing its changes raises interesting questions about what actually caused the shift. One strategy is to enrich the time-series design by collecting similar information from one or more untreated comparison groups.

Interrupted Time-Series With Comparison Group

An addition of a comparison time series of an untreated group is seen occasionally with the interrupted time-series design. This design is called *interrupted time series with comparison group* and allows the researcher to compare the dependent variable time series of the treatment group with the same series for a nonequivalent, nontreatment group.

In such a design, researchers develop a time series, interrupted by the occurrence of the independent variable, for the treatment group. In addition, they develop a concurrent time series for another group not exposed to the independent variable. If the groups are equivalent in important aspects, the addition of the comparison group can provide a check on some threats to internal validity.[38] To check the effectiveness of New York City's community policing, one might choose another major American city (Chicago, Philadelphia, or Boston) where community policing was not implemented. To control for selection as a threat to internal validity, the comparison city should have similar populations, for example, similar variations in age, ethnic background, income, and drug use. One also looks for "local history," that is, events that occurred in one city and not in the other, which could have caused the changes in crime rate.

The interrupted time series with comparison group design is diagrammed as follows:

$$O_1\ O_2\ O_3\ O_4\ O_5\ X\ O_6\ O_7\ O_8\ O_9\ O_{10}$$
$$O_1\ O_2\ O_3\ O_4\ O_5\quad\ \ O_6\ O_7\ O_8\ O_9\ O_{10}$$

Example 3.6 summarizes an interrupted time series with a comparison group design.[39] The researchers were evaluating the impact of community development initiatives on housing values. They looked at the intervention of a redevelopment effort in a declining neighborhood in Portland, Oregon. To control for other changes, including trends such as the general increase in housing values, they collected data on housing values in other low-income neighborhoods and on average housing values in all of Portland. Thus, they had two control groups.[40]

With quasi-experimental designs, having more than one comparison group, if possible, is important, since no one comparison group is as good as a control group formed through random assignment.

Example 3.6 An Application of an Interrupted Time-Series Design

> *Problem:* A nonprofit organization sponsored the redevelopment of a commercial area in a declining neighborhood, called Belmont, in Portland, Oregon. Sponsors asked if efforts to revitalize distressed inner-city neighborhoods made any difference. Research was conducted to provide an answer to this question.

Subjects: Belmont neighborhood, other low-income neighborhoods in Portland, City of Portland.

Independent variable: Commercial redevelopment in 1996 and 1997.

Dependent variable: Sales price of single-family housing units

Design: Yearly data were collected from 1988 through 1999.

Table 3.4 Yearly Data From 1988 Through 1999

Belmont:	$O_1 \ldots O_4 \ldots O_8 \, X \, O_9 \ldots O_{10} \ldots O_{11} \ldots O_{12}$
Other low-income neighborhoods:	$O_1 \ldots O_4 \ldots O_8 \quad O_9 \ldots O_{10} \ldots O_{11} \ldots O_{12}$
All of Portland:	$O_1 \ldots O_4 \ldots O_8 \quad O_9 \ldots O_{10} \ldots O_{11} \ldots O_{12}$

Threats to internal validity:

> History: data from other low-income neighborhoods and the rest of Portland controlled for a general increase in housing values and the effects of other events and factors.
>
> Instrumentation: the measurement of housing values did not change throughout the data collection period.
>
> Selection: a possible threat. Belmont may have had characteristics making it a particularly good candidate for revitalization.
>
> Maturation, statistical regression: controlled by the design.
>
> Other threats were assumed not to apply.

Possible threats to external validity: Belmont had been a well-to-do neighborhood, and the redevelopment was based on commercial concerns. A number of old homes surrounded the commercial district.

Findings: Figure 3.5 reports the median value of single-family housing for each year between 1988 and 1999. The redevelopment activities began in 1996 and were completed by the end of 1997. As the graph shows, the values of housing in the neighborhood began to increase before revitalization was completed.

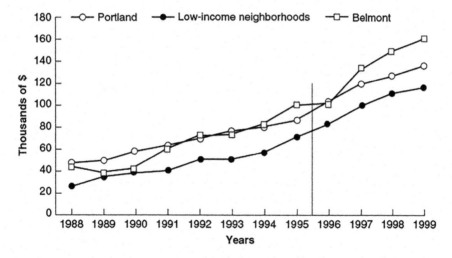

Figure 3.5 Median Sales Price of Single-Family Homes by Year

Discussion: The researchers inspected the graphed raw data and conducted detailed statistical analysis. They concluded that the graphical data demonstrated that median value of housing in the Belmont neighborhood had increased during and just after revitalization by more than had the value of housing in other low-income neighborhoods and in the City of Portland as a whole. Statistical analysis verified the conclusions based on the visual data. However, the increase was in the base value of housing. The increase after revitalization in Belmont did not continue at a higher rate than in other low-income neighborhoods or in Portland as a whole. Researchers concluded that revitalization efforts had had a positive impact.

Nonexperimental Designs

Nonexperimental designs do not control for threats to internal validity. If an intervention is implemented on randomly selected subjects, which occurs occasionally, the designs have some external validity. The value of nonexperimental designs may be overlooked because of the obvious limitation of drawing conclusions when threats to validity, including history, maturation, statistical regression, and selection, are uncontrolled.

Three nonexperimental designs are:

Single group posttest: $X\ O_1$
Single group pretest/posttest: $O_1\ X\ O_2$
Nonequivalent groups posttest only: $X\ O_1\ O_1$

Single Group Posttest

To illustrate the value and limitations of a single group posttest, we selected research descriptions from two news stories and provide quotes from those stories that illustrate the newspapers' conclusions regarding success.

New York City wanted to increase transit use by elderly from the city's Chinatown. The Transit Authority sent a news release to three Chinese-language newspapers announcing that a mobile unit would come to New York City's Chinatown to take farecard applications. A *New York Times* article describing the success of this "experiment" included the following quote: "[A]bout 400 people showed up yesterday to enroll for a special [farecard] available to senior citizens for half price" after the "Transit Authority sent a news release to three Chinese-language newspapers."[41]

A *Washington Post* article described what the *Post* saw as the success of a Welfare-to-Work Program in South Carolina. The article included the following quote: "In a single year [South Carolina's welfare-to-work program] has put 442 welfare recipients on its payroll without creating any new jobs or displacing any state workers."[42]

Although single group posttest studies do not control for any threats to internal validity, the most plausible threats are noted subsequently. In the Chinese-language news release "experiment," history and maturation are the most plausible and can be eliminated. The mobile unit had been traveling throughout the city for six months; however, prior to the press release, relatively few people had used it to apply for farecards. Thus, the press release seems to have been a success.

The success of the welfare-to-work program is less certain. People who work in low-paying, low-skilled jobs rely on welfare during periods of unemployment. The 442 jobs were concentrated in food service, maintenance, and personal services. Maturation or history cannot be eliminated as threats. The 442 jobs normally may be held by people who move on and off welfare, depending on their employment status. The state did not know how many welfare recipients it had hired in previous years; 442 may have been a relatively large number, but perhaps it was not. Economic growth may have shrunk the pool of job applicants and forced the state to develop new employee sources. The state may have hired welfare recipients because of its inability to find other employees. A good economy, rather than the welfare-to-work program, may have contributed to hiring former welfare recipients.

Even if plausible threats to internal validity cannot be eliminated, the single group posttest provides useful information. First, the design indicates whether an intervention worked as well as expected. The Transit Authority expected 50 people to visit its mobile unit in Chinatown. Its expectations were greatly exceeded. If hardly any people had shown up, the Transit Authority might have concluded that press releases in foreign-language newspapers were not helpful. If South Carolina investigators found that the state had hired few welfare recipients, they would question the effectiveness of the welfare-to-work program. Second, researchers can collect other valuable, detailed information. In South Carolina, investigators studied the jobs welfare recipients had—the type of work, whether the jobs were temporary or permanent, whether they had benefits. Such information can identify additional issues that help in understanding the relationship between welfare-to-work programs and employment. The single group pretest/posttest design does not perform better at controlling for threats to internal validity, but it does add one useful piece of information. It indicates if something changed between the pretest and the posttest. A pretest/posttest design establishes if program participants changed. Participants are "tested" at a program's beginning and end. The change in their performance may be attributed to the program, but we should ask if there are alternative explanations. Consider a 10-week course to train new staff how to answer taxpayer inquiries. Participants are tested on their knowledge of the tax code. At the beginning of the course, 30 percent had a score of 85 or better, and at the end, 80 percent had a score of 85 or better.

What factors, other than the program, might have brought about the improvement?

- If participants work as they attend the training course, the posttest scores may reflect what they learned on the job about the tax code.
- People with low pretest scores may have dropped out of the course, leaving those who are doing well as a larger proportion of the whole class.
- The pretest may have motivated people to study the tax code on their own, and their "homework," rather than the training, improved their posttest scores.
- An unfamiliar format may have lowered the performance on the pretest. Some posttest improvement may be attributed to familiarity with the test format.

Problems of design contamination may arise in studies of programs intended to change attitudes or behaviors. At the time of the posttest, participants may give answers or act in a way that suggests that they have adopted the expected attitudes or behaviors. Nevertheless, their attitudes may have stayed the same, and they may have not really have altered their behavior.

Single Group Pretest/Posttest

The single group pretest/posttest has strengths similar to those of the single group posttest. First, investigators can decide if a policy or program appears to meet their expectations. Second, they can gather data to answer specific questions about a program. Third, they can combine their knowledge of the program with other information and make a plausible case that it is effective. Fourth, they can identify what questions they may want to investigate further.

Whether an investigator examines a policy's impact with a single group posttest, a single group pretest/posttest, or interrupted time-series design may depend on the availability of data. As policy agendas change, so do data requirements. For example, apparently South Carolina at that time did not track how many former welfare recipients worked for the state. The single group posttest or single group pretest/posttest may be the best one can do.

Nonequivalent Groups Posttest Only

The last nonexperimental design, nonequivalent groups posttest only, compares two groups: a group that experienced an intervention and a group that did not. The design has the same limitations as the other nonexperimental designs. One cannot tell if differences between groups existed prior to the intervention, nor can one tell what, if any, changes took place in each group. Yet, as in the other nonexperimental designs, the nonequivalent group's posttest can provide useful information.

Example 3.7 describes a nonequivalent groups posttest design to evaluate the effectiveness of parent training. The information identified the parts of the program that were working and the parts that needed improvement. The particular study had a problem often encountered in collecting data from social service clients, for example, the inability to keep in contact with former clients. Contact information such as addresses, telephone numbers, and telephone service changes often.

Example 3.7 An Application of Nonequivalent Groups Posttest Design

Problem: A program conducts a parent education course for parents charged with child abuse. Program staff want to know how well the program is working.

Subjects: Data gathered from 22 people who had completed the parent training program and 20 people on a waiting list to attend the next training program. Although 35 people had completed the course, only 22 could be reached.

Intervention (Independent variable): Parent education course

Dependent variable: Knowledge of effective parenting strategies

Design:

$$X\ O_1$$
$$O_1$$

Findings:

Strategies for handling a child's temper tantrum:

68 percent of course participants suggested a positive strategy

40 percent of waiting list members suggested a positive strategy

Strategies for handling a child who doesn't want to go to school:

59 percent of course participants suggested a positive strategy

60 percent of waiting list members suggested a positive strategy

Threats to internal validity:

Selection: Waiting list members assumed similar to attendees, but criteria for referring to program or putting on waiting list may have changed over time.

Experimental mortality: Problems in reaching course participants; may have reached only more stable (and more successful) participants.

History: Not controlled.

Instrumentation: Controlled; investigators were randomly assigned course participants and waiting list members (which controlled for different interviewing styles).

Maturation, statistical regression, testing: Not applicable.

Discussion: The staff used information from the study to assess program components. They identified which concepts needed more emphasis and which needed less.

Participants did not seem to give socially acceptable responses. When asked how they handled stress, some parents indicated that they used alcohol or overindulged in their favorite foods. Similarly, parents said they did not agree that they should negotiate with children, even though this was taught in the course.

Source: Adapted from Combs, W. et al., "The Evaluation of the Nurturing Program," unpublished evaluation, Department of Political Science and Public Administration, North Carolina State University (Raleigh, NC,1996).

Determining Causality With Cross-Sectional Studies

Administrators analyzing a cross-sectional study may want to infer causal relationships from the data. For example, administrators may survey graduates of employment training programs to determine the effectiveness of instruction. Imagine a survey finds that persons receiving on-the-job training earn more than people who received only classroom training. Can one conclude that on-the-job training caused the higher incomes? The finding of greater earnings for on-the-job participants shows a statistical association between type of training and earnings. We can establish quickly if the training (the independent variable) occurred before reported earnings (the dependent variable). Nevertheless, the concluding piece of evidence, the ability to eliminate alternative hypotheses, is elusive.

When analyzing data from experiments and quasi-experiments, the analyst contends with a limited number of threats to internal validity. In cross-sectional studies, a seemingly unlimited number of possible, uncontrolled threats can be identified. The major uncontrolled threat is the inability to control subject selection or assignment, as the researcher cannot control who enrolls in on-the-job training and who does not. The greater earnings of on-the-job training participants may be due to their greater job readiness at the beginning of training or their higher degree of motivation and ambition. If you had more information on the dataset, you could identify other problems with internal validity. For example, if the training programs were conducted in different locations, the on-the-job training participants may be merely benefiting from a more robust economy. Investigators may suggest other plausible third variables, any one of which may demonstrate that the relationship is spurious.

When designing studies to establish causality, researchers prefer experimental and quasi-experimental designs. The deliberate introduction and manipulation of an independent variable provides much stronger evidence of causality than can a cross-sectional study. Nevertheless, if an experimental or quasi-experimental design cannot be implemented, or if the analyst has to rely on existing cross-sectional data, she may employ sophisticated analytical techniques to make plausible inferences about causal relationships. The adequacy of her analysis depends on her ability to select and use tools to accurately represent complex interactions among variables.[43]

Drawing upon theoretical knowledge and empirical experience, she should build a model, such as the schematic shown in Figure 3.6, which specifies the sequence and direction of relationships.

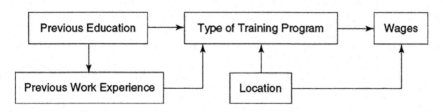

Figure 3.6 Model to Infer Training Effect on Earnings

Such a schematic representation places variables in time order from left to right. The model will guide her statistical analysis.

The model helps users and critics analyze a problem. They can debate whether the model includes the relevant variables and if the effect of included variables has been identified correctly. Whether the model offers a basis for inferring causal relationships depends on its theoretical strength. That is, the analysts should have a coherent, logical explanation justifying the linkages between variables. Statistical tools allow the researcher to control for some variables. This assumes, however, that these variables are observable, have been properly measured, and are included in the dataset.

Summary

Experimental and quasi-experimental designs provide us with useful models of how to conduct research to establish causal links between independent and dependent variables. Rather than just describing the changes in the values of dependent variables, the analyst uses these designs to help explain those changes. In an experimental design, the researcher has control over the assignment of subjects to groups, the introduction of the independent variable, and the research conditions. Quasi-experimental designs lack one or more of these controls.

With the constraints and resources of a given situation, each experimental or quasi-experimental design must be constructed to minimize possible threats to internal and external validity. An internally valid design is one that eliminates factors other than the independent variable as possible explanations of variation in the dependent variable. The common threats to internal validity are history, selection, maturation, statistical regression, testing effects, instrumentation, experimental mortality, and design contamination. Experimental designs eliminate most threats to internal validity at the onset of a research study through random assignment of subjects or a combination of matching and random assignment. Nevertheless, simply implementing an experimental design does not protect against subject attrition, nor can it prevent design contamination. If subjects drop out of a study, a researcher should document their attrition and attempt to document whether dropouts are systematically different from continuers. Contamination may be unavoidable in some studies. If participants are markedly affected by the outcome of a study, they can be expected to behave in such a way as to enhance their own interests.

An externally valid design yields findings that can be generalized beyond the test population. Unique features of a program can unintentionally become part of the treatment. A characteristic of the participants, the setting, or the time may interact with the treatment to make the program succeed or fail. Cook and Campbell have postulated[44] that replication of program findings among different populations and settings provides acceptable evidence of external validity even if these studies have poorly designed samples.

An understanding of internal and external validity and the possible threats to each helps an administrator evaluate the quality of a design and decide whether she is justified in having confidence in a study's findings. An administrator must, therefore, seek to determine whether plausible threats to validity, which could radically change the findings and her decision, have been eliminated. The experimental and quasi-experimental designs discussed here are used widely to minimize threats to validity. Nevertheless, an administrator will want to go beyond identifying the type of design and determine if serious threats were considered and controlled.

Experimental designs depend on random assignment of subjects to a treatment or a control group. The two-group design may be modified to include several treatment groups and to eliminate a no-treatment control group. This modification frequently is necessary to avoid ethical problems of withholding treatment. Researchers will often prefer the ease and efficiency of the randomized posttest design to the classical experiment. In addition to checking for experimental mortality and design contamination, the researcher should make sure that, save for exposure to the independent variable, the experimental and control groups experience similar conditions from the beginning of the experiment to its end. If program personnel conduct the study, the researcher should ensure that subjects are randomly assigned to groups and that the study is implemented as designed.

Experimentation requires researchers to control the research environment. Frequently, they cannot exert such control and must, as a result, use a quasi-experimental design. In designing a quasi-experiment, they try to control as many threats to internal and external validity as possible. The major difference between experimental and quasi-experimental designs is that the latter do not have randomly assigned subjects.[45] Often researchers use comparison groups, but they cannot eliminate the possibility of a selection bias.

Nonexperimental designs lack internal validity. Nevertheless, they can demonstrate if an intervention met expectations. Information gathered using these limited designs may be useful in developing appropriate evaluations. Existing information may act to eliminate plausible threats to internal validity. A researcher may be able only to implement a nonexperimental design because no pretest data were gathered.

In deciding what type of design to use, administrators should define clearly what they want to find out and how they plan to use the information. If they want data on variables that have received little attention, they may want a limited study before authorizing a more extensive experimental study. Whether the design should be quasi-experimental or descriptive depends on the question to be investigated and the availability of data. Occasionally, the inability to control the implementation of a program or the selection of participants requires an analyst to settle for a cross-sectional design. If a cross-sectional design is utilized, analysts must understand fully the relationships among variables and collect data on relevant variables. With a carefully specified model, analysts may use sophisticated statistical tools to infer causality by controlling for some variables representing threats to validity.

Chapter 3 addressed the validity of conclusions investigators could reach based on the characteristics of the research design used. Another type of validity, operational validity, is important for evaluating the measures of variables. The validity of measures of variables and the validity of the design used to collect data are both important in order for investigators to reach proper conclusions (see Figure 3.7).

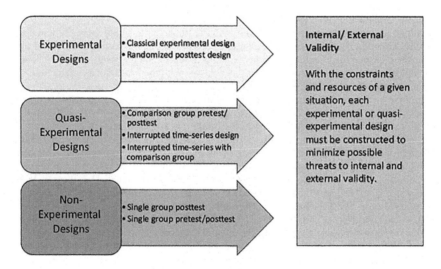

Figure 3.7 Designs for Explanation

Notes

1. See T. D. Cook and D. T. Campbell, *Quasi-Experimentation: Design and Analysis Issues for Field Settings* (Boston, MA: Houghton Mifflin, 1979), 9–36, for a discussion of the concept of causality.

2. See S. Grimmelikhuijsen, S. Jilke, A. L. Olsen, and L. Tummers, "Behavioral Public Administration: Combining Insights from Public Administration and Psychology," *Public Administration Review* 77, no. 1 (2017): 45–56; S. Jilke, S. Van de Walle, and S. Kim, "Generating Usable Knowledge Through An Experimental Approach to Public Administration," *Public Administration Review* 76, no. 1 (2016): 69–72; B. Bozeman and P. Scott, "Laboratory Experiments in Public Policy and Management," *Journal of Public Administration Research and Theory* 2, no. 3 (1992): 293–313.

3. Elizabeth Linos, "Simple Changes to Job Ads Can Help Recruit More Police Officers of Color," *Harvard Business Review*, 2018; Elizabeth Linos, "More Than Public Service: A Field Experiment on Job Advertisements and Diversity in the Police," *Journal of Public Administration Research and Theory* 28, no. 1 (2017): 67–85.

4. F. J. Graveter and L. B. Forzano, *Research Methods for the Behavioral Sciences*, 2nd ed. (Belmont, CA: Thomson-Wadsworth, 2006).

5. For two examples of quasi-experimental designs, see James E. Rosenbaum, "Changing the Geography of Opportunity by Expanding Residential Choice: Lessons from the Gautreaux Program," *Housing Policy Debate* 6, no. 1 (1995): 231–269; Susan D. Hyde, "The Observer Effect in International Politics: Evidence from a Natural Experiment," *World Politics* 60, no. 1 (2007): 37–63.

6. See Kenneth J. Meier and David R. Morgan, "Speed Kills: A Longitudinal Analysis of Traffic Fatalities and the 55MPH Speed Limit," *Policy Studies Review* 1, no. 1 (1981): 157–167.

7. Ralph Kolstoe, *North Dakota Highway Patrol 55 MPH Speed Limit Project: Final Evaluation* (Grand Forks: University of North Dakota, Bureau of Governmental Affairs, 1981).

8. Graveter and Forzano, *Research Methods for the Behavioral Sciences*.

9. See L. Christensen, R. Johnson, and L. Turner, *Research Methods: Design and Analysis*, 12th ed. (Upper Saddle River, NJ: Pearson Education, Inc., 2014) for a discussion of *statistical conclusion validity*, which refers to the inference that independent and dependent variables covary. As the name implies, the issues involved are statistical. In a later chapter, we discuss that in order for researchers to conclude that independent and dependent variables covary, they must have confidence that the relationship is a real one. Also see W. R. Shadish, T. D. Cook, and T. D. Campbell, *Experimental and Quasi-Experimental Designs for Generalized Causal Inference* (Boston, MA: Houghton Mifflin, 2002).

10. D. T. Campbell and J. C. Stanley, *Experimental and Quasi-Experimental Designs for Research* (Chicago: Rand McNally, 1966); Cook and Campbell, *Quasi-Experimentation*.

11. See Shadish, Cook, and Campbell for an expanded definition of design contamination that includes threats in addition to those discussed in this chapter.

12. See Matthew D. Shapiro and Joel Slemrod, "Did the 2001 Tax Rebate Stimulate Spending? Evidence from Taxpayer Surveys," in *Tax Policy and the Economy*, vol. 17, ed. James M. Poterba (Cambridge, MA: MIT Press, 2003), 83–110. NBER Book Series Tax Policy and the Economy, available at www.nber.org/chapters/c11535.pdf.

13. Cook and Campbell, *Quasi-Experimentation*, 55–56.

14. Any of these factors may act in combination with the treatment to produce the particular outcome of the study. Although the treatment may be present for other cases, the particular values of these factors may not be present. Hence, external validity would be affected and the results could not be generalized. Methodologists refer to the combination of these factors with the treatment as interaction.

15. Cook and Campbell, *Quasi-Experimentation*, 73, 75–76.

16. See the following for a discussion of the various procedures for matching to obtain experimental and control groups: L. Christensen, R. B. Johnson, and L. A. Turner, *Research Methods, Design, and Analysis*, 12th ed. (New York: Pearson Education, Inc., 2014), 195–198.

17. See L. S. Meyers and N. E. Grossen, *Behavioral Research* (San Francisco: W. H. Freeman and Company, 1974), 126–127. For a more detailed, statistical discussion of matching, see D. G. Killeinbaum, L. L. Kupper, and H. Morganstern, *Epidemiological Research* (Belmont, CA: Wadsworth, Lifetime Learning Publications, 1982), Chapter 18, or S. Anderson, A. Auquier, W. H. Hauck, D. Oakes, W. Vandaele, and H. Weisberg, *Statistical Methods for Comparative Studies: Techniques for Bias Reduction* (New York: Wiley Interscience, 1980). Also see A. L. Edwards, *Experimental Designs in Psychological Research*, 5th ed. (New York: Harper and Row, 1985), Chapter 9, for arguments in favor of matching.

18. See M. L. Dennis, "Assessing the Validity of Randomized Field Experiments: An Example from Drug Abuse Treatment Research," *Evaluation Review* 14 (1990): 347–373, for a bibliography on the use of randomized field experiments.

19. Alan S. Gerber and Donald P. Green, *Field Experiments: Design, Analysis and Interpretation* (New York: W. W. Norton and Company, 2012). For examples of experiments, field experiments, and studies using quasi-experimental designs in public administration, see Justin B. Bullock, Hal G. Rainey, and Andrew B. Whitford, "Collection of JPART Experiments," *Introduction-Oxford Journals,* available at www.oxfordjournals.org. For an excellent recent example, see Claudia Avellaneda, "Mayoral Decision-Making: Issue Salience, Decision Context, and Choice Constraint? An Experimental Study with 120 Latin American Mayors," *Journal of Public Administration Research and Theory* 23, no. 3 (2013): 631–661.

20. E. Brodkin and A. Kaufman, "Policy Experiments and Poverty Politics," *Social Service Review* (December 2000): 507–532.

21. A dated but dramatic example of this is a time series showing the number of monthly calls to directory assistance in Cincinnati and how they changed after the telephone company began charging for them. The number dropped immediately, and substantially, only to begin increasing again after a few months. See A. J. McSweeny, "Effects of Response Cost on the Behavior of a Million Persons: Charging for Directory Assistance in Cincinnati," *Journal of Applied Behavioral Analysis* 11 (1978): 47–51.

22. G. P. Waldo and T. G. Chiricos, "Work Release and Recidivism: An Empirical Evaluation of a Social Policy," *Evaluation Quarterly* (February 1977): 87–107.

23. Cook and Campbell, *Quasi-Experimentation,* 344–382, discuss the limitations of field experiments as well as situations when they are appropriate. P. H. Rossi, M. W. Lipsey, and H. E. Freeman discuss the limitations of randomized field experiments in *Evaluation: A Systemic Approach,* 7th ed. (Thousand Oaks, CA: Sage Publications, 2004), 252–262. See Christensen, Johnson, and Turner, *Research Methods, Design, and Analysis,* 36–39 for a brief discussion of the disadvantages of an experimental approach.

24. E. J. Posavac, *Program Evaluation: Methods and Case Studies, Studies,* 8th ed. (Boston, MA: Prentice-Hall, 2011).

25. Cook and Campbell, *Quasi-Experimentation,* 367–369.

26. Campbell and Stanley, *Experimental and Quasi-Experimental Designs,* 34.

27. Posavac, *Program Evaluation,* 167–170, discusses the problems associated with selecting comparison groups and gives examples.

28. E. Hirst and R. Talwar, "Reducing Energy Consumption in Low-Income Homes," *Evaluation Review* (October 1981): 671–685. (Copyright © 1981 by Sage Publications.)

29. Ibid.

30. G. Glaster, K. Tempkin, C. Walker, and N. Sawyer, "Measuring the Impacts of Community Development Initiatives: A New Application of the Adjusted Interrupted Time-Series Method," *Evaluation Review* 28, no. 6 (2004): 502–538; R. McCleary and R. A. Hay, Jr., *Applied Time Series Analysis for the Social Sciences* (Beverly Hills, CA: Sage Publications, 1980), Chapter 3, present the statistical analysis procedures for confirming these patterns. Cook and Campbell, *Quasi-Experimentation,* Chapter 5, discuss the types of effects, various time-series designs, and how to handle gradual implementation or delayed causation.

31. S. G. West, J. T. Hepworth, M. A. McCall, and J. W. Reich, "An Evaluation of Arizona's July 1982 Drunk Driving Law: Effects on the City of Phoenix," *Journal of Applied Social Psychology* 19 (1989): 1212–1237.

32. Cook and Campbell, *Quasi-Experimentation,* 225–232, Chapter 6.

33. "Biography of Mayor Rudolph W. Giuliani," *NYC Link,* New York City's Official Web Site, available at www.ci.nyc.ny.us.

34. Catherine Alford, *How Medical Marijuana Laws Affect Crime Rates* (Charlottesville, VA: University of Virginia, September 2014).

35. C. Spohn and J. Horney, "A Case of Unrealistic Expectations: The Impact of Rape Reform Legislation in Illinois," *Criminal Justice Policy Review* 4 (1990): 1–18.

36. Ibid., 15.

37. McCleary and Hay, *Applied Time Series Analysis,* present statistical procedures to statistically analyze time-series data. Alternatively, the reader can consult L. J. McCain and R. McCleary, "The Statistical Analysis of Simple Interrupted Time-Series Quasi-Experiment," in *Quasi-Experimentation: Design and Analysis Issues for Field Settings,* eds. T. D. Cook and D. T. Campbell (Boston, MA: Houghton Mifflin, 1979), Chapter 6. Michael Lewis-Beck provides visual examples and discusses statistical tools for evaluating changes in interrupted time series. See Michael S. Lewis-Beck, "Interrupted Time Series," in *New Tools for Social Scientists: Advances and Applications,* eds. William D. Berry and Michael S. Lewis-Beck (Beverly Hills, CA: Sage Publications, 1986), Chapter 9, 209–240. Meier, Brudney, and

Bohte discuss using regression analysis to statistically analyze the effect of an intervention on a time series. They also provide graphical examples of changes. See Kenneth Meier, Jeffrey Brudney, and John Bohte, *Applied Statistics for Public and Nonprofit Administration*, 9th ed. (Stamford, CT: Cengage Learning, 2014).

38. Shadish, Cook, and Campbell refer to this method as "interrupted time series with non-equivalent, no treatment control group time series" (Shadish, Cook, and Campbell, *Experimental and Quasi-Experimental Designs for Generalizing Causal Inference*).

39. Galster, Tempkin, Walker, and Sawyer, "Measuring the Impacts of Community Development Initiatives," 502–538.

40. Ibid., 515–525.

41. G. Pierre-Pierre, "Crowd in Chinatown Reveals an Unexploited Market for Metrocards," *New York Times*, May 16, 1997, A16, available at www.nytimes.com/1997/05/16/nyregion/crowd-in-chinatown-reveals-an-unexploited-market-for-metrocards.html.

42. J. Havemann, "Welfare-to-Work Program: A South Carolina Success Story," *Washington Post*, April 21, 1997, A7.

43. For information on making causal inferences from cross-sectional designs, see H. B. Asher, *Causal Modeling*, 2nd ed., Series on Quantitative Applications in the Social Sciences, #3 (Beverly Hills, CA: Sage Publications, 1983); O. Hellevik, *Introduction to Causal Analysis: Exploring Survey Data by Crosstabulation*, Contemporary Social Research Series, #9 (London: Allen & Unwin, 1984); Cook and Campbell, *Quasi-Experimentation*, Chapter 7. All these works include extensive discussions on statistical techniques. For a discussion that focuses on the validity issues, see L. B. Mohr, *Impact Analysis for Program Evaluation* (Pacific Grove, CA: Brooks/Cole, 1988), Chapter 10.

44. Cook and Campbell, *Quasi-Experimentation*.

45. It is important for students to remember that random assignment is not the same as random sampling. Random assignment is assigning participants to control and experimental groups by chance, but random sampling is designed to make generalizations back to the broader populations, and this is not a requirement for experiments. Experiments may use both random assignment and random sampling. However, due to the cost of doing both, this often is not done. [could we say ". . . this is seldom done."]

46. National Association of County Commissioners. Local Option Sales Tax Referendum. G.S. 105–535, available at www.nacc.org/227/ (accessed April 26, 2020):-Local-Option-Sales-Tax-Referenda.

47. The 47 counties were: Cherokee, Graham, Swain, Clay, Jackson, Haywood, Madison, Buncombe, Rutherford, Ashe, Wilkes, Surry, Alexander, Lincoln, Gaston, Catawba, Cabarrus, Stanly, Anson, Rowan, Davidson, Forsyth, Rockingham, Randolph, Montgomery, Anson, Moore, Chatham, Orange, Durham, Lee, Harnett, Robeson, Cumberland, Sampson, Duplin, Onslow, Jones, New Hanover, Greene, Pitt, Edgecombe, Halifax, Martin, Bertie, Hertford, and Pasquotank.

Terms for Review

experimental design

quasi-experimental design

internal validity

external validity

threats to internal validity

threats to external validity

reactive effects of experimental arrangements

random assignment

experimental group

control group

comparison group design

interrupted time-series design

nonexperimental designs

Questions for Review

The following questions should indicate whether you have a basic competency in this chapter's material.

1. a. What types of evidence are necessary to support a claim that one variable is the cause of a change in another?
   b. Why is each of these types of evidence necessary?
2. Which of the threats to internal validity is likely to be the most common? Why?
3. Which of the threats to internal validity is likely to cause the most trouble or harm? Why?
4. Contrast history with maturation as threats to internal validity.
5. "Internal validity takes precedence over external validity." Evaluate this statement.
6. Discuss three ways to improve external validity. Compare history as a threat to internal validity with history as a threat to external validity.
7. Consider a study on the effects of noise. Fifty subjects will enter data into a spreadsheet. The researchers will introduce random noise in the room where the experimental group is working. Data will be gathered on the number of forms each subject processes correctly in an hour.

   a. This experiment will be held in two rooms (a "quiet" room and a "noisy" room) with 25 workstations. Subjects will not know which room is which. Should the subjects decide which room they want to work in, or should they be assigned to one of the rooms? Justify your answer.
   b. This study will begin at 9 a.m. A randomly selected experimental group will work for an hour without intermittent noise, and the first hour's data will be collected. From 10 to 11 a.m., the researchers will introduce noise into the room and collect the second hour's data. The control group will work from 1 to 3 p.m. The data will be collected twice, but no intermittent noise will be introduced. Critique the internal validity of this design.
   c. This study will begin at 9 a.m. The experimental and control groups will work in separate buildings from 9 to 11 a.m. In one of the buildings, a fire alarm (not part of the experiment) goes off and everyone leaves for 20 minutes. The group returns to the room and works until 11:20 a.m. How would the fire alarm affect the experiment?

8. Does the classical experimental design guarantee that the threats to internal validity are controlled? Explain your answer.
9. What conditions should be met for a posttest-only design to work well?
10. A researcher wants to see if question wording affects how much people say they are willing to pay for a program to prevent the deaths of birds from oil spills. He asks 200 people who come into a San Francisco science museum to volunteer to answer a questionnaire. The volunteers are randomly assigned to receive one of four versions of a questionnaire.

    Version 1: How much a respondent is willing to donate for the program
    Version 2: How much a respondent is willing to be taxed for the program
    Version 3: How much a respondent is willing to donate for the program (with a reminder that other people will also be donating)
    Version 4: How much a respondent is willing to be taxed for the program (with a reminder that other people will also be taxed)

     a.   Use the *R X O* notation to summarize the study design.
     b.   Briefly assess if the author's use of a nonrandom (convenience) sample will affect the internal validity and external validity of the study.

11. Discuss the strengths and weaknesses of the comparison group pretest/posttest design. Contrast the design with the single group pretest/posttest design. Be sure to discuss how each protects or fails to protect against threats to validity.

12. Why is the interrupted time-series design used so often in administrative and policy research?

13. Summarize the advantages and disadvantages of nonexperimental designs.

14. A newspaper reported that after an "experiment" in cutting the cost of Internet access, half the residents of Blacksburg, Virginia, were "regular users of the global computer network."

     Blacksburg, with a population of 37,000 at the time, is home to Virginia Tech, a university with a strong engineering program. Infer the study's design. What threats to internal validity seem plausible?

Problems for Homework and Discussion

1.   To determine if halfway houses improve the transition from prison, 50 inmates who will be released in a month are randomly assigned to one of two groups. One group is sent to a halfway house to participate in a program called "ReEntry." The other group stays in prison until the time of release. Two months after release, data measuring "success in job placement" are gathered on the subjects in both groups.

     a.   Use the *R, X, O* notation to sketch the research design.
     b.   An analyst says that the design does not control for statistical regression since the subjects are inmates and thus represent an "extreme in their social behavior." Do you agree that the design has problems with statistical regression? Explain.
     c.   Define the terms *internal validity* and *external validity* in the context of this study.

2.   Three hundred people working in 40 state offices process X120 forms.

     a.   To evaluate a proposed change in the X120 form, would you randomly assign offices or individuals? That is, if offices are assigned, 15 offices would use the form with the proposed change, and 15 offices would continue to use the existing form. If individuals are assigned, roughly 50 percent of the individuals in the office would use the form with the proposed change, and 50 percent would continue with the existing form.
     b.   Would you allow subjects to volunteer to test the new X120 form? Justify your answer.
     c.   Would you test the new X120 form in the three slowest offices and evaluate the change in productivity? Justify your answer.

3.   Rye County Hospital's Wellness Center is testing two programs to get smokers to quit. STOP is offered on Tuesday nights, and QUIT is offered on Thursday nights. Ninety smokers volunteered to be subjects. They will be randomly assigned to STOP, QUIT, or No Treatment (neither group).

     a.   What two threats to internal validity are controlled by random assignment?
     b.   About the time the smoking study started, cigarette taxes increased sharply. Would this affect the internal validity of the study? Justify your answer.
     c.   Several subjects assigned to QUIT had a schedule conflict with their aerobics class. Before the program began, a secretary let them trade places with subjects assigned to STOP. Comment on how the secretary's action probably affected the study's internal validity.

d. STOP and QUIT begin and end in the same week. At the last session of the STOP program, data on the subjects' smoking behavior were gathered. The analysts forgot to gather the same data from the QUIT subjects. Three weeks later, the oversight was discovered, and the data were collected. Comment on how the oversight may have affected the study's internal validity.

e. What is the value of having a No Treatment group?

4. A school system conducts a workshop on child abuse and neglect. Attendance is voluntary; 30 percent of school employees indicate that they will attend. To decide whether the workshops should be continued and attendance made mandatory, the staff selects a comparison group from among employees who did not preregister. Before the workshop, both groups are given a questionnaire on the signs of child abuse and neglect, attitudes about violence, and knowledge of reporting requirements. A similar questionnaire is given to members of both groups after the workshop is over.

   a. Name this type of design. Use the *R*, *X*, *O* notation and diagram the design in symbols.
   b. Consider each threat to validity, and explain whether it is a problem in this design.
   c. Evaluate the external validity of this study.
   d. Suggest a design that would provide better evidence for the decision. Would the staff find this alternative design practical to implement? Explain.

5. To reduce crime, a police force tries team policing, which has two officers walk a beat together. If trouble arises, the team calls a patrol car for backup reinforcement.

   a. Suggest two different designs to determine the effectiveness of team policing.
   b. Indicate how well each design handles the threats to internal validity.
   c. Assess the external validity of these designs.

6. A state housing agency director wants to see if changing procedures in the emergency housing assistance program will speed the processing of claims. He selects a sample of 10 counties, then randomly selects 5 counties to adopt the new procedures and 5 to continue with current procedures. He measures the length of time required to process claims before and after the changes are adopted.

   a. Name and diagram this design in the *R*, *X*, *O* symbols.
   b. Revise the design, making it a randomized posttest design. Explain how the study would be conducted; diagram the new design. Would this design have any advantages over the original design?
   c. Revise the design, making it an interrupted time-series design with comparison group. Assume a time period of six years with data recorded every three months and the new procedures adopted at the end of year four. Draw a graph illustrating the time series if the new procedure was successful.

7. To study organic farming, researchers matched each organic farm with a nonorganic farm of the same size and soil composition. Some agricultural researchers criticized the study design, which they said should have compared small adjacent plots—one exposed to chemical fertilizers and insecticides and the other not exposed. Assess the benefits of the matched study over the classical experimental study to determine the effectiveness of organic farming.

8. A school system will conduct research prior to deciding to require ninth-graders to take Algebra I. (Currently, only 30 percent of ninth-graders take Algebra I. Very few students entering ninth grade have completed algebra; any ninth-grader who has passed Algebra I will not be required to take it again.) Three proposed research plans are outlined here:

Plan A: Select a random sample of files of students who complete ninth grade in June. Compare the math achievement scores of students who took Algebra I with those of students who have not taken Algebra I.

Plan B: Require all ninth-graders in a "typical" high school to take Algebra I in this school year. Compare students' math achievement scores at the beginning of the year with their end-of-year scores.

Plan C: Select two similar high schools. Randomly assign one of the schools to require all ninth-graders to take Algebra I in this school year. In the other school, students may choose whether to take Algebra I. Compare the two schools' math achievement scores at the beginning of the year with their end-of-year scores.

a. Evaluate the three plans, and recommend which plan the school system should implement.

b. Assume the school system requires all ninth-graders to take Algebra I. The school system wants to conduct an interrupted time series to assess the requirement's effectiveness. What features should the time series include to yield useful information?

Working With Data

Access the County Data file accompanying this text for the following assignments.

1. In North Carolina in 2007 per Article 46 GS 105–537, counties had the option to increase their sales tax by one quarter of a cent in North Carolina by holding a referendum.[46] Other options for increasing the local option sales tax could be levied through other state legislative measures, but the goal was to make it easier for counties to adopt the quarter-cent increase to improve specifically county revenue. The first set of referenda were held in November of 2007 in order to take effect in 2008. Prior to this, the majority of counties levied very similar rates. All had either 2.5 (six) or 2.75 (ninety-four). After the first year of state authorization, five counties successfully adopted the increase (Catawba, Martin, Pitt, Sampson, and Surry), and it took effect in April of 2008 (at first, most county residents were hesitant to hold a referendum and vote for the additional sales tax revenue; the majority of referenda failed). Over a decade later in 2019, the number had increased to 42 counties out of 100 counties with the additional county option rate.[47] As a result of this measure, state legislators hoped to increase tax revenue by increasing the sales tax rates for county governments, especially when they could no longer provide as much state aid as in the past. Data for both 2008 and 2019 are available in the dataset. Use the 2008 and 2019 data for the following.

a. Separate the data into four groups—those from counties with the experimental policies (added sales tax option) and those without for 2008 and 2019. The counties with the local county option sales tax (Article 46) have asterisks in the spreadsheet and are listed in the subsequent endnote. Next, compare the counties that participated in the experimental local option sales tax under Article 46 to those that did not participate based on their rates. How do the values for the two groups of counties differ regarding these two variables in 2008 and 2009. (Report high, low, median, and mean values and others you think relevant.) Do you think participating in the experimental policy led to large differences in sales tax rates? What else might account for the differences between counties that added the sales tax option over time versus those that did not adopt it?

b. Can a researcher generalize from these findings to other states considering similar experiments? Explain.

2. What type of study might you recommend for another state interested in allowing counties to adopt extra local option countywide sales tax increases to improve their overall sales tax rate to increase revenue?

3. Diagram a model of how the original experiment could have been set up with counties with experimental policies hoping to have higher sales tax rates to help mitigate budgetary shortfalls. Make your model like the schematic in Figure 3.4. Use a total time period of 11 years. Make another model according to the study you designed in Question 2. Comment on the internal validity of both studies.

Recommended for Further Reading

Cook, T. D., and D. T. Campbell, *Quasi-Experimentation* (Boston, MA: Houghton Mifflin, 1979) is a valuable resource for understanding experimental and quasi-experimental designs in field settings. The authors discuss and extend the ideas presented in Campbell, D. T., and J. T. Stanley, *Experimental and Quasi-Experimental Design for Research* (Chicago: Rand McNally, 1966).

Shadish, W. R., T. D. Cook, and T. D. Campbell, *Experimental and Quasi-Experimental Designs for Generalized Causal Inference* (Boston, MA: Houghton Mifflin, 2002) continue, modify, and expand the concepts and tools in earlier work by Cook and Campbell. This book is appropriate as a text in research design for graduate students and is advanced in parts. It is excellent and valuable for those interested in the topic.

An extensive review of the Shadish, Cook, and Campbell book is Julnes, J., "Book Review of Shadish, Cook, and Campbell," *Evaluation and Program Planning* 27 (2004): 173–185. Available at web.pdx.edu/-stipakb/download/PA555/ExperDesignBookReview.pdf

Cook, Thomas discusses certain aspects of external validity in more detail in the following: "Generalizing Causal Knowledge in the Policy Sciences: External Validity as a Task of Both Multi-Attribute Representation and Multi-Attribute Extrapolation," *Journal of Policy Analysis and Management* 33, no. 2 (Spring 2014): 527–536. Those with a strong interest in the topic will find this article rewarding.

If you are interested in reading more about experiments in public management using vignettes, please see Jilke, S., and G. Van Ryzin, "Survey Experiments for Public Management Research," in *Experiments in Public Management Research: Challenges and Contributions*, eds. O. James, S. Jilke, and G. Van Ryzin (Cambridge: Cambridge University Press, 2017), 117–138, doi:10.1017/9781316676912.007. The recommended text on the statistical analysis of time-series data is McCleary, R., and R. A. Hay, Jr., *Applied Time Series Analysis for the Social Sciences* (Beverly Hills, CA: Sage Publications, 1980).

Lewis-Beck, Michael discusses interrupted time series designs and statistical tools used to analyze data from them. One of the tools is regression analysis, which most master's-level students have studied. See "Interrupted Time Series," in *New Tools for Social Scientists: Advances and Applications in Research Methods*, eds. William Berry and Michael Lewis-Beck (Beverly Hills, CA: Sage Publications, 1986), Chapter 9.

Texts in program evaluation discuss these designs and give cases illustrating their use. See Posavac, E. J., *Program Evaluation: Methods and Case Studies, Studies*, 8th ed. (Boston, MA: Prentice-Hall, 2011); Rossi, P. H., M. W. Lipsey, and H. E. Freeman, *Evaluation: A Systematic Approach*, 7th ed. (Thousand Oaks, CA: Sage Publications, 2004); *Handbook of Practical Program Evaluation*, 3rd ed., eds. J. S. Wholey, Harry Hatry, and Kathryn Newcomer (San Francisco: John Wiley & Sons, Inc., 2010); Valadez, J., and M. Baberger, *Monitoring and Evaluating Social Programs in Developing Countries* (Washington, DC: World Bank, 1994) present 10 quasi-experimental designs with examples of their use.

4   Measuring Variables

In this chapter, you will learn:

1. How policy makers and administrators define and measure concepts and variables.
2. Why you should review the quality of measures developed and used by others.
3. Categories used to describe the level of measurement and their role in developing measures.
4. Strategies for assessing the reliability, operational validity, and sensitivity of measures.

Researchers, reporters, administrators, and citizens should understand measurement, that is, how numbers, words, or other descriptors are assigned to broader concepts, behaviors, or phenomena. This is true whether data represent environmental conditions, the state of the economy, achievement, public opinion, or something else. Data can influence policy decisions, demonstrate accountability, and classify people. Understanding measurement quality allows us to assess if data are appropriately comparing different cities or counties on a characteristic, such as poverty, or a standard, such as air quality. Consequently, researchers must take care in designing measures, and data users must understand measurement concepts.

Beginning the Measurement Process

Early on, investigators must specify and clarify the research question and the meaning of each variable. The measurement process may start with an abstract concept. We can label the variable under consideration a *concept* and its meaning a *conceptual definition*.

Concepts and Conceptual Definitions

A concept is an abstraction representing an idea, such as poverty, happiness, or economic development. A conceptual definition indicates what one means by a concept. Conceptual definitions range from brief descriptions to thorough, detailed statements.[1] How detailed a definition should be depends on the intended use of the measure. Consider the concept of alcohol abuse. The appropriate conceptual definition depends on the question at hand. A university, wanting to identify the extent of alcohol abuse among its undergraduates, may focus the definition on heavy weekend or binge drinking. An employer, wanting to identify sources of workplace problems, may define it as drinking that affects an employee's work performance. A physician looking for signs of abuse in her patients may define it as involving a combination of verifiable physical or mental symptoms.

Operational Definitions

Next, guided by the conceptual definition, the investigator finds a way to measure the concept. An *operational definition* is formulated from the conceptual definition. It details *exactly* how values will be assigned to a variable. This is a vital step because, without an operational definition that provides a practical, clear way to measure a concept, the process cannot move forward. Too often, researchers may fail to think through or may even try to skip the difficult task of determining the conceptual definitions of the concepts until the last minute. As we discuss in this chapter, this can lead to poor measures and therefore weak research.

For example, if the conceptual definition of alcohol abuse was whether a person indulged in binge drinking, the accompanying operational definition of personal alcohol abuse might be "Did you drink five or more servings of wine, beer, or hard liquor at one time during the past month?" Individual alcohol abuse is measured by determining who matches this definition. Alternatively, an operational definition may be written to gather data from behavior logs, observation, or self-reports.

Consider the concept of employee empowerment, discussed in Example 4.1. First, you may want to consider what the concept means to you. One conceptual definition states that empowered employees have four characteristics. They believe that:

1. their work is meaningful,
2. they are competent enough to do their job,
3. they have autonomy in performing their work, and
4. their work has an impact on the organization.[2]

These characteristics are combined to represent the overall concept of empowerment. This conceptual definition implies that the operational definition should measure meaning, competency, autonomy, and impact.

An Operational Definition of Employee Empowerment

The researcher's next task is to operationally define these characteristics. Example 4.1 suggests survey items that could constitute the operational definition of empowerment. Each item is linked to one of the components of the conceptual definition. Respondents should read the statements describing their work and check the appropriate response to each statement. To estimate the overall level of employee empowerment, the investigator might count the responses for each item, combine some or all items, or look for patterns across responses.

Example 4.1 An Operational Definition of Employee Empowerment

Survey Items and Responses:
Please indicate your level of agreement with each of the following statements.

1. The work I do is meaningful to me.

    _____Strongly Agree
    _____Agree
    _____Neither Agree nor Disagree
    _____Disagree
    _____Strongly Disagree

2. I know what is expected of me in my job.

    _____Strongly Agree
    _____Agree

_____Neither Agree nor Disagree
_____Disagree
_____Strongly Disagree

3. In the past two years, I have been given more flexibility to do my job.

    _____Strongly Agree
    _____Agree
    _____Neither Agree nor Disagree
    _____Disagree
    _____Strongly Disagree

4. The work I perform contributes to the organization's mission.

    _____Strongly Agree
    _____Agree
    _____Neither Agree nor Disagree
    _____Disagree
    _____Strongly Disagree

Step 1: Assign values to each response; that is:

    Assign 5 to every "Strongly Agree" response.
    Assign 4 to every "Agree" response.
    Assign 3 to every "Neither Agree nor Disagree" response.
    Assign 2 to every "Disagree" response.
    Assign 1 to every "Strongly Disagree" response.

Step 2: Total the responses to statements 1–4 to determine a respondent's perceived degree of empowerment. Note: If all items are answered, values can range from 4 (least empowered) to 20 (most empowered).

Step 3: Decide whether to group values, for example:

    Not empowered: scores 4–10
    Limited empowerment: scores 11–15
    Most empowered: scores 16–20
    Alternatively, the mean of responses may be calculated and grouped.

Source: Adapted from items in *Merit Protection Survey 2000*, U.S. Merit Protection Board. In practice, researchers would create several items to measure the separate components indicated in the conceptual definition.

Concepts and Operational Definitions

Example 4.2 draws upon an incident in which a staff analyst is charged with supporting the decision to approve of changes in military uniforms and gear. As is common practice, each year the military invites its suppliers to demonstrate improved uniforms and gear. The ultimate decision for whether there needs to be a change in the current uniform and gear in the military usually rests with the senior officer. We assume that the senior officer and the analyst did not discuss their conceptual definitions of improvement or how it would be defined. The staff officer formulated an operational definition of improvement based on his conceptual definition of better uniforms. Using his operational definition, he concluded that the new uniforms with their small increase in per unit cost seemed like a good value. The cost analyst had to figure out how to pay for the new uniforms, and when the increase in cost was calculated out over the five-year defense planning cycle, it amounted to over a $400,000 increase. Since this large sum of money would have to be reallocated from some other part of the budget that may reduce unit effectiveness, this improvement to the uniforms and gear may not have been an improvement for the Army. The second time the cost analyst went to the supplier's demonstration, he brought the cost calculations to show that cuts would need to be made elsewhere in the budget to support the changes in uniforms.

Example 4.2 Concepts and Operational Definitions

Problem: As a member of the Army Uniform Board (AUB), members recommend changes to army-provided soldier uniforms and gear to increase effectiveness of army personnel.

Senior officer's definition of effectiveness: New and improved clothing and gear.

AUB cost analyst's definition of effectiveness: New clothing and gear that does not increase costs or have more than a minimal, negative impact on the budget.

Operational definition of effectiveness (based on cost analyst definition): Cost.

Analysis: Calculate the additional cost of the clothing items for the five-year defense plan.

1. Costs and assumptions

   a. $1—Increase in cost of clothing item (relative to the previous year's clothing item that is still readily available)
   b. 80,000—Number of recruits per year
   c. 1 and 5—Years in planning horizon

2. Cost calculations

   Cost increase in first year = $1 × 80,000 = $80,000
   Cost increase over the five-year defense plan = $1 × 80,000 × 5 = $400,000

Action: Originally, the senior officer accepted the new uniform proposal because the very small increase in the cost of the clothing item seemed negligible and they were told that the clothing item was better than the previous year's clothing item. However, when the cost analyst showed them the effect on the budget over the five-year defense plan, the AUB stopped approving incremental improvements to army clothing and gear that also came with increased costs.

Result: In the first year, the increase in cost of that clothing item was $80,000 more than was in the budget. In subsequent years, the cost analyst came armed with his calculations, and the budget busting new uniform was subsequently denied.

Discussion: The senior officer's concept of effectiveness was new and improved clothing and gear for army personnel. However, this definition of effectiveness leaves out cost. The money for the cost increase in uniforms had to be found somewhere else in the budget, which may have reduced overall unit effectiveness. The cost analyst concept of effectiveness does not rule out that the uniforms can be better, but the uniform change had to have a negligible impact on the budget or make a cut that would not impact unit effectiveness.

Source: Geiger, D., "Cost Management Innovations in Federal Agencies," in *Cost Accounting in Government*, ed. Z. Mohr (New York: Routledge, 2017).

The misunderstanding in Example 4.2 should convince you of the importance of conceptual and operational definitions. Studies of the same concept, such as efficiency or effectiveness, can arrive at different, even conflicting, conclusions if they use different conceptual or operational definitions. An operational definition may be faulty because it is based on an inappropriate or inadequate conceptual definition, or it may have technical flaws. Administrators who ignore the details of defining variables may waste time and money. Studies based on faulty measures may generate misleading, incorrect, or useless data. A sound conceptual and operational definition is consistent with the intended use of the data. At a minimum, administrators should carefully review a proposed conceptual definition and how the concept is operationally defined.

Levels of Measurement and Types of Measures

The terms *operational definition* and *measurement* are similar in meaning. *Measurement* applies rules to assign values to a variable. *Operational definitions* indicate exactly how a specific variable is to be measured. Even common and simple measurements can have flaws. For example, consider the variable "number of bedrooms." The U.S. Census Bureau defines a bedroom as a room "used mainly for sleeping." This seems like a logical, straightforward operational definition. Yet the number of bedrooms in a studio apartment or other one-room unit is zero. A guest room is counted if it is reserved primarily for sleeping, even if it is used infrequently. A living room with a sofa bed is not counted, even if it is used for sleeping all the time. If a simple measure such as number of bedrooms has problems, imagine the challenges involved in *operationalizing* more complex concepts. For example, to measure air pollution, the operational definition includes precise instructions on how to select an air sample, how many samples to take, requirements for calibrating instruments, and laboratory personnel competencies.[3] To take an even more abstract concept, think about how you would operationalize community health and well-being.[4]

Exactly what value should you assign to an object or event? To determine the answer to this question, you first need to identify and understand the *unit of analysis* for your study. The unit of analysis refers to the *level* of the entity for which you are measuring the concept. For example, many studies ask, "What influences educational achievement?" Some studies use a measure of the concept educational achievement relevant to each student, such as his or her test scores or GPA. In this case, the unit of analysis is the individual. Other studies use schools as the unit of analysis, using a measure relevant to schools, such as the percent of all students passing grade-level exams or schoolwide average test scores. Yet other studies use school districts or systems, or even states or nations, as the unit of analysis. In each of those cases, the unit of analysis is a geographic entity. Once the unit of analysis for a study is in place and the hypotheses are determined, the data for the study are generally measured and gathered on the level of that unit.[5]

After determining the unit of analysis, the researcher must think about *measurement* for each variable. Most writers on measurement discuss four levels of measurement:

- nominal
- ordinal
- interval
- ratio

Each level of measurement involves categorization. Each object or event observed should be described by a value or category, and each object or event should be described by one and only one value or category. In other words, the values or categories should be mutually exclusive and exhaustive. Students need to understand the different levels of measurement because the level of measurement used for a variable affects the statistics you choose and how you interpret them.

Nominal Measures

Nominal variables can only be measured with nominal measures. Nominal measures are sometimes called categorical measures, or categorical variables, because nominal measures identify and label the categories that are relevant. For example, the categories for employment status might be working full time, working part-time, or not working. They do not have relative values. That is, you cannot rank or order the individual categories. Even though numbers are sometimes assigned to nominal categories, these numbers have no particular meaning beyond being a label or a coding system, allowing one to classify and count the number of cases in each category. One might, for example, classify town employees by department. You could devise the following nominal measure (Table 4.1):

Table 4.1 Example: Nominal Measure for City Department

1	Planning
2	Human Resources
3	Finance
4	Budgeting
5	Public Works
6	Public Safety

Note that if other numbers were assigned to the various categories, the information would remain the same. The numbers are a device to identify categories; letters of the alphabet or other symbols could replace the numbers and the measure would be unchanged. Remember, too, that values of nominal measures are not ranked. In other words, the numbering system does not imply that Finance is more or less important than Budgeting. Typical examples of characteristics measured by a nominal measure are gender, region, and hometown.

Ordinal Measures

Ordinal measures identify and categorize values of a variable and put them in rank order according to those values. They do so, however, without regard to the distance between values. Ordinal measures are characterized by determining that one case has more or less of the characteristic than does another case. If you can rank values but not determine how far apart they are, you have an ordinal measure. *Any numbers assigned to the values must be in the same order as the ranking implied by the measurement.* For example, the value represented by 3 is greater than the value represented by 2, and the value represented by 2 is greater than the value represented by 1. However, the numbers do not imply an amount of the characteristic measured, only an order of more or less. Table 4.2 shows how you could assign numbers to ordinal categories.

You can try other numbering schemes *if the numbers preserve the rank order of the categories.* For example, you could reverse the order and number "Strongly Agree" as "1" and "Strongly Disagree" as "5." Alternatively, you could skip numbers and number the categories 10, 8, 6, 4, and 2. Because you cannot determine the distance between values, however, you cannot argue that an employee who answers "Strongly Agree" to all items is five times happier than an employee who answers "Strongly Disagree" to all items.

Rankings commonly result in an ordinal measure. A supervisor may rank 10 employees and give the best employee a "10" and the worst a "1." The persons rated "10" and "9" may be exceptionally good, and the supervisor may have a hard time deciding which one is better. The employee rated with "8" may be good, but not nearly as good as the top two. Hence, the difference between employee "10" and employee "9" may be very small and much less than the difference between employee "9" and employee "8." Unfortunately, while ordinal data, by definition, do not use an equal or standard category, consumers of research may try to use ordinal results that way. They assume a "2" is twice as good (or bad) as a "1" in ordinal ranking.

Interval and Ratio-Level Measures

Interval and ratio measures are characterized by categories that are ranked with specific numerical differences between them. The distance between objects whose characteristics have been measured can be determined by using standard and equal intervals. The difference between 10 and 20, for example, measures the same amount of the characteristic as does the difference between 30 and 40.

One important difference between interval and ratio measures is that a *ratio measure* has an absolute zero; an *interval measure* does not. An absolute zero permits one to use ratios to describe relationships between measured objects. You can add or subtract the values for objects measured by both interval and ratio measures. In contrast, you can multiply and divide the values of the characteristics measured only with a ratio measure. For example, temperature is a common example of an interval measure. (In the U.S., we typically use the Fahrenheit thermometer to measure temperature.) You can state that 50°F is 25 degrees warmer than 25°F. But because zero on the Fahrenheit scale is not an absolute zero,[6] temperature is not a ratio measurement—you cannot state that 50°F is twice as warm as 25°F. The *difference* in heat between 0°F and 50°F is twice of that between 25°F and 50°F; but the *amount* of heat is not double.

In administrative research, investigators commonly work with ratio measures. Frequently used ratio measures include amount of a total budget or its components, program costs, the population size, or the number of program participants. In this text, we use a common social science convention and ignore the distinction between interval and ratio measures. We will usually refer to them both as "interval measures."

The numbers assigned to interval measures correspond to the magnitude of the phenomenon being measured. Thus, the numbers assigned could be the actual number of persons working in an agency, the number of homeless persons in a city each year, or the per capita income in a city. If the county has 100 employees and one of its municipalities has 50, we can say that the county has 50 more employees than the town. We can also note that the county has twice as many employees as the town.

In practice, the boundary between ordinal and interval measures may be blurred, especially if the ordinal measure has a large number of values. Some ordinal measures have been created to approximate an interval measure, that is, to estimate the points along an underlying interval measure. For example, ordinal measures, such as IQ scores, suggest the magnitude of differences along the measure, have many categories, and appear to be interval. One can argue that such measures do not have equal-distance intervals; for example, one cannot state with certainty that the difference between an IQ of 100 and an IQ of 110 is the same as the difference between an IQ of 120 and one of 130. Analysts frequently treat summated measures, such as the employee empowerment scale presented earlier in Example 4.1, as interval measures or interval indices.

Table 4.2 Assigning Numbers to Ordinal Scale Items

5	Strongly Agree
4	Agree
3	Neither Agree nor Disagree
2	Disagree
1	Strongly Disagree

Students often mistakenly assume that groups or categories containing interval data form an interval measure. They do not. Rather, the new measure is ordinal. For example, although age is usually measured with an interval measure, the following categories constitute an ordinal measure: under 21 years old, 21 to 30 years old, 31 to 40 years old, and so forth. The exact distance—that is, the age difference between any two people—cannot be determined. While we know a person who checks off 21 to 30 years old is younger than a person who checks off 31 to 40 years old, the age difference between them could be a few days or nearly 20 years.[7]

Alternative Terminologies

Nominal and ordinal variables may be called *categorical variables*, while interval and ratio variables may be called *numerical variables*. The values of categorical variables have labels; the values of numerical variables are numbers that indicate quantities.

Variables also may be described as *discrete* or *continuous*. Discrete variables have a limited number of values; continuous variables can take on an unlimited number of values, depending on the precision of the measuring instrument. For example, the number of divisions in an organization is a discrete variable. An agency may have 10 or 11 divisions; it cannot have 10.3 divisions. The amount of time an employee spends on a task forms a continuous variable. An employee may be said to spend 5 hours on a task, 5.4 hours, 5.46 hours, and so on.

The Role of Measurement and Types of Measures

Before data are collected, someone needs to ensure that variables are measured in a manner appropriate to the planned analysis. Some people may try to collect interval data for all variables, because a greater number of statistical tools can be applied. Nevertheless, this is not always advisable. First, a shift in measurement level may require a change in the variable. In a study of office automation, for example, you might want to learn what types of computer hardware various agencies use, a nominal measure. By asking how many different types of hardware the agencies own or the total cost of computer hardware, you could generate interval-level data. In the process, however, you would have changed the definition of the variable radically. Equipment cost is not equivalent to types of equipment.

A second reason for not attempting to create an interval variable where one does not readily exist is that this effort sometimes leads to a request for information that respondents are unwilling to provide. For example, respondents may feel that requests for information on their exact age may be combined with other information on an anonymous survey to reveal their identity. Alternatively, a request for precise information that requires the respondent to consult records may irritate a respondent. For some studies, such irritation may result in a lower response rate or made-up information. An investigator must then decide whether less-precise, grouped data are consistent with a study's purpose. Annual income is an example. Although individuals may know their approximate annual income, few have exact figures at hand. If asked, they may have to check tax records or may give a rough estimate. Questions that ask about income within ranges, such as less than $25,000; $25,000 to $49,999; $50,000 to $74,999; $75,000 to $99,999, and greater than $100,000 often obtain a better response rate.

Reliability

The critical factors in selecting a measure are reliability, operational validity, and sensitivity.

- When you ask, "Is my measurement tool accurately and consistently measuring the same thing for all subjects?" or ". . . from one time to another?" you are questioning the reliability of the measurement.

- When you ask, "Does this measure actually produce data on the concept or variable of interest?" you are questioning its operational validity.
- When you ask, "Is this measure sufficiently precise?" you are questioning its sensitivity.

Reliability evaluates the consistency of a measure. Differences over time or between subjects may be due to measurement errors. *Measurement errors* are random errors that result from the measurement process. No measurement is error free. Errors occur because of respondent characteristics, the measure itself, or the process of assigning values. Uninterested or distracted respondents introduce errors when they answer questions rapidly or stop and restart questions. Questions with ambiguous terms or inappropriate responses lead to errors. Raters may be inconsistent in how they assign scores.

For a familiar example of an unreliable measure, consider a bathroom scale. If you weigh yourself and find that you have gained five pounds, you may react immediately by reweighing yourself. Suppose the second time, the scale shows you lost three pounds, an eight-pound difference within seconds. You would consider the measurement from the scale inconsistent or unreliable. While random error cannot be eliminated completely, the error rate should be kept at a tolerable level. Perhaps a half-pound difference is tolerable in a bathroom scale. A measure that yields a lot of random error should be judged unreliable and discarded. In our example, you should throw the bathroom scale out!

Three Dimensions of Reliability

Reliability has three dimensions: stability, equivalence, and internal consistency.[8]

Stability

Stability refers to the ability of a measure to yield the same results time after time, assuming that what is being measured has not in fact changed. The scale that indicates a different weight for an object each time it is weighed lacks stability. We are also concerned about stability when we ask if an investigator assigns the same number to the same phenomenon each time. For instance, if a school counselor examines student files to identify children with learning difficulties, how consistent is she? Do her criteria, which constitute her measure of learning difficulties, change with mood changes, fatigue, degree of attention, or other uncontrolled factors? She should check a sample of her work to make sure each case was identified and recorded consistently. Did she identify the same students both times? If she finds few inconsistencies in the sample, she can assume that her measure is stable and reliable.

Equivalence

Equivalence considers:

1. Whether two or more investigators using a measure assign the same value to the same phenomenon, or
2. Whether different versions of a measure assign the same value to a phenomenon.

For example, different inspectors rating the quality of housing should give the same houses very similar scores. To check the reliability of a rating system, each inspector rates all the houses in a sample of houses. The sample should represent the range of housing types and conditions apt to be encountered in an actual survey of housing. If the rating system is reliable, the inspectors' ratings for each house should be very similar. Nevertheless, some random error must be anticipated, and a decision must be made as to how much error a reliable instrument may contain.

Occasionally, different versions of a measure must be used. This occurs most commonly in testing. Different versions of a test may be created to avoid cheating. Consider written driving tests. If each applicant took the same test, cheating would be a problem, so numerous versions are needed. Equivalence requires that any test taker receive the same or very similar score no matter which test version he takes.

Internal Consistency

Internal consistency applies to measures with multiple items. It considers whether all the items are related to the same phenomenon. For example, employee surveys may include questions measuring four concepts:

1. job satisfaction,
2. empowerment,
3. fair treatment, and
4. training.

To be internally consistent, the measures for each concept should in fact be measuring that concept. We believe that job satisfaction should be related to empowerment, fair treatment, and training. Each of these four concepts should be both related and distinct. In general, a measure's reliability increases as the number of items increases. The importance of using more than one item to measure a variable depends on the underlying concept. A single item is sufficient to measure variables such as a person's age, sex, or similar demographic characteristic. However, measuring individual achievements, attitudes, and opinions usually requires multiple items. For example, in Example 4.1, we used four items to measure empowerment.

The following summarizes the dimensions of reliability.

Stability: measure gives same result when applied to the same case more than one time; that is, different results should occur only if the underlying phenomenon being measured has changed.

Equivalence: measure gives same result when applied to the same case by more than one investigator; different versions of a test give same result when applied to the same case.

Internal consistency: all items constituting a measure are related to the same phenomenon.

Establishing a Measure's Reliability

Measures are seldom, if ever, completely free from random error. Mood, test conditions, and other factors can affect performance on any given day. We need to determine how much error to tolerate with respect to any given measure. For example, a large high school identifies one week during which students can change their class schedules. To schedule advisors during that week, a supervisor checks school files to identify patterns of schedule changes. She divides the files into four categories: no schedule changes, minor schedule changes, major schedule changes, and complicated schedule changes. With a thousand or more students, she knows that she will probably place some files in the wrong category. She may feel an error rate as high as 15 percent is tolerable. She rechecks 100 randomly selected files. To consider her data reliable, she needs to place at least 85 of the rechecked files in the same category as she did the first time.

On the other hand, if she were reviewing files to identify students who were going to be assessed for learning disabilities, an error rate of 15 percent would be too high. The impact on a child of not receiving needed services or wasting expensive and limited resources on a child who does not need services is more serious than time spent on a schedule change. The counselor could err in two ways:

1.   missing children in need of assessment
2.   incorrectly referring children for assessment

To establish a criterion for reliability, an administrator or policy analyst should consider the practical consequences of trying to increase reliability by reducing random error. Minimizing random error may seem desirable, but many measures cannot be refined to meet high standards of accuracy. Estimating and reducing the level of error may be costly. Procedures to estimate the degree of reliability have their own assumptions and limitations, further reducing our ability to establish with certainty a measure's reliability.

To estimate reliability or to apply a measure with known reliability, you should remember that reliability is not only a property of a measure. Reliability also is a function of the conditions under which it is applied. For example, most bathroom scales are unreliable at the extremes of body weight; that is, they do not give consistent weights for infants or very heavy people. Similarly, other measures are designed to work in certain settings. A driver's test may not be reliable for persons with limited knowledge of English or poor reading skills. A form to rate housing quality may not work equally well in a city and in a rural area.

Qualitative Assessment of Reliability

Investigators use qualitative and quantitative methods to estimate reliability. Qualitative methods, which consider many procedural issues, cannot estimate a measure's degree of random error precisely; hence, their importance and value may be underestimated. However, in practice, qualitative methods can pick up serious problems that will discredit a measure. *Qualitative methods* require that the investigators make sure that persons responsible for data collection, assigning values, and entering data are sufficiently trained and supervised and that the investigators review a measure to decide if:

*   terms are defined clearly and consistently
*   ambiguous items or terms have been eliminated
*   necessary information is accessible to respondents
*   multiple-choice responses cover all possible, reasonable responses
*   directions are clear and easy to follow

Consider how the items listed here can affect reliability. First, remember that an unreliable measure is one in which observed differences between subjects or in the same subject over time are due to the instrument or data collection procedures rather than to actual differences. Now consider trying to measure the number of deaths caused by fires. If you asked city clerks how many people in their towns died in fires during the past calendar year, you may assume incorrectly that each town counts fire deaths the same way. Yet some jurisdictions may include only deaths that occur within a certain time period, such as within 72 hours of a fire. Other jurisdictions may include only deaths caused by burns or smoke inhalation. Thus, a person who died from a heart attack while escaping a fire might not count as a fire death in Town A but will count as a fire death in Town B.

Similar to the problem of unclear or inconsistent definitions is the problem of ambiguity. For one reason or another, an item that seems perfectly clear to the investigators may confuse respondents. The reliability problem caused by ambiguous items may be seen easily by considering the following question and responses:

What is your marital status?

```
_____ Single
_____ Married
_____ Separated
_____ Divorced
_____ Widowed
```

As you know, many adults lead complicated lives, and if the word "current" does not appear in the question, a person can accurately check more than one response. Similarly, because many formerly married people consider themselves single, the single category should be "single, never married." If just "single" is used, some currently divorced people may check just "single," others "divorced," and others both "single" and "divorced."

If the requested information is not available to respondents, they may guess or use different rules to estimate the correct response. Consider the example of fire deaths. If towns place all fire deaths in one category and keep no records on specific causes, how will they answer a request to report how many deaths are caused by smoke inhalation?

If multiple-choice items do not include all possible reasonable responses, respondents left to their own devices may guess the item's intent. Recall your classroom experiences with multiple-choice tests. What did you do if none of the listed answers seemed remotely related to your answer? Chances are you guessed. So too will some respondents when faced with choices that do not seem to apply to them. Unclear or complicated directions may cause unreliable data because respondents misinterpret the directions or answer more or less at random because of frustration, anger, or boredom. Responses should, in general, be mutually exclusive and exhaustive, so that everyone can respond, even if the response is "I don't know" or "Other."

Persons involved in data collection must be trained and supervised to reduce problems of reliability, that is, to minimize the number of subjective and therefore potentially inconsistent decisions. Thus, if investigators are sent to fire departments to collect data, each investigator should define fire deaths the same way and use the same procedures for resolving problems of inaccessible or ambiguous information. If individual raters decide on a case-by-case basis how to handle ambiguous responses, the decisions may be inconsistent and the data more unreliable.

Mathematical Procedures

Reliability can be estimated mathematically.[9] Specific tests estimate the measurement error associated with each dimension of reliability. We will limit our discussion to the most common tests that establish internal consistency, equivalence, and stability. Administrators who construct or work extensively with job tests, achievement tests, or personality tests will want to become familiar with other methods for mathematically estimating reliability. An up-to-date textbook on tests and measurement can be very helpful with this.

TESTS OF INTERNAL CONSISTENCY

When a measure includes several items, the researcher wants items that will prompt internally consistent responses. A *test of internal consistency*, easily performed with statistical software, can demonstrate whether a measure has a high level of internal consistency or instead has extensive random errors due to some unreliable or unrelated items.[10] Example 4.3 applies a test of internal consistency to a scale of police professionalism. A *scale* or an *index* is a composite measure that combines several measures into a single variable that is used to measure a concept. (In Chapter 10, we distinguish between a scale and an index.)

Example 4.3 Establishing Internal Consistency

Problem: Establish the reliability of a measure of police professionalism.
Procedure:

1. Create a measure of police professionalism by adding respondents' answers to the following items:

   In my most recent contact with a city police officer:

	Agree	Strongly Agree	Disagree	Strongly Disagree
The officer was courteous (Courteous)				
The officer was competent (Competent)				
The officer displayed a professional attitude (Attitude)				
The officer's overall performance was good (Performance)				
As I move around the city, city police are	Often visible	Sometimes visible	Rarely visible	Never visible

2. Examine the correlation matrix (Table 4.3). The closer a correlation coefficient is to 1.00, the stronger the relationship between variables.

Table 4.3 Correlation Matrix

Attitude		Courteous	Competent	Performance	Visible
Attitude	1.0000				
Courteous	.6121	1.0000			
Competent	.6403	.6652	1.0000		
Performance	.6311	.6505	.6675	1.0000	
Visible	.0580	.0298	.0827	.0233	1.0000

Note that the variable "Visible" has virtually no relationship with any of the other items; that is, how an individual rates the attitude, courtesy, competence, or performance of the police is unrelated to his perception of police visibility.

3. Note the value of alpha. For these five items, alpha = .7795.
4. Note if alpha can be improved by deleting any one of the items. See Table 4.4.

Table 4.4 Evaluating Alpha

Item	Alpha if Item Deleted
Attitude	.6940
Courteous	.6923
Competent	.6749
Performance	.6884
Visible	.8787

5. Decide to eliminate "Visible" from the scale. Eliminating "Visible" clearly improves the reliability of the scale, that is, its homogeneity. The low correlation coefficients for "Visible" confirm that it is unrelated to the other scale items. Removing any other items would weaken the homogeneity of the scale.

6. Report the reliability test by simply noting the value of alpha. "Police performance was measured by summing respondents' perceptions of police officers' attitudes, courtesy, competence, and performance (alpha = .8787)."

The items included citizen ratings of police officers' attitude, competence, performance, and courtesy and an item about police visibility. The key statistic estimating internal consistency is alpha, a reliability coefficient. The closer alpha is to 1.0, the more reliable the measure. If alpha is close to 0.0, the measure has too few items or the items are statistically unrelated to one another. (A negative alpha indicates that the items violate the assumption of homogeneity; in other words, they do not all measure the same concept.) In the example, alpha = .78; deleting the visibility item will increase alpha to .88. Upon reflection, an investigator should realize that visibility is not related to the other measures of professionalism. The alpha has helped him think about the measure; it has not done his thinking for him.

A major benefit of the test is that it identifies items that diminish the measure's internal consistency and should be excluded from the measure. If an investigator reports the results of a test of internal consistency, the reader can then decide whether the evidence of the measure's internal consistency is weak, strong, or somewhere in between.

INTER-RATER RELIABILITY

Inter-rater reliability establishes the equivalence or consistency of the measures reported by two or more observers. Inter-rater reliability applies to a wide range of administrative data collection projects when more than one person applies a measure. If two or more staff members review and rate job applicants, they should agree on who are the best applicants. Inter-rater reliability is established by having observers apply a measure to the same phenomena and independently record the scores. The scores are then compared using one of several procedures to determine whether the level of agreement is appropriate.

Example 4.4 outlines the steps researchers took to ensure surveyors collecting data on racial profiling agreed on each driver's race. A similar procedure could be used to see if staff gave similar ratings to job applicants. The actual study on racial profiling found about 88 percent agreement in identifying an individual's race. Like using alpha values to test internal consistency, before adopting the measure, the researcher has to decide if 88 percent agreement is sufficient and what the consequences are of not requiring a higher percentage of agreement.

Example 4.4 Establishing Inter-Rater Reliability

Problem: To determine if police officers engage in racial profiling.
Key variables:

Police stops; all stops (either traffic or pedestrian) made by a police officer which do not result in a ticket.

Racial composition of transient population: Asian, Black, White, Middle Eastern, Other, Unknown.

Procedure for measuring racial composition of transient population: Two surveyors are posted at various intersections. Each identifies the race of the drivers driving in one lane of traffic.

Establish reliability of measure of racial composition:

$$reliability = \frac{A}{A+D} \times 100\%$$

where:

A = number of agreements (surveyors assign same race to an individual)

D = number of disagreements (surveyors record different race for an individual)

Notes:

1. This measure indicates how surveyors perceived a person's race—it does not measure the person's actual race;
2. The findings on the racial composition of the transient population are then compared with the racial composition of the population subjected to a police stop.

Source: Adapted from Lamberth, John C., *Ann Arbor: Police Department Traffic Stop Data Collection: Methods and Analysis Study* (Chadds Ford, PA: Lamberth Consulting, 2004).

Measuring race, including developing reliable measures, is controversial and challenging.[11] In a longitudinal study, the officially recorded race of 20 percent of the young people changed sometime during the time they were part of the study, perhaps due to how the interviewer perceived a young person's race or how the young person viewed her race.[12] Detailed analysis showed how an individual's recorded race was associated with the individual's social situation.[13] For example, if the person was unemployed or coming out of prison, the recorded race was more likely to be black.[14] If the changes were random—that is, if no consistent patterns had been found—one might ascribe it to interviewer error. In either case, the measure used was not reliable; that is, changes in a young person's race were due to how it was measured, not a change in the person.

The equivalence of two or more versions of a measure can be established by the *alternative forms technique.* Two or more versions of a measure are detailed on a survey, interview protocol, or test. Respondents are asked to answer both versions. The scores on both versions of the measure are compared. If the measures yield similar scores, the versions are reliable. However, creating comparable versions of a test is difficult. Consequently, establishing reliability with alternative forms occurs primarily in situations where tests will be given to large numbers of people over a period of time, such as tests for driver's licenses.[15]

TEST-RETEST

The *test-retest* technique establishes the stability of a measure. Test-retest requires that an instrument or test be administered to the same people at two points in time. If the results are dissimilar, the measure might be unreliable. This procedure seems direct and reasonable, but if you think about it, you may recognize its limitations:

1. The initial testing may affect the responses to the retest. Suppose you asked people to rate the adequacy of police protection in their community. Responding to this question could focus their attention on policing in their community and eventually lead to a change in their perceptions. Their retest responses would change. Such a change would not mean the

measure involved was unreliable. Conversely, they may remember their first response and repeat it. In such a case, the lack of change between Time 1 and Time 2 does not verify the measure's reliability.

2. With practice, respondents' performance may change. They may benefit from familiarity with previously unfamiliar question formats or tasks.
3. An actual change, independent of the first testing, may take place between the two administrations, or tests, of the measure. The more time that passes, the more likely an actual change has occurred.

From our teaching experience, we have learned that students:

* underestimate the problems of using test-retest to estimate reliability
* confuse test-retest with the pretest-posttest of experiments

With experiments, researchers measure the value of a variable before the intervention of the independent variable for a pretest and then after that intervention for the posttest. Only the experimental group members are exposed to the independent variable. With test-retest, there is no intervention for any participants between test administrations.

All research designs, including experiments, should use reliable measures. Administrators, depending on how much they work with data and in what capacity, vary in their concern with reliability. At a minimum, they want to affirm that data are free of gross errors. If they know a measure's operational definition, they can apply the qualitative method and identify the most serious threats to reliability. In some situations, administrators need quantitative information to estimate precisely a measure's error rate. In these cases, more attention is given to a measure's reliability. A standard rule of thumb is to understand how reliable a measure needs to be for the task at hand. In the next section, we summarize the major features of reliability.

Summarizing Reliability for Administrators

Reliability: evidence that a measure distinguishes accurately between subjects or over time. If a measure is reliable, one can assume that differences between subjects or over time are accurately and consistently measured.

Evidence of a measure's reliability:

Qualitative method: judgment used to evaluate the reliability of any measure.

Internal consistency: used to demonstrate empirically that each of several items of a single measure is related to the others.

Inter-rater reliability: used if more than one investigator collects data on a measure. Inter-rater reliability produces empirical evidence demonstrating that similar scores are assigned by different raters.

Alternative or equivalent forms: used to demonstrate empirically that two different versions of a measure are assigning similar scores to subjects. Administrative researchers, other than those who construct tests, rarely use this method.

Test-retest: used to demonstrate empirically the stability of a measure. Test-retest will not produce useful evidence of reliability if the first testing is likely to affect retest responses or performance or the phenomenon being measured is likely to have changed between the test and the retest.

When to determine reliability: Reliability should be determined prior to collecting the research data, introducing an intervention, or testing a hypothesis as part of the main research. Internal consistency may be determined after collecting data but prior to data analysis. Nevertheless, specific problems, such as ambiguous terms, may have been accidentally overlooked during initial reviews of a measure and identified only after the data have been collected or analyzed.

Relative nature of reliability: the degree of accuracy required depends on the purpose of the measure. A measure may be reliable for one purpose and not for another.

By definition, an unreliable measure contains an unacceptable amount of random error. Data produced by an unreliable measure should be discarded. Depending on the degree and source of the unreliability, the measure itself may be discarded or simply revised. And, as you'll see subsequently, demonstrating a measure's reliability is only one step in determining whether it should or should not be used in a given study.

Operational Validity

In addition to knowing that a measure is reliable, we also want to know whether it has been named correctly and that it measures what we intend to measure. Measures that measure what they are devised to measure are said to be *operationally valid*. Operational validity commonly is referred to simply as *validity*. We have found that students often confuse the validity of *measures* with the internal and external validity of experimental and quasi-experimental *research designs*. Thus, for the purposes of clarity, we prefer to use the term *operational validity* when discussing the validity of measures.

Judgment and Statistical Analysis

Accepting a measure as operationally valid is a matter of judgment, although it may also involve statistical analysis. Consider measures intended to compare and rank the quality of life in American cities. First, as discussed earlier, we would start by determining a conceptual definition. What would represent quality of life in this case? Many ideas might come to mind, but let's say quality of life is determined by living in a desirable place. In turn, we would develop measures to determine how desirable people consider a community as a place to live. If one investigator chose items that focused on social diversity, arts and cultural events, access to technology, and employment opportunities, while another selected items that focused on commute time, recreational opportunities, natural beauty, and violent crime rates, the results surely would differ. The first investigator would probably conclude that major cities were the best places to live. The second would probably assign higher rankings to towns or smaller cities. The operational validity of either measure could be challenged by arguing that it does not really measure quality of life.

Measuring the incidence of crime may seem to be a simpler assignment. At first glance, one might assume that the number of crimes reported to police would be a clear-cut measure of the crime rate. However, not all crimes are reported. As a result, officials in cities with a high number of reported crimes can argue that reported crime actually measures the vigilance of police officers rather than the actual crime rate. Officials may suggest that in cities with lower crime rates, police discourage victims from reporting crimes or that these cities do not have procedures to ensure accurate recording of crimes. Five events must occur for an actual crime to become part of the crime report:[16]

1. The act must be known or perceived by someone other than the offender; for example, you must know your home was burglarized.
2. The person perceiving the act or its consequence must define it as a crime; for example, what some might consider to be vandalism, others may think of as only a neighborhood prank.
3. Someone must report the act to the police.

4.   The police must define the reported act as a crime.
5.   The police must record the crime in the appropriate way and category.

The requirements that a reported crime must be perceived, recognized, reported, defined, and recorded correctly do not imply that the crime report constitutes an operationally valid measure of crime rate. Rather, each of the first four requirements presents an opportunity to miss a crime and undercount the crime rate. The fifth requirement means that one type of crime might be undercounted and another overcounted.

Hate crimes are an example of how the operational validity of a measure of crime rate is criticized. When the FBI issued its annual report on hate crimes in 2010,[17] advocacy organizations suggested it did not accurately reflect trends in actual hate crimes.[18] Victims might be unwilling to report such crimes. Alternatively, the police may be unwilling to report crimes as hate crimes or may not have sufficient knowledge to report them at all.[19]

A Measure's Use and Operational Definition

Policy makers and program administrators should carefully consider how a proposed measure will be used and pay attention to its operational definition. This advice applies whether users are developing a new measure or applying an existing one to arrive at findings. They start with a clear idea of what information they need. Consider the measures of poverty.[20] The U.S. Census counts a family as poor if its pretax income falls below the "poverty threshold." The measure was developed in the 1960s. To determine the poverty threshold, the estimated cost of feeding different-sized families a minimally adequate diet was multiplied by three. The multiplier was based on a 1955 finding that a typical family spent one-third of its income on food.

The poverty threshold has provided a consistent measure of the poverty rate for over a half a century, yet it has been criticized during much of that time.[21] Questions about the measure arise when it is used to determine who needs financial assistance. How do 1955 circumstances apply in the 21st century? For example, the average American household spends less than 15 percent of its income on food today.[22] Should non-income benefits be considered? What constitutes income today? The Census Bureau collects data on 15 definitions of income. Should regional variations in cost of living be considered? Does relative poverty, which considers social inequality, provide more insights into the effects of poverty? To address some of these concerns, a supplemental poverty measure was released by the Census Bureau in late 2011. However, due to what many consider political and logistical reasons, the original poverty measure from 1955 is still the official, widely used statistic.[23]

The way in which a measure is labeled affects our perceptions and even our behaviors. For example, regarding the study mentioned previously on how race was tracked in the National Longitudinal Study of Youth, researchers are now arguing that perceptions can affect how we apply even what were thought to be well-designed measures. Out of necessity, administrators deal with summarized information. Often, they have time only to read report highlights or hear brief presentations. Consequently, the label attached to a measure can seduce and mislead administrators. For example, a harried administrator may remember the findings about the "best" American city without stopping to learn what criteria were used to rank cities. If you plan to act on findings you have read or heard about, you should first learn more about the measure used. Not only will you be able to defend your own use of the measure, but you will be in a better position to decide whether your action will get the intended results.

Establishing a Measure's Operational Validity

Developing measures and amassing evidence of their validity is an ongoing process. Both the developers and users have responsibilities. The developers provide evidence supporting the measure's interpretation, and the users evaluate the evidence in the light of the measure's intended use.[24] The evidence may be based on a measure's content and its relationship to other variables.[25] Evidence provides insight into what the measure is and is not measuring. How convincing the evidence has to be depends on the potential impact of the measure. Evidence that shows a measure is inadequate or not working as expected is still a step forward. Investigators can then build on the evidence to improve the measure.[26] In the following sections, we will address evidence based on content and evidence based on the relationship to other variables.

Evidence Based on Content

An important and almost obvious piece of evidence of operational validity is demonstrating that, in the eyes of the researcher, the operational definition is consistent with the conceptual definition. Evidence of content-based validity addresses the issue of how well the researcher has translated the idea or definition of the variable into a measure. Two types of content-based evidence are *face* and *content* validity.[27] *Face validity* is an initial judgment one might have when first reading about the measure. *Content validity* more thoroughly evaluates the content of the measure. In content-based evidence of validity, the measure is not judged against other measures. It is judged against what the conceptual definition suggests should be included in the measure. Consider our previous discussion of Example 4.1, which measured employee empowerment. To be consistent with the conceptual definition, the operational definition needed to measure whether employees believe that (1) their work is meaningful, (2) they are competent to do their job, (3) they have autonomy in performing their work, and (4) their work has an impact on the organization. The data produced by the operational definition have merit only if potential users agree on the conceptual definition. In Example 4.2, the engineer and the budget director had different conceptual definitions of productivity. If the budget director had examined the content of the operational definition, he would have raised questions about its validity.

To see how investigators develop evidence based on a measure's content, consider personnel selection procedures. At the simplest level, a job description (the conceptual definition) outlines the most basic knowledge, skills, and duties required to do the job. Staff then decide how to evaluate the applicants using a selection process consistent with the description. A sound selection procedure identifies and ranks applicants based on their ability to do the stated job.

A more elaborate process may occur if positions are numerous, openings frequent, and the applicant pool large. Analysts begin by systematically identifying the elements essential to the job. What specific skills must an employee have? What detailed knowledge does she need? What is the relative importance of the skills and the knowledge? That is, are some skills or types of knowledge more important than others? After this is done, a blueprint for the selection procedures is created; the more detailed blueprint outlines the job components and their relative importance. Appropriate measures are developed for each job component, and more important skills or knowledge should receive greater weight than less important ones. The measures may be answers to an objective test, performance in a simulated exercise, or responses to interview questions.[28]

Operationally Valid Definitions

Before using a measure, you need to review the operational definition and any documentation describing its design. You need to judge its validity. An operationally valid definition measures what you want to measure; that is, it captures the essence of the conceptual definition. Note that a measure may reflect a strong relationship between your conceptual and operational definitions and still be judged invalid by others.

For example, consider measures of a nation's wealth. Such measures may count all production without regard to its social value. Some critics question whether the measures should consider the cost of depleting a nation's natural resources or if production of weapons, tobacco products, or toxic wastes should be included.[29] Others point to the possible impact on human conditions. The small Himalayan nation of Bhutan measures its well-being by gross national happiness. It has documented and posted the conceptual and operational definitions of nine domains, including health, education, and psychological well-being.[30]

Validity Over Time

A measure's validity can change over time as well. Consider the validity of official statistics. While an operational definition may stay the same, it may become less relevant over time. To estimate the inflation rate, governments track the costs of a sample of consumer goods and services. A major problem arises when consumer behavior changes. For many years, the typical American diet was heavy on red meat, eggs, and milk. Americans now eat less of these products and more poultry, fruits, and vegetables. So a measure that tracks the cost of foods eaten by a typical American family in 1950 will not reflect accurately the cost of foods eaten by a typical American family today.

Some reasons for not changing the measure of inflation:

1. First, establishing the typical American diet requires a costly nutritional survey.
2. Second, if the measure is changed, then comparability with the past is lost. To retain comparability, the revisions may be applied to older data, or both measures may be reported for a time. If a measure was changed in the year 2005, data for earlier years may be adjusted to reflect the change, or beginning in the year 2005, data for both measures may be reported.
3. Third, changes in certain measures inevitably bring charges of political manipulation.

Evidence Based on a Relationship to Other Variables

Reviewing a measure's content does not tell the whole story. Investigators gain additional insight if they correlate the measure with a criterion—a kind of double-check. *A criterion measures the concept of interest; it may be an alternative or similar measure or a measure of a future outcome.* The investigators look for a similar response pattern between the measure being validated and the criterion. Examples of measures and similar criteria include the following, as shown in Table 4.5.

Criterion-based evidence of validity can only be produced when a suitable criterion exists and can be applied to the same population. For example, analysts can compare respondents' answers to whether they voted (the measure) with actual voting records (the criterion). Finding a weak relationship between a measure and a criterion may be disconcerting, but it may start a valuable exploration of what information a measure is and is not providing. For example, the observation that more people say they voted than actually voted has led to follow-up studies.[31]

Table 4.5 Examples of Measures and Criteria

Measures	Criteria
Citizen or user satisfaction with a service	Objective measure of service quality
Employee descriptions of a supervisor's management style	Formal professional assessment of supervisor's management style
Self-report of health status	Medical records or medical examination

Researchers sometimes use one measure that is available to provide information on a characteristic that cannot be directly measured. In this case, the measure obtained must have *concurrent validity* in order for the researcher to have confidence in it. In other words, measure A is used in place of measure B, which has been previously validated. For example, psychologists use polygraph machines to measure galvanic skin response, breathing rate, and heart rate to assess stress levels. These measures in turn indicate areas of personal conflict and emotional difficulty for the individual. See Example 4.5 for another example of concurrent validity.

Example 4.5 Criterion-Based Evidence of Operational Validity: Citizen Satisfaction with Government Services

The following two studies illustrate an attempt to establish the validity of a measure with criterion-based evidence. The specific type of operational validity under consideration is concurrent validity. As you read the summaries of the studies, decide if you believe citizen satisfaction is a valid indicator of service quality.

Study 1

Units of analysis. U.S. cities with both citizen survey and performance measurement data.
Measure: 100-point scale measuring average resident satisfaction with a service. Respondents were asked to rate their satisfaction with street maintenance.
Criterion: Percent of lane-miles street engineers assessed as being in good condition
Purpose: Assess correlation between measure and criterion to determine if resident satisfaction is a valid measure of condition of streets. This will determine a type of validity known as concurrent.
Finding: Weak, but negative relationship ($r = -.36$, $n = 7$) between resident satisfaction level and city engineers' assessment of street quality. Possible explanations:

1. Citizens have different standards for road quality than engineers, or
2. The two groups consider different factors important. (Can you think of other reasons the measure and the criterion do not correspond very well?)

Conclusion: Satisfaction with street maintenance is not a valid concurrent measure of engineering assessment of street quality.

Study 2

Unit of analysis: Iowa towns, population between 500 and 10,000; $n = 99$.
Measure: Average of town's surveyed residents' response to rate the condition of streets in [name of town] as "poor," "fair," good," or "very good."
Criterion: Trained observers rated four randomly selected streets in each town based on the physical condition of their surface and curbing and whether they were free of weeds and debris.
Purpose: To determine if residents' rating of street condition is a valid measure of street condition. Validity assessment by comparison with that of trained observers as criterion.
Finding: Weak, direct relationship between the measure and the criterion ($r = .49$).
Conclusion: Citizen assessments indicate how citizens view their streets, not their quality.

What are some lessons learned from these studies?

1. Criterion-based evidence may be limited because of lack of a suitable criterion (small *n* in study 1 was due to the fact that few communities collected both survey and performance data) or feasibility (cost in having trained observers gather data in Study 2).
2. Evidence suggests that experts—street engineers in Study 1, trained evaluators in Study 2—and citizen satisfaction as obtained by surveys are not measuring the same thing.
3. Weak or contrary evidence can provide valuable insights about a measure's quality and what it is measuring.
4. These studies are part of a research literature on satisfaction measures. The body of literature confirms the observation that "even if the evidence shows a measure is not working the way it was intended, it is a step forward. The challenge is to find the way to make the evidence useful in the next iteration."

Sources: Based on studies conducted by D. Swindell and J. M. Kelly, "Linking Citizen Satisfaction Data to Performance Measures: A Preliminary Evaluation," *Public Performance and Management Review* (2000): 30–52; and M. J. Licari, W. McLean, and T. W. Rice, "The Condition of Community Streets and Parks: A Comparison of Residence and Nonresident Evaluations," *Public Administration Review* (2005): 360–368.

PREDICTIVE VALIDITY

A measure may be designed to predict something in the future, for example, who will succeed in university, be a valuable employee, or commit crimes. The criteria for these respective examples might be grade point average and graduation status, a measure of employee productivity, or arrest and conviction records. An investigator will need data on both the measure, taken at time one, and the criterion, taken at a later time. This validity is appropriately called *predictive validity*. The Graduate Record Exam, for example, is often used by graduate program admissions committees to help identify which applicants are likely to be successful in the program to which they have applied. Even if an appropriate criterion is identified, a measure's ability to correctly predict a future outcome will be limited. Events and situations concurrent with the criterion may strongly affect the outcome. For example, what happens on a job after a person is hired may explain his productivity more than the selection criterion.

Example 4.6 illustrates the use of predictive evidence to guide placements in the military. The example shows that it is important to measure the predictive validity of the measure in terms of its stated purposes. Often, testing the predictive validity of the measures is essential to ensuring that the measurement can be used in a valid scale.

Example 4.6 Criterion-Based Evidence of Validity: A Nobel Laureate Tries to Predict Service Placement in the Military

Background: Daniel Kahneman is one of the most famous social scientists of the 21st century. His ideas about how people behave under conditions of loss and uncertainty, with Amos Tversky, which are known as prospect theory, won Kahneman the Nobel Prize in Economics in 2002. Kahneman and Tversky's work is generally regarded as the foundational work that started the field of behavioral economics.

An Important Problem: Just after finishing his psychology degree, Kahneman was required to serve two years in the Israeli military. Since this was the 1950s, Israel did not have many psychologists. Kahneman was asked to develop a way to predict which branch of the military new recruits should be assigned. Military personnel directors generally believed that personality was very important for determining the branch of the military to assign a recruit. For example, a member of the infantry would have a very different personality than a pilot in the air force.

Measures: Kahneman began developing measures of soldier behavior on relevant dimensions to the military such as sociability, bravery, and so on. He thought that by measuring these attributes and then seeing which ones corresponded to performance in the different branches of the military that he would find the soldier characteristics that were most relevant to each branch of the military.

Unit of analysis: Newly enlisted soldiers in the Israeli military.

Findings: The measures of behavior were *not* associated with placement in the military with either the soldier's or their officer's evaluation of whether this was a good placement for the soldier. What the test did show was that the measures of behavior that Kahneman developed were highly valid measures of predicting a soldier's performance in *any* branch of the military. According to Michael Lewis, the measures that Kahneman developed are regarded as one of the keys in the development of the Israeli military and have been in use in at least some form ever since they were developed.

Further note: What are some lessons you can deduce from this example?

Measuring the predictive validity of scales is very important. Kahneman may have asserted that his measures of bravery, for instance, made the recruit an ideal candidate for the airborne unit. However, just measuring bravery and asserting that there is a theoretical connection to the placement in the airborne unit would have missed that this is important in other branches of the military as well. By assessing the predictive validity of the measure, Kahneman was able to show generally who would likely perform well and who was not likely to perform well, which allowed the best soldiers to be deployed to the most strategically advantageous branches of the military.

Source: Lewis, M., *The Undoing Project* (New York: W.W. Norton & Company, 2017), 52–84.

A researcher may also want evidence that a measure correlates with (1) another measure that is supposed to measure the same concept and but does not correlate with (2) a measure not intended to measure the concept. If the researcher's measure correlates with (1), the measures are said to *converge*, and the researcher has evidence of convergent validity for his measure. If the researcher's measure does not correlate with (2), they are said to *diverge*, and the researcher has evidence of discriminant validity for his measure. For example, scores on a quantitative aptitude test should be strongly related to grades in math but not to grades in English.[32]

Summarizing Operational Validity

Operational validity is not a characteristic of a test; rather, it is the user's judgment that a measure is appropriate. It means that a measure does measure what is intended. Administrators who work with data should ask themselves: How was the measure operationally defined? Will the operational definition yield useful data? What will these data indicate? What evidence supports an assumption of validity? Is it related to an alternative or similar measure? Is it correlated with theoretically linked variables and uncorrelated with unrelated variables? Whether a measure is accepted as operationally valid depends on the user's purpose and conceptual definition. If the data will drive decisions, investigators should monitor the consequences of using the measure.[33] As with reliability, operational validity should be determined prior to collecting the data, introducing an intervention, or testing a hypothesis. Because of the role of judgment, the operational definition and evidence supporting its validity should be reviewed with stakeholders.

Sensitivity

The *sensitivity* of a measure refers to its precision or calibration. A sensitive measure has sufficient values to detect relevant variations among respondents; the degree of variation captured by a measure should be appropriate to the purpose of the study. Measures that are reliable still may not detect important differences. Consider a salary survey. Suppose employees were asked:

What is your salary? (check appropriate category)

_____ Less than $25,000
_____ $25,000 to $34,999
_____ $35,000 to $44,999
_____ $45,000 to $54,999
_____ $55,000 to $64,999
_____ $65,000 to $74,999
_____ $75,000 or more

The categories included in this measure may be adequate for a survey of entry-level and mid-level city employees. They probably would not be adequate for a survey of city managers or department heads. Since most top managers will fall in the last category, the measure is likely to be insensitive to variations in city manager salaries. Thus, in constructing a measure, an investigator wants to avoid having respondents clustered in a single category. If most responses fall into one category, the measure would be considered insensitive. Insensitive measures do not allow investigators to compare respondents. There is not enough detail to show variation—and variation often drives analysis.

Measures developed to study one type of population may be insensitive to differences in other more homogeneous populations. In the previous example, the problem could be solved by creating more and narrower categories, especially at the higher income levels. With complex measures, such as job-satisfaction scales or intelligence tests, different populations may require different sets of indicators to ensure sensitivity. For example, a job-satisfaction measure developed for organizations employing unskilled and skilled laborers, clerical workers, and technical, administrative, and professional staff may be a poor choice to study a work unit largely made up of professional employees. If individual differences are of interest, then the measure would not be sufficiently sensitive to identify differences among employees in the more homogeneous group. Sensitivity in policy and administrative research is a relative concept, and no standard for an acceptable level of sensitivity exists. Sensitivity is related to operational validity, because if the measure is unable to capture the degree of variation, you probably are not measuring what you want to measure.

Summary

Measurement allows us to place cases in categories based on some value of a variable. It makes it easier to compile, analyze, and compare information on phenomena. Nevertheless, the process of measuring sacrifices the richness of a concept. No measure can fully describe quality of life, employee empowerment, or level of poverty. The information provided by measures has great value, but you also should recognize a measure's limitations.

The measurement process begins with a conceptual definition for each variable to be studied. The conceptual definition indicates what the investigator means by a concept. The conceptual definition should be consistent with the investigator's purpose in developing or selecting a measure and collecting data. The conceptual definition serves as a blueprint for the operational definition. The operational definition details exactly how a concept or variable was measured and how its values were determined. Both conceptual and operational definitions are critical for selecting the unit of analysis and the appropriate measure of a concept.

A measure's values are commonly classified as nominal, ordinal, interval, or ratio. Nominal measures categorize, but do not rank, data. Ordinal measures rank data, but the exact distance separating two categories cannot be determined. Interval and ratio measures indicate the distance between two pieces of data. Measures should be reliable, operationally valid, and sensitive. Reliable measures allow an investigator to conclude that identified differences between subjects or over time are consistent and accurate and not due to the measure or the measuring process. Reliable measures yield the same results time after time if whatever is being measured has not changed. Two or more investigators using a measure should give very similar scores to the same phenomenon. Carefully reviewing the procedures used to implement measurement and data collection can markedly improve reliability. The reviewer makes sure that directions are clear and easy to follow, that items are clearly defined, that given responses are mutually exclusive and exhaustive, and that the respondent has access to the requested information. To avoid unreliable measures, people responsible for data collection and data processing should be well trained. If knowing and limiting the amount of random error are important, an investigator should use mathematical procedures to establish reliability. Unreliable data should be discarded.

A reliable measure is not necessarily operationally valid. An operationally valid measure actually measures the concept of interest. Analysts and data users should examine evidence that the measure is appropriate for their purpose. The evidence informs the judgment of data users; it does not replace it. For example, evidence can show the measure is well aligned with the conceptual definition or that correlation exists between a measure and a similar or alternative measure. A measure should also not correlate strongly with measures of concepts that should not be related. A sensitive measure sufficiently distinguishes cases from each other so that they can be compared.

An administrator who includes quantitative information in her decision-making may make an inappropriate decision if the data are unreliable or invalid. Quantitative information may be ignored if administrators consider the measures meaningless or inadequate to support decisions. Administrators should not expect researchers to make all decisions about a measure's appropriateness, nor should they base their decisions upon data without questioning the quality of the measures that generated them.

Measurement decisions are tied to other research decisions. The level of measurement determines what statistics the investigator can use. The analysis chapters identify the statistics appropriate to data measured at the nominal, ordinal, or interval level. The data collection chapters explicitly discuss measurement. The number of items and their wording must be considered in conjunction with the data collection method. An investigator who uses existing data must determine the measures' reliability and operational validity.

In practice, investigators construct measures and samples independently, although implied links exist (see Figure 4.1). Every measure is not necessarily appropriate for all subjects. The degree of precision can affect the cost of data collection or processing; thus, sample size may limit the sensitivity of measures. Investigators may overemphasize sample construction at the cost of developing high-quality measures.

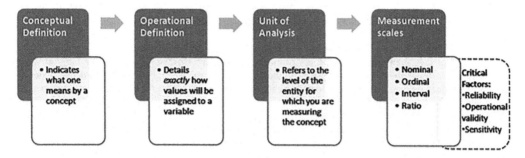

Figure 4.1 The Measurement Process

Notes

1. For an excellent example of detailed conceptual and operational definitions and the relationship between them, see J. E. Royce, *Alcohol Problems and Alcoholism* (New York: The Free Press, 1981), 15–19.
2. K. W. Thomas and B. A. Velthouse, "Cognitive Elements of Empowerment," *Academy of Management Review* 15, no. 4 (1990): 666–681, cited by S. T. Shelton, "Employees, Supervisors, and Empowerment in the Public Sector," Unpublished Dissertation, North Carolina State University, Raleigh, NC, 2002.
3. J. S. Hunter, "The National System of Scientific Measurement," *Science* 210 (November 1980): 869–874.
4. Gallup-Healthways has recently developed a "well-being" index to measure this concept. Information can currently be found at www.healthways.com/solution/default.aspx?id=1125 and www.gallup.com/poll/106756/galluphealthways-wellbeing-index.aspx.
5. Of course, the researcher can determine in the middle of the study that they have chosen an incorrect unit of analysis, and the researcher may have to go back to collect more data. This is a costly mistake and should be avoided by good planning. Another exception to the single unit of analysis occurs when the researcher carries out multilevel analyses. We do not discuss multilevel studies in this text. For information, see, for example, Harvey Goldstein, *Multilevel Statistical Models*, 4th ed. (Hoboken, NJ: John Wiley & Sons, Inc., 2011). See also Douglas A. Luke, *Multilevel Modeling*. Series: Quantitative Applications in the Social Sciences (Thousand Oaks, CA: Sage Publications, 2004).
6. Absolute zero is generally defined as the point at which the motion of particles stops and is generally thought to only be theoretically possible. Absolute zero is 0 degrees Kelvin or –459.7 degrees Fahrenheit.
7. Frequency distributions may be constructed so that the measures can be considered or assumed to be interval. In general, we do not recommend that you use interval statistics on ordinal measures unless you are sufficiently familiar with the measures and the statistic that you can interpret the statistical findings appropriately. For an excellent summary of the issues, see D. J. Hand, *Measurement Theory and Practice: The World Through Quantification* (London: Arnold Publishers, 2004), 72–83.
8. For a useful discussion, see Ellen A. Drost, "Validity and Reliability in Social Science Research," *Education Research and Perspectives* 38, no. 1 (2011): 105–123.
9. See Susana Urbina, *Essentials of Psychological Testing*, 2nd ed. (Hoboken, NJ: John Wiley & Sons, Inc., 2014), Chapter 4, for more details on the formulae and their appropriate use.
10. For more detailed discussion on internal consistency and coefficient alpha, see Anne Anastasi and Susan Urbina, *Psychological Testing*, 7th ed. (New York: Palgrave Macmillan, 1997), 91–102.
11. Margaret A. Winker, "Measuring Race and Ethnicity: Why and How?" *Journal of the American Medical Association* 292, no. 13 (2004): 1612–1614.
12. Aliya Saperstein and Andrew M. Penner, "Racial Fluidity and Inequality in the United States," *American Journal of Sociology* 118, no. 3 (2012): 676–727, especially p. 678.
13. Ibid., 698–705.
14. Ibid., 707.
15. Other terms, such as *parallel forms*, are also used to describe this type of reliability.
16. R. F. Sparks, *Research on Victims of Crime* (Washington, DC: U.S. Department of Health and Human Services, 1982), 14.
17. For an overview, see: U.S. Department of Justice—Federal Bureau of Investigation, "FBI Releases 2010 Hate Crime Statistics," November 14, 2011, available at www.fbi.gov/about-us/cjis/ucr/hate-crime/2010/Hate%20Crime%202010-Summary.pdf.

18. Mark Potok, "DOJ Study: More Than 250,000 Hate Crimes a Year, Most Unreported," *Southern Poverty Law Center*, March 26, 2013, available at www.splcenter.org/blog/2013/cb/doj/DOJ-study-more-than-250,0000-hate-crimes-a-year-a-third-never-reported.
19. Ibid.
20. This discussion draws on John Cassidy's article, "Relatively Deprived: How Poor Is Poor," *The New Yorker*, April 3, 2006, 42–47, and links found at The Census Bureau web page, available at www.census.gov/topics/income-poverty/poverty/guidance/poverty-measures.html. The Census Bureau web page includes links to poverty-related definitions, publications on measures of poverty and income, and research on alternative definitions of poverty.
21. See Scott Horsley, "Five Decades Later, Time to Change the Way We Define Poverty?" *National Public Radio*, January 8, 2014, available at www.npr.org/2014/01/08/260807955/five-decades-later-time-to-change-the-way-we-define-poverty.
22. *Consumer Expenditures in 2012* (Washington, DC: U.S. Department of Labor, U.S. Bureau of Labor Statistics, March 2014), Report 1046. In the years including and leading up to 2012, the amount of household income, on average, spent on food was 13 percent.
23. Thomas B. Edsall, "Who Is Poor?" *The New York Times*, March 13, 2013, available at http://opinionator.blogs.nytimes.com/2013/03/13/who-is-poor/. See also "Measuring America: How the U.S. Census Bureau Measures Poverty," September 2017, available at www.census.gov/library/visualizations/2017/demo/poverty_measure-how.html.
24. Urbina, *Essentials of Psychological Testing*, 161.
25. See *Standards for Educational and Psychological Testing* (Washington, DC: American Educational Research Association, 2014), 11–17, for a more detailed discussion of sources of evidence. We have confined our discussion to the sources that readers without more training in tests and measurements are most likely to produce and encounter.
26. Mark Wilson, *Constructing Measures: An Item Response Modeling Approach* (Mahwah, NJ: Erlbaum Associates, Publishers, 2005), 156.
27. William Trochim, *Research Methods Knowledge Base*, 2nd ed. (Cincinnati, OH: Atomic Dog Publishing Co, 2001), Chapter 3, 66–69; Kenneth Meier, Jeffrey Brudney, and John Bohte, *Applied Statistics for Public and Nonprofit Organization*, 9th ed. (Stamford, CT: Cengage Learning, 2014), Chapter 2.
28. M. W. Huddleston and Dennis Dresang, *The Public Administration Workbook*, 7th ed. (Upper Saddle River, NJ: Longman, 2011).
29. M. H. Maier and J. Imazeki, *The Data Game: Controversies in Social Science Statistics*, 4th ed. (Armonk, NY: M. E. Sharpe, 2013), Chapter 7, discuss problems with gross domestic product (GDP). Gross domestic product is often used instead of gross national product (GNP) as a measure of the nation's economy. For the United States, the difference between the two is slight. See Maier and Imezeki, *The Data Game*, 141.
30. See Lisa Napoli, "Gross National Happiness Measures Quality of Life," *National Public Radio*, June 9, 2010, available at www.npr.org/templates/story/story.php?storyId=127586501.
31. For a discussion of the literature and an extensive bibliography, see R. Bernstein, A. Chadha, and R. Montjoy, "Overreporting Voting," *Public Opinion Quarterly* 65 (2001): 22–44.
32. Hand, *Measurement Theory and Practice*, 133. For an in-depth discussion of patterns of convergence and divergence, see Urbina, *Essentials of Psychological Testing*, 169–181. Other discussions of convergent and divergent validity can be found in Trochim, cited previously, and at: www.socialresearchmethods.net/kb/measval.php. Also see Meier, Brudney and Bohte, *Applied Statistics for Public and Nonprofit Organization*, Chapter 2; E. Carmine and R. Zeller, *Reliability and Validity Assessment: Quantitative Applications in the Social Sciences* (Thousand Oaks, CA: Sage Publications, 1979).
33. Urbina, *Essentials of Psychological Testing*, 208–210, discusses and gives sources for the debate on the role of consequences as evidence of validity.

Terms for Review

concept
conceptual definition
operational definition
measurement
levels of measurement
measures
reliability

stability
internal consistency
test-retest
operational validity
content-based evidence of validity
criterion-based evidence of validity
sensitivity

Questions for Review

The following questions should indicate whether you have a basic competency in this chapter's material.

1. Indicate whether each statement represents a conceptual definition, part of an operational definition, or a hypothesis.

    a. The organizational capacity of not-for-profit organizations consists of their relevance, responsiveness, effectiveness, and resilience.
    b. To determine the equity of police services, we asked respondents if they thought there was less, about the same, or more crime in their neighborhoods than in the rest of the city.
    c. The more politically efficacious the respondents, the greater the probability that they will have a favorable attitude toward government services.
    d. Home health care has been defined as an array of therapeutic and preventive services usually provided to patients in their homes or in foster homes because of acute illness or disability.
    e. Controlled inventoried items are those that must be identified, accounted for, secured, segregated, and handled in a special manner.
    f. Uncontrolled inventoried items have a higher rate of wastage than controlled items.
    g. The rate of school vandalism was measured by counting the number of cracked and broken windows.

2. What type of measure—nominal, ordinal, or interval/ratio—does each describe?

    a. Counties in a state (Ash County, Beach County, Maple County)
    b. Number of participants in food stamp programs
    c. Reputations of colleges
    d. Leadership ability measured on a scale from 0 to 5
    e. Divisions within a state agency
    f. Inventory broken down into three categories: tightly, moderately, or minimally controlled

3. A job-training program measures its effectiveness by the number of persons placed in a job. Factors such as salary and length of employment are not considered. Comment on the apparent reliability and operational validity of the program's effectiveness measure. How should the program improve its effectiveness measure?

4. Suggest possible operational definitions to test the hypothesis: Parents of elementary-age schoolchildren (5- to 11-year-olds) go to public libraries more often than parents of younger or older children.

5. Evaluate the following statements:

    a. A measure can be reliable for one study but not for another.
    b. A reliable measure will assign the same value to the same subject each time the measure is used.

    c.   Test-retest to establish reliability is the same as an experiment's pretest-posttest.

    d.   An agency should establish a maximum rate of random error for all the measures it uses.

    e.   All measures should be qualitatively reviewed before mathematically estimating reliability.

6.   Consider a measure of school vandalism that includes seven indicators of vandalism.

    a.   What does a researcher mean when she says that the measure of school vandalism is reliable?

    b.   What does she mean if she finds that the measure is stable, equivalent, and internally consistent?

    c.   What does she mean if she says that the measure is operationally valid?

    d.   What evidence could she produce to demonstrate the measure's validity?

    e.   What does she mean if she says that the measure is sensitive?

7.   Consider the measure of school vandalism. In October, the vandalism at eight schools is determined using the measure. Six weeks later, the vandalism at the same eight schools is again determined using the same measure.

    a.   What procedure is being used to determine the measure's reliability?

    b.   The amount of vandalism between the two applications of the measure increased. Should the researcher assume that the measure is unreliable? Explain.

8.   A social service agency developed a measure of client satisfaction. What should another agency learn about the measure before using it to survey its own clients?

Problems for Homework and Discussion

1.   A city department plans to administer a survey to a sample of citizens to assess program quality. Department managers intend to measure citizen satisfaction with services and want to make sure that the survey instrument contains reliable and operationally valid measures.

    a.   Present one strategy that you can use to determine that the measures of program quality are reliable.

    b.   Present one strategy that you can use as evidence that the measures of program quality are operationally valid.

2.   To gather information on safety violations in factories, a state agency plans to send investigators to selected factories to obtain safety information.

    a.   How should the agency determine whether its measure of factory safety is reliable?

    b.   What content-based evidence would you look for to decide if the measure of factory safety was operationally valid?

    c.   Suggest a criterion and strategy to produce further evidence of operational validity of the measure of factory safety.

    d.   If the relationship between the criterion and measure of safety is weak, what would you recommend?

3.   Student observers measure traffic flow through a section of the city. They count the number of vehicles that pass through specified intersections. Each observer works for a four-hour period. An engineer notes that observers become careless in counting near the end of the fourth hour. Is this a problem in reliability, operational validity, or both?

4.   A planner wants to assess road quality. He sends a questionnaire to a sample of residents and asks them to evaluate road surfaces, repair quality, and amount of traffic. He finds the

results of limited value, so he assigns engineers to travel through the neighborhoods to rate the roads. The engineers are given general instructions but no written rating forms. Still dissatisfied, he develops a rating form that lists the number of potholes per mile, the depth of potholes, the thickness of asphalt, the width of the roadway, and similar measures.

a. Comment on the reliability and operational validity of each approach.
b. Which approach would you consider the most operationally valid? Comment on the ways each approach could be modified to improve its reliability and validity.

5. To develop an employment test, two versions of the exam are written and administered. In the morning, one-half of the test takers receive Form A and the other half receive Form B. In the afternoon, the groups and tests are reversed so that those who took Form A in the morning take Form B and those who took Form B take Form A.

a. What was the agency trying to do?
b. If later administrators learn that people who had passed either Form A or Form B did not correctly interpret information on spreadsheets, is the problem with the examination's reliability or operational validity? Explain.
c. If the test takers retook the tests three weeks later and their scores improved, could the administrators correctly conclude that the tests were unreliable? Explain.

6. A human resources department administers employment tests to large groups of applicants. To minimize the temptation to cheat, applicants are given different forms of the exams.

a. Why is it important that the reliability of the test be established?
b. Briefly describe how the examiners could establish the reliability of the different forms.

7. A city council wants to learn what the impact would be if it started issuing "tennis licenses" to allow residents to use the tennis court. A sample of players who use public courts are asked, "If the city charged $10 for tennis licenses, would you still use the public courts?" Discuss the operational validity of this question. Do you think that it would give an accurate indication of the effect of a tennis license?

8. Table 4.6 shows the number of people living below the poverty level as determined by two different operational definitions of poverty just after the Great Recession. Discuss how you would choose one of these definitions as operationally valid. Do you think that your decision would be the same if you were a legislator, a budget officer, or a provider of social services to clients?

Table 4.6 Identifying an Operationally Valid Measurement of Poverty

Operational Definition	Number Below Poverty Level (2010)	Percentage Below Poverty Level (2010)
Money income (adjusted by household size; $23,850 for family of 4)	46,602,000	15.2%
Money income plus capital gains (losses), less income and payroll taxes, plus value of all noncash transfers	46,094,000	16%

Source: The Research Supplemental Poverty Measure: 2010. www.census.gov/prod/2011pubs/p60–241.pdf.

9. Go to http://fedstats.sites.usa.gov/ and identify an official statistic that interests you. How are key measures defined? How are the data collected? How are they used? What information is available on reliability and validity? Note: You may find the latter information if you enter "reliability" and "validity" as search terms.

Working With Data

1. State a conceptual definition for economic status for counties. Then, using the County Data file and the County Data File Notes accompanying this text, identify three variables that might be combined to create a single measure of a county's economic status consistent with the conceptual definition. Using statistical software, check on the internal consistency of these three variables (SPSS refers to this procedure as "reliability"). What do the findings tell you about the reliability of the variables? Can you assume the variables give you an operationally valid indicator of a county's economic status? Explain your answer.

2. Choose a different set of three variables, this time seeking to measure a county's poverty. Conduct the same tests. Are the variables reliable? If these were combined, would the index be operationally valid? Given your findings, which indicator, one of economic status or poverty, do you believe is the better indicator? Do they represent the same thing? Can one be a good measure of the other? Explain your answer.

3. The variables in the GSS dataset that begin with CON are related to a person's confidence in the federal government, Congress, and the courts (CONFED, CONCONG, and CONCOURT, respectively). If you combined these variables, are they internally consistent? Do you think they would be reliable?

Recommended for Further Reading

Standards for Educational and Psychological Testing (Washington, DC: American Educational Research Association, 2014). The standards were created by a joint committee of the American Educational Research Association (AERA), American Psychological Association, and National Council of Measurement in Education and approved by their respective governing bodies. The AERA endorsement stated "we believe the Standards to represent the current consensus among recognized professionals regarding expected measurement practice. Developers, sponsors, publishers, and users of tests should observe these standards" (Standards, viii). (The standards were first published in 1999.)

Urbina, Susana, *Essentials of Psychological Testing*, 2nd ed. (Hoboken, NJ: John Wiley & Sons, Inc., 2014) serves as a basic, accessible resource on psychometrics, that is, measuring characteristics including knowledge, abilities, attitudes, and opinions.

Hand, David H., *Measurement Theory and Practice: The World through Quantification* (London: Arnold Publishers, 2004). The book has a thorough discussion of measurement theory and applications. Its chapter on measurement in the social sciences includes a discussion of performance indicators, customer satisfaction, and crime.

Maier, M. H., and J. Imazeki, *The Data Game: Controversies in Social Science Statistics*, 4th ed. (Armonk, NY: M. E. Sharpe, 2013), present an insightful discussion of social science statistics. They identify and critique health, labor, crime, and educational statistics, among others. Their case-study questions are provocative.

5   Sampling

In this chapter, you will learn:

1. The reasons for sampling.
2. Common sampling terminology.
3. How to identify, construct, and interpret common probability and nonprobability samples.
4. The guidelines for determining the appropriate sample size.
5. How to evaluate the adequacy of samples.

Every 10 years, the United States government is required to conduct a *census* of the nation's population. A census attempts to count every person in the population. Although the Census Bureau has used sampling techniques for studies since 1937,[1] the decennial census is a complete count. For the 2000 Census, the Census Bureau proposed to interview a *sample* of the population—rather than carrying out a census—to improve the accuracy of the count and to reduce expenses. Whether you study groups of 100, 1,000, or 1,000,000, drawing a sample is an economical and effective way to learn about their members. This applies if your data come from individual respondents, case records, agencies, or computerized datasets. Understanding the principles of sampling would have helped most observers appreciate the Census Bureau's proposal; however, the proposal was controversial and was ultimately rejected by Congress.[2]

In fact, many people unfamiliar with sampling may misunderstand it, mistrust it, and feel that studies should include an entire population. This chapter explains the value of sampling and explains techniques to build a sample that is representative of a larger target population, whether that population is people, case records, organizations, or some other unit.

Consider some practical reasons for sampling. For many groups, it is impossible to identify every member. Thus, you cannot contact every member. And even when you can identify them, you often cannot realistically collect data on every unit of that population. The cost of doing so would likely be prohibitive. Contacting all members of a large or widely dispersed population can require tremendous commitments of time and money. Even with a relatively small group, sampling allows one to obtain information on members in a much shorter amount of time.

Sampling Basics and Terminology

Investigators draw *sample* units from a *population*—the complete set of people or other units whose attitudes and characteristics we would like to study—in order to generalize from those units to the other units in the population that they have not studied. Readers should recognize that this involves the concept of external validity. Generalizing the results of a study beyond the study itself is defined as external validity. Generalizing the findings from a sample of units to the population is another instance of external validity. Properly drawn probability samples allow us to generalize the measures of characteristics from a sample to a larger population.

Sampling is used in a variety of settings for a number of purposes. Perhaps its most widespread use is in survey research. Sample surveys are used to provide statistical data on a wide range of subjects for research and administrative purposes. A relatively small number of individuals are interviewed in order to gather data that will allow an investigator to find out something about the larger population. Given today's widespread use of surveys, people are often surprised to learn that the sample survey has a relatively short history.[3] Even at the beginning of the 20th century, statisticians debated whether anything less than investigation of a complete population was acceptable.[4] Sampling has since become widely accepted, and a number of techniques have been developed and refined.

Sampling involves several interrelated factors. These include the type of sample, its size, the population of interest, the accuracy desired, and the confidence the investigator wishes to have in the results. We will discuss these topics in this chapter. Before describing the methods and techniques of sampling, however, we need to introduce and define a number of terms.

Defining the Population

A *sample* is a subset of units selected from a larger set of the same units. They are the units studied and provide data for use in estimating the characteristics of the larger set. For example, polling organizations, such as the Gallup Poll, use samples of about 1,500 or fewer people to describe the opinions of over 200 million Americans.

The *population* is the total set of units in which the investigator is interested, that is, the larger set from which the sample is drawn. The population's characteristics and the relationships among these characteristics are inferred from the sample data. Investigators wish to generalize from the sample units studied to the entire population of units. A population may be composed of people, but it may also consist of units such as government organizations, households, businesses, records, or pieces of equipment such as police squad cars, and so on.

The *target population* must be specified clearly. The units in a population must conform to a set of specifications, such as "all adults living in Clark County on July 1," so that analysts will know who is considered part of the population and who is not. The interpretation of the results of a study will depend on how the population is defined. Consider a sample survey to assess support for a bond referendum to build a new coliseum in a city. How should the population be defined? Who should be included? Registered voters? Taxpayers? Only people living within the city limits? What should be the minimum age of the population? An investigator may find it useful to start by defining the population as the ideal one required to meet the study objectives. This could be called the target or theoretical population. The definition would then be modified to take account of practical limitations; the modified definition is the *study population*. The study population is a set of units that the investigator can access.

Drawing a Sample

Once the population has been defined, the question arises as to how to draw a sample from it. Many types of samples require a list of all the units in the population. The specific set of units from which the sample is actually drawn is the *sampling frame*. To sample households in a county, for example, one needs a list that identifies every household in the county. Complete lists seldom exist. Whatever list is used is the sampling frame. In the absence of a list, an equivalent procedure may be used to organize the population so a sample can be drawn.

Sampling frames may include units not defined as part of the population, or they may fail to include some members. Suppose a sample of households was drawn from a telephone directory. The directory may list businesses, which are not part of the population; it will not include households with unlisted numbers or without land lines. Other potential problems with sampling frames are that some of the listings may be groups of units rather than individual units, and some units may be listed more than once.[5] A list of building addresses, for example, may include apartment buildings containing many apartments. If an investigator is interested in individual dwelling units, many would be missed with this frame. We tend to think of sampling frames as being physical or electronic lists of units; however, this need not be the case. For example, you could place grids on a map and select a sample of the grids for further study. The sampling frame would be the gridded map.

Unit of Analysis and Sampling Unit

The term *unit of analysis* refers to the type of object whose characteristics interest us and whose characteristics we want to measure and study. If data are collected on fire departments, the unit of analysis is "fire department." If data are collected on fires, the unit of analysis is a "fire." Typically, investigators measure something about the unit of analysis. Frequently, administrative studies use agencies, not individuals, as the unit of analysis. A reader who fails to identify correctly the unit of analysis or uses results based on one unit of analysis to make conclusions about a different unit of analysis may reach incorrect conclusions about a study's findings.

A *sampling unit* is that unit or set of units considered for selection at a stage of sampling. The sampling unit may or may not be the same as the unit of analysis. In some sampling procedures, the sampling unit includes several units of analysis. For example, if analysts wanted to interview high school seniors, they might select a sample of high schools across the state and interview the seniors in those schools. The seniors would be the units of analysis, but the high schools would be the sampling units.

Parameters, Statistics, Errors, and Biases

A *parameter* is a characteristic of the population. This is the primary reason for conducting the research. It is what the investigators want to learn about. The actual, "true" percent of citizens in favor of a bond issue to finance a new municipal coliseum would be a *parameter*. However, such parameters are often unobservable. Instead, a *statistic*, a characteristic of a sample, is developed from information about the members of the sample. We use statistics to *estimate* parameter values. To return to the bond issue example, the percent of citizens in favor of the bond issue, *according to a probability sample of the relevant set of citizens*, would be a *statistic*. Typically, an investigator will take a sample statistic and estimate that the corresponding parameter of the population is within a certain range of the statistic. For example, the average cost of houses in a sample of residences in the state can form the basis for estimating the average cost of all houses in the state. The investigator does not assume, however, that the statistic exactly estimates the parameter value. For the price of houses in the state, she may say that the average cost of all houses in the state is probably within a certain number of dollars of the average cost of houses in the sample.

The difference between the population parameter and the statistic used to estimate it is the *sampling error*. It is the expected error in estimating a parameter for any given sample of a specified size. The *standard error* is a measure of the sampling error and is based on a theoretical sampling distribution. This theoretical sampling distribution is a distribution of the values of a statistic that would result if we were to take an infinite number of samples of the same size and plot those values and their frequency. The standard error is the standard deviation of that theoretical sampling distribution.

In discussing sampling, we distinguish between error and bias.[6] Not all error is "bias."[7] Sample bias is a systematic misrepresentation of the population by the sample. The misrepresentation usually comes about because of a flaw in the design or implementation of a sampling procedure.[8] Sampling error is treated as random, and its size can be estimated mathematically; nonsystematic sampling error is inherent to drawing a sample.[9] In contrast, sampling bias tends to be in one direction and, while difficult to measure, is not an inevitable consequence of drawing a sample.

Sampling Fractions

The *sampling fraction* is the percent of the population that is selected for the sample. A *sampling design* is the set of procedures for selecting the units from the population that are to be in the sample. Two major types of designs are *nonprobability* and *probability* sampling designs. The distinction between them is important. With a probability sample, each unit of the population has a known, nonzero chance of being in the sample. Probability samples allow us to avoid selection biases, use statistical theory to estimate population parameters from sample statistics, and evaluate the accuracy of these estimates. Nonprobability designs do not allow the researcher to calculate the probability that any unit in the population will be selected for the sample. From a nonprobability sample, we *cannot* generalize with any accuracy to the larger population of interest. In nonprobability sampling designs, other principles take precedence.

We discuss common probability and nonprobability sampling designs in the next section. Both types of designs are useful but typically are used for different purposes and in different situations. A weakness of nonprobability designs is that they allow for an element of subjectivity, thereby preventing the investigator from using statistical theory properly. However, these designs are usually more convenient than probability designs and can be used to provide important information to administrators.

Applying the Terms

The following application illustrates some of the previously defined terms.

- *Population:* All motor vehicles owned in the state in the current fiscal year.
- *Sampling frame:* All vehicles appearing on the state list of Registered Motor Vehicles prepared July 1 of the current fiscal year by the Department of Motor Vehicles.
- *Sampling design:* Probability sampling.
- *Sample:* 300 motor vehicles randomly selected from the sampling frame.
- *Unit of analysis:* Motor vehicle.
- *Statistic:* Average distance passenger cars in the sample were driven annually: 20,000 miles.
- *Parameter:* The actual average annual mileage of all passenger cars in the state.

Probability Sampling Designs

With a probability sample, each unit in the population has some chance of being in the sample; that chance is greater than zero and can be calculated. Probability samples permit a precise estimate of parameters. If sample statistics are to be used to accurately estimate population characteristics, probability samples are required. Although the reader need not be familiar with the calculation and interpretation of probabilities, it is important to be familiar with the more common probability sampling designs and recognize the situations in which they are likely to be useful. Our purpose here is to describe these designs so that the reader can determine whether sample findings have been interpreted properly and can participate in discussions of alternative ways of selecting a sample.

We discuss four common probability sampling designs:

1. simple random sampling
2. systematic sampling
3. stratified random sampling
4. cluster sampling

In addition, we discuss multistage sampling, which can be thought of as a variation of cluster sampling. While other designs exist, they are often hybrids of these four designs. These designs demonstrate the basic principles of probability sampling.

Simple Random Sampling

Simple random sampling requires that each unit of the population have a known, equal, nonzero probability of being included in the sample. The selection of each unit is independent of the selection of any other unit. That is, the selection of one member of the population for the sample should not increase or decrease the probability that any other member of the population will also be chosen for the sample.[10] Two common methods of constructing a random sample are:

1. lottery method or
2. random number table or random number generator

Lottery method: These methods ensure against the inadvertent introduction of a pattern of systematic bias into the procedure. In the lottery method, numbered or named balls, each representing a unit in the population, are placed in a container. The balls are mixed thoroughly, and the number of balls equal to the sample size is removed. For a sample of 100, the investigator would remove 100 balls. The sample then consists of all units of the population corresponding to the selected balls.

While the lottery method is theoretically adequate, usually it is more convenient to use a random number table. To do this, the investigator assigns a number to each unit of the population and consults the table. He then randomly selects a starting place, goes through the table across the rows or down the columns, and lists the numbers as they appear on the table. Members of the population with the selected numbers constitute the sample. The investigator selects a random starting place in order to avoid any preference he may have for beginning at a particular starting spot.

A random number table is a list of numbers generated by a computer that has been programmed to yield a set of random numbers. Computers are used since the typical human is not capable of generating random numbers. For example, a person might have a tendency to list more even numbers or those ending only in 3 or 7. The numbers in a random number table have no particular sequence or pattern. The use of random number tables ensures that the selection will be random and not influenced by any selection bias of the investigator. Computer programs operating on the principle of a random number table for selecting cases are available. Many organizations have computerized records; such records constitute sampling frames, and computer routines can select a simple random sample from them. Even a spreadsheet can select simple random samples from records in a spreadsheet file.

In drawing the sample using a table of random numbers, it is possible to select a unit's number more than once. This possibility does not exist with a lottery method because, when the unit's ball was drawn, it was not replaced and so given another chance to be selected.

- The sampling method is known as *simple random sampling with replacement* if a selected unit, or its identifying number, is returned to the population and can be selected again.
- The sampling method is known as *simple random sampling* when the sampling procedures are done without replacement.

Although the selection of one unit changes the probability that other units will be selected, prior to any unit's selection, all units have the same probability, as do all combinations with the same number of units.[11] We will emphasize the method without replacement, as it is used more commonly. Example 5.1 illustrates an application of simple random sampling.

Example 5.1 An Application of Simple Random Sampling

Problem: County officials want to assess the extent of support for a proposed bond issue to fund construction of a new wing to a high school.

Population: All people eligible to vote in a referendum on the proposed bond issue.

Procedure:

1. Identify the population: The county's registered voters.
2. Identify and obtain a sampling frame: The County Board of Election's list of 40,000 registered voters. The list is on the board's computer, and names are numbered sequentially.
3. Determine sample size: 500.
4. Select a table of random numbers, pick a random beginning point, use a consistent pattern to go through the list, and list the first 500 five-digit numbers from 00001 to 40000.
5. Select each voter whose name appears next to a number selected from the table of random numbers.

Discussion: The sampling frame is an accurate list of members of the population, for example, citizens eligible to vote in a referendum on a bonding request. A simple random sample can be executed relatively easily. A computer-generated sample would be equally good and less time consuming. A computer program could have printed out a randomly selected sample of names using its own internally generated table of random numbers. Note that each listed voter had the same chance of being in the sample as every other listed voter, and the selection of any one voter did not affect the chance that any other voter on the list would be included.

Systematic Sampling

The use of the random number table to select a simple random sample is manageable but can be very tedious. It becomes more so with large populations. Although newer developments for selecting samples, such as random digit telephone dialing, can make the process easier and faster, they cannot be used in many situations.[12] *Systematic sampling* is a widely used alternative to simple random sampling that reduces the amount of effort required to draw a sample and usually provides adequate results. It requires a list of the population units.

To construct a systematic sample (see Figure 5.1), the investigator first divides the number of units in the sampling frame (N) by the number desired for the sample (n). The resulting number is called the *skip interval* (k). If the sampling frame consists of 50,000 units and a sample of 1,000 is desired, the skip interval equals 50 (50,000 divided by 1,000). Having determined the skip interval, the investigator then selects a random number, goes to the sampling frame, and uses the random number to select the first case. Next the investigator picks every kth unit for the sample. In this example, every 50th case following the starting place on the sampling frame would be chosen. If the random number 45 was selected, cases 45, 95, 145, and so on would be in the sample. Note that the sampling fraction in this example is 2 percent.

With systematic sampling, the list is treated as circular, so the last listed unit is followed by the first. It is important to go through the entire list that constitutes the sampling frame. If the items on the list occur in a regular pattern and the skip interval coincides with this pattern, the sample will be biased. For example, consider what could happen in sampling the daily activity logs of a

Investigator:
1. Determines appropriate sample size (*n*)
2. Locates list of population cases: the sampling frame
3. Finds number of cases on the sampling frame (*N*)
4. Calculates skip interval:

If *n* = 1,000 and *N* = 50,000,
Then:

$$\frac{\text{Sampling frame size } (N)}{\text{Desired sample size } (n)} = \text{Skip interval } (k) = 50$$

5. Selects a random number for a starting place
6. Goes to the sampling frame and selects first case
7. Picks every *k*th case. For example, if 45 was first case, then cases 95, 145, and so on would be selected until 1,000 case were selected.

Figure 5.1 Steps in Constructing a Systematic Sample

sheriff's department. If the skip interval were 7, the activity logs in the sample would all be for the same day, that is, all Mondays or all Tuesdays, and so forth. A skip interval of 14, 21, or any other multiple of 7 would have the same result. Experienced law enforcement officials tell us that certain days consistently have more activity. If the skip interval matched the cycle, the sample would not represent the population of days accurately.

Periodicity

One way to deal with this problem, known as *periodicity*, is to perform the procedure twice, first doubling the skip interval. In our example, we would make 100 the skip interval, randomly select a starting point, and go through the sampling frame once. This provides one-half of the sample. We would then pick another starting point and go through the frame a second time, providing the second half of the sample. Another way is to "mix up" the list before selecting the sample, although this may not be practical. The available evidence indicates that periodicity problems are relatively rare in systematic samples.[13]

Applying Systematic Sampling

In order to sample a population whose size is unknown, for example, people attending a community event or clients attending a clinic, systematic sampling may be the only feasible type of probability sample. In such cases, one can estimate the probable population size, that is, the number of visitors or patients, determine a skip interval, say 50, and pick a random beginning point, say 6. You would then sample the 6th person to arrive (or depart), the 56th person, the 106th person, and so on, until the end of the sampling period. Example 5.2 illustrates an application of systematic sampling.

Example 5.2 An Application of Systematic Sampling

Problem: The director of a county Women's Commission wishes to compile a summary describing the characteristics of registrants in the county's Job Bank. The Job Bank has been operating for 10 years.

Procedure:

1. Identify the population: 6,500 women registered with the county Job Bank.
2. Determine the sampling frame: The set of file folders, one for each registrant, kept for the 10 years of the Job Bank.
3. Decide on sample size: 500.
4. Calculate skip interval: $k = 6{,}500/500 = 13$.
5. Select a random number between 1 and 13 for a starting place. The number 7 is selected.
6. Pull files from the filing cabinets for the sample. Start by selecting the seventh file, and continue by selecting files number 20, 33, 46, and so forth, through the entire set of files.
7. The 500 files selected constitute the sample for the study.

Discussion: A systematic sample works well where files are kept. These files were kept in chronological order, providing a useful sampling frame. To go through and number each file and then match them with numbers chosen from a random number table would be tedious. The investigator has no reason to suspect periodicity, so she may pick all members of the sample by going through the sampling frame once. Note: If the skip interval is not a whole number, you may round either up or down. The most important aspects of the procedure are to pick at random a starting place and to be sure to go through the entire list or set of records.

Limitations of Systematic Sampling

Systematic sampling does not result in a truly random sample. Although each unit in the population has the same chance of being selected, the selection of one unit affects the probability of selection of other units. Unless the researcher goes through the sampling frame more than once, there is no possibility that adjacent units within the skip interval will be selected. For instance, in the previous examples, the probability that both units 1 and 2 will be in the sample is zero. There is some nonzero chance that both units 1 and 50 will be in the sample.

In order to apply sampling statistics to systematic samples, the investigator must make certain assumptions about the sampling frame. The major assumption is that the units on the list are randomly ordered, or at least approximately so, with respect to variables under investigation. If this is the case, the systematic sample can be treated as a simple random sample. Lists arranged in alphabetical order can be treated this way. Systematic sampling is used widely and works well in practice. You need not be overly concerned about the problems noted here, although it is wise to be aware of them. Systematic sampling is also called quasi-random sampling.

Stratified Random Sampling

Stratified random sampling ensures that a sample adequately represents selected groups in the population. Analysts often use stratified sampling if a group of particular interest is a relatively small proportion of the population or if they plan to compare groups. This technique assumes some knowledge of the population characteristics. Fortunately, we usually know certain things about the population being studied. Information on demographic characteristics, such as the percent in various racial groups, ages, percent employed in manufacturing, and so on, often is available.

The first step in drawing a stratified random sample is to divide or classify the population into strata, or groups, on the basis of some common characteristics such as gender, race, or institutional affiliation such as school or agency. The classification should be done so that every member of the population is found in one and only one stratum. Separate probability samples are then drawn from each stratum.

For example, an investigator studying management information systems wanted to compare public with private sector organizations and designed a stratified sampling procedure. He developed two sampling frames, one of public sector agencies and the other of private organizations, and drew separate samples from each.[14] The benefits of stratification derive from the fact that the researcher determines the number of units selected from each stratum. Homogenous populations produce samples with smaller sampling errors than do populations with more diversity. Stratification reduces the diversity in the strata. Hence, stratified samples provide for greater accuracy than simple random samples of the same size.[15]

We will discuss two types of stratified sampling, proportionate and disproportionate.

Proportionate Stratified Sampling

In *proportionate stratified sampling*, members of a population are classified into strata, and the number of units selected from each stratum is directly proportional to the size of the population in that stratum. In other words, the sampling fraction is the same for each stratum.

If, for example, an investigator wanted to compare three types of workers—professional staff, technical staff, and administrative staff—she would begin by designating each of these types as a stratum. She would then draw her samples by taking an equal percentage of members from each stratum, say 10 percent of the professional staff, 10 percent of the technical staff, and 10 percent of the clerical staff. The resulting sample would consist of three strata, each equal in size to the stratum's proportion of the total population. A simple random procedure or some other method of probability sampling is used to draw the actual sample from each of the strata.

Applying Proportionate Stratified Sampling

Example 5.3 Illustrates an application of proportionate stratified sampling.

Example 5.3 An Application of Proportionate Stratified Sampling

Problem: The director of recruitment for the State Law Enforcement Service wanted information about applicants to the service over the previous four years.

Population and sampling frame: All applicants to the State Law Enforcement Service for the past four years.

Procedure: Construct a proportionate stratified sample, using the year of application as the strata. Select a random sample of applications from the total group for each year. The same percentage of each year's total is selected. See the accompanying tabulation in Table 5.1

Table 5.1 Constant Sampling Fraction

Stratum	Number of Applications	Sampling Fraction	Number in Sample
Year One	400	15%	60
Year Two	350	15%	53
Year Three	275	15%	41
Year Four	250	15%	38
Total	1,275		192

Discussion: The analyst used his knowledge of the population to stratify along yearly groupings and took a 15 percent sample within each subgroup. The information from each year's group can be compared or combined into one sample.

Disproportionate Stratified Sampling

The proportion of the members of a stratum selected for the sample need not be the same for each of the strata in the population. In *disproportionate stratified sampling*, a larger share is taken from some strata than from others. This is a useful technique when a characteristic of interest occurs infrequently in the population, making it likely that a simple random sample or a proportionate sample may have too few members with the characteristic to allow full analysis.[16] It also is useful when the sizes of important subgroups in the population differ greatly. For example, the State Division of Human Resources wanted to have a sample of employees in all departments of state government. The largest department had over 18,000 employees, and the smallest had 50. In disproportionate stratified sampling, the investigator selects a larger percentage of members from groups likely to be underrepresented in the population. In this fashion, the investigator assumes that the number of units included in each stratum of the sample will be large enough to allow for separate analysis of each individual stratum. Since many research projects require sample estimates not just for the total population but also for various subgroups within the population, this is often a useful technique. Note that the samples from each stratum constitute subsamples that can be analyzed separately. Since a higher percent of the total in some strata than from others have been selected—they have been oversampled—the analyst must weight the results to compensate for oversampling before combining the subsamples to form one sample. To weight the subsamples correctly, the analyst must know the size of each stratum.

Applying Disproportionate Stratified Sampling

Example 5.4 illustrates a disproportionate stratified sample and a weighting procedure. One disadvantage of disproportionate stratified sampling is that it may require a larger sample to achieve the same level of accuracy as simple random or proportionate stratified sampling.

Example 5.4 An Application of Disproportionate Stratified Sampling

Problem: Identify citizens' transportation needs for a county transportation plan.
Population: The county's 27,500 residents.
Sampling frame: List of households from county tax records.
Procedure:

1. Stratify the population by the county's three townships: Tax records are filed by township, and each township differs in its demographic makeup.

2. Note the number of households in each township.
3. Determine sample size: Approximately 500.
4. Decide between proportionate and disproportionate sampling: Note that a 5 percent sampling fraction would yield a 480-household sample, with only 26 households from Southwest Township. Because of the importance of location in developing the plan, analysts decide to oversample Southwest Township.
5. Individual households for the sample are randomly selected from the tax records for each township. One resident in each household is interviewed.

See the accompanying tabulation (Table 5.2).

Table 5.2 Illustrating Unequal Sampling Fractions

Stratum	Number of Households	Proportion of Total	Sampling Fraction	Sample Size
North Township	5,760	.60	4%	230
Southeast Township	3,330	.35	4%	133
Southwest Township	510	.05	10%	51
Total	9,600			414

Discussion: Normally, the three groups would be analyzed separately and compared to each other. If the planners wanted to combine the subsamples into one sample, they would weight the subsamples. Specifically, the results of each subsample would be weighted by the proportion of the total population that that strata constituted. The results would then be combined. For example, the average age of residents in North Township would be multiplied by that group's corresponding proportion in the total, .60, Southeast Township by .35, and Southwest Township by .05 to determine the average age of the total population.

Comparing Proportionate and Disproportionate Stratified Sampling

Next is a visual illustration of differences between proportionate stratified sampling and disproportionate stratified sampling using the example of professional, technical, and administrative staff noted earlier. Suppose that an organization consists of 1,700 staff members. Of these 1,700, 200 are professional staff, 1,000 are technical staff, and 500 are administrative staff. Figure 5.2 represents the shares of staff in the organization.[17]

If the organization wants to draw a sample of its employees, it might decide to sample 10 percent of the employees of each type. This would be an example of *proportionate stratified sampling*. The last column of Table 5.3 reports on the number of employees of each type that will appear in the resulting sample. Figure 5.3 represents the breakdown of staff in this sample. Notice that it is identical to Figure 5.2.

Notice that in the sample that results from proportionate stratified sampling, there are only 20 professional staff. If the organization wanted to study the responses of the professional staff in detail, more cases would be needed. The organization could decide to draw larger, equal proportions of staff from each stratum. Or the organization could "oversample" professional staff by drawing a larger share of staff from this stratum than from the others. For instance, the organization could sample 25 percent of the professional staff members while continuing to sample 10 percent of technical staff members and 10 percent of administrative staff members. This would create a disproportionate stratified sample. The last column of Table 5.4 reports on the number

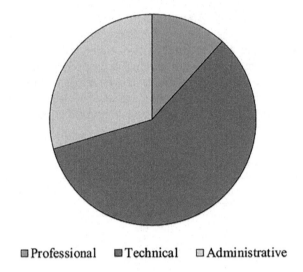

Professional   Technical   Administrative

Figure 5.2 Share of Each Staff Type in the Organization

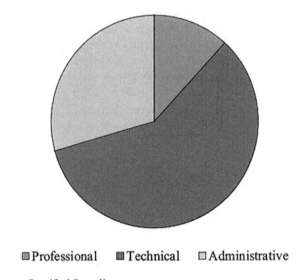

Professional   Technical   Administrative

Figure 5.3 Proportionate Stratified Sampling

Table 5.3 Share of Staff, Population, and Sample

Staff Type	Number in Organization	Proportion Sampled	Number in Sample
Professional	200	0.10	20
Technical	1,000	0.10	100
Administrative	500	0.10	50

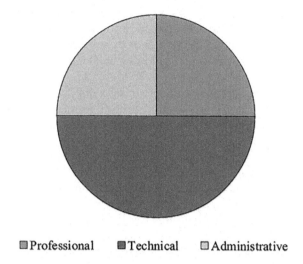

 Professional ■ Technical □ Administrative

Figure 5.4 Disproportionate Stratified Sampling

Table 5.4 Population and Sample, Disproportionate Sample

Staff Type	Number in Organization	Proportion Sampled	Number in Sample
Professional	200	0.25	50
Technical	1,000	0.10	100
Administrative	500	0.10	50

of employees of each type that would appear in the resulting sample. Figure 5.4 represents the breakdown of staff in this sample. Notice that this pie chart does *not* resemble Figure 5.2. If the organization wanted to use this sample to obtain information representative of the organization as a whole, it could use weights to do so. With 50 professional staff in this sample, however, the organization would also have more cases available to analyze this subset of staff members on their own.

Cluster and Multistage Sampling

In many sampling problems, the units of the population exist in groups or *clusters*. For example, if we wanted to survey state residents on their use of certain state services, the unit of analysis would be individual residents. But they could be grouped in clusters in various ways. Each county, city, planning district, or even city block would contain a cluster of residents. We could also view the city block as a unit, in which case each city would have a cluster of city blocks, and each city block would have a cluster of residents. In cluster sampling, we select a sample of groups from the sampling frame and obtain information on all the individual units in the groups selected. If a sample of units is taken from each selected cluster, the design is called a *multistage cluster sample*.

Drawing a sample of a population, such as state residents, can be difficult. You may lack a sampling frame of the population, or logistic problems may preclude drawing a sample directly from the sampling frame. A more efficient approach is to design a multistage cluster sample.

Assume that investigators wanted to obtain data on the use and need for mental health facilities in a five-state region. They might design the following multistage cluster sample.

1. Stage One: The investigator takes a sample of large units containing clusters of smaller units. In this case, counties would be a likely choice for the large units.
2. Stage Two: A sample of smaller areas—townships—is selected from the previously chosen counties.
3. Stage Three: A sample of yet smaller geographic areas, designated by population size, is selected from the townships chosen in Stage Two. Census tracts might be used.
4. Stage Four: Select a random sample of dwelling units—houses, condominiums, and apartments—from the census tracts chosen in Stage Three.
5. Stage Five: At this stage, the investigators would go to these selected dwelling units and interview a resident. Of course, they would need some procedure, established in advance, for choosing which resident to interview, since many of the dwelling places will have more than one adult resident.[18]

The units selected at each stage are called *sampling units*. The sampling unit may not be the same as the unit of analysis. In this example, the unit of analysis was the resident. However, different sampling units were selected at each stage of the process. Figures 5.5a–c illustrate cluster and multistage sampling and the differences between them.[19]

Figure 5.5a Illustration of Cluster and Multistage Sampling

Clusters: Note that researchers are often initially unable to identify the units inside of each of the 10 clusters.

Figure 5.5b Illustration of Cluster and Multistage Sampling

Cluster Sampling: Clusters in dark gray are sampled. Clusters in grey are not part of the sample. People (units) within the sampled (dark gray) clusters are identified. Then, all people inside each sampled cluster are interviewed.

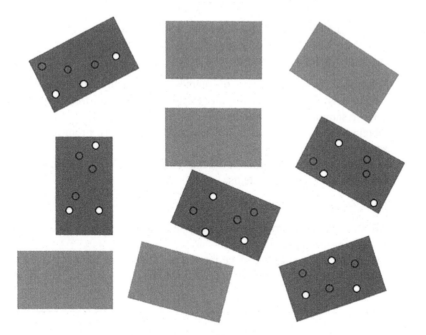

Figure 5.5c Illustration of Cluster and Multistage Sampling

Multistage Cluster Sampling: Clusters in dark gray are sampled. Clusters in grey are not part of the sample. People (dots) within the sampled (dark gray) clusters are identified. Then, a sample of one-half of the people within each sampled cluster (designated as shaded dots) are contacted for interviews.

When to Use Cluster Sampling

Cluster sampling is recommended for studies involving a large geographic area. Without the ability to limit the sample to discrete areas, that is, to selected clusters, the costs and logistics would make probability sampling of many widely dispersed populations difficult, if not impossible. If face-to-face interviewing is to be used to collect data, a sample spread thinly over a large area will be extremely expensive to reach. Cluster sampling can help reduce this cost.

Cluster sampling helps to compensate for the lack of a sampling frame for the units of analysis. Using cluster sampling in combination with other sample designs allows the investigator to proceed by only developing a sampling frame to draw the specific sample members. In the earlier description, for example, the investigators did not develop their sampling frame until Step Four, when they listed the addresses of all dwelling units in the selected census tracts. They chose their sample from this list. It would have been very costly and cumbersome if they had begun the process by attempting to find or develop a sampling frame that included addresses of all residents in this five-state region.

Although cluster sampling reduces travel time and costs, it requires a larger sample than other methods for the same level of accuracy. Note that in the multistage process, probability samples are selected at each stage. Each time a sample is selected, there will be some sampling error. Since the sampling error accumulates at each stage, a cluster sample requires a larger sample than the other probability samples. Example 5.5 illustrates a four-stage cluster sample. The research staff applied random sampling techniques four times to select the members of the sample.[20]

Example 5.5 An Application of Multistage Cluster Sampling

Problem: A state health agency needs to obtain information from high school students living in rural areas.
Population: All high school students attending school in rural areas of the state.
Sampling frames: List of counties, high schools, classes, and students.
Procedure: Construct a five-stage cluster sample.

1. Select a simple random sample of counties from all state counties considered rural.
2. Obtain lists of high schools in all selected counties and, from these, randomly select a sample of high schools.
3. Obtain a list of all classes in the selected high schools. From these, randomly select a sample of classes.
4. From the class lists for the selected classes, randomly select a sample of students.
5. Interview the selected students.

Justification for sampling technique:

1. No obtainable sampling frame of all students attending high school in rural areas existed for the entire state.
2. The study required face-to-face interviews. Cluster sampling allowed research staff to concentrate interviews in fewer locations, thus avoiding costs of traveling all over the state.

Addressing Clusters With Different-Sized Populations

What happens if the initial clusters to be sampled have very different-sized populations? If one of the clusters contains a significant proportion of the population, a multistage sample could easily miss this cluster and hence a large part of the population. To adjust for this situation, investigators sometimes use a technique called "probability proportional to size" or PPS. A typical application of cluster sampling is a study in which the investigator wants to obtain a statewide representative sample of adult residents. If an accurate list of the number of adults in each county is available, multistage cluster sampling would be appropriate. A probability sample of 10 counties could be the first stage, with a sample of residents selected from these counties for the second stage. However, the investigator would want each adult in the state to have an equal chance of being chosen. If one or a very few large counties contain a significant proportion of the state's population, as is the case in many states, the investigator needs to adjust for this. The adjustment must be done in stage one, or the probability of missing the largest county and much of the population would be high.

The investigator must weigh each county by its population. If random numbers are assigned to the counties for the first-stage selection, they would be assigned to the counties in proportion to their population: more numbers assigned to the more populous counties than to the less populous ones. The larger counties would then have a greater chance of being selected than the smaller ones. Weighting would ensure that a resident of a large county has the same chance to be selected as a resident of a small county.[21]

A sample taken from a larger group of units, which is itself a sample, is legitimate. For example, students at research universities may have access to databases containing survey responses from a large sample of individuals. Someone wanting to do research on a much smaller sample could instruct the computer to select a random sample to use. This, in effect, would constitute another type of multistage sample.

As noted early in the chapter, the sampling unit can be almost anything. One of the authors supervised a project in which the student researcher investigated newspaper coverage of a county's Work First program.[22] The student identified the news story as the unit of analysis and as the sampling unit. She used a stratified sampling design to select her sample. All local newspapers had been filed by month, so the student calculated a sampling fraction and selected the proper number of newspapers from each month.

Probability sampling designs, when used in conjunction with sound statistical methods, allow us to generalize from the sample to the population. The distinguishing characteristic of probability samples is the random selection of the units for the sample. Randomization eliminates biases that may affect the selection of units. It also allows for the correct use of sampling statistics. The principles underlying simple random sampling form the basis for other probability designs. Although other designs are widely used, they are attempts to provide a close approximation to simple random sampling when simple random sampling is either difficult or impossible.

Nonprobability Sampling Designs

Investigators also use a variety of nonprobability sampling designs. Nonprobability samples can work well for many types of studies, particularly exploratory studies and those used to generate hypotheses to be more fully tested by further research. The usefulness of the information contained in a nonprobability sample depends on the investigator's purpose, the criteria for selecting the units for the sample, and how well those units seem to represent the population of interest. These samples are also useful if it is not important to obtain accurate estimates of population characteristics. A visit to one or two homeless shelters, for example, can provide useful information to housing administrators seeking a change in policy governing the provision of services to the shelters. Since nonprobability samples do not rely on random selection, however, the results cannot be generalized to a larger population with any accuracy.

And it thus follows that one also *cannot estimate parameters from sample statistics*. Obviously, investigators attempt to do this in some subjective fashion. However, with nonprobability samples, statistical theory cannot be applied to make these estimates or to evaluate their accuracy. The adequacy of the nonprobability sample can be evaluated only by subjective means; no mathematical evaluation is possible. With nonprobability samples, one cannot determine the chance that any unit in the population will be selected for the sample. It is therefore impossible to determine, mathematically, how representative the sample is of the population.

Convenience Sampling

The various types of nonprobability samples are named according to the primary criterion used in drawing the sample. *Convenience sampling* involves sampling available units. For example, a researcher may ask a classroom of students to fill out an opinion survey. In so doing, he may hope to find out something about a larger population, such as American youth, but in order to do so, he has to assume that his sample is in fact representative, as there is no way to demonstrate this. Convenience samples are inappropriate for generalizing with any degree of certainty. They can, however, provide illustrative case material or serve as the basis for exploratory studies. For example, one could use the survey of the classroom to determine whether any items on the questionnaire were difficult for the students to understand.

Examples of convenience sampling include volunteers for a study, interviews conducted in convenient locations such as shopping centers, the respondents to a questionnaire in a newspaper, and telephone calls to a radio talk show. Convenience samples also are known as accidental samples.[23] Although the risk of bias for convenience samples is high, some will be worse than others. Particularly troublesome are those for which subjects select themselves for the sample and in which we would expect only people with strong views to respond. Because of the high risk of bias with convenience samples, it is unwise to use them to make inferences about general populations. Nevertheless, if the purpose is to identify issues of potential concern to a larger population, to pretest forms to be used by an agency, or some similar aim, convenience samples may be economical and appropriate.

Purposive Sampling

A second type of nonprobability sampling is *purposive sampling*.[24] The main criterion for selection of any unit from the population using this sampling procedure is the investigator's judgment that the unit somehow represents the population. Because of this, it is also known as judgment sampling. The probability that any unit will be selected is unknown because it depends entirely on the judgment of the investigator. Often units for this type of sample are selected on the basis of known characteristics that seem to represent the population. The investigator assumes that the units selected will represent the population on unknown characteristics as well. But even after collecting the data, an investigator cannot verify the representativeness of the sample on the unknown characteristics. He must therefore be skeptical about the accuracy of estimates.

Local governments use purposive samples when they seek information from cities and counties with a reputation for excellent administration. They may interview the personnel of these governments to ask about their experiences with outsourcing services, what performance measures they use, and how they monitor citizen satisfaction. Similarly, investigators may study school systems that have had the most success and those that have experienced failure in implementing charter schools. Investigators may also be interested in unusual cases, that is, those differing the most from others, and so select them for study. *Expert sampling* is a type of purposive sampling that involves selecting persons with known experience or expertise in an area. A panel of experts is selected to provide the views of persons with specific knowledge and skills.[25] Example 5.6 illustrates the use of purposive sampling.

Example 5.6 An Application of Purposive Sampling

Problem: A county manager plans to purchase a new computer system and wants to find out what experiences similar counties have had with computer systems, specifically their applications, costs, installation, and maintenance problems.

Population: All counties similar to that of the manager.

Procedure: The manager decides to talk at length with personnel in counties similar to hers that have computerized administrative activities.

1. Based on her knowledge of the area, the manager decides to contact five counties similar to her own in population size and makeup, budget, and services provided.
2. County managers and administrative staff in the five counties are interviewed and asked:

   a. What equipment is owned, leased, or shared?
   b. What alternatives were examined? Why was the existing system chosen?
   c. What does the computer system do?
   d. How well does the system perform? How often does it break down?

Sample characteristics: The representativeness of the sampled counties' population size and makeup, budget, and services are known. The representativeness of their computer-related decisions are not known and cannot be determined by this sampling technique.

Discussion: The manager, familiar with her county and its operation, the counties contacted, and the reputation of possible vendors, may reasonably believe that a limited nonprobability sample will provide adequate information for a good decision.

Quota Sampling

A third type of nonprobability sampling is *quota sampling*. In this technique, the investigator attempts to structure a sample that is a cross-section of the population. It is less costly and easier to administer than a comparable probability sample. Particularly with the decline in response to telephone surveys, the use of online quota samples, known as online panel surveys, has become much more common. Quota samples and online panel samples are often much less costly than random samples and can quickly generate a high number of responses. However, these samples also have limitations that are worth considering.

To conduct a quota sample, the researcher attempts to select a quota of individual units with defined characteristics in the same proportion as they exist in the population. If she thinks that gender and age are important characteristics, the population is 55 percent male, and 35 percent are over 40 years old, she will require that 55 percent of the units in the sample be men. She will also require that 35 percent of the sample be over 40. The characteristics used to guide the selection of units for the sample may be independent, or they may be interrelated.

For example, a researcher seeking to interview members for a quota sample of agency employees may be assigned the following quotas:

- 10 males under the age of 40, 5 males over the age of 40
- 6 women who have been with the agency for 5 years or less
- 7 women who have been with the agency for more than 5 years

Quota sampling is sometimes used in surveys in which the interviewer selects the specific individuals to be interviewed. Quota sampling has the advantage of not requiring a sampling frame, and it reduces the need for callbacks. With quota sampling, if an eligible person is unavailable when the interviewer calls, the interviewer simply proceeds to the next dwelling or the next eligible respondent.

One purpose in assigning quotas is to reduce the risk of selection biases that might result from giving the interviewer a totally free hand. The biases of the interviewer can reduce the representativeness of a sample, since, given a choice, interviewers are likely to interview people with whom they feel most comfortable, who are convenient, or who are most willing to be interviewed. Quota samples cannot overcome these biases, however, and as a result are likely either to overrepresent or underrepresent traits not specified in the quotas. It is difficult in practice to emphasize more than a few traits when engaging in quota sampling. For example, in 2016, the *LA Times*/U.S.C. Dornsife poll broke out its panel into very small segments that included 18–21-year-olds and race. Particularly underrepresented groups such as 18–21-year-old African American males were weighted by as much as 30 times. Because there was one 19-year-old African American man in Illinois who repeatedly said that he was going to vote for Donald Trump, he had disproportionate sway on the poll, which sometimes caused the predicted percent voting for Trump to swing by double digits within the category and even caused the entire poll to shift by as much as 1%.[26]

For these reasons, stratified sampling is far preferable to quota sampling if it is important that the sample contain an adequate proportion of members of identifiable groups. At one time, quota sampling was used widely and is returning because of problems with conducting traditional surveys. Although it may appear to be comparable to stratified sampling, it is not. In stratified sampling, the various strata, or subgroups, are identified in advance, and sampling units are selected randomly from each strata.

Snowball Sampling

Snowball or referral sampling is used when members of a population cannot be located easily by other methods and where the members of a population know or are aware of each other. We may want to sample members of professional groups who form informal networks or other elites. Or we may want to sample very small populations who are not easily distinguishable from the general population or who do not want to be identified, for example, drug users. In snowball sampling, each member of the population who is located is asked for names and addresses of other members. A bias in snowball sampling is that the more times a given person is mentioned, the more likely that person is to be included in the sample. Those who are most well known may be least typical of the population.

Nonprobability sampling designs have the advantage of being easier, quicker, and usually cheaper than probability sampling. If the purpose is to make accurate generalizations to a larger population, then probability sampling is necessary. If, however, the purpose is to undertake some exploratory investigation, then nonprobability sampling is likely to suffice. If the sample is going to be very small, then a nonprobability sample is likely to be as accurate as a probability sample, and the added inconvenience and cost of a probability sample are not warranted. Here again, this is likely to be the case at the exploratory stages of an investigation.

Sample Size and Related Issues

One of the first questions administrators ask about sampling is about appropriate sample size. Those unfamiliar with sampling procedures and theory often assume that a major determinant of sample size is the size of the entire population. But to determine the appropriate sample size, the sampling expert needs to know other information.

Our intuition tells us that larger samples are likely to give better estimates of population parameters. Generally, this is correct. However, additional units also bring additional expense, and increasing the size of the sample beyond a certain point results in very little improvement in our ability to generalize about the population. Thus, in deciding on a sample size, the administrator must balance the need for accuracy against the need to keep costs reasonable. Further, with very large samples, the quality of the data actually may decrease—for example, when a small staff must supervise the interviewing of a large number of people. With a large sample, staff will find it difficult to make many callbacks when respondents are difficult to contact. Each additional case also increases the opportunity for errors in transcribing data.

It is important to note that the following discussion refers only to probability sampling. All of the factors to be discussed here are relevant only when one proposes to use the sample data to estimate population characteristics. Since this cannot be done with nonprobability sampling, the determination of sample size in such cases must be governed by other considerations. In nonprobability sampling, for example, taking a larger sample will not eliminate bias; it may, in fact, make it worse.

Investigators expect that a statistic will not estimate the parameter exactly. Each possible estimate is likely to be somewhat off its mark. However, if numerous samples were selected, an investigator would expect most of the estimates to be within a specified range of the parameter; otherwise, the estimate would be of little value. The larger the samples, the closer we would expect these estimates to be. This difference between sample estimate and population parameter is called sampling error. It is one of the important factors involved in determining sample size.

Sampling and Nonsampling Errors

Both *sampling* and *nonsampling errors* contribute to the difference between the value of the statistic and the parameter that it estimates. Sampling error occurs because sampling has taken place; the difference between a parameter and its estimate is random and due to the probability of selecting one unit rather than another.[27] Because of sampling error, every possible sample is as likely to underestimate as it is to overestimate the value of the parameter. Nonsampling error, also called *bias*, is usually due to some flaw in the design or in the way the sampling design is implemented. If nonsampling error is present, all samples selected in the same way will either underestimate or overestimate the parameter. The sources of nonsampling error are discussed later in this chapter.

The size of the sampling error affects the accuracy of the estimate. Since the investigator usually does not know the exact value of the parameter—if he did, he would probably not need to draw a sample—the size of the sampling error must be estimated. This is done from the sample itself. The formula for doing so is illustrated in the "Calculation Examples for Sampling Error and Sample Means" section at the end of the chapter.

Sample Size and Sampling Error

The appropriate sample size depends on how large a sampling error an investigator is willing to accept, since accuracy is an important factor in determining sample size. Greater accuracy usually can be obtained by taking larger samples or by accepting less restrictive values for two other factors important in determining the sample size. These are the confidence the investigator wishes to have in the results and the variability within the population from which the sample is selected. Both are related to accuracy.

Determining Sample Size:
    The formula for calculating the necessary sample size for a study uses sampling error, confidence level, and variability.
    Proportions

The formula shown subsequently is specific for a dichotomous variable, that is, a variable with only two values, such as a respondent being either *for* or *against* a bond issue. The *p* would usually refer to the proportion "in favor of" or "for."

$$\sqrt{n} = \sqrt{p(1-p)} \times (z\text{-score for confidence level}) / \text{accuracy}$$

where:

n = sample size. (Note: A capital *N* often is used to indicate population size. The factor to adjust for population size is not included here.)

p = proportion of population in one category of a dichotomous variable. The term $\sqrt{p(1-p)}$ is the standard deviation for proportions and measures the population variability.

z-score = standard score corresponding to the appropriate confidence level. This is taken from the theoretical distribution of all sample statistics. For a 95 percent confidence level, the *z*-score would be 1.96, or approximately 2.

A similar formula is used to calculate sample sizes for studies with interval variables as well. Examples later in the chapter illustrate using the formula for sample size for nominal and interval variables. The examples will show the calculation for sample size when the variability in the population is known from previous research or is assumed. Researchers often use a pretest or pilot test survey to obtain an estimate of the population variability. Calculations for determining sample sizes for means are illustrated with examples in the last section of this chapter prior to the Summary.

The properties of sample statistics may be derived theoretically by considering what the results would be if all possible samples of a given size were drawn. A certain percentage of these possible samples would estimate the population parameter very closely, some would be less accurate, and a few would be very inaccurate. Figure 5.6 illustrates possible outcomes if several samples were drawn from the population.

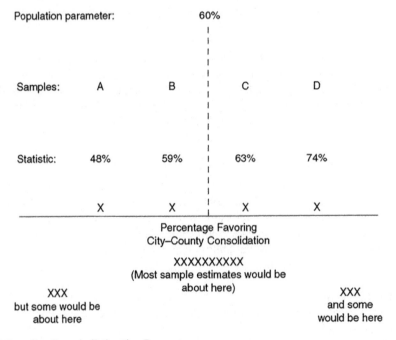

Figure 5.6 Sampling Error in Estimating Parameters

The parameter is 60 percent, and the statistics from four samples are shown. If a large number of samples were selected, the majority of their estimates would be near the parameter. However, as the figure shows, a few would greatly underestimate and some would greatly overestimate the value of the parameter. An investigator, though, draws only one sample. There is always a risk that any single sample will be one of those whose estimate of the parameter is not very accurate.

Confidence Levels, Accuracy, Population Variability, and Sample Size

The *confidence level* refers to the confidence the investigator has that the selected sample is one that estimates the population parameter to within an acceptable range. This confidence usually is expressed as the probability that a parameter lies within this range of the sample statistic. The range is called the *confidence interval* and is usually expressed in terms of units of the standard error. The higher the degree of confidence desired by the investigator, the larger the required sample size if no other factors are changed.[28]

Confidence levels usually are discussed in terms of how confident we are of our estimates. When investigators say that they are 95 percent confident that the parameter falls within a certain range of the statistic, this means that 95 times out of 100, using the same sampling procedure, the population parameter will be within that range of the sample statistic. Conversely, 5 times out of 100, the population parameter will be outside the range specified.

Accuracy of a Sample

The *accuracy* of a sample is how close the sample statistic is to the population parameter. A measure of accuracy is the standard error. If the standard error is small, then the sample estimates based on that sample size will tend to be similar and will be close to the population parameter. If the standard error is large, then the sample estimates will tend to be different, and many will not be close to the population parameter.

If we construct a confidence interval[29] of −1.96 to +1.96 standard errors around the sample statistic, we can use probability theory to state our level of confidence that the interval contains the population parameter. That is, we are 95 percent confident that the sample statistic falls within −1.96 to +1.96 standard errors of the population parameter. There will be a 5 percent chance that we are wrong, that is, that the interval does not contain the population parameter. The confidence level is the extent to which we are confident that it does.[30] Table 5.5 illustrates that as the confidence interval is narrowed, that is, the accuracy desired increases, the investigator's confidence that the interval captures that parameter decreases. Note that the parameter in the population remains the same; it is the confidence interval that changes depending on the confidence level selected and on sample size.[31]

If the investigator is not willing to accept a 5 percent risk of being incorrect, he can use a different confidence interval. The chance that the parameter will be within −2.58 to +2.58 standard errors of the sample estimate is 99 out of 100. But note that the sample estimate will not be as accurate with the same size sample.

Table 5.5 The Trade-Off Between Confidence Levels and Accuracy

Confidence Level	Accuracy as Shown by Confidence Interval
99%	−2.58 − +2.58
95%	−1.96 − +1.96
90%	−1.65 − +1.65
50%	−.68 − +.68

Note: These figures assume that the sample size remains constant.

The investigator will be 99 percent confident that the estimate is no more than 2.58 standard errors (SEs) off its mark rather than no more than 1.96 standard errors off. The width of the confidence interval can be narrower if the researcher is willing to accept a larger risk of being unable to apply the results to the population. For the same size sample, the investigator can have more accuracy but will be less confident in the results. The researcher could use an interval of −.68 to +.68 standard errors (a very accurate estimate) and have only a 50 percent chance of being correct in assuming that the population parameter is within the interval. To be more accurate without running a larger risk, the investigator needs to take a larger sample.

Usually investigators will say that they want to be at least as accurate as some figure or within a certain number of standard errors of the parameter with their estimate and have a certain level of confidence in their results. That means that a certain size sample is necessary.

Interpreting Confidence Levels

The interpretation of confidence levels is difficult because it involves thinking not only about the actual sample but about other samples that could have been drawn. For a 95 percent confidence level, if 100 samples of the same size were selected, about 95 percent of them should produce confidence intervals that include the population parameter. The other 5 percent would not.

Example 5.7 illustrates these concepts. The information shows that members of a sample had an average annual salary of $16,000. The standard error of this estimate was $250. The confidence level indicates how likely it is that the population's average salaries are within 1, 2, or 3 standard errors of the sample's average salaries. For the 95 percent level of confidence, the population average would be within 1.96 standard errors of the sample average. Note that the standard error takes on the same unit of measure as the sample statistic; in this example, it is measured in dollars. The standard error goes in both directions; that is, the sample average may be below or above the true population average.

Example 5.7 Application of Sampling Statistics to a Probability Sample

Problem: State officials want to measure the average earnings of participants in a job training program after they complete the training program.

Population: All participants who have completed the job-training program.

Strategy: Construct a probability sample of 100 participants and contact members of the sample to learn each member's annual salary.

Finding: The average (mean) salary of the sampled participants is $16,000 yearly. The standard error is estimated from the sample data to be $250.

Interpretation: The state officials can be 95 percent confident that the average salary of all program participants is between $15,510 and $16,490 (1.96 standard errors below the mean and 1.96 standard errors above the mean).

Explanation: If all possible samples of 100 job-training participants were drawn, the average salaries found in 95 percent of them would fall between 1.96 standard errors below and 1.96 standard errors above the true average.

If the parameter is a proportion, such as the percent of trainees who had completed high school, the standard error is reported as a percent. If 60 percent of the sample had finished high school and the standard error of this estimate was calculated to be 1.5 percent, then an investigator could report that he was 95 percent confident that the span of 57 to 63 percent captures the true proportion of all trainees who had finished high school (within 1.96 standard errors on either side of 60 percent). The specific range depends on the confidence level selected. You would choose the 68 percent confidence level in order to be within 1 standard error of the parameter. To be within 3 standard errors, you would choose the 99 percent confidence level. The standard error of 1.96 corresponds to the 95 percent confidence level. Sometimes 1.96 is rounded to 2. You may be familiar with this measure of accuracy from seeing newspaper reports of the results of polling organizations. Typically, they report the accuracy for a 95 percent confidence level.

Population Variability and Sample Size

If all the units in a population were exactly the same, a sample size of one would be adequate to characterize that population! For example, if everyone in a city earned $45,000 a year, a sample of 1 would be enough to estimate average salary. However, the units within a population usually are somewhat different from each other.

Population variability is the extent to which members of a population differ from each other on the variables that we want to investigate. The greater this variability, the larger the sample size required to estimate population parameters. A common measure of population variability is a statistic called the *standard deviation*. This measure is used in the formula for determining sample size.

Sample size, then, is a function of the accuracy desired, the confidence level desired, the population variability, and, to a lesser extent, the size of the population. These factors are related to each other. The relationship of each to sample size, if the others remain unchanged, is as follows:

1. Accuracy: The greater the accuracy desired, the larger the sample needs to be. This also means that the smaller the sampling error that an investigator is willing to accept, the larger the sample needs to be.
2. Confidence level: The more confidence desired, the larger the sample required.
3. Population variability: The more diversity among the members of the population, the larger the sample size needed.
4. Population size: The larger the population, the larger the required sample size. However, the population size is only a concern if the population is very small. At a certain point, adding more cases to the sample does not add information.

One common misconception about sample size is that a sample must include some minimum proportion of the population. This implies that if the size of the population is larger, the sample size must be increased by a corresponding amount. *This is not the case.* As you might expect, however, the members of larger populations usually are more diverse than those of smaller populations. In this indirect way, population size affects sample size. Statisticians use the finite population correction (fpc) factor to adjust for population size. If the population is at least five times the size of the sample, however, this factor usually has little impact on the sample size and often is omitted from the formula for calculating sample size. For samples drawn from very small populations, however, the population correction factor is very important.[32]

The second example presented in the section on calculating the standard error illustrated this for small populations. As with calculating the standard error for samples from small populations, the finite population correction factor should be used in the formula for calculating the necessary sample size. The formula for sample size then becomes:

$$\sqrt{n} = (\text{Variablity}) \times (\text{reliablity factor}) / (\text{accuracy}) \times (\text{fpc})$$

Table 5.6 illustrates the sample sizes necessary for small populations for selected accuracy and confidence levels.

Table 5.6 Sample Size for Small Populations by Selected Accuracy and Confidence Levels

For 95% Confidence Level

Population Size

Accuracy Desired	100	500	1,000	5,000	10,000	20,000
±1%	n.a.	n.a.	n.a.	n.a.	*	*
±2%	n.a.	n.a.	n.a.	1,248	1,825	2,113
±3%	n.a.	n.a.	n.a.	839	953	1,010
±4%	n.a.	n.a.	*	528	564	582
±5%	n.a.	*	237	355	369	377
±10%	*	78	87	94	95	96
±15%	24	40	41	43	43	43
For 99% Confidence Level						
±1%	n.a.	n.a.	n.a.	n.a.	n.a.	*
±2%	n.a.	n.a.	n.a.	*	2,427	3,287
±3%	n.a.	n.a.	n.a.	1,163	1,503	1,673
±4%	n.a.	n.a.	n.a.	822	930	983
±5%	n.a.	n.a.	*	575	619	641
±10%	n.a.	111	139	161	163	165
±15%	*	63	68	73	73	74

Note: Where variability is at the maximum for a dichotomous variable.
n.a.: Not Applicable.

*Some statisticians report that in the situation illustrated here, a probability sample equal to 50 percent of the population will provide the required accuracy or greater. See, for example, Taro Yamane, *Elementary Sampling Theory* (Englewood Cliffs, NJ: Prentice-Hall, 1967), 398–399. Yamane also suggests that some formulas used for sample size do not apply if sample size, *n*, is more than 50 percent of population size, *N*.

Sample Size, Accuracy, and Confidence

In order to be more accurate and more confident, an investigator needs to take a larger sample. To increase accuracy without increasing the sample size, the investigator must settle for a lower confidence level. To increase confidence and keep the sample size the same, some accuracy must be sacrificed.

The calculation of sample size for simple random samples is relatively straightforward. Determining sample sizes for cluster and stratified samples requires more complex calculations, although the determination is based on the same factors. We suggest that you seek expert advice if you are using either type of sample to estimate population parameters.

Although the accuracy and confidence level can be improved by increasing sample size, the amount of improvement realized with additional units becomes less and less. It is a classic case of diminishing returns. When the sample size is small—say 100—an increase to 400 will greatly improve accuracy and confidence. However, an increase in sample size from 2,000 to 2,300 will bring little improvement, although the additional costs are likely to be the same in both cases. Beyond some sample size, the improvements in accuracy and confidence level from increasing the sample size usually are not worth the cost.

Formula and Calculation Examples for Sample Size

An example using the formula for calculating sample size follows with information for a dichotomous variable. An investigator attempting to determine the proportion of a population supporting a bond issue, for instance, would analyze a dichotomous variable. The responses would be categorized as either "In Favor" or "Not In Favor." The population parameter would be expressed as the percent "In Favor."

The example shows the calculations required for a variable in which 50 percent of the population favor an issue and 50 percent oppose it; the confidence level desired is 95 percent, and the accuracy desired is ±4 percent.

Calculate sample size

$$\sqrt{n} = \sqrt{(.50)(1-.50)} \times (1.96)/.04$$
$$= (.5) \times (49) = 24.5$$
$$n = 600$$

Table 5.7 shows the relationship of accuracy, confidence level, and sample size for variables that have only two values. The estimates of required sample sizes shown here are conservative; that is, the size shown may be larger than actually needed.

Computing equation:

$\sqrt{Sample\,Size}$ = Population variability × z-score for confidence level × 1/degree of accuracy. Where variability is at the maximum for a dichotomous variable, $p = .50$ and the population is large.

In these calculations we assumed that 50 percent of the population was in one category and that 50 percent was in the other. These percentages are expressed as proportions (.50 and .50) and are used to measure the population variability. This is the maximum variability. A population with a smaller percent in one of the categories, say 30 percent, has less variability and therefore requires a smaller sample size to achieve the same degree of accuracy at a given confidence level. If the investigator has little or no information about the population, the easiest and most conservative approach is to assume that the population is split 50/50.

If the population split is anything other than 50/50, the population variability will be less, resulting in a smaller required sample size. Note that a commonly used standard of 4 percent sampling error and 95 percent confidence level requires a sample of 600 units, whereas to have a sampling error of only 1 percent and a 99 percent confidence level requires over 16,000 units. The figures in Table 5.7 also assume that the population is large enough that its size can be ignored as a factor in determining sample size. The reader may wish to compare the sample size figures in Table 5.6 with those in Table 5.7.

Sample Size for Nominal and Ordinal Variables

This approach is also used when calculating sample size for nominal and ordinal variables with more than two categories. The investigator treats the variable of interest as if it were a dichotomy

Table 5.7 Sample Sizes for Various Degrees of Accuracy and Confidence Levels

Confidence Level			
Desired Degree of Accuracy	*99%*	*95%*	*90%*
1%	16,576	9,604	6,765
2%	4,144	2,401	1,691
3%	1,848	1,067	752
4%	1,036	600	413
5%	663	384	271
10%	166	96	68
20%	41	24	17

and chooses two proportions for the formula, guessing that a certain percent are in one category and the remainder in all the others. For example, say the director of the employment division wants to conduct a study in which one of the variables is occupation. This variable includes the following categories: professionals, managers, and skilled workers. To calculate the appropriate sample size, her statistical consultant assumes that 34 percent of the population is in the professional category and that 66 percent is in the other two categories.

The box showing the Calculation of Sample Size for Means illustrates a calculation of sample size for an interval-level variable. Note that the appropriate sample is relatively small compared to those in Table 5.7, which are for nominal and ordinal variables. The variability of interval and ratio variables can be measured more precisely than the variability of nominal and ordinal (categorical) variables, allowing for smaller samples to estimate parameters measured at this level.

Response Rate

The projected *response rate* also affects our choice of sample size for surveys. Response rate is the percent of those selected to receive the questionnaire or be interviewed who actually complete and return the questionnaire or agree to be interviewed. In sample surveys, not all of the units selected for the study respond. A large percent of those receiving mailed questionnaires do not return them. Some respondents contacted by telephone or in person will refuse to be interviewed. The calculations in Table 5.6 assume that data will be gathered from every unit selected for the sample. Unfortunately, if some units cannot be found or if individuals refuse to respond, the sample's accuracy will be affected in a way that cannot be determined accurately. Thus, factors such as the anticipated nonresponse rate need to be considered and the sample size adjusted accordingly. If a researcher expects to have a high nonresponse rate, he would want to increase his sample size to make sure he has the necessary usable responses. Of course, the amount of time and money available also may set limits on the number to be sampled.[33]

The administrator should keep in mind a study's purpose. Unless the purpose of a study is well defined and there is reason to be confident that the validity of the measurement of the variables has been established, a preliminary study with a small sample may be more efficient than spending resources on a larger sample. Exploratory studies can be very useful. In quantitative research, just collecting and examining the data frequently give the researcher and the client an opportunity to understand and modify their models. Data collection has its moments of frustration. One is well advised not to spend resources on a large-scale study in the absence of a reasonably well-developed model whose critical measures have been tested for reliability, validity, and sensitivity.

Another important factor in determining sample size is the analysis to be undertaken. Specifically, small samples will not withstand extensive statistical exploration; some statistical procedures require a large sample. If, for example, there are a large number of independent or control variables, working with a small sample may result in the analyst's looking at individual cases rather than true samples. Dividing the sample into many small groups will have the same result. A Southeastern state annually conducted a statewide survey of about 1,300 residents. This size sample is more than adequate for estimating characteristics of the population of the state as a whole. However, the sample is not large enough to allow a meaningful analysis at the county level. For example, too few cases exist for analysts to compare the employment experience of males without college degrees in all 100 counties of the state.

Sample Size for Different Types of Studies

The minimum sample size necessary for credible research differs from study to study. Public-opinion surveys and epidemiological studies rarely use fewer than several hundred subjects, and sample sizes of over 1,000 are not uncommon. Surveys of the general population spread over a large geographic area tend to have the largest samples. Studies to discover rare events or conditions also tend to require large samples. Sample sizes for other types of studies, such as controlled experiments, can be much smaller.

Samples for Experiments

In experimental studies, sampling seldom is done in order to be representative of the population at large. An important consideration in experiments is that subjects represent the characteristics relevant to the experiment. Units are randomly assigned to the experimental and control groups to ensure that these groups are equivalent at the beginning of the experiment. The main interest is in having a sample large enough so that any difference between the control and experimental groups can be attributed to the independent variable.

Adequate experimental designs can be conducted with fewer subjects than cross-sectional designs for several reasons. A rule of thumb for experimental studies is to have at least 30 subjects for each experimental condition, for example, 30 subjects in the experimental group and 30 subjects in the control group.[34] Traditionally, researchers using experimental designs did not use probability sampling to represent a general population. Rather, they randomly assigned members to experimental and control groups. They applied many of the principles of probability sampling to assure a sufficient number of members in each group allow them to attribute a change in the dependent variable to the independent variable. More recently, however, some experimental researchers are combining survey sampling of populations with experiments. The researchers select a representative sample of the relevant population to use in an experiment. This allows them to apply laboratory findings to a larger population with more confidence in the external validity of the findings.[35]

In experimental designs, fewer groups are compared; subjects are usually assigned to no more than four groups, and often two groups are sufficient. With data from cross-sectional studies such as sample surveys, the analyst may want to compare more subgroups or look at the characteristics of a group meeting certain criteria. For example, he may want to compare the educational backgrounds of employed people over 50 years of age with those who are unemployed. The experimenter physically controls for more variables, whereas the survey researcher uses statistical controls in the analysis after the data are collected. The survey analyst divides the obtained sample data into many subgroups whose members are equivalent on one or more important variables and compares the different subgroups. In our example of the statewide survey, if the analyst wanted to compare employment experience for various age, sex, and education groups for different counties, she would need a much larger sample than if she wanted only to analyze employment experience for all respondents.

If possible, experiments should use more than the minimum sample size. If the difference between groups that is expected at the end of the study is small, it might not show up if samples are small. Statistical techniques can be used to estimate required sample sizes for experimental studies. Using these requires knowing or assuming some facts about the population, such as the difference expected between groups and the variability of the dependent variable in the population. A preliminary study can be most helpful in obtaining initial information on these factors.

Sample Size and Statistical Power

Power refers to the ability of a statistical test to detect, from a sample, the strength of a relationship between two variables, a difference in means between two groups, or another measure of effect size and do so correctly.[36] It is also the probability of obtaining statistical significance and therefore rejecting the null hypothesis if the research hypothesis (the "alternative hypothesis") is true.[37] The smaller the difference between means or the weaker the relationship between variables, the more difficult it will be to detect differences. Tables showing the sample size needed for various levels of accuracy, confidence, and power are available.[38] When testing a hypothesis, the investigator wants to include enough subjects in the sample to demonstrate that the hypothesis probably is true. For example, a sample of 50 may detect a moderate effect of an independent variable on a dependent variable only 46 percent of the time using conventional statistical criteria, whereas, using the same criteria, a sample of 1,300 will detect a very slight effect 95 percent of the time. On the other hand, an investigator does not want to waste resources by sampling many more subjects than needed. Consider the following two hypotheses:

H_1: City managers are paid higher salaries than county managers with the same length of experience.

H_2: The more hours of training employees receive, the fewer accidents they will have on the job.

To learn whether the hypotheses are probably true in the population of interest, that is, among city and county managers or employees, the investigator may begin by selecting a sample out of the appropriate groups. To decide how many subjects to sample, he needs to estimate how great an effect the independent variable is likely to have on the dependent variable.

Let's assume that the investigator anticipates that the difference in pay between city and county managers is relatively small, but he wants to confirm this difference. He also may believe that the impact of longer training on the number of accidents is relatively great, and he may not be interested in results that show only small improvements. In other words, he would not consider his hypothesis supported if increased length of training brings about only small improvements. In the case of detecting pay differences, then, he will need a larger sample than he will in detecting increased benefits of longer training. Information on the power of a test will help in selecting the proper test and sample size.

Other Sampling Issues

Although many sampling issues relate to sample size, other issues come into play as well. Some of these are discussed in the following sections.

Nonsampling Errors

Sampling errors are errors that come about because we have drawn a sample rather than studied the entire population. They are due to mathematical chance—the probability that we will select those cases not exactly estimating the parameter. If a probability sampling method has been used, the size of the resulting error can be estimated with the use of statistical theory. This is one of the powerful advantages of probability sampling. But other types of errors can cause a sample statistic to be inaccurate. *Nonsampling errors* usually result from a flaw in the sampling design itself or from faulty implementation. These errors distort the estimate of the parameter that cannot be evaluated using statistical theory. Unlike sampling errors, nonsampling errors are likely to cause the difference between the statistic and the parameter to be distorted in one direction. This also is called *bias*. Sometimes we are aware of bias but are not able to estimate its effect, at least not mathematically. Sometimes we are unaware of it. An example of an inappropriate sampling frame causing bias would be the use of the telephone book as a sampling frame for voters. Upper-income individuals are less likely to have a listed number than are those with middle or lower incomes. Younger individuals are less likely to have a land-line telephone (and therefore less likely to have a listed number) than older people.[39] The characteristics related to income and age may also be related to how someone votes.

Nonsampling errors are serious, and their impact on the results of a study cannot be easily or directly estimated statistically. Taking a larger sample may not decrease nonsampling error; such error may, in fact, increase with a larger sample. For example, if coders rush to code and transcribe data, they may make more mistakes than they would have with fewer forms. Similarly, a larger sample size may result in fewer attempts by interviewers to reach respondents who are not at home, thus increasing the nonresponse rate. In short, a larger body of data generated from a larger sample can adversely affect the quality of data gathering and data processing.

People who do not respond to surveys and questionnaires contribute to the nonsampling error of studies in which they are asked to participate. Some people routinely may refuse requests for information; others ignore questionable or poorly conceived surveys. A researcher may be unable to contact a member of a sample; the person may have moved, missed a telephone call, or be otherwise unavailable.

The greater the proportion of sample members who do not respond or who do not give usable responses, the greater the nonsampling error.[40]

If members of the sample who respond are consistently different from nonrespondents, the sample will be biased. Other nonsampling errors include unreliable or invalid measurements, mistakes in recording and transcribing data, and failure to follow up on nonrespondents. If a sampling frame excludes certain members of a subgroup, such as low- or high-income individuals, substantial bias may be built into the sample. And, as we have already said, taking a larger sample will not necessarily reduce these problems. Proper selection and construction of the sampling frame and other steps in the investigation require and merit careful effort and attention.

Assessing Sample Quality

Administrators should feel confident in their ability to evaluate the quality of samples encountered in their work. They must frequently rely on studies involving limited samples, such as small samples drawn from a narrowly defined population. For example, a researcher studying depression in elderly people might study only those in a geriatric facility because of the lack of resources or contacts necessary to study those seeking care from private doctors or hospitals. However, as long as one is aware of the limits of a given sampling design, information thus gathered may prove very useful, and one is not likely to make inappropriate generalizations from it.

In evaluating a given sample, an administrator should consider its size, how representative it is of the population of interest, and the implementation of the design.[41] The size of the sample must be evaluated in light of the purpose of the study. Small samples may be reasonable for a relatively homogeneous population, especially if the investigators do not want to generalize with great accuracy to a larger population. Nevertheless, sample size partially depends on the extent of analysis; a small sample cannot withstand intensive analysis involving investigation of the joint relationships of many variables. If a population parameter is to be estimated from a sample with accuracy and confidence, a larger sample is needed. Similarly, if an analyst intends to divide the sample into many subsamples, the sample must be larger than otherwise. Consider, for example, the *U.S. Current Population Survey*. This survey, which provides statistical estimates of population characteristics for the nation and each state, uses a sample of over 60,000 households. Still, the sample is too small to estimate the characteristics for a single city or county.

The importance of generalizing from the sample to the population depends on the sample and how an administrator plans to use it. A sample representing just one school will be adequate for the principal. It will be of less interest to the head of the state department of education, unless it uncovers something amiss or intriguing about that particular school. Awareness of a study's population and its sampling frame may alert an administrator to findings that cannot be generalized to a similar but unsampled population.

Another important consideration is how well the study was implemented; poor implementation of the sample design will cause avoidable nonsampling errors. Evidence of careless data collection, poor staff supervision, inadequate quality control, and low response rate can seriously undermine the most carefully drawn design. Many people understandably refuse to participate in studies that appear poorly conceived and developed. A well-designed sample cannot overcome the flaws in a poorly designed questionnaire. Although sampling is an important part of a study, an administrator should not assume that a well-designed probability sample will alone ensure a quality study.

Calculation Examples for Sampling Error and Sample Size for Means

The three examples here show the calculation of the standard error when the sample size and the standard deviation (a measure of variability) of the variable in the population are known.

Sampling error is estimated with the standard error formula. One component of the formula is the z-score corresponding to the confidence level desired. For a 95 percent confidence level, the corresponding z-score is 1.96 (often rounded to 2). This also is the most common level of confidence used in administrative and managerial work. For 99 percent and 68 percent confidence levels, the appropriate z-scores are 3 and 1, respectively.

Standard Error Calculations

1. **Standard Error for Proportions**

    The formula for calculating the size of the standard error of proportions is:

    $$SEp = \sqrt{p(1-p)/n}$$

where:
SEp = standard error for proportions
p = the proportion of the population in one category of the variable of interest
$\sqrt{p(1-p)}$ = the formula for the standard deviation, the measure of variability of the population
n = the sample size

The reader should note the relationship between this formula and the one for determining sample size presented in the text.

As an example, consider the following situation: A survey was taken to determine the support for city–county consolidation in a city in the Southeast. The following information was developed.

p = the proportion of citizens supporting consolidation = .53
$1 - p$ = the proportion not supporting consolidation = .47
n = sample size = 590

$$SEp = \sqrt{(.53)(.47)/590} = .0205 \text{ or } 2.05\%$$

2. **Standard Error for Means**

    The formula for calculating the size of the standard error for means is: $= SE_{\bar{x}} = s\sqrt{n}$

where:
$SE_{\bar{x}}$ = standard error of the mean proportions
s = standard deviation of the sample; this measures the variability of the population
n = sample size

As an example of the use of the standard error of the mean, consider a sample of county managers drawn in a three-state area to estimate the average age of county managers in these states. The following information was developed.

s = 4.5 years
n = 35

$$\text{SE}_{\bar{x}} = 4.5 / \sqrt{35} = 4.5 / 5.92 = .76 \text{years}$$

3.   Standard Error for Samples From Small Populations

The formula for estimating the standard error when sampling from small populations includes the finite population correction factor. The following example is from a study in which the investigator took a sample out of a population of 100 counties for a study on nursing home care.

$$SE_p = \sqrt{p(1-p)/n} \times (N-n)/N-1$$

where:
SEp = standard error for proportions
p = the proportion of the population in one category of the variable of interest
$\sqrt{p(1-p)}$ = the formula for the standard deviation, the measure of variability of the population
n = the sample size
N = the population size
$(N - n)/(N - 1)$ = the finite population correction factor
   To facilitate comparison with the previous example, assume the following:
p = .53
$1 - p$ = .47
n = 60
N = 100

Carrying out the calculations, we have

SEp = (.53)(.47)/60 3 × (100 − 60)/(100 − 1)
   = (.064)(.404) = .026 *or* 2.06%

For very small populations, investigators may gain little by sampling, even though the finite population correction factor has quite an impact. One might wonder if studying all 100 counties wouldn't have been as convenient as taking a sample of 60. In the research on which this example was based, however, the work involved in studying more counties made the reduction worthwhile.[42]

Calculation of Sample Size for Means

The example in this illustration shows the calculation for sample size when the variability in the population is known from previous research or is assumed. Researchers often use a pretest or pilot test survey to obtain an estimate of the population variability.
   A study is planned by the Center for Public Affairs. The investigators want to calculate the sample size needed for the study. They wish to be 95 percent confident that they have estimated the average age of county managers to within one year of the actual average of all county managers in the area of study. A 95 percent confidence interval can be placed around this estimate by adding and subtracting a number equal to 1.96 standard errors from the mean estimate. (The

z-score corresponding to the confidence level desired is also called the *reliability factor*. In this example, it is 1.96.)

The amount of error in the estimate is 1.96 times $SE\bar{x}$.

$$\bar{X} \pm 1.96 \times SE\bar{x}$$

The problem for the investigators is to determine how large the sample must be to reduce this error to one year.

The formula for calculating sample size is the one given in the chapter:

$$\sqrt{n} = (\text{Variablity}) \times (\text{reliablity factor}) / (\text{accuracy})$$

The variability is measured by the standard deviation; the population standard deviation is estimated by sample data or previous knowledge or is assumed to be a certain value.

Using the information from Example 5.2, the sample size for this problem can be calculated as follows:

$$\sqrt{n} = (4.5 \text{ years}) \times (1.96) / 1 \text{ year}$$

$n = 77.79$ or 78

The investigators will need a sample of 78 managers to reach the desired level of confidence and accuracy.

Summary

Sampling is an efficient and effective method of studying a population, whether it consists of people, files, agencies, or other units. Probability sampling normally is the preferred sampling method because with it, an investigator can apply statistical methods to estimate parameters, and it helps to control biases that inevitably enter into the construction of samples. The four common probability sample designs are simple random, systematic, stratified, and cluster. For simple random sampling, the investigator uses a lottery method or list of random numbers to draw the sample. Each member of the sample has an equal probability of being selected, and the selection of one member does not affect the probability that another member will be chosen. In systematic sampling, the population size is divided by the desired sample size to determine a skip interval, a random starting point in the list of the population members is selected, and cases separated by the skip interval are chosen.

In stratified sampling, the investigator divides the population into classes or strata and randomly selects units from each class for the sample. Stratified samples may be either proportionate or disproportionate. Proportionate samples include members from the strata in the same proportion as they exist in the population. Disproportionate samples include a higher proportion of the members of some strata than of others. Oversampling of some strata may occur in order to provide sufficient numbers of units from that stratum for detailed analysis. In cluster sampling, the investigator randomly selects units that contain members from the population of interest. If sample members are randomly selected from the chosen clusters, the entire sampling is known as multistage cluster sampling.

Nonprobability samples are suitable for exploratory studies. Purposive samples rely on an investigator's judgment; convenience samples rely on ease of access to sampling units. These samples may be adequate if they are free of obvious bias; however, sampling statistics cannot be appropriately applied to them. Quota samples, which specify the number of units with certain characteristics to be selected, are less satisfactory and have little to recommend them. In snowball sampling, each respondent in the sample is asked to refer the investigator to another member of the population. Although this method has important biases, it can be useful in reaching a population difficult to contact by other methods.

The appropriate sample size for any given study is a function of the desired degree of accuracy, the population's variability, the desired degree of confidence in the results, and the analysis to be done. Population size is sometimes a factor but usually affects sample size only indirectly. Increased precision, that is, a high degree of accuracy and confidence in the data, requires a larger sample size. However, this also will increase a study's cost and may increase the number of nonsampling errors.

Although sample surveys are perhaps the most well-known use of probability sampling, many other endeavors utilize the principles of probability sampling. Experimental designs seldom use samples to represent a general population. Traditionally they randomly assign members to experimental and control groups. By relying on the laws of probability, they can ensure that the groups are large enough allow the researchers to attribute change to the independent variable. With recent developments in methods, however, researchers are combining laboratory experiments with survey sampling of larger populations.

Samples are prone to two types of errors. Sampling errors are a product of the sampling process. One cannot reasonably expect any one sample to be a completely accurate representation of the population, nor can one tell with absolute certainty that the sample has only a low degree of error, but statistical theory provides a basis for estimating the extent of sampling error. Nonsampling error also affects the quality of the sample. Common nonsampling errors include nonresponse bias, careless data collection, and careless data processing. Proper implementation of the sample design is as important as is the design itself in reducing errors (see Figure 5.7).

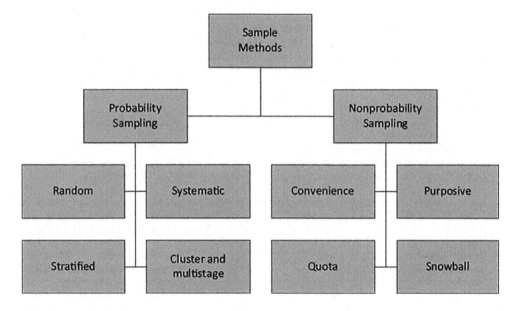

Figure 5.7 Sampling Methods

Notes

1. See U.S. Census Bureau, "Developing Sampling Techniques," available at www.census.gov/history/www/innovations/data_collection/developing_sampling_techniques.html.
2. Again, see Ibid. Also see Jennifer D. Williams, *Census 2000: The Sampling Debate, CRS Report for Congress* (Washington, DC: Congressional Research Service, 2000), available at http://research.policyarchive.org/353.pdf. In Chapter 9, we discuss the decennial census and related surveys the Census Bureau conducts.
3. Graham Kalton, *Introduction to Survey Sampling*, Sage University Paper Series on Quantitative Applications in the Social Sciences, 07–035 (Beverly Hills, CA: Sage Publications, 1983), 5.
4. L. O'Muircheartaigh and S. T. Wong, "The Impact of Sampling Theory on Survey Practice: A Review," *Bulletin of the International Statistical Institute* 49 (1981): 465–493; Kalton, *Introduction*.
5. Leslie Kish, *Survey Sampling* (New York: Wiley Interscience, 1995), 53–59.
6. Other authors make this point using somewhat different language. See, for instance, Robert M. Groves, Floyd J. Fowler, Jr., Mick P. Couper, James M. Lepkowski, Eleanor Singer, and Roger Tourangeau, *Survey Methodology* (Hoboken, NJ: John Wiley & Sons, Inc., 2004), 57. Groves et al. distinguish "sampling variance" from "sampling bias," defining them both as forms of "sampling error" (p. 57).
7. Note that the distinction between error and bias applies to aspects of survey design beyond sampling. See Groves et al., *Survey Methodology*, Figure 2.5 (p. 48), for an overview of the many places within a survey that error occurs. For instance, see Groves et al., *Survey Methodology*, p. 53, on the difference between "response bias" and "response variance," pp. 55–56 on the difference between "coverage error" and "coverage bias," and pp. 58–59 on the difference between "nonresponse error" and "nonresponse bias." See also p. 60.
8. Groves et al. indicate that "[s]ampling bias arises when some members of the sampling frame are given no chance (or reduced chance) of selection" (*Survey Methodology*, 57). Note that sampling error is *not* the same thing as "coverage error." Groves et al. note that coverage error "exists *before* the sample is drawn and thus is not a problem arising because we do a *sample* survey. It would also exist if we attempted to do a census of the target population using the same sampling frame" (p. 55; see also p. 83).
9. Groves et al. make this point in noting that "sampling variance" is error that is not bias; it simply "arises because, given the design for the sample, by chance many different sets of frame elements could be drawn" (*Survey Methodology*, 57).
10. Kish, *Survey Sampling*, 20–21.
11. Groves et al. note that "[o]ne way to think about a simple random sample is that we would first identify every possible sample of size *n* distinct elements within a frame population, and then select one of them at random" (*Survey Methodology*, 99). Also see Groves et al., *Survey Methodology*, 99–100, on the "finite population correction" applied to samples drawn without replacement.
12. Because no new principles of sampling are included in random digit dialing (RDD), we discuss it in the chapter "Using Questions and Questionnaires to Collect Data" along with other data collection methods.
13. Seymour Sudman, *Applied Sampling* (New York: Academic Press, 1976), 56–57.
14. Stuart Bretschneider, "Management Information Systems in Public and Private Organizations: An Empirical Test," *Public Administration Review* 50 (September–October 1990): 536–545.
15. Sudman, *Applied Sampling*, 110, 130.
16. M. Hansen, W. Hurwitz, and W. Madow, *Sample Survey Methods and Theory* (New York: Wiley Interscience, 1953), 40–48.
17. For a different example, with tables similar to Tables 5.2 and 5.3 to illustrate the difference between proportionate and disproportionate stratified sampling, see Tables 5.2 and 5.3 in Chapter 5 of Johnnie Daniel, *Sampling Essentials: Practical Guidelines for Making Sampling Choices* (Thousand Oaks, CA: Sage Publications, 2012).
18. For an example of multistage cluster sampling used in combination with other sampling methods, see: Appendix A in "Religion in Latin America" (Pew Research Center: Religion in Public Life), available at www.pewforum.org/wp-content/uploads/sites/7/2014/11/Religion-in-Latin-America-11-12-PM-full-PDF.pdf.
19. Similar diagrams are used by others. For instance, see The European Social Survey's explanation of cluster sampling, particularly Figure 2.3, available at http://essedunet.nsd.uib.no/cms/topics/weight/2/6.html, from Norwegian Social Science Data Services. See also diagrams on cluster sampling from Scott D. Campbell, available at http://www-personal.umich.edu/~sdcamp/up504/questionnairedesign.html.
20. Survey Research Center, *Interviewers Manual* (Ann Arbor: University of Michigan, Institute for Social Research, 1969), Chapter 8, describes a seven-stage cluster-sampling process used in a nationwide survey.

21. Kalton, *Introduction*, 38–47; E. Terrence Jones, *Conducting Political Research*, 2nd ed. (New York: Harper and Row, 1984), 61–63, contains a relevant and clear illustration. Also see Earl Babbie, *The Basics of Social Research*, 6th ed. (Belmont, CA: Thomson-Wadsworth, 2014), 219–220.

22. Work First was a program to move people off welfare by providing assistance in obtaining job training and in finding employment. For more information, see www.ncdhhs.gov/divisions/social-services/work-first-family-assistance.

23. Kalton, *Introduction*, 90; Kish, *Survey Sampling*, 19. These samples also are called "fortuitous" or "haphazard."

24. William Trochim classifies all nonprobability sampling designs as either accidental or purposive. See William M. K. Trochim, *The Research Methods Knowledge Base* (Cincinnati, OH: Atomic Dog Publishing Co, 2001), 55–59.

25. Trochim, *The Research Methods Knowledge Base*, 57.

26. See Nate Cohen, "How One 19-Year-Old Illinois Man Is Distorting National Polling Averages," *The New York Times* (October 12, 2016), available at www.nytimes.com/2016/10/13/upshot/how-one-19-year-old-illinois-man-is-distorting-national-polling-averages.html.

27. Again, see Groves et al., *Survey Methodology*, 57.

28. Geoff Cummings, *Understanding the New Statistics: Effect Sizes, Confidence Intervals, and Meta-Analysis* (New York: Taylor & Francis, 2012), discusses these topics. See Chapters 3 and 4 for confidence intervals.

29. The discussion of confidence intervals assumes that the samples were selected by a probability sample. Practitioners of quota sampling often use a measure related to the confidence interval called the "credibility interval." However, computing a credibility interval requires the use of Bayesian statistics that is beyond the scope of this textbook. For more information, see "AAPOR Statement: Understanding a 'Credibility Interval' and How It Differs from the 'Margin of Sampling Error' in a Public Opinion Poll," *American Association of Public Opinion Research*, October 7, 2012, available at www.aapor.org/AAPOR_Main/media/MainSiteFiles/DetailedAAPORstatementoncredibilityintervals.pdf.

30. Groves et al., *Survey Methodology*, 97.

31. For another way of saying this, see David S. Moore, George P. McCabe, and Bruce A. Craig, *Introduction to the Practice of Statistics*, 6th ed. (New York: W. H. Freeman and Company, 2009), 358, 359. Moore, McCabe, and Craig describe the "margin of error" around an estimate of a mean as an indication of "how accurate we believe our guess is, based on the variability of the estimate, and how confident we are that the procedure will catch the true population mean μ" (358). Moore, McCabe, and Craig describe the "confidence level" of a confidence interval as giving the "probability of producing an interval that contains the unknown parameter" (359).

32. See Kalton, *Introduction*, 13, 14, 82–84; William F. Matlack, *Statistics for Public Managers* (Itasca, IL: F. E. Peacock, 1993), 132–135, for discussions of the "finite population correction factor" and when to use it. This correction is important if the sample is a large fraction of the population, say 10 percent or over. This is likely when small populations are sampled. If the population is large and the sample is a very small fraction of the population, the correction factor can be ignored with very little effect.

33. Norman M. Bradburn, "A Response to the Nonresponse Problem," *Public Opinion Quarterly* 56 (1992): 391–397. We discuss the issue of response rate and ways to increase it in Chapter 6.

34. L. R. Gay, G. E. Mills, and P. W. Airasian, *Educational Research*, 10th ed. (Boston, MA: Longman-Pearson, 2012), 137.

35. Diana Mutz, *Population Based Survey Experiments* (Princeton, NJ: Princeton University Press, 2011).

36. H. C. Kraemer and S. Thieman, *How Many Subjects? Statistical Power Analysis in Research* (Newbury Park, CA: Sage Publications, 1987), 22–29. Also see the second edition of this book, authored by H. C. Kraemer and C. Blasey, 2015.

37. Cumming, *Understanding the New Statistics*, 12, Chapter 12.

38. R. V. Krejcie and D. W. Morgan, "Determining Sample Size for Research Activities," *Educational and Psychological Measurement* 30 (1970): 607–610.

39. John Tarnai, Danna L. Moore, and Marion Schultz, "Characteristics of Cell Phone Only, Listed, and Unlisted Telephone Households," *Survey Practice* 2, no. 7 (2009).

40. C. A. Mosher, *Survey Methods in Social Investigation* (London: Heineman, 1969), 139–144. Also see pages 246–250 for further discussion of sources of nonsampling error. G. T. Henry, *Practical Sampling* (Newbury Park, CA: Sage Publications, 1990) discusses formulas to adjust for nonresponse.

41. See Sudman, *Applied Sampling*, p. 27, for a more detailed discussion of these points.

42. Gary Rassel, "Nursing Home Facilities in North Carolina Counties," Unpublished Research, The University of North Carolina at Charlotte, Charlotte, NC, 1999. Also see Kalton, *Introduction*, 13–16, 82–84.

Terms for Review

sample
population
sampling frame
parameter
statistic
sampling error
standard error
sample bias
sample design
sampling fraction
probability sample
nonprobability sample
sampling unit
simple random sampling
systematic sampling
skip interval
stratified random sampling
proportionate stratified sampling
disproportionate stratified sampling
cluster sampling
multistage sampling
convenience sampling
panel survey
purposive sampling
quota sampling
selection bias
snowball sampling
sampling error
nonsampling error
confidence level
confidence interval
accuracy
population variability
standard deviation
unit of analysis
power

Questions for Review

The following questions should indicate whether you have a basic competency in this chapter's material.

1. What are the advantages of probability sampling over nonprobability sampling? Disadvantages?
2. Explain how systematic sampling violates the principles of simple random sampling.
3. Stratified sampling is said to reduce the variability of the population and hence provide for a smaller required sample size. Explain.
4. What is the appeal of quota sampling? Why is it not possible to assume that a quota sample is as representative of the target population as a probability sample?

5. Contrast single-stage cluster and multistage cluster sampling.
6. Describe the factors that are used to determine sample size. Why is population size often not important?
7. How are the factors used to determine sample size related to sample size and to each other?
8. Distinguish among sampling error, nonsampling error, and bias. What are some common sources of nonsampling error for an investigation using government documents such as police crime reports?
9. Identify the sampling unit in each of the stages of sampling in Example 5.5.
10. Comment on how generalizing from a sample to a population is an example of external validity.

Problems for Homework and Discussion

1. From each description, identify or infer the target population, the sampling frame, the unit of analysis, and the type of sample. Discuss whether the sampling strategy will allow the researcher to form inferences about the target population based on the sample.

   a. Problem: To study factors related to a diagnosis of depression in elderly individuals. Sample: A random selection of 150 of the 300 residents diagnosed with depression in one year while in a particular geriatric facility and 340 randomly selected from the 850 residents in the same facility who were not diagnosed with depression during that year.

   b. Problem: To obtain information on routes and types of transportation used by people traveling with the city. Sample: The Council of Governments has a computerized dataset of 110,000 trips made in one year; 10 percent of the trips are randomly selected for intensive analysis.

   c. Problem: To see whether employees changed behavior after attending a training program. Sample: Randomly selected five training sessions held in 2003, identified the trainees in the selected sessions, and compared performance reviews of the trainees recorded before the sessions with those developed after.

   d. Problem: To determine the effects of a change in publisher on a newspaper. Sample: Randomly selected seven U.S. cities and identified and studied all newspapers published in these cities over the past century.

   e. Problem: To determine the total amount of assistance received by welfare clients. Sample: Thirteen states that represented geographic regions and a full range of payments were selected. One to three counties within a metropolitan statistical area within each of these states were selected. In each selected county, the welfare director gathered data on each of the first 50 clients coming in to recertify eligibility for benefits.

2. You have been contracted by the Office of Continuing Education to construct a sample in order to determine reaction to a public service announcement on your university's night courses. The announcement is intended to attract more adults into night classes. What publicly available information do you think you would you need to carry out the following? Assume that this information is available.

   a. Design a simple random sample. Identify your population and sample frame. Recommend a sample size and justify it. Give specific directions on how members of the sample would be selected.

   b. Design a systematic sample. Identify your population and sample frame. Recommend a sample size, indicate the skip interval, and select the first 10 members of your sample.

   c. Design a cluster sample. Identify your population and sample frame. Recommend a sample size. Give specific directions on how members of the sample would be selected.

    d. For all three samples, discuss the adequacy of the sample frame.

    e. What sampling design would you recommend? Explain why.

3. Suppose that researchers investigated the impact of reductions in force in state government salaries by studying a random sample of abolished positions. The researchers determined whether the salary of the abolished position was below or above the average for the class of employee. Suppose that the researchers reported that the sample yielded "an accuracy of 7 percent at the 95 percent confidence level." Interpret the researchers' observation.

4. Select a confidence level and accuracy level for the studies in Examples 5.1 and 5.2. Justify your choices. As part of your recommendation and justification, you should comment on the cost to sample each person or file.

5. For Example 5.1, what size sample would the county need in order to have the findings within 5 percent accuracy at the 95 percent confidence level if the population split about evenly on the proposition? If the community normally splits 60–40 on bonding issues, how could this information modify the choice of sample size?

6. For Example 5.1, what parameter is the study designed to estimate?

7. The local chapter of the National Urban League wants to find out which type of training or education would be most useful to the city's low-income citizens to enable them to obtain higher-paying jobs. The chapter executives want to have focus groups. Design a snowball (referral) sampling procedure to obtain a sample for the focus groups. Identify the populations and describe the sampling procedure.

8. Construct with your classmates a hypothetical population of 50 people. Indicate the income of each member of the population. Calculate the mean income of the population and its standard deviation. Randomly select five samples of 10 members each and five samples of 20 members each. For each sample, compute the mean income and the standard error of its estimate (given by the equation: standard error = standard deviation/n). For each sample size, indicate how many samples come within ±1, 2, and 3 standard errors of the mean. What can you conclude?

9. An intern with the city's division of employment and training is designing a study to evaluate whether three training programs impact future wages of Job Training Partnership Act (JTPA) participants differently. She plans to contact between 25 and 40 participants of each program and is concerned because she knows that sample surveys usually have many more study subjects than this.

    a. Write a memo to her explaining why she may have enough subjects for her study. Do you believe that she does have enough? Explain.

    b. The intern has decided that if the difference among the postprogram wages of the participants in the different groups is less than 15 percent, the difference is not important. Will this affect her sample size? Explain.

    c. Discuss how the size of the differences among the postprogram wages of the participants and the importance of these differences is an issue of the power of a statistical test.

    d. Sketch out a design that the intern might use for the study.

10. Summarize the sampling procedure in the Bretschneider article (Endnote 14). Identify each of the different types of sampling used. Identify or describe each of the following: population, sampling frame, sample size, response rate.

Working With Data

To see how sampling statistics work, access the County Data file accompanying this textbook. Use a statistical software package or a spreadsheet to obtain a frequency distribution for public school enrollment and average SAT total scores for the 100 counties. Request the mean, standard deviation, and standard error of the mean for each variable. Identify or construct the confidence interval for each variable. Then use the software to get a 20 percent sample from the dataset and determine its frequency distribution for the same variables. Again, determine the mean, standard deviation, and standard error for each variable. Also identify or construct the confidence interval. Compare these measures for each group.

1. For both the population and the sample, which variable has the most variability? Which has the least? Describe any differences you observe.
2. Compare the means from the 20 percent sample with the means for the entire population. For each variable, determine if the 20 percent sample means come within 1 standard error, within 2 standard errors, or within 3 standard errors of the mean of the entire population.
3. Compare the means from your sample with the means from samples gathered by three or more classmates. For each variable, determine how many sample means come within 1 standard error, within 2 standard errors, and within 3 standard errors of the mean of the entire survey sample.
4. Do the same procedure as previously for the variables age and education in the GSS dataset. Which variable has the most variability? When you take a 20 percent sample, which of the sample means is closer to the original mean? Does this make sense? Why or why not?

Recommended for Further Reading

For a clear and practical discussion of many actual applications of sampling designs, see Sudman, Seymour, *Applied Sampling* (New York: Academic Press, 1976). The author discusses the strengths and weaknesses of the various designs in different circumstances.

Groves, Robert M., *Survey Errors and Survey Costs* (New York: Wiley Interscience, 1989) is recommended for an excellent treatment of methods to reduce bias and increase response rates and some of the trade-offs involved.

Groves, Robert M., Floyd J. Fowler, Jr., Mick P. Couper, James M. Lepkowski, Eleanor Singer, and Roger Tourangeau, *Survey Methodology* (Hoboken, NJ: John Wiley & Sons, Inc., 2004) is also an excellent resource on survey design. On sampling, see in particular Chapters 2 through 4.

Henry, Gary T., *Practical Sampling* (Newbury Park, CA: Sage Publications, 1990) is a short, readable treatment of sampling with well-developed examples from applied research.

Kalton, Graham, *Introduction to Survey Sampling*, Sage University Paper Series on Quantitative Applications in the Social Sciences, 07–035 (Beverly Hills, CA: Sage Publications, 1983), provides good descriptions of various sampling designs, including combinations of those presented here, and discusses the calculations for determining sample size and sampling error for them.

Kish, Leslie, *Survey Sampling* (New York: Wiley Interscience, 1965), provides an excellent discussion of sampling designs and sampling statistics.

Kraemer, Helena, and Sue Thieman, *How Many Subjects? Statistical Power Analysis in Research* (Newbury Park, CA: Sage Publications, 1987) discuss the power of statistical tests and provide tables showing sample size for various tests and power. Although much of the book is very technical, Chapter 2 has an excellent general discussion of power for the nonstatistician. A new edition of *How Many Subjects?* by Helena Kraemer and Christine Blasey became available in 2015.

Williams, Jennifer, *Census 2000: The Sampling Debate* (Washington, DC: Congressional Research Service, 2000). This reprint discusses several of the issues surrounding the debate about whether sampling should be allowed for the 2000 Census. The author cites other reports, relevant court cases, and statutes.

6　Contacting and Talking to Subjects

In this chapter, you will learn:

1.　How to conduct and use Internet, telephone, mail, and in-person surveys.
2.　How to deal with low response rates.
3.　Understanding random digit dialing.
4.　The value of intensive interviewing and focus groups.
5.　Basic requirements for conducting interviews and focus groups.
6.　How new technology has affected surveys, interviews, and focus groups.

After investigators identify the purpose of a study and outline a model, they design a data collection strategy. The design consists of (1) a plan for contacting subjects and obtaining data from them and (2) a questionnaire or other data collection instrument. Data collection decisions affect the quality of research. Furthermore, in practice, the relationship between the research design and data collection is dynamic. An investigator may change the original research design because of data collection problems. For example, the characteristics of samples affect the feasibility of conducting face-to-face interviews, telephone calls, or mailed surveys. Question wording determines the reliability and operational validity of measures.

Various disciplines have their own way of looking at things and rely on a few data collection methods to answer most empirical questions. Administrative and policy questions can touch on almost any discipline. Consequently, students studying administration may conclude that they need a background in many methodologies. In practice, administrators and policy analysts may rely heavily on collecting their own data or analyzing methods such as survey research, interviews, and focus groups. These methods produce evidence to answer diverse administrative and policy questions.

This chapter discusses common methods of data collection: Internet-based, mail and telephone surveys, and face-to-face interviews, including focus groups. Each method has its benefits and drawbacks that affect access to the target group, quality of information, response rate, and costs. Investigators should be conversant with each method so they can select the most appropriate one for a given study and combine methods as needed.

We first consider the challenges of contacting potential subjects and encouraging them to provide the needed information and questions when planning a study. We then discuss each data collection method and concentrate on the quality of responses, response rates, and costs. Two data-gathering techniques, random digit dialing and unstructured interviewing, have a direct link to other methodological considerations. Random digit dialing was developed to overcome problems of constructing a probability sample for a telephone survey.

Survey Research

Survey research refers to cross-sectional studies in which investigators use questionnaires, forms, or interviews to gather data from individual subjects. In the majority of surveys, the researcher's unit of analysis is the individual. Less frequently, the unit of analysis is the organization or location where the respondents live or work. *Experiments* and *quasi-experiments* refer to studies in which subjects are exposed to deliberately manipulated treatments or interventions that may occur under the researcher's control. For quasi-experiments, however, the intervention might not be under the researcher's control. *Economic analysis* refers to studies in which investigators gather and analyze cost or other economic data; its technical details lie outside the scope of this text. Nevertheless, economic analysis also depends on reliable, operationally valid, and sensitive measures, all of which are discussed in this text.

Rather than concentrating explicitly on any one research method, we emphasize the processes of organizing data collection, contacting subjects, eliciting their cooperation, writing questions, and designing research instruments. These skills, largely developed by survey researchers, can benefit any empirical investigator. Furthermore, these skills aid even the most basic data collection efforts, such as asking employees to fill out forms reporting organizational and individual performance data.

Basic data collection involves numerous techniques. Many decisions and steps are involved in empirical research. These include knowledge of how to contact subjects and collect data from them, question writing, and questionnaire design. All researchers must be aware of the ethical concerns regarding collecting data from, and about, human subjects. This applies to survey research as much as to any other research approach.

Survey research has been a common data collection method for over a century. However, the 20th century saw numerous and dramatic changes as sample surveying replaced census enumeration. Even more changes took place in the late 20th and early 21st centuries. Technological, legal, lifestyle, and cultural changes all affected how surveying has been conducted.

Contacting Subjects

Once an investigator has identified the target group, she must convince each potential subject to participate and provide information. Think of your own experiences. Has an invitation to take a survey popped up on your computer screen or smart phone? Have you been asked to fill out a form after attending an event? Or been asked to rate a product or seller of something you recently purchased? Do you participate in regular online surveys for rewards such as airline miles, gift cards, or magazines? In each case, you decided to participate or not. You may have declined because you were busy, or you were suspicious of the surveyor's motives. You may have participated because you were interested in the topic, you trusted the person who asked you, or you were offered an incentive. If you received a letter, telephone call, or e-mail asking for information, you may have wondered why you were contacted. This is especially true if you were asked about a sensitive topic such as your experiences in a drug rehabilitation program.

Administrators who hire researchers to collect information from program participants, employees, or stakeholders should think through how to contact subjects and who will contact them. To avoid violating privacy, administrators themselves may want to make the initial contact with their clients, employees, or stakeholders, seeking their permission to be later contacted by a survey team. In both cases, the agency explains the purpose of the survey and how respondents' privacy will be protected.

Contacting and interviewing subjects may occur simultaneously. Alternatively, researchers may contact individuals and arrange a time and place for a personal interview. Potential subjects may receive a letter or e-mail informing them that they will be called and interviewed soon or that they will receive a questionnaire by mail. They may receive a telephone call asking for a time for them to be re-contacted and interviewed. Or they may receive an e-mail with a link to a web-based survey.

Low Response Rates

Researchers must locate potential subjects and convince them to participate. A low *response rate* may suggest problems with the study's design or implementation. If a large percentage of potential subjects cannot be found, or if they refuse to cooperate, the research findings may be unrepresentative of the target population of interest. To determine the response rate and reasons for nonresponses, researchers need the following data:

- total subjects in the original sample
- number of subjects not located (nonexistent address or phone number, undeliverable mail, discontinued phone service, unoccupied housing unit)
- number of subjects not contacted (not at home, unable to understand English or other survey language, illness)
- number of subjects ineligible (sample member does not belong to target population)
- number of subjects refusing to provide information

Deciding how to categorize an uncontacted respondent can be difficult. An unanswered e-mail, ringing phone, or unanswered door does not necessarily mean that the e-mail address still works or that the house is occupied. Persons who were not reached may or may not belong to the target population. Conventions for computing response rate vary. A conservative strategy computes the response rate by dividing the number of subjects who provide data by the total number of subjects minus ineligibles. This calculation assumes the sampling frame and the data collection effort were adequate.[1]

New Technologies

The development and widespread use of new technologies such as smart phones and the Internet raise new issues, challenges, and opportunities for survey researchers. Many homes no longer use landlines and have switched to cell phones. The number of wireless-only households was 54.9 percent in 2018.[2] Consequently, survey organizations have modified their traditional data collection techniques to assure representative samples.

Household access to the Internet shows similar patterns. In 1997, 18 percent of households had the Internet available at home; by 2013, this had increased to 75 percent of households. In 2019, 90 percent of households use the Internet.[3] As of 2017, 73 percent of adults over 65 use the Internet (the lowest age group), whereas people age 18–29 reported 100 percent Internet usage (the highest).[4] Widespread access to the Internet makes new ways of reaching respondents available. Surveys can be administered by e-mail or posted on web sites. With e-mail, the survey link is sent in the body of an e-mail message or as an attachment, and the respondent completes and returns it either online or by regular mail. Web surveys are the most popular and can handle more complex designs, accommodate skip patterns, and post data directly to a database. Audio and video clips can be included. Respondents may have more confidence in the anonymity of the web survey since they can go to the link to complete the survey and are always considered voluntary since they can terminate the survey at any point in time or even ignore it (anonymity is different from confidentiality. Anonymity demands that a specific individual cannot be linked to a specific questionnaire or set of data. Confidentiality means that someone on the research team can link an individual to an item in a dataset, but the individual's identity is protected and known only to the researcher.) With an e-mail survey, the form is sent to the respondent, who may feel that confidentiality is compromised. With web surveys, intended respondents are often contacted by e-mail with a direct link to the questionnaire or directions for accessing the web site hosting the questionnaire.

Planning for a Survey

Planning for and conducting a survey require many of the same steps and considerations generally conducted for any study. Certain components of surveys, however, should receive additional attention. The steps for planning a survey include deciding on the goal of the survey and identifying the target population; how the data will be collected, analyzed, and reported; and other considerations. The process of survey research involves a set of interrelated decisions and activities. These are summarized here as questions an administrator or researcher must answer. The answers will affect activities to be carried out later in the process.

The Process of Survey Research

1. Conceptualizing:

   What is the purpose of the survey?
   What does the researcher want to find out?
   To whom will the results apply?

2. Study design and data collection method:

   What is the nature of the sample?
   What data collection method is most appropriate and economical? Will respondents be surveyed more than once?
   Are there privacy issues, and is institutional review board (IRB) review required?[5]

3. Data analysis and management:

   How are the data to be managed and analyzed?
   Are staff and resources available to do the necessary analysis? (These issues should be considered early in the process, not put off until after the data have been collected.)

4. Instrument design:

   What will be the general layout of the survey instrument and its specific questions—their type, order, and wording?

5. Sampling:

   How many respondents are necessary to achieve the study's purpose? (If precise estimates are important, a statistician may be consulted to determine the sample size and design.)
   How does the sample size affect the feasibility and cost of data collection?

6. Training:

   Who will collect the data?
   How will they be trained?

7. Pretesting:

   When will the survey be pretested?
   Who will follow up on pretest findings?

8. Pilot study:

   When will the pilot study, also called a "dress rehearsal," be conducted?
   What subjects will be involved?
   What will be done with their data?
   Who will follow up and act on the responses?

9. Administration:

    What is the plan for collecting, compiling, and documenting the data and information?

10. Data entry and analysis:

    Who is responsible for organizing, entering, verifying, and analyzing the data?

11. Reporting:

    What reports will be produced?
    To whom will they be communicated? How?

Mailed Questionnaires and Internet Surveys

Mailed surveys, along with face-to-face interviews, dominated the early days of survey research.[6] Consequently, extensive research exists on their design, use of incentives, and response rates. Much of the information obtained has been incorporated in other methods, primarily Internet surveys.

Internet surveys cost less than mail, telephone, and in-person surveys and cause minimal inconvenience to the respondent. Using Internet surveys, like mail surveys, the researcher can collect more detailed, thoughtful data, especially if the questions require reflective answers or perusal of files and records.

Disadvantages of Mail and Internet Surveys

Designing an unambiguous Internet questionnaire; sending it via e-mail addresses or launching from a web site; and having it delivered to the appropriate person, answered, completed, and successfully past spam filters takes time. Internet surveys now achieve lower response rates when compared to in-person surveys but now have higher response rates when compared to mail or telephone surveys. This is particularly true if the investigator has not taken steps to encourage responses, such as indicating the value of the information, offering incentives, and having clear directions and questions. While re-contacting nonrespondents may improve the response rate, each follow-up adds more time and increases costs and may begin to anger respondents.

Investigators conducting Internet surveys (like mail surveys) cannot tell why a recipient does not respond. A poor sampling frame, for example, one with out-of-date addresses, may lower the response rate. The recipient may not have received the survey or may have forgotten about it or ignored it. It may not have applied to the respondent.

Advantages of Internet Surveys

Internet surveys do not require live interviewers, eliminating the costs of training, supervision, travel, and salary of staff to administer the survey. If a respondent must consult records to answer the questions, additional time must be allowed. The timing of the release of the survey also must be considered; for example, a questionnaire sent to hospital administrators during the coronavirus pandemic will not receive immediate attention or be considered a high priority. In the United States, the holidays and other year-end activities probably make the last two weeks in December an unproductive time to launch Internet surveys to either organizations or individuals.

An effective Internet survey should be easy to understand and answer. Unfamiliar words, confusing directions, and unsuitable choices may lead respondents not to answer a question on the survey or to answer randomly.

Poorly worded questions affect all respondents. In addition, not all populations react in the same way to a survey. Educated groups or people accustomed to forms handle Internet surveys most easily. They can understand and follow directions and deal accurately with a variety of question-and-response formats. Some people routinely express their ideas and opinions in writing, whereas others feel uncomfortable expressing their opinions or answering open-ended questions. Aged or less well-educated individuals may find mailed questionnaires difficult to read and understand. They may dread making mistakes and appearing foolish. Thus, characteristics of the sample can affect response rate as much as the survey itself.

Administrators, managers, and other professionals may prefer Internet surveys to the more intrusive telephone interview. Professionals are busy and occupied with the details of their jobs. The researcher may find reaching them by telephone a time-consuming and costly exercise in logistics. Furthermore, filling out forms, responding to written requests, and writing memoranda constitute normal parts of administrators' work routine, and they can fit in answering questions with their other tasks. An e-mailed questionnaire works well to gather specific information from an agency, such as the number of clients served, their characteristics, and the services received. The addressee may answer parts of the survey and pause it and give other sections to staff members with the needed information or knowledge.

Response Rates

A respondent may fail to reply to a survey for several reasons. First, the respondent may start out willing to answer the survey but later simply forget to do so. Frequent reminders may alleviate some of this. Second, respondents may ignore surveys that ask for inaccessible or hard-to-obtain data. Perhaps no one has time to compile the needed data. Perhaps the respondents resent and resist the expectation that they will gather the data. A recent survey sent about bicycles and pedestrian infrastructure to local planning officials that ran through the month of March in 2020 was disrupted to a certain degree because the last week of March, the governor gave stay-at-home orders for all nonessential employees, causing many offices to close or have employees work remotely during the coronavirus pandemic. One of our colleagues had a survey returned because emergency conditions caused many workers to be furloughed.

Third, respondents may ignore surveys that seem biased. For example, some Internet surveys assume a state of affairs inappropriate to the respondent. Consider a survey that asks respondents why they have left a job. If the available responses stress job problems or lifestyle changes, then people who changed jobs to take advantage of opportunities for professional and personal growth may be reluctant to associate their job changes with the listed factors. Fourth, recipients may not know or may not trust the survey sponsor and may question how their information will be used or if it will be used. The more recipients believe that the survey will benefit them, the more likely they are to respond—even if they have to spend time compiling data and writing out answers. Conversely, if recipients question whether the survey information will benefit them, poorly designed or lengthy surveys or ones requiring looking up information will add to their disinclination to respond.

The sponsorship of a survey affects the rate of responses. In conducting a study for an agency, we have found that an e-mail from the agency head endorsing the survey and encouraging the recipient to respond helps to establish the legitimacy of the survey. The endorsements also seem to improve response rates. Some of the authors once sent a questionnaire to university faculty. The questionnaire's cover letter was from the chancellor. The survey received a 95 percent response rate. One reviewer of an earlier edition of this text noted that he got a 99 percent response rate from Massachusetts's school superintendents when he included a cover letter from the dean of Harvard's School of Education.

Sponsorship also can affect the direction of responses. We suspect that this is a particular problem with surveys of work conditions. If an employer conducts or sponsors a survey to learn about employee satisfaction and other components of the work environment, employees may doubt the confidentiality of their responses, and they may distort their responses in line with their perceived self-interest.

The failure to respond is not a trivial problem. Nonrespondents contribute to nonsampling error, therefore undermining the investigator's ability to infer from the sample. In considering response rate, we have focused on the purpose of the survey, its sponsor, and quality. In addition, there are "tricks of the trade" that can lessen people's resistance. The length of the survey; the ease in filling it out; respondent interest in the subject matter; the format and design; and a range of other factors, including the color of the font and the background color, have been linked to response rates.

Improving Response Rates

An example helps to illustrate how a set of procedures can improve response rate. An agency survey received a disappointing response rate of 34 percent, leading the survey sponsors to conduct a study on ways to improve it. The respondents were police officers, members of a profession notorious for its suspicion of researchers.[7] The questionnaire was complicated. It had a matrix, and embedded in each item were five separate questions. It had not been pretested, nor was a follow-up of nonrespondents conducted. In contrast, the revised questionnaire incorporated many recommended characteristics, as did the administration of the questionnaire. These included checklists rather than complicated questions, prior review and pretest by other police officers, and one follow-up e-mail and reminder to nonrespondents. The revised survey, sent to different police officers, had a 76 percent response rate.

The problem with the initial questionnaire in this example happens all too often. Too many surveys ignore sound procedures. In preparing to do a mail survey, beginning researchers are well advised to read a book that details the procedure, either one of the books cited in the bibliography or a similar text; arrange for a variety of people, including members of the target population, to review and comment on the questionnaire; pretest it; and conduct a pilot study.

Follow-up e-mails and even phone calls will almost always improve response rates, and usually a second follow-up will improve the rate even more.

Factors Affecting Response Rate

The following list was generated based on the study discussed previously and other sources.[8]

A. Sampling frame

    1. Accuracy: Sample members may not receive a survey if the sample is drawn from an outdated sampling frame.
    2. Relevance: Survey recipients may not respond if they believe they do not belong to the target population.

B. Questionnaire design

    1. Questionnaire length: Research shows mixed results, but when salience and follow-up are held constant, shorter surveys have higher response rates than longer ones.
    2. Item content and sequence: Items should be logically arranged by topic.
    3. Questionnaire layout and format: The questionnaire should be attractive and on sturdy, good-quality paper; items should be numbered sequentially with ample space between items.

C. Delivery

1. Prenotice: Sending a message in advance alerting the intended respondent that he or she will receive a questionnaire or be contacted about a survey usually improves response rates. If surveying an organization, identify the name and correct contact information for the appropriate respondent.

2. Cover letter in an e-mail: Response rates improve if the cover letter (a) indicates the importance of the study and the value of the respondent's participation or (b) offers to provide feedback.

3. Delivery method: Response rates from organizations may improve if the e-mail account generating the survey denotes official status in an organization.

4. Follow-up: Two or more follow-ups are recommended to get higher response rates; including a questionnaire with the follow-up request seems to yield a higher response rate than just an e-mail follow-up.[9]

5. Incentives: Including money or another incentive, for example, entry in a lottery, donation to a charity, or a token item, with the initial questionnaire markedly improves response rates.[10]

Strategies for Improving Response Rate

Example 6.1 shows a postcard reminder that evaluators sent to alumni of a graduate program. They did not send a similar reminder to subjects of a university-sponsored study, since some administrators felt it detracted from the seriousness of the survey. Postcards can be used as inexpensive reminders. Nevertheless, either on the first or second follow-up should take place. Nonrespondents can be contacted by telephone and asked if they received a questionnaire. An additional survey may be sent, or they may be surveyed over the telephone or sent a paper questionnaire.

Example 6.1 Postcards to Use as Follow-up Reminders

Situation: A graduate program director needed to send an inexpensive follow-up reminder to people who had not responded to an alumni survey.

Solution: The director decided that a humorous postcard would serve as a first reminder to alumni.

Notes:

1. Postcards were inexpensive to reproduce and mail. They were not wordy or heavy handed.

2. Alumni who did not respond to the postcard reminder were sent a short letter with a duplicate questionnaire. Although not necessary, the next step would have been to telephone nonresponding alumni and to ask key questions.

Research shows that incentives *included with* a survey have a "substantial positive" effect. One study used in developing the previous list found that groups that received $.01 to $5 cash had a 19 percent higher average response rate than the control groups.[11] In that study, a nonmonetary incentive increased the response rate about 8 percent. The promise of sending an incentive, either monetary or nonmonetary, after a completed survey was received had little or no effect.[12] And two other studies found that monetary incentives plus reminders increased the response rate more than either of these used separately.[13]

An investigator should check the assumption that nonrespondents are similar to respondents. She can contact a sample of nonrespondents and make a special effort to persuade

---

them to answer the questionnaire or at least to provide information on key characteristics. This will allow her to determine if the respondents and nonrespondents differ significantly. With information on key characteristics she can compare respondents' and nonrespondents' demographics, for example, their age, sex, and region of the country, to see whether she can detect a possible bias.[14] Plotting nonresponses on a map can also help identify any geographic bias. Table 6.1 summarizes the advantages and disadvantages of mailed surveys.

Table 6.1 Overview of Mailed Surveys

	Mail
Costs	Relatively low cost (printing, mailing, and postage, incentives for responding)
Sampling issues	Requires a good sampling frame; can distinguish nonrespondents from nonrecipients if undelivered questionnaires are returned
Response rate	Low (follow-up required)
Turnaround time	Slow
Respondent issues	Convenient, especially for hard-to-reach professionals; poor if subjects are elderly or poorly educated
Content	Highly motivated respondents can look up information, write detailed answers; other respondents require clear, easily answered questions

Internet Surveys

Internet surveys are usually posted on the web but can be sent as part of an e-mail message. They are faster and cheaper than telephone or mail surveys. No interviewers must be hired or trained, no postage or printing bills must be paid, and investigators can avoid having to enter data from paper questionnaires. In other respects, Internet surveys are similar to those sent by traditional mail. The researcher needs a sample from a sampling frame of e-mail addresses, unless he wants to send out a blanket mailing. Unlike traditional mailed surveys, however, e-mail surveys raise greater concerns about confidentiality, security of responses, and viruses.[15]

In early uses of the Internet to conduct surveys, the survey itself was included in the e-mail or as an attachment. Those included within the e-mail message could be tedious to complete unless they were well designed. A survey sent as an attachment required the respondent to download it, fill it out, and return by regular mail. Such surveys are now seen rarely. Current practice is rather to set up a survey using a web-based survey program such as SurveyMonkey, Qualtrics, or Survey Share.[16] The researcher sends an e-mail to explain the survey, establish the identity and legitimacy of the organization asking for the survey to be completed, and offer a link to the survey. Sometimes an access code is required to take the survey. Usually the link or the access code either expires after one attempt or allows the survey taker to resume the survey later if he/she abandons it partway through. Assigning a password that must be used to access the web survey is also sometimes a way to control duplicate or other unwanted responses.[17]

Challenges of Internet Surveys

Because Internet-based surveys are still a relatively recent phenomenon, questions about their quality are an ongoing area of research. Specifically, researchers are only beginning to study ways to improve samples, response rates, and questionnaire design.[18] Sampling problems are particularly challenging.[19] If a sampling frame with e-mail addresses exists, investigators can

Table 6.2 Overview of Internet Surveys

	Internet
Costs	Low cost
Sampling issues	May not provide a good sample of general public, although probability samples are possible; a good sampling frame may allow access to specific populations
Response rate	May be difficult to determine; if general population is being surveyed, it may have lowest response rate of all
Turnaround time	Quick
Respondent issues	Requires Internet access, awareness of survey, motivation to access survey
Content	May have some of the benefits of computer-assisted interviewing; may encounter privacy concerns; respondents limited in their ability to communicate if questions seem ambiguous or responses inappropriate

design a probability-based sample. If a list with e-mail addresses does not exist, accessing the target population becomes more difficult, as does designing a probability-based sample. The investigators may publicize the survey through Listservs, newsletters, newsgroups, links from relevant web pages, or similar communications. Yet, in such cases, the researcher will only be able to obtain responses from a nonprobability-based sample—most likely a convenience sample of the people who choose to take the survey. To deal with such issues, Knowledge Networks is one organization that delivers online surveys to probability-based samples of respondents.[20]

Improving Internet Survey Response Rates

To see if the lessons learned from mail surveys increased the response rate of e-mail surveys in the late 1990s, a group of researchers sent a prenotice, letter and survey, reminder, and replacement surveys to randomly selected groups of university faculty. One group received all these materials by e-mail and another group by paper. The response rate for the two groups was virtually identical. However, e-mail respondents were more likely to complete the survey, to leave fewer items unanswered, to answer open-ended questions, and to give longer answers to open-ended questions.[21]

Other features associated with effective surveys should apply to Internet surveys. Like mail surveys, Internet surveys should be easy to access, answer, and return. As is true of any survey, ease of answering and the clarity of the questions should be established through pretesting. A respondent cannot ask an interviewer to clarify a question or write in how she interpreted the question.

Research on the design of web surveys shows that the format and design of the survey affect the response rate and data quality. Many of the characteristics affecting the traditional mail survey apply to e-mail and web surveys. For example, questionnaire length and the ease with which the questionnaire can be followed and completed will also affect response rate and data quality.[22] It is possible for a researcher to exercise a great deal of control over how a survey is presented when building a survey in a program such as Qualtrics. As noted, several vendors provide web-based survey software. Before designing their own web survey, researchers may want to investigate one or more of these.[23] Table 6.2 offers a review of the advantages and disadvantages of Internet surveys.

Telephone Surveys

Telephone surveys involve special considerations with respect to drawing a representative sample, designing the questionnaire, and ensuring confidentiality. Let's look at these three issues one at a time.

Telephone Samples

The first questions about telephone surveying center on sampling. How adequate are telephone subscribers as a study population? How can one assemble a representative sample when sampling frames are quickly out of date, exclude persons with unlisted numbers, and include households with multiple listings? How does one make sure that the sample units contacted coincide with the intended units of analysis and target population? What has been the nonresponse rate associated with telephone interviewing? How does the fact that so many people rely on cell phones affect survey results? The answers to these questions have changed greatly over the last several decades.

In the past, the value of telephone survey data was limited because a sizeable portion of households did not have telephones, nor could investigators reach households with unlisted numbers. Consequently, surveyors could not expect to telephone a representative sample of individuals.

Yet, the conditions for telephone surveying changed dramatically during the 20th century. By the end of the 20th century, census data indicated that roughly 94 percent of U.S. households had regular telephone service. Households without phone service tend to have lower incomes, to move frequently, and to have fewer group memberships and community attachments. Some households, about 3 percent, have intermittent phone service. These households, which discontinue phone service from time to time because of money problems or a move, seem to be like nonphone households.[24]

Cell Phones and Land Lines

More recently, the use of cell phones has become widespread, as discussed earlier. The percentage of adults with cell phone service only varies by income level and region of the country.[25] The legality of contacting people on a cell phone number to complete a survey is a topic of debate.[26] Cell phone users may be charged for incoming calls.[27] The national- and state-level telephone polls conducted in 2004 were quite accurate, even though cell phone numbers were not included in the sampling frame.[28] It remains to be seen if this situation changes as the percent of citizens of voting age who rely on cell phones continues to increase.

A directory may be an appropriate sampling frame to sample organizations and professionals. Directories of organizations or professionals may be relatively stable and provide few problems. If an investigator plans to draw a sample of the general public from a local telephone directory, however, the sampling frame becomes problematic. A telephone directory does not include cell phone numbers, households with unlisted or unpublished numbers, families who have recently moved into a community, or those who have changed their telephone numbers. Random digit dialing can overcome many of these problems. In addition, random digit dialing reduces the clerical work required to draw a random sample.

Random Digit Dialing

Random digit dialing techniques, which produce samples of randomly generated telephone numbers, are another reason that telephone-based surveys will likely persist even though more individuals move away from land lines. Random digit dialing overcomes the limitations of telephone directories as sampling frames. Random digit dialing techniques have been developed and refined so that investigators can contact households with unlisted telephones. Investigators can be reasonably assured of telephoning a representative sample, including people whose telephone numbers are not listed in a current directory.

Random digit dialing produces a sample of randomly generated telephone numbers. The sample is an equal-probability sample of all area telephone numbers—listed and unlisted numbers, residential and nonresidential numbers. Pure random digit dialing generates a sample of telephone numbers from seven-digit random numbers. To reduce the frequency of nonworking numbers, investigators identify all working telephone exchanges in the target geographic area. The investigators assign four-digit random numbers only to working exchanges. If individuals or households make up the study population, any nonresidential exchanges such as city or state government offices, universities, and businesses are omitted before constructing the sample.

After identifying the appropriate exchanges, the investigators select an exchange and complete the telephone number by assigning a four-digit random number between 0000 and 9999. Researchers often work with telephone companies to obtain the number of exchanges and estimates of the number and proportion of telephones assigned to each exchange that are residential. With this information the investigator can create a proportional sample, as illustrated in Example 6.2.[29]

Example 6.2 Constructing a Random Digit Dialing Sample: An Illustration

Problem: Surveyors want to telephone 500 households in a community with three telephone exchanges (111, 112, 555).

Strategy: The telephone company gives the surveyors the number of residential telephones in each exchange.

Table 6.3 Residential Telephones in Each Exchange

Exchange	Residential Telephones	Percentage of Total
111	2,000	20%
112	3,000	30%
555	5,000	50%

1. Select 2,000 four-digit random numbers.
2. Assign 20 percent of the random numbers to exchange 111. If the first random number is 8,752, the calling number is 111–8752.
3. Assign 30 percent of the random numbers to exchange 112.
4. Assign 50 percent of the random numbers to exchange 555.
5. Select replacement numbers for nonworking or business telephones from the list of remaining random digits.

Notes:

1. Surveyors expect a large portion of the numbers generated in a random digit sample to be nonworking numbers or business telephones. We have assumed that a surveyor needs four calling numbers to successfully reach one household.
2. If we assigned each exchange the same number of calling numbers, 667 numbers, 1 out of 3 households in exchange 111 would be called, and 1 out of 7.5 households in exchange 555 would be called. Thus, a household in exchange 111 is 2.5 times more likely to be in the sample than a household in 555. Any unique features of the 111 exchange population will be overrepresented in the sample.

Considerations for Random Digit Dialing

Survey researchers often contract with telephone companies or a private consulting firm to draw a sample meeting certain specifications. For example, the firm may construct telephone samples for specific ZIP codes or census tracts, enabling investigators to study certain population mixes.

Although random digit dialing includes cell phone numbers, the procedures to follow through with the survey itself are usually modified. The Pew Research Center, for example, selects land line and cell phone numbers to yield a combined sample of 35 percent land line and 65 percent cell phones. Since cell phone use varies by age group and other factors, Pew uses this ratio to balance costs with demographics. Since in traditional telephone surveying with land lines, women are more likely to answer the telephone and do the interview, interviewers calling cell phone numbers ask one half of those answering to speak to the youngest male and one half to speak to the youngest female.[30]

Random digit dialing solves the problem of unlisted numbers but not of multiple telephone numbers within the same household. A household with more than one telephone or listing has a greater probability of being chosen for a sample. To handle multiple listings, interviewers may ask how many telephone numbers the household has. Later, analysts will weigh the household's responses. For example, responses from a household with two telephone numbers may receive a weight of 0.5. Investigators may decide to ignore this problem.[31] Nevertheless, one should be aware that households with multiple listings are more likely to be selected, and such households may have distinctive characteristics. Also, area codes are no longer valid indicators of where someone lives as people have shifted from landlines to cell phones.

Computer-Aided Telephone Interviewing

Another factor supporting the continued widespread use of telephone interviewing is that face-to-face interviewing may be quite costly, although technologies like Skype, Google Hangout, and other videoconferencing methods may reduce these costs.[32] In instances where survey interviews are conducted in person, expenses such as interviewer time and travel can be expensive. Furthermore, in some neighborhoods, residents and interviewers worry about safety. Potential subjects refuse to open their doors to strangers, and interviewers also may feel unsafe. People often cannot be contacted face to face during business hours. Telephone surveys can also cover a wide geographic area relatively quickly. Interviewers can also call from a central location, allowing closer supervision of their work than is possible with face-to-face interviews.

Computerization of surveying processes further supports the continued use of telephone interviewing. *Computer-aided telephone interviewing* (CATI) systems simplify the administration of complex interviews and survey monitoring. The computer can be programmed to dial the numbers selected. When the respondent is contacted, the interviewer reads items from a terminal screen and keys in the responses. The computer paces the interview, makes branching decisions for contingency questions, and checks for inconsistent responses. In addition, the computer keeps track of the number of calls made, refusals, completions, and similar data so that, at any time, the investigators know the survey's status.

Households and Individuals

Telephone surveys have the household as the implied unit of analysis. Yet individuals may be the actual unit of analysis. If the surveyor intends to contact individuals, he should not automatically interview whomever answers the telephone. For example, researchers have found that women answer a disproportionate number of telephone calls. For some telephone surveys, the interviewer will have the name of the sample member and will ask to speak to that person. Or the surveyor may be looking for a characteristic, such as the person who had most recently seen a doctor.

When interviewers can theoretically speak to any household member, they need some way to ensure a representative sample. Researchers have tested various techniques for selecting a specific household member for the interview. One nonintrusive method is to ask to speak to the member of the household who most recently celebrated his or her birthday, while others ask for the youngest male over age 18, since young men are least likely to be home when surveyors call.[33]

Telephone Response Rates

The last sample-related question has to do with response rates. As noted earlier, response rates for many types of survey have declined. The rates for many annual surveys conducted by the Survey Research Center (SRC) at the University of Michigan have declined substantially during the last 30 years. This decline has accelerated in recent years. Researchers have been unable to determine how much of the decline is due to an increasing number of refusals to participate and how much is due to failure to contact respondents; however, the research has concluded that nonresponse bias is not as bad as had been feared.[34] The decline in response rates, however, increases the costs of telephone surveying.

In general, telephone surveys have better response rates than mail surveys, although mail surveys with multiple mailings may do nearly as well.[35] Unlike mail surveys, telephone surveyors can distinguish between those they cannot reach (the inaccessible) and those who refuse to participate (the unwilling). However, because of lifestyle changes, people may be harder to reach than they were in the past. They spend less time at home and rely on an answering machine, voice mail, or caller ID to screen calls. Response rates for cell phone contacts also tend to be lower.[36] Just because someone screens calls, however, does not mean that she will refuse to answer a survey.[37] Surveyors record machine-answered calls as "not at home," although they may leave a message stating the purpose of the call and the plan to call back.[38]

A major issue affecting survey costs is whether extraordinary efforts to convince respondents to participate bias their responses. A recent study found that younger and better-educated households are harder to reach.[39] According to Robert Groves, an expert in survey research methods, a high rate of cooperation requires a population with characteristics that are not common in a nationwide sample.[40]

Questionnaire Design

We now turn to the second set of questions about telephone surveys: those about questionnaire design and questioning. How do telephone questionnaires differ from mailed and other self-administered questionnaires? What types of questions can be asked? How long can telephone surveys be? Unlike written or Internet-based surveys, telephone surveys depend strictly on auditory cues. Questions cannot be so verbose that the respondent forgets the question contents before arriving at the answer categories. Similarly, a respondent is likely only to be able to keep a limited set of answer categories in mind at one time. An interviewer has to entice subjects to participate in the study and stay on the telephone to answer the questions. She must rely on her voice and ability to convey key information succinctly.

Introducing a Telephone Survey

In telephone surveys, most refusals to participate occur after the introduction and before the first question is asked; therefore, the introduction must be clear and compelling. Consider the typical telephone survey situation. A householder in the middle of other activities is interrupted by an unexpected call by an unknown caller. The interviewer must persuade the subject that it is worth his time to delay whatever he is doing to answer some questions. The interviewer begins by introducing herself and the survey organization. Next, the interviewer explains how the subject was selected, the purpose of the survey, and its approximate length. See Example 6.3 for wording presenting this information. The interviewer should be prepared to answer questions about the survey organization, the survey, and the confidentiality of the answers.

Example 6.3 Introduction for a Telephone Survey

Situation: A sample of agency employees who participated in the DECIDE management training program were selected for a telephone interview to learn whether they used DECIDE in their work and how they used it.

Step 1: To avoid questions about the legitimacy of the interviewers, a training manager sent a memorandum to the sample explaining that the survey was to find out employee training needs:

To: xxx

From: Joe Doe, Training Manager

Subject: Training Survey

The Human Resources Development Unit will be conducting a phone survey November 7–14, regarding Management/Professional Development courses.

You have been selected in a random sample to participate. Expect a phone call from one of our interviewers during this period. You will be asked to answer a series of questions regarding training you completed, and we ask that you be candid in your responses. The interview should take no more than 10–15 minutes.

I know each of your schedules is full, but to continue to provide quality training, we need your feedback.

Step 2: Three days after the memorandum was sent, the interviewers began calling.

Each interviewer used the following introduction (printed on each survey copy):

Good morning (afternoon). I am John Cooper. I am calling in conjunction with a memo that you should have received from Joe Doe, the Human Resources Development Manager, regarding a training survey. As the memo stated, the interview should take approximately 10–15 minutes and relates primarily to training you completed with Human Resources. We appreciate your taking the time to work with us. Unless you have any questions, I would like to start.

Notes:

For information on DECIDE, see K. L. Guo, "DECIDE: A Decision-Making Model for More Effective Decision-Making By Health Care Managers," *Health Care Managers* 27, no. 2 (2008, April—June): 118–127, at www.ncbi.nim.nih.gov.

The introduction does not mention DECIDE, since interviewers did not want to introduce a bias by mentioning it at the beginning. (An early question asked respondents what management decision-making tools they used. If the respondents knew the survey was to study DECIDE, they might have been more likely to mention it.)

Interviewer mentioned the probable length of the interview, based on pretest times. This was to prevent being turned down by busy respondents.

Introduction was relatively short (as was the memorandum), but interviewer gave respondents a chance to ask any questions.

Note: Example 6.3 draws on work with the DECIDE study by authors of earlier editions of this text. The example reproduces the memorandum sent to members of the sample and the introductory statement interviewers used. An employee from the training division sent a memorandum to the trainees telling them that they would be interviewed. Note that the interviewer mentions the memorandum in his initial comments.

Determining Survey Questions

In constructing an instrument for a telephone interview, the writer considers the length of the interview and the order, length, and complexity of questions. The questions and possible responses must be clear so the respondents can correctly figure out what they are being asked. The initial questions should reinforce the purpose of the survey stated by the interviewer. The "wrong" questions may arouse respondent suspicion. As the questions proceed, the respondent should sense the underlying logic; otherwise, he may become confused, lose interest, or question the legitimacy of the study.

Unlike in a mail survey, the respondent cannot "read ahead," so investigators should consider how question order will affect later responses. The questionnaire should proceed smoothly so that the interviewer feels and sounds comfortable moving from topic to topic and the respondent can answer easily. In this regard, all telephone interviewers should be well trained and practice the telephone interview before making the first real contact.

Although the interviewer can repeat questions and responses and clarify the meaning of words, an overdependence on the interviewer burdens both the interviewer and the respondent. The interviewer does not have visual clues to indicate respondent understanding. The person writing the questions should write short, easily understood questions. The writer must watch for wording that may lead a respondent to misconstrue a question. The investigators should not assume that a respondent will ask to have a question repeated or to say that he does not understand what the interviewer is asking.

Response categories must be carefully considered because the respondent does not have the leisure to read over and consider the responses. The respondent may not remember long lists of alternatives; he may give the last stated alternative just out of convenience or stop the interviewer mid-list to choose an option. Asking the respondent to rank options does not work well, insofar as he has to keep several options in mind prior to ranking them. Instead, a respondent may be asked to rate items on a scale, for example, a scale ranging from a low of 1 to a high of 10.

Open-ended questions work moderately well in telephone surveys if the interviewers can probe beyond a quickly constructed, superficial answer. Open-ended questions work poorly if the interviewers are not well trained about what to listen for, how to probe, and what to record. On occasion, we have tried to analyze completed telephone surveys and have found ourselves bewildered by the cryptic wording of the interviewers. This problem is related to problems of reliability, where what is recorded varies with who conducted the interview.

Length of Instrument

The length of the survey instrument may affect the decision on how to collect data. Research on telephone surveying has found that few respondents terminate the interview after the first questions. Nevertheless, a telephone interview cannot go on indefinitely. Respondents may tire as it drags on. Remember that the typical survey involves question after question, with the pace controlled by the interviewer. Tired or bored respondents will give unreliable answers. Similarly, respondents may become anxious to get off the telephone and start giving terse, truncated answers.

With longer interviews, costs, including decreased data quality, may make telephone interviewing less satisfactory. Nevertheless, as with mail questionnaires, no absolute rule on questionnaire length can be formulated. The length depends on the study population, the nature of the questionnaire, the motivation to respond, and the skills of the interviewers. One can imagine keeping a professional respondent on the telephone for much more than 20 minutes answering questions relating to her area of expertise.

Table 6.4 Overview of Telephone Surveys

	Telephone
Costs	Higher cost (personnel, equipment, telephone charges)
Sampling issues	Provides a reasonably good initial sample of general public with random digit dialing; must have a mechanism for selecting respondents within a household to assure generalizability. Sample coverage constrained by increased reliance on cell phones
Response rate	Higher (may need to vary calling times to get a representative sample)
Turnaround time	Quick
Respondent issues	Most intrusive; allows easy access to dispersed populations
Content	Avoids biases caused by respondent reading ahead; allows for probing and elaboration of answers; may give socially acceptable answers; respondent fatigue if too long

Confidentiality

A third question underlies the other two; that is, how confidential are telephone surveys? Confidentiality is the last aspect of telephone surveying we consider. Random digit dialing approximates anonymity; however, respondents reached through random digit dialing are not more likely to disclose sensitive information.[41] If one is calling from a list, there is a good chance that the caller knows who the respondent is. If we plan to telephone individuals who are agency clients or employees, we can have the appropriate agency official write to the individuals explaining the purpose of the study. The letter may discuss provisions for maintaining the confidentiality of the responses. For both clients and employees, the preinterview contact minimizes suspicions about why they were called. Preinterview contact should reduce refusals and avoid lengthy explanations about the purpose of the study and procedures for analyzing and protecting the data. Table 6.4 offers a review of telephone surveys. (See Belle County Survey for an example of a telephone survey.)

In-Person Interviewing

In-person, or face-to-face, interviews allow researchers to obtain large amounts of data, perform in-depth probing, ask more complicated or sensitive questions, or contact difficult-to-reach populations, for example, homeless people. Two major types of in-person interviews are:

1. structured interviews, where all interviewers ask each respondent the same closed-ended or short-answer questions in the same order
2. intensive interviewing, where interviewers ask general, open-ended questions

In structured interviewing, several interviewers may be involved, and each must ask the same questions in the same way. In intensive interviewing, the interviewers must be experienced and often have expertise in topical areas covered in the interviews. Between these two contrasting types we find semi-structured interviews, in which the questions to be asked are the same but the interviews have flexibility in the order they are asked and in following up on each question. The interviewer's efforts to establish rapport, to interpret questions, and to elicit usable answers may also influence a respondent's answers. Telephone interviewers working in a central location under supervision are less likely to bias responses. In-person interviewers work largely on their own, with limited supervision. Thus, the probability is higher that they will bias an interview. Because of this, in-person interviewers need to be more highly trained than others.

Structured Interviewing

Structured interviews refer to surveys in which all respondents are asked the same questions in the same order by all interviewers. The investigators want different answers to reflect differences among subjects' experiences with the topic, not differences among interviewers. To minimize the effect of interviewers on the answers, investigators hire and train interviewers to be consistent in how they do the following:[42]

- explain the purpose of the study
- ask questions
- handle incomplete or inappropriate answers
- record answers
- deal with interpersonal interactions

In-person structured interviewing works well for studies with lengthy or complex instruments. Researchers have noted that respondents become fatigued if they are kept on the telephone too long, whereas in-person interviews can successfully last over an hour in some cases. Most in-depth interviews are 30–40 minutes, But it is important to respect the respondent's time. It is a valuable resource, and appointment times need to be honored. Keep your eye on the clock, and watch for non-verbal cues of fatigue or disinterest. The interviewer may elicit the respondent's trust, thus setting the stage for asking sensitive questions. Make sure to reassure the respondent of confidentiality and the purpose of the interview. Do not tape record if this makes the respondent nervous or you do not have permission in writing; take notes and write them up directly after the interview. Interviewers can use visual clues to decide whether the respondent understands a question and whether to continue probing the answers to open-ended questions. The interviewer can also use visual aids as part of a question. For example, in-depth interviews with the elderly may require visual aids when ranking items, or the respondent may need extra time.

Structured interviewing depends on standardized interviewers who give the same explanation of the study's purpose, ask questions without changing the wording, probe without directing answers, and record answers verbatim. Furthermore, interviewers focus on the survey and avoid talking about themselves or sharing their opinions about the study or the questions. This interview style can be used in qualitative case analysis if structured properly.

Computer assistance is an important component of most large-scale in-person surveys. It allows for more complex instruments, standardized interviews, reduced data entry costs, and decreased time between data collection and data analysis. Interviewers using computer-assisted personal interviewing (CAPI) read the questions presented on the computer screen and enter the data. Interviewers using computer-assisted self-interviewing (CASI) have respondents read the questions or listen to them. Respondents may key in their answers or reply orally. CASI helps in collecting sensitive information and maintains the confidentiality of a respondent's answers. However, unlike personal interviews, asking subjects to discuss meaning and clarify answers with probes is not typically used. The interviewers give instructions and answer questions; the computer presents the question, and the interviewers take care not to observe or overhear a respondent's answers.[43]

With the widespread use of tablets and other mobile devices, their use in survey research is likely to increase. They are used by market researchers and others for field studies. They are especially useful when respondents are in rural areas.[44] Typically the survey is given in an electronic link or the respondent downloads an app (application) and the surveyor or the respondent fills out the survey or questionnaire on the tablet or other mobile device. The data can then be transferred electronically into software such as Excel, Stata, R, or SPSS. Some survey applications allow the mobile device to be tracked by GPS from a central location.

Intensive Interviewing

Another type of in-person interviewing is more similar to a conversation, with the interviewer working with a list of general questions. The questions asked, their wording, and their order can vary from interview to interview. This type of interviewing is called *intensive interviewing* or, as some experts call it, *responsive interviewing* to indicate that the interviewer may change the questions and their order depending on how those interviewed respond.[45] Researchers may apply intensive interviewing techniques when interviewing a group. One type of group interview is the focus group, discussed later in this chapter.

For administrative researchers, both structured and intensive interviews provide rich data to improve a research design, elaborate on the statistical findings of a study, or triangulate with another method. Before discussing the intensive interviewing process further, we illustrate its value to designing a research study through an example.

This example helps illustrate how interviews can help identify appropriate measures, especially in program evaluations. A colleague once participated on a team studying the effects of discontinuing free transportation on low-income patients receiving renal dialysis. The administrators wanted to know "if anyone had died as a result." Interviewers constructed a sample and called patients or their families to learn whether patients had missed dialysis sessions and their current health status. They also interviewed staff at the dialysis centers. One employee mentioned that health status was the wrong outcome measure, "since the patients would do anything to avoid dying." She argued that dietary status and adherence to medication schedules were more valid measures of the impact of termination. Her observation led investigators to reconsider how they designed program evaluations; they began to include interviews with program constituents and other stakeholders as part of a design.[46]

In this example, interviews helped program evaluators to learn about the details of the program they were going to study:

- background
- objectives
- processes
- accomplishments
- failures

Whether used to plan another study or to obtain the primary information desired, effective interviewing requires practice and careful preparation. Specific populations may present special challenges. One colleague, for example, tells about interviewing members of a white supremacist group. He arranged to meet them on street corners after going through extensive questioning to assure them that he was not conducting a criminal investigation. Furthermore, the respondents insisted on anonymity, which required elaborate procedures for conducting the interviews.

Interviewer Skills

You may occasionally read studies based on intensive, unstructured interviews. Typically, the interviewers have training in clinical interviewing, for example, in counseling, social work, psychology, or a similar therapeutic discipline. Nevertheless, the interviewer borrows some techniques, such as good listening skills, practiced by professionals such as counselors and psychiatrists. The ability to get the respondent to talk openly and comfortably is also important. The interviewer also needs to have a good understanding of the subject matter of the interview.

Finally, the interviewer needs to have had a sufficient amount of interview experience and practice to avoid wasting a subject's time, overlooking important material, or distorting responses. Interviewers conducting intensive interviews must be more well trained and more experienced and possess more information about the subject matter of the interview than do those doing structured interviewing.

Whom and When to Interview

Whom to interview and when to interview them depend on the study and logistics. Some simple guidelines apply to most interviews.

The interviewer should not speak with the most important respondents first.

By most important, we mean the persons whose information, opinions, and insights are expected to be most valuable. Normally, interviewers can identify important respondents. They may be more difficult to reach and more protective of their time. Interviewers do not expect them to provide basic descriptions or information. The rule not to interview important respondents first arises from the logic that one does not want to throw away an important interview by asking the "wrong" questions. Initial interviews also often suggest additional items to be included.

Factual questions covering widely known or easily attainable information constitute a type of question that may be "wrong" for some respondents. Evaluators often start with knowledgeable program staff or constituents willing to describe program operations and accomplishments. These early interviews bring the interviewer up to date on the program, its operations, and current issues.

After she has practiced her interview skills for the specific project, identified current issues, and begun to formulate hypotheses, the interviewer can learn more from later respondents. She can test out her hypotheses with these respondents. Furthermore, she may bring insights and observations to the interview that the respondent will find valuable. We often forget that an interview can benefit the respondent. Frequently, in their day-to-day work, managers have little time to reflect upon their programs, why they developed as they did, how they work, and how they could be changed. The first interviews serve as a learning period during which the researcher seeks respondents who will be sympathetic to gaps or errors in her information. Delaying interviews with difficult respondents too long also has its costs. During the initial interviews, the interviewer is most open and flexible. Later, as the interviewer becomes more convinced of the correctness of her hypotheses, people with a different point of view may be considered "exceptions," and their insights may be ignored or overlooked. Interviewing respondents with a diversity of views is especially recommended to identify unexpected outcomes of or problems with programs or policies.

Good Listening Skills

Interviewers need good listening skills to tap the insights of respondents. Otherwise, they may misinterpret respondent comments or override them with their own points of view or agendas. Common listening errors include reaching a premature conclusion about the correctness of a respondent's observations. An interviewer who does this may stop listening. Alternatively, the interviewer may become anxious during periods of silence and either begin to fire questions rapidly or put words in the respondent's mouth. In both cases, the interviewer loses opportunities to learn from the respondent. Other common complaints about interviewers are that the interviewer did not know enough, the interviewer did not use the right vocabulary, and the interviewer interrupted with irrelevant comments or had to have too much explained. These errors tend to put respondents off, and they will become less involved with the interview.

Example of an Effective Interview

Example 6.4 summarizes a project based on interviews with female police officers. One of us worked with a colleague to learn how female officers participated in a police department:

- How did they learn about their jobs?
- In what policing activities were they involved?
- What were their career goals?
- How were they planning to achieve them?

Example 6.4 Interviews With Female Police Officers

Problem: Academic researchers wanted to learn about the socialization of female police officers, particularly their attitudes toward using force and the military.

Procedure:

1. Obtained permission from the police chief to interview officers.
2. Arranged to interview two or three officers together for approximately 1 hour.
3. Developed a set of interview questions to ask the police officers:

   a. How are women different from men in their approach to police work?

      1. Different emphases in law enforcement, service, community relations?
      2. More/less likely to deviate from department rules?
      3. More/less committed to a career in police work?
      4. More/less integrated into the social aspects of police work as a profession?

   b. Once you were out of the academy, who helped you become experienced as a police officer?

      1. What kinds of partnership arrangements did you have?
      2. What role did other women play?
      3. Was your immediate supervisor involved?
      4. What kinds of problems arose most frequently? Who showed you how to handle them?
      5. Did you sense antagonism to women as police officers?

   c. What have been your major problems as a police officer?

      1. Work problems—dealing with the public, other officers, supervisors, handling physical or other aspects of assignments.
      2. Personal problems—family, social image.

   d. The researchers found that in a survey of this department, most officers who answered said that force was one of the things about policing they liked least. Some people who have studied police, however, have found that the best police officers are those who do not avoid the use of force and are not uncomfortable about the fact that it is sometimes necessary.

      1. Can you explain why the officers in this department answered that force was the part of the job they liked least?
      2. Have you ever had to use force in your work?

3. Do you think that men handle this part of the job better?
4. Is this a problem for policewomen?

e. How will this job make a difference in your total career?
f. In your police training and in your work, does the subject of being "professional" come up? What do they mean by that term?
g. What is most attractive/least attractive about police work for women?

1. Scheduled interviews: placed specialized and senior officers midway in the interviews; scheduling depended on officer availability.
2. Conducted interviews:

a. The two researchers introduced themselves and the purpose of the study.
b. Interviews were largely conversational.
c. Both researchers took notes during the interview and wrote up their notes immediately following.

The interviewers met the women in small groups to encourage free discussion and interaction among the officers. The group interviews seemed to work well for this exploratory study. Interviewers introduced themselves and explained the purpose of the study at the beginning of the interviews. They mentioned their criminal justice experience. With most of the officers, the interview proceeded smoothly. Interviewers asked the basic questions and followed up with specific questions as needed. Generally, before asking the next question, they glanced at the interview schedule to be sure all the planned topics had been covered. Occasionally they interjected with comments such as, "Could you explain that?" or "Can you give me an example?" to encourage the officers to expand upon their answers. Also, to ensure against projecting their own professional experiences into the answers, the interviewers often repeated what the women had said. For example, one interviewer might say, "Let me repeat what I think you said to make sure I got it right."

Table 6.5 Overview of In-Person Interviews

	In-person
Costs	Highest cost (personnel, travel)
Sampling issues	No distinctive issues
Response rate	Highest (problems with not-at-homes)
Turnaround time	Slow (depends on sample size and dispersion)
Respondent issues	Convenient, if appointment scheduled; allows access to hard-to-reach populations
Content	Allows for in-depth, probing questions; good source for qualitative detail; valuable for exploratory studies; respondent–interviewer interaction may introduce bias; use of self-administered interviewing may encourage disclosure of embarrassing or sensitive information

Taking Notes

Before interviewing, an investigator must decide how to take notes. A method that the interviewer is comfortable with works with most subjects. For example, interviewers who use tape recorders report little resistance. Nevertheless, an interviewer who uses a tape recorder may want to jot down the important points of the interview. Transcribing or even listening to an entire interview can be costly in money or time. Most interviewers still take notes and leave plenty of time immediately after the interview to amplify and organize their notes. Unless done promptly, the details fade quickly. The interviewer should not conduct back-to-back interviews. The interviewer may become fatigued, and the content of one interview may merge into the next. Some interviewers work in teams of two, with one person conducting the interview and recording it on tape and the other taking written notes. Table 6.5 provides an overview of interviewing.

Focus Groups

Investigators assemble focus groups to get in-depth information on and reactions to a few topics in any one study. Survey researchers, program evaluators, and program planners often construct focus groups. Surveyors use focus groups to identify appropriate survey questions or to supplement survey findings. Program evaluators use them to delve into the experiences of program clients. Program planners use them to get reactions from potential service recipients. The hallmark of focus groups is group interaction—members of the group can respond to answers given by others in the group. Investigators believe this interaction produces data and insights that would not be available if data were collected from separate individuals.

No one focus group represents a population; however, most studies include several focus groups to ensure that their insights and observations extend beyond any one group. Recall that the best evidence of external validity is replication, that is, achieving similar results under somewhat varying conditions. Similarly, if several focus groups express similar attitudes or experiences, the investigators may persuasively argue that the groups represent the opinions and experiences of a larger population.

A focus group study involves five major phases: preparation, forming groups, conducting interviews, analyzing notes and records of interviews, and reporting results (see Figure 6.1). Steps 1 through 3 are discussed here. Analyzing notes and reporting results are discussed in a separate section following that.

Figure 6.1 Phases of Focus Group Study

Step 1: Preparation

Preparing for a focus group is like preparing for any other empirical research project. The investigators clarify the study's purpose, what information is needed, why, and by whom. They write a limited number of open-ended questions for the group to discuss. Everyone in a focus group should have a chance to discuss each question and to react to each other's comments. Thus, the questions must be carefully thought out to ensure that they will spark interest and develop the needed information.

Step 2: Forming a Group

Focus groups bring together a small group, ranging from 4 to 15 members, of unacquainted people who agree to meet for a focused discussion. Generally, each member should have some common characteristic related to the topic of the study. The group should be small enough for everyone to have a chance to participate but large enough to provide some diversity of opinion. In larger groups, some members may dominate the discussion; some members may have little or no chance to talk because there never is a pause long enough for them to speak. In small groups, members may hesitate to voice contradictory opinions and fail to move beyond listless conversation. Experts differ in their recommendations regarding the maximum size of groups. However, the optimal size depends on the topic, the participants, the skill of the moderator, and the purpose of the study.[47] For example, complex topics presented to knowledgeable participants would require and work well with smaller groups.

Researchers recruit focus group members who are homogeneous on some important characteristics. They may have the same occupation, have the same income level, receive services from a public agency, or have a common experience, such as recently losing a job or having heart surgery. Group members must sufficiently encourage open discussion. If a group represents widely different socioeconomic levels or lifestyles, social dynamics may inhibit free or productive conversations. People with markedly different experiences may not work together effectively in a group. For example, large-scale farmers and family farmers may have an interest in the implementation of new farm legislation, but their different perceptions, concerns, and needs may cause such a group to spend most of its time clarifying and overcoming differences.

Focus group members are selected from lists of persons with the required characteristics for a particular study. Potential members are contacted by letter or telephone and are asked if they wish to participate. Potential members are again screened, often in more depth, to make sure that they meet the criteria for participation. For example, each focus group of farmers may consist of farmers who are actually responsible for running the farm, who work on similar-sized farms, and whose farms generate similar amounts of revenue annually.

Step 3: Conducting the Focus Group

The staff needed to conduct a focus group consists of a moderator and one or two persons to take notes and address logistics such as setting up any recording equipment. Other researchers may observe, either behind a one-way mirror or off to the side of the room. The moderator should be familiar with the study's topic; however, she may be less knowledgeable than an intensive interviewer. She may be "brought up to speed" by learning about the questions she will ask and their purpose.

A focus group usually meets for less than two hours. The meeting begins with the moderator's brief description of the study. Then, in a somewhat spontaneous fashion, she leads the group through the list of questions. Her job is to keep the group going and to prevent it from bogging down on unproductive conversations. As soon as the focus group session ends, the moderator and any observers meet to discuss and record their impressions of the group and the interview.

Focus Group Example

Example 6.5 highlights the focus-group segment of an evaluation of a program offering college courses in prisons. The program director wanted to learn what the inmates thought of the program. Focus groups seemed to be the best way to develop this information. (For an additional example of focus group use and reporting, see the "Focus Group Report" accompanying this electronic textbook.)

Example 6.5 Focus Groups With Prison Inmates

Problem: The director of a prison program to provide college education to inmates wanted to know if the program was meeting their expectations. The evaluators decided to survey the prisoners who had taken college courses in the program and to supplement the surveys with information collected in focus groups.

Procedure:

1. Survey forms were created, focus group questions were written, and necessary permissions and authorizations were obtained.
2. The evaluators met with inmates at three prisons. At each prison, the evaluators met with the group at one location. The evaluators introduced themselves and the purpose of the study, then the inmates filled out a questionnaire.
3. Randomly selected inmates were asked to stay for a discussion of educational programs.
4. A research team member acted as a moderator and led the inmates through a discussion based on the following questions:

   a. What do you think about the survey?
   b. What do you consider to be the value of education?
   c. What do you think will be the value of the courses you are taking?
   d. What benefits do you expect from taking courses here?
   e. How do prison administrators view the program?
   f. What kinds of courses would you like to see offered?
   g. How would you explain to "John Q. Public" that his taxes help pay for someone who committed a crime to take college courses? The discussions proceeded smoothly, with inmates commenting on or reacting to each other's remarks. Security restrictions prevented recording the focus group at the maximum-security prison; the other groups were recorded.

5. The research team met after each session to discuss and record the information; the tapes also were transcribed for further study.
6. The focus group data were included in both the written and oral evaluation reports. Some of the comments supplemented survey data. Discussions on the value of education and the types of courses desired delved into concerns that could not be easily addressed or ascertained by the questionnaire. Comments on the hostile attitudes by the custodial staff addressed an issue that was completely ignored by the survey items.

Discussion: This study violated the focus group model that the participants should not be acquainted. The focus groups elucidated inmates' opinions of the program, and group interaction yielded information that would not have been picked up by a survey or individual information. The survey helped to focus the discussion.

The team working on this project had no previous experience in running focus groups or in prison research. They found that the focus groups provided far superior information than they would have obtained with just a survey. Their judgment mirrors our experience that focus groups add immeasurably to investigators' knowledge of how a program is perceived and how well it is working.

Steps 4 and 5 (analyzing data and reporting results) are discussed next.

Analyzing and Reporting Data From Intensive Interviews and Focus Groups

Since intensive interviews and focus groups primarily provide qualitative information, this seems to be an appropriate place to include a few comments about analyzing qualitative information. Much of the quantitative analysis is done with the aid of computer programs. Similarly, several computer programs are available for use in managing and analyzing qualitative data. We discuss these in this section as well.

Analyzing

Analysis of information from focus groups and intensive interviews depends on the quality of the investigator's records. The investigator should have thorough and accurate notes or recordings of the interviews. He may want to have both. In addition, he should leave time after each interview or focus group to write down his observations and to fill in any gaps in his notes. An individual interview or focus group session produces extensive data. When he reviews his notes or recordings, the investigator looks for trends and patterns that reappear among various participants or groups. The quality of the analysis and final report will depend on the analyst being familiar with the data. He should read all notes and transcripts and listen to recordings of the proceedings, preferably more than once.[48]

The analysis must be systematic and verifiable. Systematic analysis follows a prescribed sequential process. The analysis is verifiable if another researcher can use the same process, information, and documents to reach a similar conclusion.

Analysis of data from focus groups and intensive interviews is driven by the purpose of the study. It also is directly related to the questions asked. Throughout, the analyst must keep the purpose of the study in mind. The objectives should guide the analysis. Researchers clarify and state objectives and ask questions to obtain information according to the purpose of the study. The questions elicit the material that will be analyzed. Clear, well-stated questions related to the study's purpose will make the analysis easier. Responses to complex or confusing questions will be difficult, if not impossible, to analyze. The analyst typically examines each question separately and summarizes the information produced in response to that question. He also should consider the context in which comments were made and what triggered them.

Interpreting and Reporting

To begin with, the analyst may focus on general trends and patterns. He may then do a more detailed analysis in which he codes topics and even phrases and groups those that are similar. He also will consider the range and diversity of information that he heard. In addition to focusing on themes, the analyst needs to note the intensity of statements.[49] After the information is summarized, the analyst interprets it. Of course, the quality of the interpretation relies heavily on the analyst's insights, intellectual talents, and previous experience with the subject matter and with analyzing qualitative data.[50]

In the final report, the analyst should review the main purpose and objectives of the study and review how those have been met. He should also note the individual questions and write at least one sentence regarding each but typically does not report at length on each. Researchers often include quotes from the interviews to more fully interest the reader and to illustrate the discussion.[51]

Computerized Analysis

There are computer programs that are available to aid in analyzing qualitative data. These range from standard word-processing programs to very specialized software. They vary in the range of capabilities and price. Some accomplish numerous tasks and will handle a large set of data, whereas others will do only the basics, such as identify themes and create categories. The size of the dataset that can be handled also varies from program to program.

The U.S. Centers for Disease Control and Prevention has available a free program, CDC EZ-Text. Similar to most text analysis programs, it is "a software program developed to assist researchers [to] create, manage, and analyze semi-structured qualitative databases."[52] This program can be downloaded free of charge. The publisher of a popular statistical package, Statistical Package for the Social Sciences (SPSS), also has a program for qualitative analysis. This package is IBM SPSS Text Analytics. It works best with short-answer open-ended items.[53] Two popular programs with a large capacity are HyperRESEARCH and NVivo.

Potential users are encouraged to investigate different programs on the Internet. Each has a web site.[54] Researchers should assess what kind of analysis they want done and what they need from a program. Some of the programs are easier to use, and some take time to learn. The user should assess what technical help is available. In addition to information on qualitative data analysis programs, see the previous reference to Rogers and Goodrick for general information on qualitative data analysis. The book by Lyn Richards is recommended for information on qualitative data analysis programs.[55]

Using Social Media

Social media has exploded in recent years. Given its widespread use, we might ask what role it plays in research. Is social media used to collect data? Recruit subjects? Analyze data? Although some researchers have used social media to collect data, its greatest benefit for researchers appears to be to enable them to work together on projects more effectively and to communicate results to a wider audience more easily.[56]

Types of Social Media

The term *social media* encompasses numerous digital communication programs. The number and style of programs that can be included has also grown rapidly. It may not be possible at any one time to list them all. Facebook, YouTube, and Twitter may be the most commonly known and used social media. However, many other digital programs also qualify. For example, Dropbox and Google Docs allow us to collaborate and save changes in conjunction with others. Blogs are also used by researchers. While many social media programs are used to communicate small amounts of information and messages between and among researchers, some are used as depositories for sharing large documents.

Two authors, Kaplan and Haenlein, categorize social media programs by the type of content usually involved, the type of user, and how people use it. One very common use is for researchers to collaborate on projects. Kaplan and Haenlein note that many researchers also use social media to seek information and sources of information. They state that social media is "a platform to facilitate information exchange between users."[57]

Using Social Media

Minocha and Petre's book, *Handbook of Social Media for Researchers and Supervisors*,[58] is primarily about ways that researchers can use social media to better communicate about research. In it, they present several innovative resources for developing and maintaining a social media strategy for research dialogue. They also discuss ways to communicate with the public and sponsors using social media. They report a comment by G. Small that the real value of social media for scientists is in teaching them to communicate concisely and in forcing them to think about how to share ideas and results with a broader audience.[59]

Minocha and Petre identified and obtained information from over 100 researchers in the United Kingdom, United States, Europe, and Asia using mailing lists, discussion forums, and the researchers' presence on social media such as blogs, Twitter, and Facebook. In addition, they conducted a more traditional survey using a web-based survey program. As noted earlier, Minocha and Petre were not interested in the researchers' use of social media for conducting research but rather in how they communicated about research. They found that researchers used e-mail extensively to communicate about research. They concluded from their results that social media helped expand literature searches; Skype and Elluminate were important for web conferencing. Dropbox, using cloud computing, was popular for sharing files and folders.

Much of the material they gathered was textual and did not fit into a quantitative analysis program. It required thematic and other qualitative analysis.[60] A limitation of using social media sources as data is the lack of computer programs that can be broadly applied to them. Most of these data are textual and require qualitative analysis. One source noted that "realizing the value of social media requires innovative computing."[61] To use qualitative programs requires an additional step of moving the data from original source documents to a single file or database that can be used to manage the data. Researchers wishing to obtain a mix of quantitative and textual data are likely to use a survey, in which case a web-based survey with available software is likely to be used. The data can then be recoded easily into a data analysis program.

As researchers rely more on cloud-based and other online methods for communication, sharing research, and storing data, we can expect social media communications to be used as source documents for data. In other words, we may see more research similar to that of Minocha and Petre. The development of more analytic programs is sure to follow. Even today, experiments using social media as part of the process have been designed and implemented. This is likely to spur greater use of social media for data collection in the future.

Selecting a Method of Data Collection

The data collection method selected will depend to a great extent on the nature of the study. The study topic and the required sample interact to suggest the appropriate method. For example, an investigator should consider mailing a complicated questionnaire, or one with many open-ended questions, only to an articulate and highly motivated sample. Studies with inarticulate or socially isolated populations require face-to-face interviews. If the research question requires extensive qualitative information, intensive interviewing or focus groups work best. To explore how stakeholders view program plans or how clients react to an existing program, focus groups may be most effective. Table 6.6 compares characteristics of the various methods of data collection.

Comparing Mail, Telephone, Internet, and In-Person Surveys

Researchers have compared structured data collection methods, that is, mail, telephone, Internet, and in-person surveys. Their findings suggest that no one factor alone, with the possible exception of cost, automatically points to the desirability of one method over another. Problems such as

Table 6.6 Characteristics of Data Collection Techniques

	Mail	Internet	Telephone	In-person
Costs	Relatively low cost (printing, mailing, and postage, incentives for responding)	Low cost	Higher cost (personnel, equipment, telephone charges)	Highest cost (personnel, travel)
Sampling issues	Requires a good sampling frame; can distinguish nonrespondents from nonrecipients if undelivered questionnaires are returned	May not provide a good sample of general public, although probability samples are possible; a good sampling frame may allow access to specific populations	Provides a reasonably good initial sample of general public with random digit dialing; must have a mechanism for selecting respondents within a household to ensure generalizability; sample coverage constrained by increased reliance on cell phones	No distinctive issues
Response rate	Low (follow-up required)	May be difficult to determine; if general population is being surveyed, it may have lowest response rate of all	Higher (may need to vary calling times to get a representative sample)	Highest (problems with not-at-homes)
Turnaround time	Slowest	Quick	Quick	Slow (depends on sample size and dispersion)
Respondent issues	Convenient, especially for hard-to-reach professionals; poor if subjects are elderly or poorly educated	Requires Internet access, awareness of survey, motivation to access survey	Most intrusive; allows easy access to dispersed populations	Convenient, if appointment scheduled; allows access to hard-to-reach populations
Content	Highly motivated respondents can look up information, write detailed answers; other respondents require clear, easily answered questions	May have some of the benefits of computer-assisted interviewing; may encounter privacy concerns; respondents limited in their ability to communicate if questions seem ambiguous or responses inappropriate	Avoids biases caused by respondent reading ahead; allows for probing and elaboration of answers; may give socially acceptable answers; respondent fatigue if too long	Allows for in-depth, probing questions; good source for qualitative detail; valuable for exploratory studies; respondent–interviewer interaction may introduce bias; use of self-administered interviewing may encourage disclosure of embarrassing or sensitive information

response rate, questionnaire length, complexity, or sensitivity can be resolved to accommodate the constraints of the data collection method. In our presentation, we identified some general advantages and disadvantages of each type of data collection technique. We have summarized these characteristics later in the chapter.

Trends in Data Collection

Prior to selecting a data collection technique, the investigator should carefully consider the appropriateness of conducting a survey. The rate of refusals of the general public has increased since 1950.[62] Refusal rates as high as 50 percent for telephone surveys are not uncommon now. The ability to successfully contact respondents may further decrease response rates. Overnight and one-day polls may have response rates of less than 30 percent.[63]

Another trend researchers have noted is the incidence of multiple participation; that is, a respondent who participates in more than one sample survey a year. A few years ago, one of the authors was asked to participate in three telephone surveys within a month. One was a political survey, and the other two were marketing surveys. Her opinion of how long a telephone survey should take changed when she saw how an interview can interrupt family routine.

A 1984 study on survey participation found that 23 percent of the respondents interviewed had been surveyed during the year; over half of them had been interviewed more than once during the year. The interview subjects were more likely to be female and younger, wealthier, and better educated than subjects who were not interviewed. The investigators concluded that these differences reflected sampling techniques designed particularly by market researchers to capture the opinions of certain segments of the population. They suggest that oversurveying may increase both refusal rates and multiple participation.[64]

The problems with multiple participation have not been explicitly studied. Investigators wonder whether cooperative subjects overrepresent some characteristics measured in a survey or if multiple participants evolve into "professional respondents," thereby affecting the representativeness of the data.

From our own experience, an analogous case can be made with organizational studies. Just as market researchers target certain populations, trends in administrative studies focus on certain organizations and professions. Imagine how many questionnaires are sent to city managers, municipal police departments, and public administration faculty by students and faculty doing academic research. Add to this professional associations and government agencies that compile statistical data or conduct studies to answer specific questions. Furthermore, city managers and police departments have to contend with inquiries from the public, public officials, and journalists. You may begin to see how repeated requests for information may overwhelm and annoy the most cooperative individual.

Before deciding to conduct an original survey, investigators, especially academic researchers, should attempt to identify and locate existing databases. If no appropriate existing data can be found, the decision to survey should proceed only after carefully pretesting the data collection instrument. The less experience the investigator has with the problem at hand, the more modest the research effort should be. If investigators limit their survey efforts to well-thought-out and carefully designed instruments that respect the time and experience of respondents, refusals to respond by organization members and by members of specific samples should decrease, and the quality of the data should improve.

Summary

The data collection strategy is integral to implementing a research plan. How an investigator contacts potential subjects and obtains data from them may determine the success of a project (see Figure 6.2). Mail, telephone, Internet, or in-person surveys each have strengths and weaknesses.

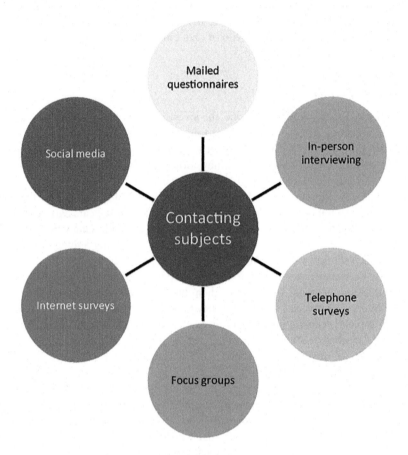

Figure 6.2 Contacting Subjects and Obtaining Information

Investigators should consider both the study and its population and choose the data collection method best suited to meet the study objectives.

Mail

Mail surveys generally cost the least of any traditional survey method, although Internet surveys can also be quite low in cost. The benefit of lower cost must be balanced against the time required for data collection. Mail surveys allow motivated respondents time to look up requested information and to provide detailed answers to open-ended questions. Mail surveys require the investigators to take time to make sure that the final questionnaire is free from confusing or ambiguous items, because no one will be there to answer the respondent's questions about what information is wanted. Complex, multipart questions also must be kept to a minimum.

Mail surveys work best with educated respondents who are used to working with forms. These respondents will not be intimidated by open-ended questions where they have to express their ideas in their own words. Mail surveys seem appropriate for many organizational studies in which data must be located and more than one person may be involved in answering the survey. Furthermore, in organizational research, the mail survey can be squeezed into the respondents' schedule, whereas telephone or in-person surveys may be disruptive and hard for the investigator to schedule.

The major weakness of mail surveys has been their low response rates. Investigators have found that even in surveys of the general public, who are the least motivated to respond to most surveys, mail surveys perform as well as telephone or in-person surveys if the researchers follow up with nonrespondents. To check the assumption that the nonrespondents are similar to respondents, researchers should compare the known demographic characteristics of the nonrespondents and the respondents. Also, with mail surveys, a random sample of nonrespondents may be telephoned or visited and asked to answer the questionnaire. Their answers may be compared with those of other respondents to assure the investigators that respondents and nonrespondents are not systematically different.

Internet

Internet surveys may work well in surveying a discrete group with easy Internet access. The low cost of Internet surveys may be offset by the more common problems of finding a good sampling frame, making potential respondents aware of a survey, and motivating them to answer it. These surveys are becoming more popular and can work well, especially for an identified sample whose members can be sent notices or follow-up contacts. Internet surveys are often conducted with software that provides immediate results on simple items.

Telephone

Telephone surveys are becoming more difficult to conduct given the movement of households to cell phones and spam-blocking software. People often no longer answer calls from an unknown number. While in the past, random digit dialing solved the inability to reach homes with unlisted telephone numbers and reduced the cost of selecting a representative sample, this is becoming less and less the case. However, technology that assists telephone surveys can provide rapid turnaround in data collection time when live interviewers do not have to be available. This also reduces costs, but they are likely harming the response rate because people are more likely to hang up on a computer than a live interviewer. Live interviewing can also be run more efficiently now than in the past because it relies on the computer to help administer a survey. The interviewer keys in the answer; the computer then selects the appropriate next question, reducing the logistical problems associated with making sure the interviewer finds the right follow-up questions. In both computer interviewing and live interviewing, the computer keeps track of inconsistent responses, stores the data, monitors the interviewer's performance, and often uses software that can provide immediate statistical results. The cost of acquiring the equipment, programming each survey, and training interviewers makes computer-assisted interviewing feasible only for organizations that constantly conduct large telephone surveys.

Other limitations of telephone surveys are their length and inability to elicit information on sensitive subjects. Although most nonrespondents have already refused to participate in a survey before the questioning commences, length is the most common complaint about telephone surveys. Long surveys may lead a respondent to refuse later requests for information. In addition, answer quality may degenerate toward the end of a long survey, or the respondent may quit answering altogether.

In-Person Interviews

In-person interviews, which approximate telephone interviews, have limited use, insofar as costs and improvements in telephone surveying methods have made them less desirable. Nevertheless, in-person interviews seem to work best for complex studies where interviewer–subject rapport is necessary.

Structured interviewing depends on standardized interviewing. Without standardized interviewing protocols, a researcher does not know whether the answers obtained are a product of respondent characteristics or of those of the interviewer. To conduct structured interviews, interviewers must be consistent in how they explain the study's purpose, ask questions, handle inadequate answers, and record information. Above all, the interviewer should not draw attention to his own experiences and opinions.

Intensive interviewing, a type of in-person interviewing, and focus groups can greatly improve administrative studies. The interviews allow investigators to consider the adequacy of their models and the appropriateness of their measures. Furthermore, interviews produce important background information necessary for understanding a problem, how it came to be, what solutions have been tried, and how they have worked.

Effective intensive interviewing requires practice. Some simple steps can minimize common errors. The first interviews usually are informational and focused on factual information. The interviewer normally avoids asking the "most important" subjects for relatively common information and waits to interview key subjects until she has a fuller understanding of the program or the problem at hand. Similarly, the interviewer avoids limiting her early interviews to people who share her own point of view. Unless people with another viewpoint are interviewed early, their insights and information may be inadvertently filtered out.

The interviewer must listen carefully to the subject. Interviewer training often works toward eliminating listening habits that cause interviewers to ignore or misunderstand what a subject is saying. Similarly, interview preparation includes familiarizing oneself with the topic of the interview and the applicable terminology or jargon; otherwise, the subject tends to spend too much time explaining.

Focus Groups

Focus groups have the same benefits as intensive interviewing. They also foster group interaction and discussion, which can provide even more information for program evaluation and program planning. Focus group research requires more than one investigator; the responsibilities of moderating and recording should not fall on one person. A major challenge in conducting focus groups is to identify a discrete number of open-ended questions that will elicit discussion consistent with the study's purpose. Because focus group studies involve relatively few individuals who agree to participate, the findings cannot be unambiguously generalized to a larger population. Probably more important than selecting a method of data collection is asking the question, "Are original data needed?" As the number of researchers grows and the value of research findings spreads, people are asked for more and more information. Market researchers target certain groups known to have more discretionary money to spend. Academic and administrative researchers have analogous target groups. The result of all this survey activity may be multiple requests for information, which raises questions as to whether multiple respondents are truly representative and whether in the long run they will continue to answer questions willingly. We assume that the problem is not how many times one is asked but how worthwhile respondents feel their participation is. If a subject questions the worth of a study or his participation in it, we assume that he either will not answer or will give cursory answers.

Notes

1. This categorization of potential sample members is drawn from M. A. Hidigoglou, J. D. Drew, and G. B. Gray, "A Framework for Measuring and Reducing Nonresponse in Surveys," *Survey Methodology* 19 (1993): 81–94. See pages 82–84 for a discussion of components and calculations of various rates.
2. www.cdc.gov/nchs/data/nhis/earlyrelease/wireless201812.pdf.

3. www.pewresearch.org/fact-tank/2019/04/22/some-americans-dont-use-the-Internet-who-are-they/.

4. www.pewresearch.org/Internet/chart/Internet-use-by-age/.

5. Organizations receiving U.S. government support must have an Institutional Review Board to review and approve of research conducted by members of the organization. Most, if not all, universities have Institutional Review Boards, as do teaching hospitals.

6. Robert M. Groves, "The Three Eras of Survey Research," *Public Opinion Quarterly* 75 (2011): 861–871.

7. *Establishing and Sustaining Law Enforcement-Researcher Partnerships—Guide for Researchers*. Final Report. U.S. National Institute of Justice, 2005, available at www.theiacp.org/Portals/O/documents/pdfs/establishing-and-sustaining-law-enforcement-researcher-partnerships-guide-for-researchers/.

8. For another list of "Design Features to Reduce Unit Nonresponse," see pp. 189–195 of Robert M. Groves, Floyd J. Fowler, Jr., Mick P. Couper, James M. Lepkowski, Eleanor Singer, and Roger Tourangeau, *Survey Methodology* (Hoboken, NJ: John Wiley & Sons, Inc., 2004).

9. See F. J. Yammarino, S. J. Skinner, and T. L. Childers, "Understanding Mail Survey Response Behavior: A Meta-Analysis," *Public Opinion Quarterly* 55, no. 4 (1991): 613–639 for a more thorough discussion of research on response rates.

10. A. H. Church, "Estimating the Effect of Incentives on Mail Survey Response Rates: A Meta-Analysis," *Public Opinion Quarterly* 57, no. 1 (1993): 62–79.

11. See Ibid., 62, 68.

12. Ibid., 73.

13. J. M. James and R. Bolstein, "Large Monetary Incentives and Their Effect on Mail Survey Response Rates," *Public Opinion Quarterly* 56, no. 4 (1992): 442–453; J. M. James and R. Bolstein, "The Effect of Monetary Incentives and Follow-Up Mailings on the Response Rate and Response Quality in Mail Surveys," *Public Opinion Quarterly* 54, no. 3 (1990): 346–361.

14. For a discussion of nonresponse in surveys and how to handle the problem of nonresponse, see N. Bradburn, "A Response to the Nonresponse Problem," *Public Opinion Quarterly* 56 (1992): 391–397. See R. M. Groves, *Survey Errors and Survey Costs* (New York: Wiley Interscience, 1989/2004) for a description of methods for weighting the responses of a sample of "converted" nonrespondents. I-Fen Lin and N. C. Schaeffer evaluate two models for estimating information from nonrespondents in "Using Survey Participants to Estimate the Impact of Nonparticipation," *Public Opinion Quarterly* 59, no. 2 (1995): 236–258.

15. Lin and Schaeffer, "Using Surveys."

16. www.surveymonkey,com; https://siue.qualtrics.com; www.surveyshare.com.

17. C. Hewson, P. Yule, D. Laurent, and C. Vogel, *Internet Research Methods: A Practical Guide for the Social and Behavioural Sciences* (London: Sage Publications, 2003), especially Chapter 5.

18. D. A. Dillman, J. D. Smyth, and L. M. Christian, *Internet, Phone, Mail and Mixed-Mode Surveys: The Tailored Design Method*, 4th ed. (New York: John Wiley & Sons, Inc., 2014).

19. This section relies on the observations and findings reported by B. K. Kaye and T. J. Johnson, "Research Methodology: Taming the Cyber Frontier: Techniques for Improving Online Surveys," *Social Science Computer Review* 17, no. 3 (1999): 323–337.

20. For an accessible overview, see David Stanton, "Connecting Polls with Quality: An Interview with Sunshine Hillygus," *GfK Knowledge Networks* (2008), available at www.knowledgenetworks.com/accuracy/spring2008/hillygus.html. For an example of research using Knowledge Networks, see D. Sunshine Hillygus and Simon Jackman, "Voter Decision Making in Election 2000: Campaign Effects, Partisan Activation, and the Clinton Legacy," *American Journal of Political Science* 47, no. 4 (2003): 583–596, see especially p. 585, note 4.

21. The study described in this paragraph is D. R. Schaefer and D. A. Dillman, "Development of a Standard E-Mail Methodology: Results of an Experiment," *Public Opinion Quarterly* 62, no. 3 (1998): 378–397.

22. M. P. Couper, M. W. Traugott, and M. J. Lamias, "Web Survey Design and Administration," *Public Opinion Quarterly* 65 (2001): 230–253. Also see Hewson, Yule, Laurent, and Vogel, *Internet Research Methods*, Chapter 5, "How to Design and Implement an Internet Survey," which addresses several of the issues discussed here.

23. Hewson, Yule, Laurent, and Vogel, *Internet Research Methods*, Chapter 5.

24. Information on nonphone households and households with intermittent service is from S. Keeter, "Estimating Telephone Noncoverage Bias with a Telephone Survey," *Public Opinion Quarterly* 59 (1995): 196–217. See also T. W. Smith, "Phone Home? An Analysis of Household Telephone Ownership," *International Journal of Public Opinion Research* 2 (1990): 369–390, cited by Keeter, "Estimating Telephone Noncoverage Bias with a Telephone Survey," 198. The estimate of 3 percent of households with intermittent phone coverage is based on 1992–93 Current Population Surveys data cited by Keeter, "Estimating Telephone Noncoverage Bias with a Telephone Survey," 199.

25. Drew DeSilver, "CDC: Two of Every Five U.S. Households Have Only Wireless Phones," *Pew Research Center*, available at www.pewresearch.org/fact-tank/2014/07/08/two-of-every-five-u-s-households-have-only-wireless-phones/.

26. A. Kempf and P. Remington, "New Challenges for Telephone Survey Research in the Twenty-First Century," *Annual Review of Public Health* 28, no. 1 (2007): 113–126, doi:10.1146/annurev.publhealth.28.021406.144059.

27. See www.donotcall.gov/ and www.consumer.ftc.gov for information on the National Do Not Call List. The number of calls from telemarketers can be greatly reduced by using this list. Legitimate survey organizations are still allowed to call.

28. Keeter, "The Impact of Cell Phone Noncoverage Bias on Polling in the 2004 Presidential Election," *Public Opinion Quarterly* 70 (2006): 88–98.

29. A popular random digit dialing method is described in J. Waksberg, "Sampling Methods for Random Digit Dialing," *Journal of the American Statistical Association* 73, no. 361 (1978): 40–46. A random number is constructed by adding a two-digit random number to a phone bank (a phone bank is the first eight digits of a phone number, starting with the area code). The number is called. If it is a residential number, several other random numbers using the same phone bank are called. If the first phone number is not residential, the phone bank is discarded. J. M. Brick et al., "Bias in List-Assisted Telephone Samples," *Public Opinion Quarterly* 59, no. 2 (1995): 218–235, identify commercially available lists that categorize phone banks according to whether they contain any residential numbers. The Pew Research Center's description of how it applies random digit dialing to include cell phones is a good illustration.

30. Pew Research Center, "Our Survey Methodology in Detail," available at www.people-press.org/methodology/our-survey-methodology-in-detail/.

31. See Charles Herbert Backstrom, Gerald D. Hursh-Cesar, and James A. Robinson, *Survey Research, Literary Licensing* (Whitefish, MT: LLC, 2012), 115; S. Voss, A. Gelman, and G. King, "The Polls—A Review: Pre-Election Survey Methodology: Details from Eight Polling Organizations, 1988 and 1992," *Public Opinion Quarterly* 59, no. 1 (1995): 98–132 gathered data from eight organizations that do presidential polling. Only two organizations reported weighting by the number of phone lines, and both organizations used a weight of 1 for homes with only one phone line and 5 for homes with more than one phone line.

32. While in theory reducing costs, the feasibility of videoconferencing use is low. T. Weinmann, S. Thomas, S. Brilmayer, S. Heinrich, and K. Radon, "Testing Skype as an Interview Method in Epidemiologic Research: Response and Feasibility," *International Journal of Public Health* 57, no. 6 (2012): 959–961, doi:10.1007/s00038-012-0404-7.

33. See James H. Frey, *Survey Research by Telephone*, 2nd ed. (Newbury Park, CA: Sage Publications, 1989), 78–85, for a discussion and evaluation of strategies for selecting persons to interview. See Voss et al., "The Polls—A Review: Pre-Election Survey Methodology: Details from Eight Polling Organizations, 1988 and 1992," 110–112, for information on practices of specific polling organizations.

34. This discussion draws from Richard Curtin, Stanley Presser, and Eleanor Singer, "Changes in Telephone Survey Nonresponse over the Past Quarter Century," *Public Opinion Quarterly* 69 (2005): 87–98. The Survey Research Center conducts high-quality social science research, including a number of national studies. Among its regular studies are the Survey of Community Attitudes and the National Election Studies. Also see Pew Research Center.

35. D. A. Dillman, *Mail and Telephone Surveys: The Total Design Method* (New York: Wiley Interscience, 1978), 248.

36. Pew Research Center, "Cell Phones," available at www.people-press.org/methodology/sampling/cell-phones/.

37. M. W. Link and R. W. Oldendick, "Call Screening: Is It Really a Problem for Survey Research?" *Public Opinion Quarterly* 63 (1999): 577–589.

38. Backstrom, Hursh-Cesar, and Robinson, *Survey Research*, 115, suggest that more than three callbacks are inefficient and not worth the benefit of an increased response rate. Dillman, *Mail and Telephone Surveys*, 47–49, has an excellent discussion on substituting sample members in telephone surveys. For how to handle answering machines, see R. W. Oldendick and M. W. Link, "The Answering Machine Generation: Who Are They and What Problem Do They Pose for Survey Research?" *Public Opinion Quarterly* 58 (1994): 264–273; T. Piazza, "Meeting the Challenge of Answering Machines," *Public Opinion Quarterly* 57 (1993): 219–231; M. Xu, B. J. Bates, and J. C. Schweitzer, "The Impact of Messages on Survey Participation in Answering Machine Households," *Public Opinion Quarterly* 57 (1993): 232–237. The Piazza study, based on California data, considers the probability of getting through, the best times to get through, and the effect of multiple callbacks.

39. S. Keeter, C. Miller, A. Kohut, R. Groves, and S. Presser, "Consequences of Reducing Nonresponse in a National Telephone Survey," *Public Opinion Quarterly* 64 (Summer 2000): 135–148.

40. Groves, *Survey Errors*, 238, Chapter 5, for a thorough discussion on refusals to cooperate, efforts to increase cooperation, and research on noncooperation.

41. W. S. Aquilino and D. L. Wright, "Substance Use Estimates from RDD and Area Probability Samples: Impact of Differential Screening Methods and Unit Nonresponse," *Public Opinion Quarterly* 60 (1996): 563–573.

42. F. J. Fowler, Jr., *Survey Research Methods*, 5th ed. (Thousand Oaks, CA: Sage Publications, 2014), Chapter 8.

43. Research on survey mode and willingness to report sensitive information found that respondents using self-administered surveys seem more likely to disclose embarrassing information. CASI administered in school settings may compromise the accuracy of the responses if the respondents are physically close to one another. See R. Tourengeau and T. Smith, "Asking Sensitive Questions: The Impact of Data Collection Mode, Question Format, and Question Context," *Public Opinion Quarterly* 60 (1996): 275–304, which also summarizes previous research relating survey mode to disclosure of sensitive information; D. L. Wright, W. S. Aquilino, and A. J. Supple, "A Comparison of Computer-Assisted and Paper-and-Pencil Self-Administered Questionnaires in a Survey on Smoking, Alcohol, and Drug Use," *Public Opinion Quarterly* 62 (1998): 331–353; Y. Moon, "Impression Management in Computer-Based Interviews: The Effects of Input Modality, Output Modality, and Distance," *Public Opinion Quarterly* 62 (1998): 610–622.

44. Craig Leisher, "A Comparison of Tablet-Based and Paper-Based Survey Data Collection in Conservation Projects," *Social Sciences* 3, no. 2 (2014): 264–271, doi:10.3390/socsci3020264. "New Study Shows Using a Tablet Computer and QuickTapSurvey Cut Time Needed for a Survey by 46% and Cost by 74%," available at www.quicktapsurvey.com/blog/2014/05/27/research-data-collection-methods-paper-vs-tablet. Also see www.dooblo.net for information on SurveyToGo.

45. Herbert Rubin and Irene Rubin, *Qualitative Interviewing: The Art of Hearing Data*, 3rd ed. (Thousand Oaks, CA: Sage Publications, 2012).

46. See E. O. Sullivan, G. W. Burleson, and W. Lamb, "Avoiding Evaluation Cooptation: Lessons from a Renal Dialysis Evaluation," *Evaluation and Program Planning* 8 (1985): 255–259.

47. Richard A. Krueger and Mary Anne Casey, *Focus Groups: A Practical Guide for Applied Research*, 5th ed. (Thousand Oaks, CA: Sage Publications, 2014).

48. "Analyzing Interviews and Open-Ended Questions," in E. O'Sullivan, G. Rassel, and J. D. Taliaferro, *Practical Research Methods for Nonprofit and Public Administrators* (Boston, MA: Longman-Pearson, 2011), Chapter 10. This short chapter has much practical advice for analyzing interview data.

49. See Richard A. Krueger and Mary Anne Casey, "Focus Group Interviewing," in *Handbook of Practical Program Evaluation*, 3rd ed., eds. J. Wholey, H. Hatry, and K. Newcomer (San Francisco: Jossey-Bass, 2010), Chapter 17, pp. 396–402. They present a detailed outline for analyzing focus group data.

50. For additional information and examples on analyzing and reporting results from focus groups, see Krueger and Casey, *Focus Group*, Chapters 6 and 7.

51. The University of Texas, Instructional Assessment Resources, Evaluation Programs, "Reporting Qualitative Results," available at www.utexas.edu/academic/ctl/assessment/iar/programs/report/focus-Report.php.

52. Centers for Disease Control and Prevention, "CDC EZ-Text," available at www.cdc.gov/hiv/library/software/eztext/.

53. Patricia Rogers and Delwyn Goodrick, "Qualitative Data Analysis," in *Handbook of Practical Program Evaluation*, eds. J. Wholey, H. Hatry, and K. Newcomer (San Francisco: Jossey-Bass, 1994), 429–453, Chapter 19.

54. For HyperRESEARCH, see www.researchware.com/. For NVivo, see www.qsrinternational.com. For SPSS, see www.ibm.com/software/products/en/SPSS-text-analytics-surveys.

55. Lyn Richards, *Handling Qualitative Data*, 3rd ed. (Thousand Oaks, CA: Sage Publications, 2014).

56. S. Minocha and M. Petre, *Handbook of Social Media for Researchers and Supervisors* (2012), available at www.oro.open.ac.uk.

57. Quote from page 60 of Andreas Kaplan and Michael Haenlein, "Users of the World, Unite! The Challenges and Opportunities of Social Media," *Business Horizons* 53, no. 1 (2010): 59–68, available at www.sciencedirect.com.

58. Minocha and Petre, *Handbook of Social Media for Researchers and Supervisors*; italics added by authors for emphasis.

59. G. Small in *Nature*, as reported in Minocha and Petre. See the Small blog post itself, available at www.nature.com/naturejobs/science/articles/10.1038/nj7371-141a.

60. D. R. Thomas, "A General Inductive Approach for Analyzing Qualitative Evaluation Data," *American Journal of Evaluation* 27 (2006): 237–246. Also see Virginia Braun and Victoria Clarke, "Using Thematic Analysis in Psychology," *Qualitative Research in Psychology* 3 (2006): 77–101.

61. From Ben Shneiderman, Jennifer Preece, and Peter Pirolli, "Realizing the Value of Social Media Requires Innovative Computing Research," *Communications of the ACM* 54, no. 9 (2011): 34–37.

62. Research covering several decades confirms a continuing decline. See C. G. Steeh, "Trends in Nonresponse Rates, 1952–1979," *Public Opinion Quarterly* 45 (1981): 40–57; P. Farhi, "How Would You Answer This One?" *Washington Post*, April 14, 1992, A1, A4; R. Rothenberg, "Surveys Proliferate," A1, A6; T. W. Smith, "Research Notes: Trends in Non-Response Rates," *International Journal of Public Opinion Research* 7 (1995): 157–171.

63. Farhi, "How Would You Answer?" A4.

64. S. Schleifer, "Trends in Attitudes Toward and Participation in Survey Research," *Public Opinion Quarterly* 50 (1986): 17–26.

Terms for Review

response rate
nonresponse rate
random digit dialing
computer-aided telephone interviewing (CATI)
Internet surveys
structured interviewing
intensive interviewing
focus groups
social media
blog

Questions for Review

The following questions should indicate whether you have a basic competency in this chapter's material.

1. Why would a low response rate to a survey concern a researcher?
2. What is the value of being able to distinguish between the response rate and refusal rate? Can researchers distinguish between response rates and refusal rates on mail surveys? On telephone surveys? On Internet surveys? In in-person interviews? Explain.
3. Assess the advisability of investigators' assuming that respondents are similar to nonrespondents. What is the value of telephoning a sample of the nonrespondents to a mail or Internet survey?
4. Explain what random digit dialing is and its value for telephone surveys.
5. Compare and contrast information you would collect to pretest a mail survey, a telephone survey, or an Internet survey.
6. When should structured, in-person interviewing be used instead of telephone interviewing?
7. What is the value of intensive interviewing? What type of people would you most likely target for an intensive interview? Why?
8. Some students tend to "wing it" if they have to do intensive interviewing. Assess the dangers of this strategy.
9. Contrast the advantages of using focus groups to mail surveys to learn about each of the following. In each case, state which method you would recommend.

    a. alumni opinion about the program you are studying
    b. parents' reactions to a proposed countywide after-school program

10. Suggest and justify a data collection method for each of the following proposed studies:

    a. to find out how many agency employees are attending continuing-education classes
    b. to understand how parole board members make parole decisions
    c. to identify the demographic characteristics of city managers in a state

Problems for Homework and Discussion

1. Examine an issue of a journal such as *Public Administration Review* or *Administrative Science Quarterly*. Select an article based on data collected from individuals or organizations and, for that article, identify and report:

    a. who the research population was
    b. how the subjects were contacted (mail, telephone, Internet, in person)
    c. what the response rate was
    d. how the investigators handled subjects who did not respond to the first contact, that is, how did they follow up and what were the number of follow-ups? Compare your results with your classmates'. Can you make any generalizations linking the nature of the study population to the method of contacting subjects and linking the number of follow-ups to the response rate?

3. Two Internet surveys are designed. Survey 1 asks state, city, and county managers what benefits they offer employees. Survey 2 asks public employees about the benefits they receive from their employer. How would you contact respondents for each survey?

4. You want to learn more about research: how projects are selected, models built, measures operationally defined, and new data collected.

    a. Outline what you would ask in an unstructured interview of a faculty member at your university.
    b. Your instructor may assign each student to interview a faculty member about their research. If so, exchange your interview report with one of your classmates and critique that interview.

5. You decide to construct a study of the effectiveness of your academic program. What data collection method reviewed in this chapter would you use? Whom would you interview or survey (give general characteristics of the potential interviewees or respondents)? How many people would you include? How would you select them? If you tried to cover different groups, say students and faculty, in what order would you interview or survey them? What questions would you ask? Sketch out a research plan for this study. Compare and discuss the research plans in class. Would you base the entire study on your chosen method? If not, what alternative data collection methods would you use, and when would you use them?

Working With Data

Two files containing examples of data collection instruments or studies discussed in this chapter accompany this text. The first file contains a survey instrument used in the General Social Survey. The second file, Focus Group Report, reports the results of focus groups used in a larger study. The following items require you to use the information in those files.

1. Open the General Social Survey file and review the survey item named Fear. Imagine you are interested in learning more about why people fear walking in their neighborhood at night beyond the quantitative data. Working with two of your classmates, describe a

procedure you would use to develop a focus group to pair with the survey data. Include the following:

    a. Develop a list of no more than six questions or items to be discussed.

    b. How you would recruit members for the focus group?

    c. Information you would need from the survey data prior to designing the focus group.

2. Open the Focus Group Report file and read through the report. Respond to the following:

    a. What changes would you make in the list of questions?

    b. What additional information would you want in the report?

    c. Based on the report, what, if anything, would you have done differently with the focus groups?

Recommended for Further Reading

Many how-to-do-it books on survey research are available. The following two go through the entire process from designing a study to analyzing the results: Rea, Louis M., and Richard A. Parker's, *Designing and Conducting Survey Research: A Comprehensive Guide*, 4th ed. (San Francisco: Jossey-Bass, 2014); Arlene Fink, *How to Conduct Surveys: A Step-by-Step Guide*, 5th ed. (Thousand Oaks, CA: Sage Publications, 2012).

Also recommended for public administrators is Foltz, David, *Survey Research for Public Administration* (Thousand Oaks, CA: Sage Publications, 1996). Another excellent source is Dillman, D. A., J. D. Smyth, and L. M. Christian, *Internet, Phone, Mail and Mixed-Mode Surveys: The Tailored Design Method*, 4th ed. (Hoboken, NJ: John Wiley & Sons, Inc., 2014).

The following is recommended as a volume that brings together chapters on how to do surveys and practical advice with theoretical underpinnings of survey research: *Survey Methodology*, 2nd ed. (Hoboken, NJ: Wiley Interscience, 2009) by R. M. Groves, F. Fowler, M. Couper, J. Lepkowski, E. Singer, and R. Tourangeau.

An excellent resource book examining and comparing errors in mail, telephone, and in-person interviewing is Groves, R. M., *Survey Errors and Survey Costs* (New York: Wiley Interscience, 2004).

For information on unstructured interviewing, a classic text is Dexter, L. A., *Elite and Specialized Interviewing* (Evanston, IL: Northwestern University Press, 1970); Rubin, Herbert, and Irene Rubin, *Qualitative Interviewing: The Art of Hearing Data*, 3rd ed. (Thousand Oaks, CA: Sage Publications, 2012) is also informative.

For a how-to-do-it guide on focus groups, see Krueger, R. A., and Mary Ann Casey, *Focus Groups: A Practical Guide for Applied Research*, 5th ed. (Thousand Oaks, CA: Sage Publications, 2014).

Readers interested in focus groups also should consult *The Focus Group Kit*, eds. David L. Morgan and Richard A. Krueger (Thousand Oaks, CA: Sage Publications, 1998). The entire kit contains six volumes covering all aspects of focus-group research, including planning, question development, moderating, and analyzing and reporting results, among other topics. The volumes are available individually.

For up-to-date research findings on surveying strategies, see *Public Opinion Quarterly*, the journal of the American Association of Public Opinion Research, published by the University of Chicago Press. *Social Science Computer Review*, published by Sage Publications, provides timely information on web surveying and data collection software. An excellent source to read about the potential for using social media is *Social Media, Sociality, and Survey Research*, eds. Craig Hill, Elizabeth Dean, and Joe Murphy (John Wiley & Sons Inc., 2013). Undoubtedly more experts will address this topic in the future.

Political Analysis is the journal of the Society for Political Methodology and the American Political Science Association's Political Methodology Section and is published by Oxford University Press. It routinely carries articles on topics such as how to analyze massive quantities of text using automated methods.

7 Collecting Data With Questions and Questionnaires

In this chapter, you will learn:

1. About the role of model building and the model in designing a survey.
2. About the various types of questions for questionnaires.
3. The proper roles for open-ended and closed-ended questions.
4. How to properly word questions and structure questionnaires.
5. How question content and wording affect reliability and operational validity.
6. Ways to avoid asking unnecessary questions and forgetting important ones.
7. Pretesting methods to prevent wasting resources on surveys.

Writing questions and designing questionnaires and surveys are important research skills for administrative investigators. Agencies use questionnaires in program planning, monitoring, and evaluation. Surveys help agencies plan programs by identifying problems, assessing support for policy alternatives, and learning what services are needed. Administrators may use questionnaires to monitor programs by compiling information on client characteristics and what services clients use, when, and how often. Evaluators use questionnaires to gather information from clients about their satisfaction with a program, its actual practices, and its effectiveness. Social scientists rely on questionnaires to collect a wide range of data on individuals and organizations.

Writing questionnaires may seem easy. Yet, as we implied in our discussion of cross-sectional designs, poor-quality questionnaires are common. Poor questionnaires may result in low response rates, unreliable or invalid data, or inadequate or inappropriate information. In our classes, we have observed that students identify a problem for which they need data and begin immediately to write a questionnaire. Potential questions flow quickly as they brainstorm. We believe that similar behavior occurs in offices when someone wants information about other people or organizations.

What is wrong with a questionnaire that comes out of such a session? The questionnaire may be inappropriate for the study's purpose. It may ask too few questions or the wrong questions. Unless it is adequately reviewed and pretested, serious errors affecting reliability, such as ambiguous questions or confusing directions, may not be detected. The underlying concepts may not be clarified, resulting in invalid measures. So while this may be an important step in questionnaire development, much more is required for a good survey.

Quality questionnaires require well-worded questions, clear responses, and attractive layouts. In addition, investigators must use systematic procedures to decide exactly what to ask. First, the investigators must clarify the study's purpose, ascertain that the proposed questions are consistent with this purpose, and determine whether the administrators believe that the survey information will be adequate. Second, the investigators want to make sure that the questions and the overall data collection strategy will yield reliable and operationally valid data. Third, they want to verify that the data meet the needs of the intended users. Investigators may meet with intended users to confirm that the study will produce appropriate and adequate information for their needs.

Creating Quality Questionnaires

A city may contract for a citizen survey to find out what services citizens most want and who wants them. Staff who plan to use the survey data should meet with the survey designer to discuss how they want to use the data and to review the survey draft. If a survey designer fails to ask respondents where they live or how old they are, staff will not know if demands for better recreational facilities vary by neighborhood or age group. Alternatively, a designer may ask about potential services that the city has no intention of providing. The questions may falsely imply planned changes in city services, irritating those who feel the city is doing too much and misleading those who think the city is not doing enough.

This chapter examines strategies for developing questionnaires that yield useful data. It begins by considering strategies for making sure that an instrument asks the right questions. It presents techniques for writing effective questions and questionnaires. It suggests a strategy for mapping out the questionnaire to confirm that it is complete. The chapter discusses how to pretest questionnaires. Investigators cannot accurately assess the clarity of the questionnaire or its completeness without a pretest. This chapter stresses the necessity of thoroughly pretesting all questionnaires before using them to collect data and conducting a pilot study to see how a questionnaire and the data collection strategy actually work.

Questionnaire and Question Content

A questionnaire with short, easily answered questions may seem to have required little effort to write. But, in fact, questionnaire writing can be tedious, involving several drafts and more than one pretest. Questionnaire writing involves deciding what variables to measure, writing questions that accurately and adequately measure the variables, assembling the questions in a logical order on a questionnaire, pretesting the questions, and conducting a pilot study.

In deciding what to include in a questionnaire, researchers need to identify the variables they want to measure, the type of questions that will measure the variables, and the number of questions needed to ensure reliability and operational validity. Many analysts, including ourselves, have labored on questionnaires only to discover later that they asked questions that were not used in the analysis. Some questions were not needed; others were useless because the respondent misunderstood them. Researchers also fail to ask questions critical to the study.

Taking care to construct the questionnaire properly can help minimize the occurrence of such events and is well worth the time and effort.

Survey Models

The first component that defines a survey's content is the model. We are certain that many surveys are written with no reference to a model. It may have faded into memory, but more likely, no one ever constructed an explicit model for the study. Model building requires investigators to identify and link variables needed to answer a question or solve a problem. Prior to building a model, investigators need to know why it is being built, that is, its purpose. Similarly, investigators need to know a survey's purpose. They also should be convinced that a study is necessary and that a survey is better than alternative methods of data collection.

The investigators identify those variables for which data are needed. They must find or develop measures for each variable in the model. Other variables should be included only if they serve a clear purpose. Next, the investigators decide what linkages among variables need to be studied. Explicitly listing the variables and the linkages of interest should guide the debate about whether a specific question is necessary.

As useful as we find model building, for some projects, developing a coherent, integrated model may be inefficient. Some users may actively resist efforts to build a fully articulated model. Investigators conducting a study for a specific user, for example, an administrator or a legislator who wants the information to help make a decision, may prefer to focus on the specific questions that the user wants to have answered. Michael Patton, in a classic program evaluation text, suggests that investigators first identify the purpose of the study and its primary users. Next, the investigators ask the primary users to indicate how they plan to use the study to answer their questions or to solve a problem.[1] The discussion between the investigators and primary users should detail exactly what information they need, how they will use it, and whether it will be adequate. With this information, questions can be formulated to yield the needed data.

Decisions on what questions to include may extend over the planning period. Later in the chapter, in Example 7.7, we show a procedure to track how each question in a questionnaire will be used. The procedure should prevent asking unneeded questions and failing to include necessary questions. The investigators also should analyze the data gathered during pretests or pilot studies. The data analysis may help the investigators and users to evaluate the adequacy of the questionnaire. However, in general, the data gathered in pretests should not be included in the final analysis of data gathered by the survey. The questions may have changed after the pretest, or the respondents completing the pretest may not have been selected in the same fashion as other respondents.

Types of Questions

The second component of content is the type of question being asked. Knowledge of question type can help you decide the operational validity of a measure. For example, survey analysts may categorize questions as gathering information on:

- facts
- behaviors
- opinions
- attitudes
- motives
- knowledge

Factual questions elicit objective information from respondents. The most common factual questions are demographic questions, such as sex, age, marital status, education, income, profession, and occupational history. Factual questions may be asked about anything on which someone wants information and may ask about the present or the past, such as:

- How much money does a participant in a charity walk-a-thon raise?
- How long does it take a city to issue a building permit?
- How many miles of highway does a state pave annually?
- Has the respondent ever had the measles?

Behavior questions, which are a type of factual question, ask respondents about things they do or have done. For example, residents may be asked how far they travel to work or whether they have ever attended a public hearing. They may be asked if they have volunteered and, if so, for how many hours a month.

Opinion questions ask people what they think about a specific issue or event. Opinions are said to be the verbal expressions of attitudes.

Attitude questions try to elucidate more stable, underlying beliefs or ways of looking at things.[2]

Motive questions ask respondents to evaluate why they behave in a particular manner or hold certain opinions or attitudes.

Knowledge questions, which are similar to test questions, determine what a person knows about a topic and the extent or accuracy of the information. Knowledge questions may act as filters to eliminate the answers of uninformed respondents. Knowledge questions commonly appear in evaluations of training programs to find out whether subjects learned what was taught. Knowledge questions can measure what a population knows about a policy or issue. They can indicate whether citizens know that a particular service or program exists.

Choosing the Correct Question Type

Investigators who know a model's purpose and understand how different types of questions produce different information avoid collecting the "wrong" data. Consider a problem in program evaluation. Participants may be asked their opinions of a program. For example, trainees may be asked to rate the quality of a training program they had completed. Administrators, however, may want to know whether training increased participants' knowledge or changed their behavior on the job. The administrators may question maintaining a program that does not produce these results. They may find opinions interesting, even helpful. Respondent opinions may lead to future program changes. Still, the opinions do not indicate whether the program has performed as desired.

Question wording, emphasis, and order may have a greater effect on opinion and attitude questions than on factual questions (see Table 7.1). Yet for any type of question, the respondent must understand what information is sought. For example, in a housing survey, respondents may report fewer bedrooms than their houses contain because they do not think of a den, a playroom, or a home office as a "bedroom." If the questionnaire writer wanted to count such converted rooms as bedrooms, she may be stuck with unreliable data.

Table 7.1 Types of Questions

Questions	Identify the Question Type	What Is the Purpose of the Question?
How much did you pay for your first house?	Factual questions	To elicit objective information from respondents
Have you ever attended a public hearing?	Behavior questions	To ask respondents about things they do or have done in the past
Do you think the city government should spend money for a professional sports arena?	Opinion questions	To ask people what they think about a specific issue or event
Do you consider yourself more of an optimist or pessimist?	Attitude questions	To elucidate more stable, underlying beliefs or view of the respondents
Why do you contribute 10 percent of your income to your church every year?	Motive questions	To ask respondents to evaluate why they behave in a particular manner or hold certain opinions or attitudes
What is your U.S. senator's position on reducing tax deductions for charitable contributions?	Knowledge questions	To determine what a person knows about a topic and the extent or accuracy of the information

Number of Questions

The third component of content is the number of questions to be asked to measure each variable. The number of questions or indicators affects a measure's reliability. To avoid burdening respondents, we should keep the number of questions to a minimum. Thus, we have to consider the trade-off between lower reliability and higher response rates.

The number of questions also affects operational validity. Many variables cannot be adequately measured with only one indicator. For example, employee empowerment can be measured by four variables: employees' belief that:

1. Their work is meaningful.
2. They are competent to do their job.
3. They have autonomy in performing their work.
4. Their work has an impact on the organization.

Similarly, examining satisfaction with a police department may involve asking citizens to rate the department's response time; its ability to prevent crime and to solve crimes quickly; how fairly officers treat all citizens, especially young people and minorities; and the department's general competency. Measures of variables, such as empowerment or satisfaction with police services, requiring multiple questions or other items to determine the variables' breadth and depth, are called composite measures.

The number of questions and operational validity problems are of special concern when gathering organizational data. To determine organizational or program effectiveness requires more than a single indicator. Imagine if the only information collected on the performance of a job-training program was the percentage of trainees placed. Relying on this one piece of data may inaccurately measure program effectiveness. Information on the quality of trainees' jobs—for example, pay rate, skills required, and stability of employment—gives a more complete and valid picture. Furthermore, measures that have inadequate content may seriously distort staff behavior. Staff members may focus on the one measured indicator, for example, the percentage placed, and accept only trainees with a high placement potential, or they may place graduates in low-paying or temporary jobs to keep the placement rate up.

Questionnaire Structure

Questions may be either open ended or closed ended. *Open-ended* questions require the respondent to answer in her own words. *Closed-ended* questions ask the respondent to choose from a list of responses. For example, respondents may check whether they "strongly agree," "agree," "disagree," or "strongly disagree" with a statement, or they may indicate a number representing the most appropriate answer.

Open-Ended Questions

Researchers ask open-ended questions for at least five reasons.

1. They help a researcher identify the range of possible responses.
2. They avoid biases that a list of responses can introduce.
3. They yield rich, detailed comments.
4. They give respondents a chance to elaborate on their answers. Just as "a picture is worth a thousand words," a comment can add immeasurably to an investigator's information.

5. Finally, respondents can more easily answer some questions with a few words rather than by selecting an answer from a long list of possible responses. For example, the question, "In what state do you live?" is easier to answer by writing a state name than by looking for the state on a list.

Open-ended questions have two major drawbacks.

*   *First, they are often unanswered.* Respondents with little time or limited communication skills may ignore them.
*   *Second, they complicate data compilation.* Categorizing and counting the answers can be a formidable task. In March 2008, the American National Election Studies (ANES)—which conducts surveys on a range of political attitudes and behaviors—released a report on problems that had been uncovered with open-ended questions on its surveys used to measure political knowledge.[3]

DECIDE: A Management Training Program

Open-ended questions are most valuable during the early stages of questionnaire design. For example, one of the text authors was once a member of a team studying how those enrolled in DECIDE, a management training program, used their training.[4] The team did not know how the trainees used the decision-making techniques taught in the class, nor did they know if trainees had found unique uses for the techniques. The team asked trainees for examples of how they applied DECIDE. Some respondents could not give a work-related example, but they did give examples of using DECIDE to make personal decisions. If the team had written a list of responses before asking the trainees for examples, they would not have included personal applications and would have overlooked one reason DECIDE was popular with employees. Using the examples, the team categorized and summarized how employees used DECIDE and then wrote appropriate closed-ended questions.

For example, one question asked:

*   How often do you use DECIDE to help you make budget request decisions?
*   How often do you use DECIDE to help you make purchasing decisions?
*   How often do you use DECIDE to help you make personnel allocation decisions?
*   How often do you use DECIDE to make a decision regarding something at home?

The team selected specific decisions from the list of common uses in constructing the questions.

Political Polling and Policy Alternatives

In political polling, the movement from open-ended to closed-ended questions is well established. Long before American political parties and voters express any formal preference for a presidential candidate, pollsters may ask respondents to name the person they prefer for the next president. This procedure allows pollsters to identify candidates with early support. It avoids the possibility that respondents will simply choose a familiar name, in which case the poll is measuring name recognition rather than support. As Election Day approaches, open-ended questions will have disappeared, replaced by closed-ended questions that focus on actual, active candidates.

A similar pattern occurs as decision makers study policy alternatives. Administrators may ask constituents to indicate policies that they want adopted. Example 7.1 presents several open-ended questions from a state health planning survey of county health directors.

Example 7.1 Application of Open-Ended Questions

Problem: State planners are developing policies and programs to address the problem of teenage pregnancy. County health agencies will administer the programs.

Strategy: To make sure planned programs are acceptable, the planners write a series of questions to get needed information and survey county health directors. The following questions are included:

1. Describe the problem of teenage pregnancy in your county.
2. In your opinion, what programs or policies would help with the problems of teenage pregnancy? Make any additional comments on the strengths and weaknesses of these solutions. Your comments will help us formulate statewide plans.
3. What is the size of your annual budget?

Discussion: Initially, planners informally reviewed the answers to select the most feasible programs and policies. The planners also compiled the answers and presented them systematically in a report. The answers to Question 1 disclose how the directors view teenage pregnancy; the question avoids distortions introduced by a set of responses. The answers to Question 2 may reveal innovative programs or policies; later closed-ended questions and responses may be developed from the answers. Categorizing the answers can be tedious. Reliability problems may occur if directors either describe similar programs with different terms or use similar terms to describe different programs. Question 3 was open-ended because the planners did not know the range of possible responses. A list of responses could have been either insensitive to actual variations or too long. The answers are relatively easy to categorize and present quantitatively. State health planners asked the directors to suggest programs and policies to address the problem of teenage pregnancy. Question 1 asks for a description of the problem of teenage pregnancy in the director's county. From the descriptions, the planners can learn about local conditions and programs, agency attitudes about teenage pregnancy, and acceptable programs. For example, what information is selected to show that teenage pregnancy is a problem? What, if any, causes of teenage pregnancies are cited? Are the needs of young teenage parents distinguished from the needs of older teenage parents?

Question 2 asks directors to suggest programs or policies to address problems of teenage pregnancy in their counties. From the comments, the planners can hypothesize which programs will be acceptable and which ones will be unacceptable. Innovative solutions or programs may be suggested, and unanticipated benefits or costs of possible programs may be mentioned. The answers to Questions 1 and 2 help the planners find out more about the policy environment. They may learn how much time county directors have spent thinking about and learning about teenage pregnancy programs, whether any programs have been planned or implemented, and what local conditions would affect the success of a program.

Benefits to Open-Ended Questions

Including open-ended questions during the process of questionnaire design may reduce biases introduced by the responses attached to a closed-ended question. Respondents may stick to the available responses, or they may use characteristics of the response to decide the most appropriate answer. In the case of DECIDE, a list of work-related uses would not have captured personal uses. Questions asking respondents to rate effects of climate change might encourage them to rate effects that had never occurred to them and ignore unlisted concerns that bothered them a lot.

Answers to open-ended questions provide the rich detail that puts a mass of collected data into context. If a policy is being planned or evaluated, a comment may add immeasurably to an investigator's understanding about the policy and how the public views it. The answers may help explain the findings and how the respondent interpreted a question. Respondents may be invited to explain their answers to a preceding question. For example, respondents may be asked whether they are doing better, worse, or about the same economically this year as last. Then, they will be asked to explain why they answered the way they did. From the explanations, investigators learn how respondents evaluate their economic situation and what factors they believe contribute to their economic well-being. Comments from open-ended questions are normally incorporated into a report. The comments add to its readability and keep the attention of readers who are less interested in the quantitative analysis.

An open-ended survey gives the respondent the opportunity to transmit information directly to the investigator or study sponsor. The respondent may wish to comment on related issues and bring them to the attention of others. The research team that carried out the survey of DECIDE trainees ended the survey by asking respondents whether they had comments that they wanted to pass on to the training department. Several did; one respondent had a lengthy complaint about problems with computer training. In community surveys, respondents can comment on the policies being considered, comment on other policies, or mention related concerns they have about local government or agency operations.

The answers to broad, open-ended questions may add to the administrator's general information. They may contribute to respondents' positive feelings about surveys. Soliciting a respondent's opinion may serve as minimal compensation for his time, effort, and cooperation. Finally, an open-ended question at the end of a questionnaire could ask the respondent about any shortcomings in the questions or provide additional thoughts. If a researcher has the space and time in the survey to do so, a catch-all question of this type can be very valuable in interpreting the results.

Problems of Open-Ended Questions

Still, you can infer the problems of open-ended questions from Example 7.1. First, a respondent must be motivated to spend the time and effort to answer an open-ended question. It is typical that many respondents skip long or open-ended questions. In online survey software, you can require responses before the respondent can move on to the next question. You can also require a minimum number of characters be utilized prior to the respondent moving to the next question. But researchers should be cautioned by overusing these techniques; they may cause more respondents to terminate the questionnaire early in an online survey where respondents are voluntary. Institutional review boards may require that the respondent receive their incentive up front (if using one) and also require the researcher to explicitly state the respondent is free to quit the survey at any point. It is important to note that open-ended questions can be quite valuable in online surveys because they can serve as attention checks to make sure the respondent is actually thinking through their answers to questions.

In Example 7.1, we may assume that the survey about teenage pregnancy was important to the directors and that they took the time to write thorough, thoughtful answers. The directors were probably motivated by more than the content of the survey. Since the survey was sponsored by a state health department, the directors may have assumed that their answers would affect state policy. More often than not, however, respondents will not be so motivated.

Respondents may answer open-ended questions more readily if they are asked over the telephone or in person than online, and this may allow the interviewer to probe for more clarification. It is essential interviewers be properly trained. Otherwise, they may add their own interpretation to respondents' answers, amplify the question in a way counter to the investigator's intentions, or fail to record the comments accurately.

Using Open-Ended Information Effectively

The other problem of open-ended questions is how to use the information effectively. If relatively few people are surveyed, the answers may be carefully read, analyzed, and quoted. If a large number of people are surveyed, the probability of carefully reading the answers diminishes. A sample of answers could be selected, but that approach has disadvantages. Sampling of responses is useful in some contexts but may not be appropriate, since in qualitative research, looking at the outliers, not just the majority, is important. Sampling may miss interesting and important comments that researchers could explore later when resources allowed.[5] Therefore, it is important to determine the need for the question to begin with. If the responses will not receive the appropriate analysis, it may be best to omit the question.

Categorizing and Counting Open-Ended Responses

Categorizing and counting open ended responses is difficult and time consuming and requires thoughtful attention. A perusal of the questionnaires by someone familiar with the study's model should help create categories. The difficulty rests in assigning responses accurately and consistently. To ensure reliable data based on open-ended questions, the investigators should apply a test/retest procedure to a random sample of cases. If several analysts are categorizing responses, inter-rater reliability also should be established. To do so, each analyst categorizes the same sample of cases. Then, an investigator compares the work of the analysts to see if they reported similar results.

Closed-Ended Questions

Closed-ended questions come in several formats. All closed-ended questions require a respondent to select one or more appropriate responses from a list. The reliability and operational validity of closed-ended questions partially depend on the list provided.

A list may include several options with instructions to "check all that apply." While such a list may burden respondents less, they may not actually check all that apply. An unchecked item (1) may not apply or (2) may have been overlooked. More items may be reported if a respondent is asked to check an appropriate response for each item, for example, "yes" or "no."[6]

Including an "other" category does not guarantee that additional appropriate responses will be mentioned. Some respondents may limit themselves to the list. They may do this out of laziness, because they have identified a desirable response, or because the list has led them to accept the investigators' frame of reference. Even if they answer the "other" category and write in a response, the investigator may be stuck interpreting and categorizing these responses.

Forced-Choice Questions

Some response lists do not include "other," "no opinion," "not sure," or "not applicable" as choices. Alternatively, the question may include an even number of possible responses. This prevents a respondent from taking a more neutral position. Both types of questions are called *forced-choice questions*. Of course, respondents cannot be forced to choose, and some will pen in their own answers or ignore the question. In general, respondents should be given the option to express "no opinion" or indicate that the question isn't applicable. Otherwise, the investigator does not know why a question was skipped. If a web-based survey does not allow a respondent to proceed without answering, we assume that she may quit or randomly select an answer. Similarly, including a neutral response, such as "neither agree nor disagree" may increase reliability. Research has found that the ratio between positive and negative responses is the same whether or not a neutral response is provided.[7]

Common Formats of Closed-Ended Questions

Example 7.2 takes a selection of questions from a survey that a state employees' association (SEA) sent to its members.

Example 7.2 Application of Closed-Ended Questions

Situation: A state employees' association wants to survey its members to learn their opinions about SEA and state employment.

Strategy: Develop a closed-ended questionnaire to survey members. Format should be chosen to facilitate quick responses by members and easy tabulation. The questions include:

1. What is your opinion of why some employees do not join SEA? Check all that apply.

   _____ Money (dues)
   _____ Dislike organizations
   _____ SEA doesn't adequately represent employees
   _____ Don't consider their jobs permanent
   _____ Don't care
   _____ Don't know

2. How much influence do you believe the average SEA member has in SEA?

   _____ Very much
   _____ Quite a bit
   _____ Some
   _____ Very little
   _____ None
   _____ Don't know

3. Number the following in order of priority, 1 being highest, 2 next highest, etc. SEA should work for

   _____ payment of sick leave upon separation from the service
   _____ collective bargaining
   _____ payment of time and a half for overtime
   _____ reduction in retirement contributions
   _____ dental insurance
   _____ higher salaries
   _____ increased mileage and per diem rates

The questions illustrate some common formats used in closed-ended questions. You should visualize how easily an investigator can compile the responses. At the same time, you may think of how the responses could be misinterpreted. For example, some respondents may answer the questions on why others do not join the association by reflecting their own dissatisfaction with SEA or a similar organization. Members who have tried to recruit others to join may have more accurate perceptions. As the questionnaire is written, the analyst cannot separate the two types of respondents, and their answers will be grouped together.

Another type of closed-ended question asks the respondent to rank or rate items. Look at Question 3 in Example 7.2. Respondents are asked to rank organization activities proposed for an employee association. Inevitably, some respondents select several items and rank each item with a "1." A few may appear to reverse the rating scale. In the example, they would give a "7" to the highest-priority item and a "1" to the lowest-priority item. Some will rate all items; others will ignore all but a few items. Similar to ranking questions are questions that ask people to assign percentages to how much time they spend on a list of activities. A surprising number of people assign values that add up to well over 100 percent.

Note also that Question 3 does not include an "other" option. Consequently, respondents who wanted the organization to pursue on-site day care or flexible work schedules did not have an opportunity to voice their opinion.

Limitations of Closed-Ended Questions

The limitations of closed-ended questions are best understood by thinking of the strengths of open-ended questions. Closed-ended questions encourage the respondent to accept the investigator's response categories. Investigators may assume that they and the respondents interpret the questions and response choices the same way. A surveyor once asked agency heads about the nature of volunteers. The question asked respondents to note how many volunteers in their agency had certain characteristics. Included among the response categories was "Native Americans." When the surveyor wrote Native Americans, she was referring to American Indians, Hawaiians, Eskimos, and Aleutians. Only after she received several surveys indicating that "all" volunteers were Native Americans did she realize that her terminology was interpreted differently by the respondents.

To minimize misinterpretation of responses and inadequate or inaccurate response lists, closed-ended questions should be carefully pretested. Investigators also should satisfy themselves that their testing included procedures to review item reliability and operational validity.

Choosing the Question Type

The decision as to whether to choose an open-ended question over a closed-ended question depends on the type of information needed, who is supplying the information, the data collection method, and the time available for completing the study. Open-ended questions are important in the first stages of questionnaire design. They help with question wording and creating appropriate response categories. The type of respondent and the data collection method may affect the choice. Questions that require longer, thought-out answers, such as in Example 7.1, may work best if the survey is sent to a targeted sample of interested respondents. Less-interested and less-motivated respondents may avoid answering open-ended questions that require more than a few words. In-person surveys may be more successful in getting complete responses to open-ended questions; however, unless the interviewers are well trained, the information may be unreliable or invalid. Closed-ended questions can be compiled and analyzed far more quickly. Investigators should not waste respondents' time by asking open-ended questions unless they have budgeted time for the proper analysis of the answers.

Contingency Questions

Contingency questions identify respondents who should answer follow-up questions and direct the respondent or the interviewer to the appropriate parts of a questionnaire. Also known as skip patterns in online survey software, contingency questions are sometimes called filter questions. These questions are nearly always closed-ended. Some contingency questions are used to identify potential respondents who should not continue with the survey. In other words, the question filters out those who do not meet certain criteria. For example, a survey on voting behavior might ask: "Are you registered to vote in this district?" Those answering No would not continue with the survey. A contingency question in a housing survey might be:

Do you own or rent the house in which you now live?

_____ Own (answer Question 10).
_____ Rent (skip to Question 11).

Can you imagine one or more questions that apply to homeowners but not renters? For example, questions on mortgages, interest rates, and property taxes would apply to owners but not directly to renters.

An effective format leads a respondent to the questions he should answer. It should prevent him from missing questions that apply to him and from answering questions that do not apply to him. The Own/Rent contingency question has written directions next to each response category, telling respondents the next question they should answer. Visual coding is especially effective; for instance, see Example 7.3. Arrows may direct the respondent to the appropriate questions. In designing contingency questions, the investigator must take care to keep the questionnaire simple and the instructions clear. In Internet-based surveys, it is possible to design contingency questions to take a respondent automatically to the next appropriate question(s), given his or her answer. This process is called adding *skip logic* to the survey.

Example 7.3 Visually Coded Contingency Questions

Situation: A state survey of citizens includes questions on educational policy. Questions check compliance with requirements for periodic testing, contact between parents and teachers, and support for education-related policy.

Strategy: Questions on testing apply only to families with children in certain public school grades, questions on parent/teacher relations apply only to families with children in school, and questions on education-related policy apply to all respondents. The survey is conducted over the telephone. The arrows save time as the interviewer selects the appropriate questions based on a respondent's answer.

1. Were any children in your household enrolled in public school grades 1, 2, 3, 6, or 9 last year?

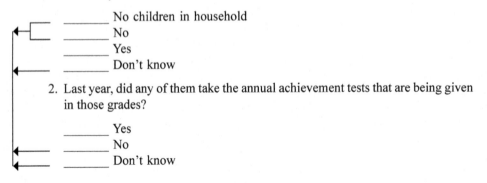

   _____ No children in household
   _____ No
   _____ Yes
   _____ Don't know

2. Last year, did any of them take the annual achievement tests that are being given in those grades?

   _____ Yes
   _____ No
   _____ Don't know

3. Were the appropriate persons in your household given the results of the tests by the school system?

_____ Yes
_____ No
_____ Don't know

→4. Do you think that free breakfast should or should not be provided for all students?

_____ Should
_____ Should not
_____ Don't know

Source: Questions from North Carolina Citizen Survey, Fall 1979.

Question and Response Wording

Respondents must see the value of a question. Respondents providing information for performance appraisals, program evaluations, or other management studies may resist spending time collecting and recording data that they do not value. They may distort data that they believe can adversely affect them. They may estimate or make up data to conform to management expectations or to get a form off their desk. Administrators and others who are frequently surveyed may also show resistance. They may start to answer a questionnaire and stop if it is too time consuming or if they perceive it will produce worthless information. We suspect that respondent perceptions are shaped as they read the specific questions.

Question Wording and Formatting

Each question should be worded so that the respondent accurately understands its meaning and answers honestly. An investigator wants to make sure that she and the respondent view a question as asking for the same information. Imagine a question asking, "Where do you like to shop?" Are you asking where you do your grocery shopping? Or do you prefer to shop online or in a physical store? And what do you mean by "like"? Are you asking if the store is customer friendly or has the best bargains? This type of ambiguous question confuses the respondent. Survey questions need to be precise for reliability and operational validity.

Questions should be clear, short, and specific. Words in common usage and not subject to different interpretations add to question clarity. Long questions tend to be complicated and confusing; they may be easily misinterpreted. Frequently, a respondent will consider part of a long question in his answer and ignore the question's broader meaning.

Specific questions leave little room for interpreting the time, place, or amount involved. To estimate clinic usage in a town, investigators may ask the following question:

"How many visits did you make to a medical clinic in Middletown between September 1 and December 31 of this year?"

If "Middletown" did not appear in the question, respondents might count out-of-town clinic visits. If the question said "recently" instead of "between September 1 and December 31," each respondent could answer the question with a different time interval in mind. Some respondents may think of the number of visits within the last month; others may think of the number of visits within the past year.

Challenges With Short, Specific Questions

The advice to keep questions short and specific may conflict. A vague term may confuse respondents and lead to unreliable data. Asking teenagers how many hours they worked or asking adults how many hours they exercise may raise questions about what constitutes work or exercise. The researchers must know exactly what they want to measure. They must know if their conceptual definition defines baby-sitting as work, or if working in the yard constitutes exercise. To solve this problem, the definition may be incorporated into the instructions. Alternatively, the definition may be incorporated into the body of the question. Example 7.4 illustrates:

1. including definitions in the instructions
2. including a definition in the question

In practice, Strategy 2 would include parallel questions about the behavior of top administration and the work group.

**Example 7.4 Two Strategies for Defining Ambiguous
Terms in a Survey**

Strategy 1: Key terms are defined as part of the instructions.
Actual wording and format included on cover page:
The following terms are used in the survey:

TOP ADMINISTRATION means the president/CEO and vice presidents or senior
  administrators
SUPERVISOR means the person to whom you report on a daily basis
WORK GROUP means the people you work with on a daily basis

Strategy 2: Define terms as part of the question.
Actual wording and format included in question:

"How often, if at all, does your supervisor, that is, the person to whom you report on a
  daily basis, overrule the findings or conclusions of a report and require that changes
  be made?"

1[ ] Never, or almost never
2[ ] Some of the time
3[ ] About half the time
4[ ] Most of the time
5[ ] Always, or almost always

Source: Strategy 2 item adapted from *Federal Personnel Management: OPM Reliance on Agency
Oversight of Personnel System Not Fully Justified* (Washington, DC: General Accounting Office,
GAO/GGD—93—94, December 1992), 51.

Improving Question Validity

Writing separate questions that include different components of the conceptual definition may improve a measure's operational validity. Example 7.5 illustrates a list of separate questions used to ask employees to rate their satisfaction with employee benefits.

Example 7.5 Dividing a Question Into Separate Questions

Strategy: Rather than ask one broad question, the investigator may divide a question into a series of questions.

The following questions ask respondents to indicate their level of satisfaction with various employee benefits. The survey directions told respondents to X out the number that corresponded to their response. The directions included a sample item with an X covering one of the numbers.

Actual wording and format: In these questions, please indicate how satisfied you are with each of the following employee benefits

(1 = very dissatisfied; 2 = dissatisfied; 3 = slightly dissatisfied; 4 = indifferent; 5 = slightly satisfied; 6 = satisfied; 7 = very satisfied).

How Satisfied Are You With:

Medical—Employee coverage	[1]	[2]	[3]	[4]	[5]	[6]	[7]
Medical—Family coverage	[1]	[2]	[3]	[4]	[5]	[6]	[7]
Dental insurance	[1]	[2]	[3]	[4]	[5]	[6]	[7]
Life insurance (Employee)	[1]	[2]	[3]	[4]	[5]	[6]	[7]
Dependent life insurance	[1]	[2]	[3]	[4]	[5]	[6]	[7]
Long-term disability	[1]	[2]	[3]	[4]	[5]	[6]	[7]
Pension plan	[1]	[2]	[3]	[4]	[5]	[6]	[7]

Notes: This question did not include a "Not applicable" category, which may affect the accuracy of the results. Employees who have neither dependent medical insurance nor life insurance may check "indifferent" or not answer the question. Either response is subject to misleading interpretations.

Response Statements

An alternative and preferred format has respondents react to specific statements.[8] For example, employees may indicate their degree of agreement with statements such as:

- "I am satisfied with the medical benefits I receive on my job."
- "I am satisfied with the life insurance I receive on my job."
- "I am satisfied with the pension benefits I receive on my job."

Possible responses may range from "strongly disagree" to "strongly agree." Asking people about their satisfaction with various components of their benefits package yields data that better represent overall satisfaction with benefits. Nevertheless, at some point, the cost of obtaining more accurate information must be weighed against the benefits. An example of erring by seeking too much specific information occurred in the Nationwide Food Consumption Survey. Respondents were asked to keep a two-day record of what they ate and whether they ate it at home or away, how it was prepared, and the size of the portion. Data were collected on extensive checklists, which included 350 foods. By putting the responses together, the investigators could learn that one day Jane Doe ate at home and had a boneless, roasted slice of chicken breast, 2-by-1 1/2-by-1/4-inch. The survey's low response rate of 34 percent was partially blamed on the burdensome questionnaire.[9]

In constructing a list of responses, investigators need to list all probable responses. Recall that this is a qualitative indicator of reliability. Respondents may resent response categories that seem to exclude something that applies to them. For example, listed responses may not allow a young adult to report that her job is only temporary or a disabled respondent to indicate that he depends on public transportation. The respondents may become frustrated or annoyed; they may toss out the questionnaire. Agency directors may react similarly. For example, an agency director whose volunteers deliver client services may be put off if a surveyor lists fund-raising and office work as volunteer tasks but does not mention delivering services.

The Importance of Being Specific

The question or the list of responses should indicate how precise an answer is expected. For example, a county official wanted to estimate how user fees would affect the use of county recreational facilities. He surveyed recreation department heads and asked them, "How much did the implementation of user fees impact the use of recreation facilities and programs?" The answers received were "very much," "very little," and so forth. A closed-ended question that gave the possible responses such as "less than 10% decrease in use of facilities," "10–25% decrease in use of facilities," would have avoided this problem.

Related to amount specificity is the need to ask a respondent for information she has or can easily get. A demand for specific information may require a respondent to calculate or search out data with which she could give a reasonably accurate estimate. For example, surveys asking for exact amounts of household income are not likely to get that information due to privacy concerns or the stigma attached to divulging salary and income information. The better approach often is to provide several categories, each with a range of incomes. The respondent chooses the one that is the closest to his or her household income. The questioner needs to keep in mind the purpose of the research and the trade-off between precision and the cost of information gathering.

Writing Questions for Self-Reports of Behavior and Opinions

Question and response wording are especially important in surveys that ask people to report their behavior, attitudes, or motives. Clearly defined questions requesting accessible information can produce reasonably accurate organizational information, for example, the number of full-time staff or the size of the budget. Furthermore, investigators can verify the accuracy of such information. Self-reports by individuals are harder or even impossible to verify; consequently, investigators have far less confidence in their reliability and operational validity.

Misreporting

People have models of how they should act, and their answers may be more in line with how they think they should act than with how they actually act. Alternatively, the answers people give may depend on how they store and retrieve information. Consider information on who votes. From time to time, researchers check voting records to establish the validity of self-reports of voting. They have found that roughly 15 percent of the people surveyed after an election misreport whether they voted, and a higher percent of those polled after an election report having voted for the winning candidate than indicated by actual votes.

The direction of misreports is not random. Typically, nonvoters "erroneously" report voting. Three reasons have been proposed to explain why misreporting occurs. First, those misreporting may wish to give a socially desirable answer. Second, they may be people who usually vote, and they forgot that they did not vote in the most recent election. Third, voting records may be in error. For example, a check of records may miss a voter whose name on the voters' rolls differs from the name she gave to the surveyor.[10] The accuracy of voting information has been studied extensively because self-reports can be validated and because the relatively large number of misreports may have distorted knowledge of American voting behavior. Because an act as straightforward and uncontroversial as voting is consistently misreported, we recommend interpreting all self-reported information with caution.

Similar problems of social desirability and misreporting occur with opinion questions.

Uninformed or disinterested respondents may nonetheless express an opinion. Including "don't know" as a response choice may alleviate this problem; however, investigators cannot infer what a "don't know" indicates. Respondents may answer "don't know" because they have problems answering the question. They may be ambivalent; that is, they may have thought about the issue but have yet to reach a firm opinion.

Some respondents may become distracted while hearing a question or have trouble understanding it. They will answer "don't know" to avoid the trouble of having the question repeated or trying to decipher its meaning. Thus investigators must examine any question that generates a high proportion of "don't knows" to make sure it is not ambiguous or otherwise flawed.

To get a snapshot of the public's opinion, the distinction between disinterested and ambivalent respondents may be less necessary.[11] If an investigator wants to identify disinterested respondents, she may ask a filter question. For example, respondents may be asked: "How interested are you in the bond referendum to build a civic center?"

An alternative solution, illustrated subsequently, asks a dichotomous question and follows up with an intensity question. Since asking a series of questions may tax respondents' patience, respondents may be simply reminded that "not interested" is an appropriate response.

1. Do you support or oppose the bond referendum to build a civic center?

    _____ Support
    _____ Oppose
    _____ Don't know

2. How interested are you in the bond referendum to build a civic center?

    _____ A lot
    _____ Somewhat
    _____ Not much
    _____ Not interested

Among the respondents who answered "don't know" to Question 1, investigators may assume that disinterested respondents answered "not much" or "not interested" to Question 2, and ambivalent respondents answered "a lot" or "somewhat" to Question 2.

Questions asking respondents how often they do something should cover a limited time span. Someone may accurately remember how many times he used a library in the past month. If he is asked about the past year, he may multiply the number of visits in the past month by 12. He is unlikely to adjust his estimate to indicate that he used the library more than usual or less than usual. A response list may serve as a frame of reference, and a respondent may alter his answers if they seem atypical.[12]

Avoiding Set Responses

If a measure includes a series of questions with similar answers, such as "agree" and "disagree," the respondent may answer all questions in the same way regardless of content. Serious distortions may occur if a respondent guesses the purpose of the questions.

For example, a survey to identify opinions about community development may have a number of questions about types and levels of development. A respondent who is generally opposed to development may answer all questions with a "disagree" and ignore the content of the specific questions. This problem can be reduced if all questions do not go in the same direction. In other words, for some questions a "disagree" would be a pro-development choice and for others a "disagree" would be an anti-development choice. Nevertheless, the respondent may become annoyed or confused with a questionnaire in which questions change direction capriciously. Some questions may be misread or misunderstood simply because the respondent missed a key word such as "not."

The items here, taken from a client satisfaction survey, illustrate how questions are worded to avoid set responses. Investigators conducted the survey to see how agency clients rated support staff. The respondent ranked each statement on a scale ranging from "strongly agree" to "strongly disagree."

1.   The agency staff can answer my questions.
2.   The agency staff is polite when I telephone.
3.   The agency staff is accessible by telephone.
4.   The agency staff is rude when questions are asked.
5.   The agency staff interrupts me when I talk with them.

A person who is pleased with agency staff should agree to items 1 through 3 and disagree with items 4 and 5. The shift from "agree" to "disagree" is referred to as a change in direction. The shift is achieved without resorting to awkward wording. A respondent should quickly and accurately grasp the meaning of each question. Still, no solution ever works all the time. One group of researchers grouped similar items together on a web survey. They found respondents were more likely to give the same answer for each item, including those items worded in the opposite direction.[13]

A major problem with changing the direction of items may occur when the data are compiled. The direction should not be changed when the data are coded and entered into a computer file. By keeping the original direction, the analyst reduces the risk of data-entry errors. Instead, she can easily change the direction later as part of the computer programming.

Biased Questions

A *biased question* is worded so that it encourages respondents to give one answer over another, eliciting inaccurate information. Although questions may be intentionally worded to lead the respondent to give a specific answer, biases also may be introduced unintentionally. A response list is often a source of question biases. Responses should cover the range of plausible options. One error is to use a rating scale with a disproportionate number of positive (or negative) ratings.

For example: a four-point rating scale with responses of "excellent," "good," "satisfactory," and "poor" has three positive responses out of four possible responses. This increases the probability that a person will express some degree of positive satisfaction. The failure to include a particular response on a list of responses also may create biases. Instead of writing in the appropriate information, a respondent may settle for checking off the next-best answer.

Loaded Questions

Loaded questions are worded so that a respondent gives an acceptable answer. Some adjectives or phrases tend to have a positive or negative value, thus leading a person to ignore the major content of the question. For example, labeling a proposed policy as "liberal," "permissive," or "bureaucratic" will tend to measure respondents' reactions to these words and not their general opinions about the policy. Also, questions may include an assumption affecting answers.

If we ask people whether they agree or disagree that the "state should build more prisons in order to decrease violent crimes," we are assuming that building prisons will decrease violent crime. The question may unwittingly lead respondents concerned about violent crime to support building more prisons.

Asking Two Questions in One

Another common, unintentional bias is to ask two questions in one. These questions (often referred to as double-barreled questions) are easily identified by the conjunction "and." Consider the question, "Are you in favor of tightening the state's DUI (Driving Under the Influence) laws and making it illegal for parents to serve liquor to minors in their homes?" A person may favor tightening up the DUI laws but not agree with making it illegal for parents to serve liquor to their minor children. How should the person accurately report her attitude?

Lack of Fit Between Questions and Responses

Another source of bias occurs if the available responses do not fit the question asked. For example, an author of this text once received a questionnaire with the following item:

Do you think that state annexation statutes need to be changed, or do you think that they are appropriate as they are?

_____ Yes
_____ No

The question writer was following the professional practice of wording the question so that it did not favor either changing the statutes or leaving them alone. Unfortunately, the responses were not appropriate. There was no way to accurately report an opinion short of writing out an answer. This problem is easily solved by altering the response wording.

The two responses to this item could be:

_____ Yes, statutes need to be changed.
_____ No, statutes are appropriate as they are now.

Other Considerations

Whereas inexperienced question writers unintentionally write biased questions, special-interest groups may intentionally word items to encourage a particular answer. A *New York Times* article reported on a study contrasting an interest-group question with questions written by professional surveyors.[14]

An example of a leading question and a balanced version of the same question is:

Version 1: "Should laws be passed to eliminate all possibilities of special interests giving huge sums of money to candidates?"

Version 2: "Should laws be passed to prohibit interest groups from contributing to campaigns, or do groups have a right to contribute to the candidate they support?"

The differences in wording should leap out at you.

- The first question has charged phrases "special interests" and "huge sums of money."
- It also lacks the feature of stating a contrasting choice in the question in order to keep the item neutral.

Of course, Version 2 would not "work" if the response categories were "yes" and "no." Rather, response choices included "laws should be passed to prohibit interest groups from contributing to campaigns" and "groups have a right to contribute to the candidate they support."

We cannot contrast the responses to the two questions because the first question was answered by a voluntary sample. Version 2, which was administered to a random sample, found 40 percent favored prohibiting contributions. Another version of the question that asked, "Please tell me whether you favor or oppose the proposal. . . . The passage of new laws that would eliminate all possibility of special interests giving large sums of money to candidates," found 70 percent favored prohibiting contributions. The example suggests how question wording can affect the response pattern and how fragile one question is at measuring attitudes and opinions.[15]

Questions that ask respondents how much they would be willing to pay for something are particularly sensitive to wording. If the respondent feels that her answer will affect service cost, she may cite an amount lower than what she would pay. Consider a question from a city's recreation department that asks people whether they would pay $10 for a tennis license to play on the city's public courts. A tennis player, who may be quite willing to pay $10, might try to influence the policy outcome by answering "no." Alternatively, people asked how much they would pay for various environmental programs are likely to overestimate their willingness to pay.[16]

Some studies involve sensitive subjects, and one must be aware of the probability that a respondent will distort his or her answers. Questions on drug use and other activities that are illegal or widely criticized may bring about inaccurate responses. A naive schoolchild may report drug use to make herself appear knowledgeable, whereas a drug user fearful of the consequences of accurately reporting her behavior will deny or underreport her drug experience.

Each of us has feelings about what we have accomplished and how we have failed. Surveys to learn about individuals undoubtedly will have questions that make some respondents uncomfortable. Questions that seem likely to create respondent discomfort should be carefully considered. First, such questions may raise ethical concerns; that is, the research may harm the subjects. Second, the questions may cause respondents not to participate in a study and may even result in a negative attitude toward research in general. Third, the questions may lead respondents to give inaccurate information.

Knowing which questions are potentially threatening depends on the subjects. Observers should not assume that their perceptions of what is a sensitive subject or a threatening question agree with a respondent's perceptions. The pilot test may identify sensitive questions. Alternatively, investigators familiar with the subject population may be aware of inappropriate questions. Such considerations are especially important if researchers plan to survey people in other countries, where different norms may prevail about what constitutes a personal or sensitive subject.

Stating a question so that it suggests to the respondent that any possible answer is acceptable, not surprisingly, helps reduce threat. For example, such questions may be introduced with the phrase, "Some people find . . ." Question order also can reduce the discomfort associated with a question. Potentially threatening questions should not be asked at the beginning of a questionnaire or interview, where they may make a respondent suspicious and less willing to cooperate. On the other hand, a sensitive question should not be placed at the end, either, where it may cause a person to finish a survey feeling anxious and wondering if the survey had a hidden purpose. Finally, respondents are more likely to share sensitive information on self-administered questionnaires, including those that a respondent answers on a computer or with paper and pencil while an interviewer waits.

Questionnaire Design

Obviously, one cannot put questions on a form in random order. Questions should be logically ordered. The initial questions can affect a respondent's willingness to answer a questionnaire. If the early questions are confusing, threatening, or time consuming, a respondent may not complete the survey or cooperate with the interviewer. Self-administered questionnaires and interview schedules follow a general sequence.

Survey Introduction

An introduction states the nature and purpose of the survey. It identifies the person or organization conducting the survey. The introduction should be short and to the point. An overly long explanation of the purpose of a survey may bias results. Introductions to web-based surveys that extend over one screen may immediately seem burdensome and consequently discourage participation.

Subjects must be told what, if any, risks may be associated with participating in a study. Subjects must understand that their participation is voluntary, they may decline to answer any question, and they can withdraw from a study at any time. Regulations on "protection of human subjects" cover requirements for informing subjects of their rights and the risks of participation.

All researchers conducting investigations sponsored by government must be aware of these regulations and ethics and submit the proposed research for review.

Introduction Examples

Example 7.6 illustrates two questionnaire introductions.

Example 7.6 Examples of Survey Introductions

Sample 1: The introduction is from a survey given to town employees. The surveys were given directly to employees, and they could mail their responses back to the investigators.

Actual wording: This questionnaire is designed to find out how you and others feel about the Town of Oaks as a place to work. The data collected will provide information needed to better understand how people feel about the quality of working life in the organization.

If this questionnaire is to be useful, it is important that you answer each question frankly and honestly. There are no right or wrong answers to these questions. We are interested in what you think and feel about your life at this organization.

Your answers to these questions are completely confidential. All questionnaires will be taken to State University for analysis. No one in the Town of Oaks organization will ever have access to your individual answers.

For analysis and data collection purposes, a number has been put on this questionnaire that can be matched with your name on a list at the university. It would be appreciated if you would leave this number intact.

Thank you in advance for your cooperation and assistance.

Sample 2: This survey introduction is from a telephone survey of community residents. The purpose of the survey was to learn their opinion of the town police force.

Actual wording:

> Introduction: Hello, my name is (interviewer name), and I'm calling from the Town of Oaks.
>
> PURPOSE: In conjunction with State University, we are calling about 500 homes in Oaks to find out how you feel about police services in Oaks. (TO INTERVIEWER: IF CALLING BETWEEN 5 and 6:30 p.m., ADD: Is this a convenient time to call you, or should I call back later this evening?)
>
> CONFIRMATION: Do you live within the town limits of Oaks? (If the answer is no, END THE CALL.) Is this a home phone or a business? (If a business, END THE CALL.)

The first is from a survey given to town employees. The second is from a phone survey of the general public. Each introduction indicates the purpose of the study. Each introduction encourages respondents to participate by indicating the importance of their answers. Note that both introductions failed to mention how long it will take to respond. As people become oversurveyed, this information may be critical in deciding whether to participate. Each introduction implies voluntary participation. In the case of mail or telephone surveys, voluntary participation is often implied, because subjects usually feel no compunction about not returning the survey or hanging up the phone.

The introduction to the survey sent to employees fully detailed how the investigators planned to protect respondents' confidentiality. Such surveys are particularly sensitive because employees are suspicious about what will be done with the information. The other survey informs the respondent that the survey is anonymous.

Recall that sampling frames are not perfect. A sample may include inappropriate respondents who should be screened out immediately. The introduction or the first question should establish whether the respondent should complete the survey. In Example 7.6, Sample 2, the respondents were asked whether they lived in the town. In Example 6.1, an enclosed postcard allowed inappropriate respondents to quickly excuse themselves from replying. Respondents who do not fall within the target population are usually asked to return the questionnaire uncompleted. An interviewer will have a parallel set of directions that also tell him if and how he can replace an ineligible respondent.

First Questions

Following the introduction, relevant, easily answered questions appear. These involve the respondent in the study. During an interview, this section builds rapport between the interviewer and the respondent. In a self-administered questionnaire, the questions draw the respondent into making a "psychological commitment" to complete the questionnaire.

For the recipient of a mailed or web-based survey, the first questions keep her from putting the survey aside and forgetting it; however, if the survey becomes unduly complex or confusing, the benefits of good first questions may be lost. In a telephone survey, the first questions may be important in getting cooperation from a wary respondent who may suspect that the caller is really trying to sell him something.

In the questionnaire sent to employees of the town of Oaks, the first three questions, each answered on a strongly agree to strongly disagree scale, were:

1. I get a feeling of personal satisfaction from doing my job well.
2. I work hard on my job.
3. If I had the chance, I would take a different job within this organization.

Major Questions

The distinction between the initial questions and the major questions may be virtually non-existent. Many questionnaires ask a few introductory questions, then group questions in sections addressing the major issues of the survey. In the sections, the question order may be important. In-person interviewers start with more general questions, followed by specific questions. Care must be taken to avoid sequences that can bias answers.

The introduction of the DECIDE study told respondents that the survey's purpose was to learn more about management training needs. The questionnaire started with questions about respondents' management responsibilities. The next grouping of questions asked respondents about management decision-making tools: What tools were they familiar with? What tools did they commonly use? What were some of the problems with the tools?

The following section asked about DECIDE as a decision-making tool. In Question 5, respondents were asked, "What decision-making tools do you commonly use?" A note to interviewers told them that they might need to prompt the respondent. The first mention of DECIDE was in Question 7, after respondents had been asked about management decision-making tools. Nearly 90 percent of the respondents who said they used DECIDE several times a month had mentioned it earlier as a decision-making tool. If we had mentioned DECIDE early in the study, we would have been more suspicious of the level of reported use.

Sequencing to avoid bias is not as effective when using self-administered paper-based questionnaires. The respondent can read the entire questionnaire before answering any questions, or he can go back and change answers. This concern is less applicable in Internet-based surveys, where the order in which respondents receive questions can be controlled if the survey designer so chooses, despite the survey being self administered.

Demographic Questions

The final section typically asks personal questions about income, age, race, education, or employment. Analogous questions are asked in organizational surveys. By the end of the interview, the respondent may feel less reluctant to answer personal questions. Personal questions at the beginning of a survey may make a person wonder about its real purpose. Regardless of placement, however, it is best to limit the personal questions asked to those necessary for the study. Often researchers include demographic questions such as race, gender, or age even though that information is not germane to the study. Too many personal questions may disturb respondents who worry about the confidentiality of their answers.

Physical Design

The physical layout of the questionnaire or interview schedule affects its utility. In organizing a questionnaire, the designer must consider its impact on the person filling it out and the person compiling or coding the information. Question wording and sequencing are elements of questionnaire design; they affect the clarity and validity of the responses.

An interviewer needs a form that is easy to read while speaking to a respondent and recording responses. A respondent needs a questionnaire that is easy to follow. A questionnaire that can be easily and quickly answered reduces the possibility that it will be put aside and forgotten. Factors such as the design of the pages also may affect the response rate and the quality and quantity of the information obtained. Furthermore, a well-designed questionnaire may communicate to the respondent the seriousness of the research effort and favorably influence her inclination to respond.

Following is a partial "instrument checklist" to guide you in constructing a survey. You may wish to add to this checklist as you read other materials or review various data collection instruments.

A Checklist for Constructing Surveys

Self-Administered Surveys

- The purpose is clearly stated in the introduction.
- Directions on how to answer are clear.
- Recipients who do not belong to the target population are quickly identified—for example, by the use of a screening question—and are told what to do with the survey.
- Survey instrument deadline date and return instructions should appear on survey instrument.

Interviewer-Administered Surveys

- Interviewer's introduction clearly indicates the purpose.
- Directions on how to ask questions and record answers are clear.
- Recipients who do not belong to the target population are quickly identified, and interviewer instructions indicate if and how he should look for a replacement respondent.

On All Surveys and Data-Collecting Instruments

- Critical terms are defined.
- Abbreviations are not used.
- Conjunctions such as *and* are avoided, because an answer may not apply to both parts of the question.
- Response choices must be adequate, appropriate, and mutually exclusive.
- Requested data must be easily accessible.
- Item groupings are logical.

For Opinion Questions

- Question wording should be neutral, such as, "Do you favor or oppose?"
- Responses should be balanced; for example, there should be an equal number of positive and negative responses.

Questionnaire Efficiency

Trying to make certain you need every question you ask and ask every question you need is difficult. The more you can visualize the information a survey will produce and how you will use it, the better off you are. Working with a model serves as a beginning point in deciding what questions to ask and how to ask them, yet it may be insufficient. Identifying the output the survey will produce should improve its efficiency.

Before fielding a survey, researchers should think through how they, or those for whom they are conducting the survey, plan to analyze the resulting data. Then they should link the plan with the questionnaire to double-check that each question is needed. The survey design team can review the plan with the study users to show them what information the survey will produce. To help users visualize what information they will receive, they may be shown formatted tables or charts with hypothetical survey data. The review should help users identify gaps in the planned survey as well as information not needed. After the data are collected, the researchers can refer to the plan to guide their data analysis.

Example 7.7 shows a partial outline produced to help design a questionnaire to learn how agency managers trained in DECIDE used the technique. The DECIDE training program required two licensed trainers to conduct weeklong training seminars of 16 agency managers. During the week, teams of trainees systematically analyzed problems, identified and evaluated possible solutions, selected a solution, and developed an implementation plan. The agency needed the information to help it decide whether the benefits of the training justified the costs.

Example 7.7 Using Planned Output to Guide Questionnaire Content

Situation: Investigators plan to survey managers who received DECIDE training and ask them how they used the training in their jobs.

Step 1

Describe planned analysis. Indicate what outputs will be produced, identify the questions that will be included to produce each output, and indicate the value of each output to the study. Note the number of categories that are needed to perform analysis.

Outline of Planned Output:

Univariate Output: Trainees' use of DECIDE
Frequency of use (Question 3)
Impact on present job (Question 5)
Types of use (Question 4)
Specific examples of use (Question 4; categorize and analyze open-ended responses)

Bivariate Output: Factors associated with how often trainees use DECIDE, how they use it, and problems with the use. Tables will be created to examine the following relationships:

1. Frequency of use by
    a. years since training
    b. management functions
    c. position
    d. division
    e. education

2. Type of use by
    a. years since training
    b. management functions
    c. position
    d. division
    e. education

3. Problems with use by
    a. years since training
    b. management functions
    c. position
    d. division
    e. education

Value and use of bivariate output on use: If patterns of use are related to years since training, the pattern may be attributed to decay (training forgotten over time), changes in the types of persons trained over the years, or changes in trainees' jobs.

If patterns of use are related to management functions, the number of employees super-vised or position pattern may show whether variations in use are associated with either a specialized or generalist management focus. Information could be used to limit training to groups identified as frequent or effective users of DECIDE.

If patterns of use are related to work unit, the pattern may suggest that DECIDE is more suited to the work of specific units or that some unit managers are more supportive of the technique. Information could be used to limit training to units identified as frequent or effective users of DECIDE. If supervisor training in DECIDE is related to frequent or effective use, managers may suggest the need to train supervisors before training their staff.

Step 2 Check outline against draft questionnaire to make sure it contains all items needed for analysis; consider deleting questions that are not noted on outline.

Step 3 Review questionnaire and outline with study users to see if they can identify unnecessary or overlooked information.

Note: As noted in the endnotes, for a version of DECIDE, see: K. L. Guo, "DECIDE: A Decision-Making Model for More Effective Decision-Making by Health Care Managers," *Health Care Manager* 27, no. 2 (2008, April–June): 118–127, at www.ncbi.nlm.nih.gov/

Planning, Pretesting, and Piloting Surveys

Good data collection instruments that yield useful and used information emerge from an iterative process. The first stage, which has been the focus of the chapter, involves planning and writing the instrument.

Planning

As ideas are fleshed out to form an actual survey, investigators better understand the purpose of the study and what it can find out or accomplish. The planning process brings critical actors together to decide what information they want, whether a survey is needed, and if it will yield the desired information. From the various discussions, the investigators should be able to identify a list of variables. With this list, they can develop measures and design an instrument. Finally, the contents of the questionnaire are outlined, as in Example 7.7, to evaluate if all the proposed items are needed and if all needed items have been included. Once it is outlined, further feedback may be solicited from critical actors.

Next, the questionnaire should be reviewed by researchers and their colleagues. They should determine if the questions seem reliable:

- Are there any ambiguous items?
- Are the response choices appropriate?
- Is the information easily available to respondents?

Writing questions takes time and effort. Survey researchers need a thick skin, because good questionnaires are more likely to emerge if exposed to a lot of criticism and argument. Good questionnaire design takes place among equals, with everyone's comments getting consideration. In our experience, not only do participants act as equals, but they become intense and pointed in their comments. To illustrate the process of question writing, we have selected just one question from the DECIDE survey. Example 7.8 shows the changes that question underwent through various drafting sessions.

Example 7.8 The Evolution of a Question

Situation: Questionnaire on DECIDE training was designed to learn how trainees felt it affected their jobs.

Stage 1

A group of three wrote a questionnaire. The first questionnaire had several questions on DECIDE's effect and use on the job. Questions 1 and 2 were both open ended.

1. Were DECIDE course objectives related to your job needs?
2. Do you use any DECIDE processes on the job? If not, why?

Stage 2

Questionnaire map drawn up and questionnaire redrafted by another team member.

1. In general how would you rate DECIDE's impact on your present job? (responses: excellent, good, fair, poor, no impact, unsure)
2. About how often do you use any DECIDE process on your job? (responses: one time a week, several times a month, about once a month, several times a year, rarely or never, don't know)
3. Let me know if DECIDE helps you think about any of the following: (responses included: setting priorities, setting objectives, identifying alternatives)
4. Give me a specific example of how you have used the DECIDE approach on your job.
5. In applying the DECIDE process, what, if any, problems have you encountered?

Stage 3

A team of seven reviewed the draft and made the following changes:

1. In general, how would you rate DECIDE's effect on your job performance? (changed "no impact" response to "no effect")
2. Question kept the same. Placed "rarely and never" responses into separate categories. Respondents who answered "never" were directed to a later question.
3. Question kept the same. Deleted two response categories.
4. Give me a specific example of how you have used the DECIDE process on your job.

Stage 4

Questionnaire pretested. Sample of 10 called. Frequency of responses to each question reviewed. The only change in the previous questions was to further reduce the number of response categories for Question 3.

Note: As noted in the endnotes, for a version of DECIDE, see: K. L. Guo, "DECIDE: A Decision-Making Model for More Effective Decision-Making by Health Care Managers," *Health Care Manager* 27, no. 2 (2008, April—June): 118–127, at www.ncbi.nlm.nih.gov/

Pretesting

After the questionnaire is redrafted, it should be *pretested*. (Note that this use of *pretest* refers to evaluating how questions actually work. It is different from the use of *pretest* to refer to data gathered prior to an experimental intervention.) To conduct a pretest, one asks a small group to answer the proposed questionnaire. These pretest subjects should represent common variations found in the target population. For example, a survey of city employees should be pretested on employees ranging from receptionists to senior management. The investigators also may want pretest subjects who represent various city departments, such as public works, public safety, planning, and finance.

Pretests normally involve face-to-face contact with the subjects. The investigators determine if subjects understood and could answer the questions and if their answers showed enough variation for the planned analysis to be conducted. The investigators should record how long it took individuals to answer the survey and if it held their interest and attention. If a survey does not engage respondents, the information may be less reliable, incomplete questionnaires may be more common, or, in the case of mail and Internet questionnaires, the response rate may be lower.

After answering the questionnaire, the pretest subjects may be interviewed or debriefed about their answers and reactions to the questions and the survey as a whole. The investigators use their own observations, the subjects' comments, and their responses to the survey questions to determine how coherent the questionnaire is:

- Are directions clear?
- Is the question order logical?

The survey should be redrafted based on the pretest findings. Common changes include changing unclear terms, rewording or dropping ambiguous questions, changing response categories, and shortening instruments.

Piloting

The final stage is the *pilot study*, or dress rehearsal. Numerous works on questionnaire design and survey research emphasize the importance of conducting the pilot test.[17] Even a lack of time should not justify skipping a pilot test.

The pilot study involves conducting the entire study as planned on a small sample representing the target population. The questionnaire is administered as planned, that is, either in person, through the mail, or over the telephone. The data from the returned surveys are analyzed according to plan. Any special procedures to be used in collecting or compiling data, for example, computer-assisted interviewing, are used during the pilot study.

By the time of the pilot study, questionnaire problems should be minimal. Investigators pay closer attention to the feasibility of the sampling, data collection, and analysis procedures. The dress rehearsal should identify any problems in contacting sample members and getting their cooperation. The investigators can estimate the average amount of time required to answer a survey, to organize the information, and to prepare the data. The adequacy of data preparation and analysis plans can be checked. Previously unnoticed, unreliable, or invalid measures may be detected. Problems encountered in the pilot study should be resolved prior to implementing the final study design.

In modest studies, the pilot study and the pretest may overlap, but at a minimum, investigators should test the instrument on people resembling potential subjects, check that items have sufficient variation, and conduct minimum analysis to assess whether the survey will produce useful information. Pretesting and pilot testing are most effective at detecting technical problems associated with questions and questionnaires, for example, questions that are misunderstood. They do less well at detecting measurement problems, especially operational validity. Focus groups, asking subjects to talk aloud as they answer questions, and other techniques are being developed to better capture what questions are actually measuring.[18]

Using poorly designed or inadequate questionnaires can waste large amounts of resources; this can be avoided with thorough pretesting and pilot studies. In addition, those conducting survey research should be sure to commit adequate resources to the management, analysis, and reporting of the results when the questionnaires are returned. All too often, these important steps are overlooked.

Earlier in this chapter, we mentioned the Nationwide Food Consumption Survey, which had a response rate of 34 percent. The problems with that survey illustrate two common situations. First, identifying problems in a pretest or dress rehearsal is not enough. An effective solution to the problems identified in the pilot study must be put in place.

Second, without a dress rehearsal, serious problems may be missed or underestimated. Two pretests of the food consumption survey had been conducted. Interviewers and subjects complained of fatigue, with a survey that took as long as five and one-half hours to complete. As a result, the computer-assisted interviewing procedures were supposed to be improved. The pretest of the improved computer-assisted interviewing methods found that interviewers required rigorous training. To save money, the planned dress rehearsal was canceled. Note that the dress rehearsal would have evaluated the effectiveness of the interviewers' training. The long survey was not shortened, and interviewer training was not sufficiently rigorous. As a result, a study that cost $7.6 million yielded seriously flawed data.

Summary

Questions and questionnaires constitute basic elements of data collection. Writing questions and designing questionnaires require one to understand how these aspects of a survey affect the implementation of a research design. Each question or series of questions has implications for reliability and operational validity (see Figure 7.1).

Outlining

A first step in questionnaire construction is to outline its content. An investigator wants to avoid a questionnaire that asks trivial questions or fails to ask important ones. To prevent a mismatch

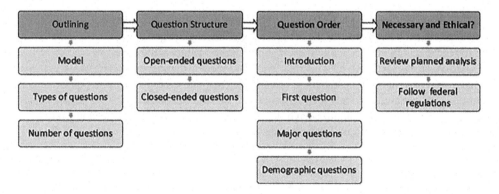

Figure 7.1 Data Collection

between the research purpose and a survey instrument, each question's role in the research should be explored and justified. Put differently, each question should be designed to generate a measure of a variable that appears in the intended model. One also should determine whether the types of questions asked elicit the desired information. Asking the wrong type of question—for example, asking an opinion question where a behavior question is needed—affects operational validity. Common question types ask about facts, behaviors, opinions, attitudes, motives, and knowledge.

Question Structure

A second step is deciding on whether to ask open-ended or closed-ended questions or to include contingency questions. Open-ended questions require respondents to write their answers. These questions work well during questionnaire development, when investigators want to learn the range of possible responses. Open-ended questions provide rich, detailed information; they may be asked in conjunction with closed-ended questions. Because open-ended questions are difficult to quantify, may unnecessarily tax respondents, and may lower the response rate, researchers prefer closed-ended questions. Closed-ended questions have a respondent select an appropriate response from a list. These questions work well if the questionnaire has been carefully pretested to ensure that the questions are understood and the response categories are appropriate.

Contingency, filter, or screening questions direct subgroups of respondents to appropriate follow-up questions. For example, renters may answer one set of questions and homeowners another set. On mailed questionnaires, arrows may direct respondents to applicable questions. Interviewers may use the same tools, or they may use computer software that automatically goes to the appropriate follow-up questions.

Question wording is a major component of measure reliability and operational validity. In general, questions should be clear, short, and specific. The respondent should understand what information the question is asking for. Questions can be biased, either intentionally or unintentionally. A biased question increases the probability that a respondent will choose one answer over another. Words that respondents perceive as positive or negative may bias answers. Response lists that have disproportionately more positive or negative alternatives also introduce bias. Questions on sensitive subjects must be handled with special care. Respondents may distort their answers; furthermore, such questions can raise ethical concerns if confidential information is not adequately protected.

Question Order

The order in which questions are asked affects response rate and the operational validity of the questions. Researchers must consider how the style of survey administration that they choose— also known as the "mode"—will impact their ability to control the order in which respondents receive questions.[19] Respondents answering self-administered paper-based questionnaires may be able to read ahead or change their answers, reducing the value of the sequence.

Questionnaires start with an introduction. Any needed demographic questions are usually placed at the end. Some researchers get in the habit of routinely asking a set of demographic questions, without considering which demographic variables are needed for the planned analysis. Asking needless questions should be avoided; furthermore, demographic questions may raise the distrust of respondents who wish to remain anonymous.

Necessary and Ethical?

Researchers should consider whether other data sources are available before pursuing constructing their own survey. (Various web sites such as FedStats identify accessible databases from government agencies and public opinion archives such as iPoLL, Inter-university Consortium for Political and Social Research, and Pew Charitable Trusts.) If the data do not exist, then the survey should proceed only after a pilot study. A pilot study rehearses the research plan, including the analysis. In analyzing a pilot study, researchers may focus on question wording, the sensitivity of the response categories, the response rate, and the time required to complete the survey. Researchers may be tempted to ignore the opportunity to evaluate their planned analysis. Yet a review of the planned analysis may be most helpful in avoiding unnecessary surveys or questions.

Gathering data from human subjects raises ethical concerns. Several key incidents in previous research projects that observers felt crossed ethical boundaries led to federal regulations and professional guidelines for protecting human subjects. The regulations now dominate data collection. The requirement for informed consent dictates what potential research subjects must be told about a planned study and its effects. Furthermore, the provision that research participation must be voluntary challenges researchers to design studies that respect respondents' privacy and entice their participation.

Notes

1. M. Q. Patton, *Utilization-Focused Evaluation*, 4th ed. (Thousand Oaks, CA: Sage Publications, 2008).
2. This is not to say that attitudes never change. For more on this topic, see Diana C. Mutz, Paul M. Sniderman, and Richard A. Brody, eds., *Political Persuasion and Attitude Change* (Ann Arbor: University of Michigan, Institute for Social Research, 1996).
3. See Jon A. Krosnick, Arthur Lupia, Matthew DeBell, and Darrell Donakowski, "Problems with ANES Questions Measuring Political Knowledge," (March 2008), available at http://electionstudies.org/announce/newsltr/20080324PoliticalKnowledgeMemo.pdf.
4. For a version of DECIDE, see: K. L. Guo, "DECIDE: A Decision-Making Model for More Effective Decision Making by Health Care Managers," *The Health Care Manager*, 27, no. 2 (April–June 2008): 118–127, available at www.ncbi.nlm.nih.gov/.
5. To reduce the work, researchers can look at the words, phrases, or themes occurring most often. This is not optimal, however. Sampling may miss interesting and important comments that researchers could explore later when resources allowed. Often these are the beginnings of a paradigm shift.
6. K. A. Rasinski, D. Mingay, and N. M. Bradburn, "Do Respondents Really 'Mark All That Apply' on Self-Administered Questions?" *Public Opinion Quarterly* 58 (Fall 1994): 400–408; Jolene D. Smyth, Don A. Dillman, Leah Melani Christian, and Michael J. Stern, "Comparing Check-All and Forced-Choice Question Formats in Web Surveys," *Public Opinion Quarterly* 70 (Spring 2006): 66–77.
7. N. Bradburn, S. Sudman, and B. Wansink, *Asking Questions: The Definitive Guide to Questionnaire Design*, rev ed. (San Francisco: Jossey-Bass, 2004), 141–142, 162–163. For discussion on neutral responses and their placement, see Don A. Dillman, *Mail and Internet Surveys: The Tailored Design Method*, 2nd ed. (Hoboken, NJ: John Wiley & Sons, Inc., 2007), 58–60.
8. Dillman, *Mail and Internet Surveys*, 102–104.
9. *USDA's Nationwide Food Consumption Survey* (Washington, DC: General Accounting Office), GAO/RCED—91–117. The GAO report on why the survey yielded poor-quality data is highly recommended; it vividly points out several errors that can occur in a major survey project. For more detailed and recent information on nutrition surveys and a link to articles examining their reliability and validity, go to the web page of the Food Surveys Research Group: www.ars.usda.gov/northeast-area/beltsville-md-bhnrc/beltsville-human-nutrition-research-center/food-surveys-research-group/docs/past-surveys/.
10. Most of the information on validating voting data is from S. Presser and M. Traugott, "Little White Lies and Social Science Models: Correlated Response Errors in a Panel Study of Voting," *Public Opinion Quarterly* 56 (Spring 1992): 77–86. For a discussion on voter records as a source of error, see P. R. Abramson and W. Claggett, "The Quality of Record Keeping and Racial Differences in Validated Turnout," *Journal of Politics* 54 (August 1992): 871–880.
11. Researchers have reported that the ratio of positive to negative responses remains the same whether or not "don't know" is offered as a response alternative. For further discussion of "don't know," see L. F. Feick, "Latent Class Analysis of Survey Questions That Include Don't Know Responses," *Public Opinion Quarterly* 53 (1989): 525–547; J. M. Converse and S. Presser, *Survey Questions: Handcrafting the Standardized Questionnaire*, Sage University Paper Series: Quantitative Applications in the Social Sciences 63 (Beverly Hills, CA: Sage Publications, 1986), 35–39; M. Gilljam and D. Granberg, "Should We Take Don't Know for an Answer?" *Public Opinion Quarterly* 57 (Fall 1993): 348–357.

12. Sudman, Bradburn, and Wansink, *Asking Questions*, 103–104.
13. R. Tourangeau, M. P. Couper, and F. Conrad, "Spacing, Position, and Order: Interpretative Heuristics for Visual Features of Survey Questions," *Public Opinion Quarterly* 68 (Fall 2004): 368–393.
14. D. Goleman, "Psychologists Offer Aid on Bias in Polls," *New York Times*, September 7, 1993, B5, B7. A version of the article is available at www.nytimes.com/1993/09/07/science/pollsters-enlist-psychologists-in-quest-for-unbiased-results.html?pagewanted=2&pagewanted=all. The article relies heavily on information included in an article in the June 1993 issue of *The Public Perspective*: David M. Wilber, "H. Ross Perot Spurs a Polling Experiment (Unintentionally)," *The Public Perspective* (May–June 1993): 28–29. The two versions of the survey questions quoted are discussed in that article; see page 29.
15. See Wilber, "H. Ross Perot Spurs a Polling Experiment (Unintentionally)," 29.
16. See P. Passell, "Polls May Help Government Decide the Worth of Nature," *New York Times*, September 6, 1993, for an example of a question to ask how much a respondent would be willing for the government to pay to clean up an oil spill.
17. For a thorough discussion on pretests and pilot studies, see Converse and Presser, *Survey Questions*, 51–75; Floyd J. Fowler, Jr., "Presurvey Evaluation of Questions," in *Improving Survey Questions: Design and Evaluation* (Thousand Oaks, CA: Sage Publications, 1995), Chapter 5. Fowler discusses the use of focus groups, intensive interviewing, and field pretests.
18. Stanley Presser, Jennifer M. Rothgeb, Mick P. Couper, Judith T. Lessler, Elizabeth Martin, Jean Martin, and Eleanor Singer, *Methods for Testing and Evaluating Survey Questionnaires* (Hoboken, NJ: Wiley Interscience, 2004), provide in-depth coverage of various techniques. A shorter discussion of the various techniques is found in Robert M. Groves, Floyd J. Fowler, Jr., Mick P. Couper, James M. Lepkowski, Eleanor Singer, and Roger Tourangeau, *Survey Methodology* (Hoboken, NJ: John Wiley & Sons, Inc., 2004), Chapter 8.
19. For more on survey modes, see Chapter 5 of Groves et al., *Survey Methodology*.

Terms for Review

open-ended questions
closed-ended questions
forced-choice questions
contingency questions
biased questions
loaded questions
pretest
pilot study

Questions for Review

The following questions should indicate whether you have a basic competency in this chapter's material.

1. List the steps required to write a questionnaire.
2. Recommend a procedure for deciding what content to include in a questionnaire.
3. Imagine you are writing a questionnaire to evaluate training given to new employees. (The training introduces employees to the agency, its purposes, and major policies.)

   a. Write a factual, behavior, opinion, motive, and knowledge question that you could include on the questionnaire.
   b. How valuable do you imagine each type of question (factual, behavior, etc.) would be in determining the quality of the training?
   c. On reflection, would you include all the questions you listed in part a?

4. A young researcher believes that questionnaires tend to be too long and adopts a rule of asking only one question to measure a variable. Evaluate the advisability of such a rule.

5. Compare and contrast the value of open-ended questions and closed-ended questions.
6. In general, would you recommend asking respondents ranking questions? Justify your answer.
7. What characteristics of question wording affect reliability? Explain.
8. Should you always include questions asking for demographic information? Why or why not? If you choose to include them, where would you place them in the survey?
9. Compare and contrast pretesting and piloting a questionnaire.
10. An investigator has written a questionnaire asking police officers about their career satisfaction. She pretests the questionnaire on students in an undergraduate criminal justice class. Comment on the adequacy of her pretesting strategy.
11. You are interested in neighborhoods and communities. You find a questionnaire that was used in metropolitan New York. The questionnaire covers topics that you are interested in studying in a moderate-size Midwestern city. What would you do in order to decide that the questionnaire was appropriate for your study?
12. Review Example 7.3 illustrating visually coded contingency questions. Explain what would be done differently if the interviewer were using computer-assisted telephone interviewing. Explain what would be done differently if the survey were a web-based survey designed through a program such as Qualtrics, SurveyMonkey, or Surveyshare.

Problems for Homework and Discussion

1. The following is from an actual town survey. Identify the different parts of the survey. Identify the type of questions asked. Consider the value of findings from this questionnaire. Assess the wording of the questions.

   1. Questions a–d could be answered "yes," "no," or "no opinion." Questions f–h are open ended.
   2. Hi, I am _____. I am helping to conduct a census for Wind Valley. Would you mind answering a few questions? It will take only a few minutes.

      a. Are you a resident of Wind Valley? (if yes, continue; if not, thank him/her and leave)
      b. Would you support redevelopment of the downtown business area?
      c. Would you favor once-a-week garbage collection if doing so would help to reduce service charges?
      d. Would you be willing to accept fewer services for a reduced tax rate?
      e. If city services had to be reduced in one area, which would you choose? (responses were: streets, fire, sanitation, parks and recreation, police, library, other, none)
      f. How long have you lived in Wind Valley?
      g. How many people are living in this household?
      h. How many people are over 60 years of age?

2. The following questions are taken from a draft of a survey to identify stereotypes that older public-sector workers may have about millennials they work with. Review each question and then the five questions together. Assess the wording of the questions. Consider whether the questionnaire seems to risk a set response pattern problem. Note: All questions were to be answered "yes" or "no."

   a. Do you put down millennials (the generation that includes people born between 1980 and 2003) for their lifestyle?
   b. Are millennials a bunch of dreamers??
   c. Does society in general label millennials as poorly prepared and lazy?
   d. What is it to you to be a millennial today?
   e. Do you provide training on what millennials need?

3. Evaluate questions a and b from a "Job Turnover Questionnaire" sent to public administrators. What changes would you recommend?

Most important reasons for leaving your last job (check two only):

_____ Lack of job satisfaction (e.g., lack of achievement, recognition, growth potential)
_____ Disagree with agency policy/administration
_____ Poor interpersonal relations on job
_____ Salary insufficient
_____ Poor physical conditions
_____ Discrimination
_____ Workload
_____ Sexual harassment
_____ Continue education
_____ Family reasons (e.g., spouse moving, birth of child)
_____ Termination (end of contract, position defunded, reorganization)
_____ Other, specify

Type of job (check one only):

_____ Administration
_____ Counseling/social work
_____ Research/technical
_____ Secretarial/clerical
_____ Teaching
_____ Other; specify

4. A community survey sent interviewers to residents' homes. The respondent, any adult living in the household, was asked to supply the following information for each person in the household. Evaluate this question.

Table 7.2 The Major Source of Income

	Person 1	Person 2
Salary		
Hourly wages		
Investments		
Retirements/pensions		
Parent support/inheritance		
Welfare		
Self-employment		
Refusal		
Don't know		

5. Evaluate the following question for use on a survey: "Do you oppose measures promoting the abolition of the death penalty?" What aspects of the question may be problematic from the perspective of measurement? What changes in wording, if any, would you recommend?

6. Evaluate the following question and response categories on a survey given to nursing home residents. What changes in wording, if any, would you recommend?

1. How much do you use or participate in the following services or activities?

   a. The Beauty or Barber Shop
   _____ Often
   _____ Occasionally

_____ Rarely
_____ Never

b. Crafts

_____ Often
_____ Occasionally
_____ Rarely
_____ Never

c. Card and Game Night

_____ Often
_____ Occasionally
_____ Rarely
_____ Never

d. Religious Services

_____ Often
_____ Occasionally
_____ Rarely
_____ Never

7. Draft a survey to determine needs and preferences of part-time students enrolled in your degree program. The survey should identify changes the program could make to better serve the needs of part-time students. Draft a series of questions to obtain this information. Would you include a list of possible changes that could be made to include in the survey instrument? Explain.
8. Contrast the value of (a) finding questionnaires that have questions you can adapt and (b) writing questionnaires with a group of colleagues.
9. Investigators are often tempted to add questions to a survey that they may or may not get around to analyzing within the project for which the survey is designed. Evaluate this strategy.
10. Review and evaluate a software package used to design questionnaires. For example, obtain information on SurveyMonkey, Qualtrics, or Surveyshare. SurveyMonkey can be used free of charge (www.surveymonkey.com/). Your institution may subscribe to Qualtrics or Surveyshare.

Working With Data

1. Limitations of a questionnaire and the resulting dataset may become obvious only as you analyze the data. To get a feel for this role of analysis, this exercise asks you to use the County Data file.

   Imagine that the governor of the state examined in the County Data file is planning a series of statewide town meetings to become more familiar with the issues concerning residents. To be well prepared, he/she would like to know how population characteristics vary by region of the state. He/she needs the information immediately, and your office only has the information in the County Data file dataset on hand.

   a. Compare the three regions in the state with regard to average population density, social security beneficiary population, Medicaid-eligible population, and crime indices.

    b.   Based on your analysis, write a one-page, single-spaced memo to the governor presenting your key findings. Make appropriate recommendations for topics the governor may wish to discuss or avoid.

    c.   Based on your analysis, what information should be added to the office database for the staff to be better prepared for such a request were it to come again? Write a memo to your office director summarizing your suggestions (one page, single spaced).

2.   Use the General Social Survey to complete the following exercises and answer the questions.

    a.   Identify one example of each of the following type of question: factual, behavior, motive, opinion, knowledge, closed ended, open-ended, filter.

    b.   Select five closed-ended questions from the GSS that a county manager may want to have answered. Of those, identify two to which you could add an open-ended option. Explain your answers.

    c.   If you needed to recommend to the county commissioners reducing or terminating a program, would you recommend using the survey data or a focus group to get information to aid in the decision? Write a one-page memorandum justifying your decision.

Recommended for Further Reading

Several excellent books on survey research and questionnaire design are available. We recommend that you review one of the following or a similar book before writing a questionnaire.

De Vaus, D., and D. de Vaus, *Surveys in Social Research* (London: Routledge, 2013).

Dillman, Don A., J. Smyth, and L. Christian, *Internet, Phone, Mail and Mixed-Mode Surveys: The Tailored Design Method*, 4th ed. (Hoboken, NJ: John Wiley & Sons, Inc., 2014).

Fowler, Floyd J. Jr., *Survey Research Methods*, Sage Applied Social Research Methods Series, 5th ed. (Thousand Oaks, CA: Sage Publications, 2014).

Professional associations and researchers publish guides for designing questionnaires and conducting surveys. Two guides that review major points covered here and provide additional examples are Heumann, L. F., "Citizen Surveys: How to Do Them, How to Use Them, What They Mean," *American Planning Association. Journal of the American Planning Association* 67, no. 4 (2001): 485. Citizen Surveys for Local Government: A Comprehensive Guide to Making Them Matter.

Miller, Thomas I., Michelle A. Miller, Michelle Miller Kobayashi, and Shannon Elissa Hayden, *Citizen Surveys for Local Government: A Comprehensive Guide to Making Them Matter*, 3rd ed. (Karachi: ICMA, 2009).

Other how-to-do-it books can help refresh your memory of key points; for example, see Patten, Mildred L., *Questionnaire Research: A Practical Guide*, 4th ed. (Los Angeles: Routledge, Pyrczak Publishing, 2001), and a very helpful basic brochure series called *What Is a Survey?* provided by the American Statistical Association's Survey Research Section.

Rea, Louis, and Richard Parker, *Designing and Conducting Survey Research: A Comprehensive Guide*, 4th ed. (San Francisco: John Wiley & Sons, Inc., 2014), provide a clear, thorough guide from start to finish of the process for conducting a survey. Sudman, S., N. M. Bradburn, and N. Schwarz, *Thinking About Answers* (San Francisco: Jossey-Bass, 1996), provide a good review of survey research findings.

To keep up to date with current research on survey research topics such as question wording, questionnaire design, data collection, and data analysis, see *Public Opinion Quarterly*, the journal of the American Association for Public Opinion Research, published by the University of Chicago Press.

Also helpful are classic readings on survey research on response accuracy and meta-analysis. See Westland, Ellen J., and Kent W. Smith, *Survey Responses: An Evaluation of Their Validity* (San Diego: Academic Press, 1993); Marquis, Kent, "Survey Responses: An Evaluation of Their Validity," *Public Opinion Quarterly* 58, no. 4 (1994): 636–636.

8 Protection of Human Research Subjects and Other Ethical Issues

In this chapter, you will learn:

1. About the major cases informing ethical practices in conducting research on human subjects.
2. Elements of ethical practice in conducting research on human subjects.
3. Requirements for obtaining informed consent from a research subject.
4. What to consider in protecting confidential information.
5. Federal requirements for protecting human subjects.
6. What administrators should consider before permitting research on their employees or agency's clients.

The nature of research involving humans creates a special relationship between researchers and their subjects. Researchers require the cooperation of humans to conduct investigations. At the same time, subjects rely on researchers to treat them respectfully and ethically. Nevertheless, some research projects may expose participants to risks, including physical injury, psychological discomfort, or loss of privacy.

While the subjects of administrative or policy research seem unlikely to experience physical harm or life-threatening effects, the risks may be more than trivial. Innocuous-appearing studies may leave participants feeling angry, upset, humiliated, or otherwise worried. They may then be less willing to participate and tell others of their experience, further decreasing support for research. Sound, ethical research practices can greatly reduce such negative effects.

Administrators should be sensitive to the issues surrounding research with human subjects and recognize how participation can create distress. Otherwise, administrators may fail to adequately protect their agency, its employees or clients, or themselves. Administrators in hospitals, schools, prisons, and social services agencies should be particularly vigilant, since their employees, patients, students, or clients often interest researchers. Employees may also conduct research. An administrator is responsible for protecting the rights of employees and clients, whether she is conducting the research or contracting or cooperating with other researchers. If she agrees to provide investigators access to employees or clients, she should ensure that a study conforms to ethical research practice and that results are reported objectively and accurately.

The environments within which scientists and other researchers work and learn have changed over the years. New government, professional, and institutional regulations issued in response to research misconduct require protection of human subjects and provide guidance in the responsible conduct of research. The Department of Health and Human Services Office of Research Integrity has posted cases of research misconduct on its web site, https://ori.hhs.gov/case_summary. The large number of cases suggests that not all researchers act ethically and responsibly. Many in the scientific community believe that investigators should have an understanding of the ethical responsibility they have in conducting research. This approach means that research ethics should go beyond compliance to include conscience, responsibility, and integrity.[1]

This chapter begins by highlighting some well-known dramatic cases; however, they were not necessarily aberrations from research practices when conducted. For one reason or another, they achieved notoriety and shaped current thinking about ethical research on human subjects. Although the last cited case occurred in 1963, from 1993 to 2001, four different medical studies resulted in the deaths of eight research subjects. As a result, bio-ethicists began to review policies intended to protect research volunteers.[2] Ethical concerns are not limited to medical research. In 1999, parents filed a lawsuit claiming that a student survey which asked questions about teenagers' lifestyles violated their children's privacy.[3]

The second part of the chapter focuses on principles of ethical treatment of human subjects, for example, obtaining informed consent. The administrator familiar with informed consent will have a firm foundation for deciding on the appropriateness of research involving agency personnel or clients. The chapter considers privacy and confidentiality, presents strategies to protect confidentiality, and summarizes current federal regulations on protecting human research subjects. This part of the chapter also discusses new requirements by the National Institutes of Health (NIH) and the National Science Foundation (NSF) regarding training researchers to conduct research responsibly. The chapter concludes with a discussion of administrative concerns, which should guide a decision on whether to cooperate with a research effort.

Illustrative Cases

As you read the cases, put yourself in the place of the judges, hospital administrators, or others who have a role in approving research. Put yourself in the position of the patient, juror, or an observed citizen. Imagine how you would react to a survey asking about your management style. As you think about these different scenarios, you may realize that protection of human subjects can directly involve you even if you never conduct research in your professional career.

Tuskegee Syphilis Study

The Tuskegee Syphilis Study is perhaps the best-known U.S. example of an egregious abuse of human subjects.[4] Begun in 1932 by the U.S. Public Health Service, its researchers monitored the health of two groups of African American males. One group had untreated syphilis; the other group was free of syphilis symptoms. The research documented the course of untreated syphilis. At the time the study began, treatments for syphilis were potentially dangerous, so denying treatment might have been rationalized. However, from the mid-1950s on, penicillin was known to be an effective treatment for syphilis and was widely available. Yet by 1973, when the study was discontinued, the participants had not received penicillin and had been actively discouraged from seeking treatment elsewhere.

The failure to treat the subjects was particularly disturbing because the study continued unchallenged despite the findings of the Nuremberg Trials and a later lawsuit against the Jewish Chronic Disease Hospital. At the end of World War II, disclosure of Nazi atrocities included reports of abuses committed by doctors and scientists performing human experiments. The Nuremberg Military Tribunal judgment against these doctors and scientists listed 10 principles of moral, ethical, and legal medical experimentation on humans. The principles, referred to as the "Nuremberg Code,"[5] formed the basis for later regulations protecting human subjects.

The Tuskegee study did not comply with the Nuremberg principles. The violated principles included free and informed consent from the subjects, the researcher's obligation to avoid causing unnecessary physical suffering, the subject's ability to terminate his or her participation at any time, and the researcher's obligation to discontinue an experiment when its continuation could result in death.

Brooklyn's Jewish Chronic Disease Hospital

In 1963, the issue of informed consent again received public attention. A lawsuit and investigation questioned whether 22 patients at Brooklyn's Jewish Chronic Disease Hospital (JCDH) had given informed consent when they agreed to be injected with live cancer cells.[6] The patients were asked if they would consent to an injection for research on immune system responses. They were not told that the experiment was unrelated to their disease or its treatment. They were not told that the injection contained live cancer cells. The investigation concluded that asking a patient to consent to a vaguely described procedure could not be considered informed consent. Still, the publicity surrounding the JCDH case did not change the course of the Tuskegee study, which continued until 1973, when an ad hoc advisory committee to the U.S. Secretary of Health found that the study did not meet requirements for informed consent.

Laud Humphreys

In a well-known deceptive study, a doctoral student, Laud Humphreys, studied men who engaged in casual, anonymous sexual activities with other men. Humphreys offered to act as a lookout for men who were using a public restroom for sexual encounters. He recorded the license plate numbers of the unsuspecting—and nonconsenting—subjects. He traced the licenses and linked them to the subjects' names and addresses. A year later, he changed his hairstyle and manner of dress and took on the role of surveyor. He visited the homes of the previously identified subjects and asked them to participate in an anonymous public-health survey.

Critics questioned whether the benefits of Humphreys' research outweighed the costs.[7] Even if the risks of disclosing the subjects' identities were negligible, the extent of other risks was unknown and inestimable. Given the publicity surrounding the study, one can imagine the men figuring out that they were the subjects, experiencing a loss of privacy, and feeling used and betrayed by Humphreys.

Some social scientists argued that Humphreys had a right to pursue knowledge, that his strategy was appropriate to the question at hand. Others believed that deceptive practices were morally wrong and demeaning to subjects.[8] Some critics worried that knowledge of deceptive research studies might result in a general reaction against social science research.

Stanley Milgram

Another deceptive research study illustrates the possibility that research subjects may gain unwanted information about themselves. Psychologist Stanley Milgram designed a study to see how ordinary people could be induced to obey authority. The research question was sparked by his interest in understanding how the Holocaust happened. The subjects were told that the experiment was to study learning theory. As part of the experiment, the subjects were told to administer what they believed were electric shocks to other participants. The other participants were actors, who vividly acted out the pain of the (non-existent) "shocks."

Some subjects refused to continue administering the shocks and withdrew from the study. Others continued but were clearly distressed at their participation. The ethical problem that critics noted was that the subjects had learned that they would follow orders and seriously harm others. This self-knowledge may have never been gained without their participation in the experiment.

In addition, without their knowledge, subjects were forced to choose between two strong and competing values. On the one hand, they had an implied contract with the researcher to carry out the research as designed. On the other hand, they held the value of not harming another person.

Milgram was conscious of the study's potential to disturb the subjects, and his study included a careful debriefing procedure. A sample of participants was interviewed by a psychiatrist, who was experienced in outpatient treatment; he reportedly found no evidence of injurious effects.[9] Milgram's own and others' follow-up research found that the self-knowledge had no lasting ill effects, and that some participants reported benefiting from the insight. Nevertheless, the case stands as a landmark in the history of deceptive research.

Ethics and Privacy

The knowledge of Nazi medical experiments, the Tuskegee study, and other reported cases has had a lasting effect. Most notably, withholding beneficial treatment from control-group subjects is considered unethical. Several controlled studies have been discontinued when a marked improvement occurred in the experimental group. For example, a five-year clinical trial was stopped after two and a half years. The researchers learned that the study drug was decreasing the rate of breast cancer recurrence by nearly half.[10] Appropriate treatment was then offered to all subjects. Alternatively, if some treatment may be beneficial, but the most beneficial treatment remains unknown, the control-group subjects are assigned to a form of treatment. For example, in a study of depression, all subjects were assigned to some form of treatment. Each form of treatment was believed to be better than no treatment, but the relative effectiveness of the treatments was unknown.

Racial and ethnic minorities are underrepresented in medical research and less likely to donate their organs for transplant.[11] Some researchers attribute this outcome to a distrust of many medical practices by African Americans due to the Tuskegee study.[12] Other studies, however, while finding that minorities are underrepresented in medical trials, did not find that it was due to distrust in medical research.[13] Santos found distrust of medical researchers among racial and ethnic minorities but also found that other groups at times were also underrepresented. These included women and HIV-positive drug users.[14]

The lack of informed consent is a theme that runs through most of the reported abuses of human subjects. The following case illustrates other themes that have helped define ethical research practices. In 1954, researchers from the University of Chicago Law School secretly recorded jury deliberations. The investigators wanted to learn more about the jury processes, processes that were known only by anecdotes, posttrial interviews with jurors, and jury simulations. The researchers believed that open recording would affect and distort the deliberations. They conscientiously developed procedures to obtain undistorted information while protecting the rights of the parties involved. Only juries deliberating civil cases were studied, and the presiding judge and the attorneys representing the litigants gave their permission for the recording. After the recordings were made, they were kept with the judge until the case was closed.

As word of the jury study spread, a U.S. Senate subcommittee held a hearing to learn more. Despite the researchers' precautions to protect the jury members and other parties to the case and to avoid influencing the judicial processes, the research compromised the secrecy of jury deliberations. A comment by the subcommittee's chairman identified the potential harm: "Would a member of a jury hesitate to frankly express his opinion if he thought there might be a microphone hidden, taking down what they said? . . . [H]ow is he going to know whether in that particular case there is a microphone or not?"[15]

Research participants may lose their privacy. *Privacy* refers to an individual's ability to control the access of other people to information about himself. A loss of privacy may occur in deceptive research, where a participant is not told a study's real purpose. While few public administrators become directly involved in deceptive research, it is an issue associated with social science research.

Principles of Ethical Treatment of Human Subjects

In response to these and other reported abuses, the Belmont Report, written by the National Commission for the Protection of Human Subjects of Biomedical and Behavioural Research, identified three basic ethical principles:

1. respect for persons
2. beneficence
3. justice[16]

Respect for persons requires that "subjects enter into research voluntarily and with adequate information."[17] Beneficence requires maximizing possible benefits and minimizing possible harm.[18] Any risk of possible harm can only be justified if it is outweighed by the expected benefit. Justice requires that research subjects not be selected "simply because of their easy availability, their compromised position, or their manipulability, rather than for reasons directly related to the problem at hand."[19]

To implement these principles requires that subjects give informed consent, that benefits and risks be identified and weighed, and that selection of subjects be fair.[20] Informed consent demonstrates respect for persons, assessing risks and benefits demonstrates beneficence, and fairly selecting subjects demonstrates justice.[21]

Informed Consent

Informed, voluntary consent is a cornerstone of ethical research practice. Voluntary consent is based on respect for individual autonomy and personal dignity.

Potential subjects must be given adequate information so they can make an informed, voluntary decision to participate.

1. Potential subjects need to know the general purpose of the study. This information provides a foundation for a subject's assessment of the costs and benefits of participating.
2. Federal guidelines and common ethical practice require that subjects be informed about the procedures, their purposes, and possible risks, including risks that are unknown. In other words, subjects should be told what they will be expected to do and what will be done to them. This information on risks should include mention of possible discomfort, anxiety, unwanted information about oneself, or inconvenience. In low-risk studies, subjects may be more concerned about the time involved than they are about the procedures.

Deceptive studies, such as the Milgram experiment, misinform subjects about a study's purpose or procedures. Deceptive research pits two values against each other:

- the potential value of a study's findings
- respect for individual autonomy (an individual's right to make choices)

Although the debate about the acceptability of deceptive research continues, existing procedures can minimize informed consent concerns. For example, a subject may be told that deception is part of the research and that he is being asked to agree to participate without knowing the full details of the study's purpose.[22]

3. Potential subjects should be told why they were selected. While this is not required or even implied in the federal regulations, this practice strengthens the subject's understanding of a study's purpose, his importance to the study, and the fairness of the selection process.[23]
4. Potential subjects should be told what will be done with the collected information and any limitations on its confidentiality. Who will receive the information from the study? What type of information will be disseminated? What steps will be taken to protect the identity of the subject? What will happen if a researcher learns of an illegal act or identifies a health risk? If photos, movies, recordings, or similar research records are being produced, the subject should know what will be done with them. We know of one student who was shocked to learn that an interview tape produced as part of an experiment she took part in was being shown to classes at her college.

Understanding Risks and Benefits

Providing information about a study's purpose, procedures, and risks is only part of ensuring voluntary participation. Other factors can compromise the voluntary nature of research participation. First, the subjects must understand the potential risks and the probable benefits. The way the risks and benefits are communicated must be appropriate to the subject population and the study. For example, in a research project that puts a subject at risk, oral consent is not adequate. Nor is just signing a statement adequate. Rather, the researcher needs to select words and techniques to ensure that subjects fully appreciate what is being asked of them.

Second, the relationship between the researcher and the potential subject may cloud the subject's judgment. For example, imagine a teacher who asks students to participate as research subjects. The teacher may state that participation is voluntary and will have no effect on class grades. Nevertheless, students may volunteer because they imagine that declining to participate may influence the teacher's feelings about them. The question is not whether a decision to participate or not to participate affects the teacher's grading or other class-related behaviors. Rather, if a student feels that his or her academic progress could be affected, then the decision to participate may not be voluntary.

Third, specific circumstances may impede a subject's ability to make a voluntary decision. Prisoners, members of the military, and schoolchildren, all of whom are in controlled settings, may interpret requests for information or participation as commands. Research involving prisoners is closely scrutinized. Current federal policy limits research involving prisoners to studies that address questions associated with criminal behavior or incarceration.[24] The regulations remind researchers that even modest incentives can compromise a prisoner's ability to assess the risks of participation and may preclude a voluntary decision to participate.

Patients may mistakenly believe that their research participation will have a therapeutic benefit. Voluntary research participation requires clear, realistic information on the benefits and risks of participation. Participants should be told how research interventions may affect their present treatment. Still, the ability of seriously ill persons to give informed consent, no matter what they are told, is questionable. A *New York Times* article summed up the problem as follows:

> [P]otential participants are often desperately ill and may grasp at any straw—even signing a document without reading it. For this reason, many say there is no such thing as informed consent, only consent.[25]

Fourth, care must be taken in working with vulnerable populations, such as children, aged people, and mentally disabled people, who may not be fully capable to make an informed decision or to protect their own interests. In general, researchers try to get informed consent from such subjects and from a legal guardian.

Finally, voluntary participation requires the ability to withdraw from a study at any time. The potential subject must be told this as part of informed consent. Furthermore, potential subjects must be told that other benefits they are entitled to will not be affected by their decision to participate or to discontinue their participation. For example, a client receiving public assistance must be told that his continued eligibility for assistance does not depend on participating in a research study.

Informed Consent Forms

Informed consent and an informed consent form are not one and the same. A subject must be free to make her own decision about whether she wishes to participate. To make this decision, she must be informed about the study and give her consent to participate. A signed informed consent merely documents what information was given, that the participant received and understood the information, and that she consented to participate. The greater the risk to a subject, the more extensive the informed consent procedures required. To obtain informed consent for online surveys, subjects may read a statement describing the research and then indicate their willingness to participate by clicking on an "accept" button. For projects where a subject experiences no risks beyond the risks of everyday life or ordinary professional responsibilities, signed statements may be reasonably viewed as unnecessary. Just because the recipient of a mail or telephone survey does not sign an informed consent form does not relieve the surveyor from giving the subject sufficient information so that she can give her informed consent to participate.[26]

Identifying and Weighing Costs and Benefits

People may not agree on the costs and benefits of research participation. Individuals' commitment to research and their educational, social, and professional backgrounds contribute to what they see as risks and benefits and their importance. Consequently, informed consent requires that researchers, impartial reviewers, and potential subjects separately assess a proposed study and decide if the benefits outweigh the risks. Researchers may incorrectly assume that a proposed study presents minimal or no risk. They may overestimate the benefits of a study. Researchers are expected to be especially vigilant if potential subjects represent a distinctly different population from themselves. For example, researchers may underestimate the potential harm in studying recent immigrants. In such cases, the researcher is more likely to misjudge what constitutes a risk or a benefit for a participant. She may erroneously assume that the way she requests consent is unbiased and informative.

To minimize the potential of overlooking risks, overestimating benefits, and assuming consent is informed and voluntary, an investigator should solicit the opinion of others. Most university-based research or biomedical research is reviewed by an institutional review board that determines if a proposed project adequately protects its human subjects. (IRBs are discussed in more detail later in this chapter.)

Sometimes the research involves a treatment that may relieve a physical or psychological problem. Sometimes the research may seem to benefit a group that a potential subject values. For example, alumni may agree to participate in research on the effectiveness of their education because they believe that the findings will help future students. For some studies, perhaps most, the subject may participate because the research question seems somewhat interesting and the inconvenience is minimal.

For some studies, remuneration is a valuable benefit. We know of a few graduate students who subsidized their incomes by participating as subjects for biomedical research projects. Paying subjects for the inconvenience of participating is not unethical, unless the remuneration is so large that it may be considered a bribe or questionable inducement to participate. What distinguishes reasonable reimbursement from "undue inducements"? Federal regulations offer no guidance, and opinions vary as to whether participants should be paid more than their direct costs for participating[27]. For example, see the National Institutes of Health 2009 update on the requirement for instruction in the responsible conduct of research, NOT-OD-10–019. Additional information can be located at https://grants.nih.gov/grants/guide/notice-files/not-od-10-019.html. The Food and Drug Administration requires that paid subjects who withdraw from a study receive a prorated payment.[28]

The most commonly cited risks are physical harm, pain or discomfort, embarrassment, loss of privacy, loss of time, and inconvenience. A study could potentially uncover illegal behavior. Other risks, alluded to in the illustrative cases, include undermining confidence in public institutions and lessening interpersonal trust.

As part of informing a subject, he must be told what risks may occur during the study or as a result of the study. If the risks are unknown, or if researchers disagree on the risks, this information must be communicated to the subject. Terminally ill patients may agree to participate in a study that promises to lead to a future cure. Nevertheless, a researcher must inform potential subjects only of benefits that can be reasonably expected. Theoretically, a study may be groundbreaking; however, most studies are not. Consequently, a subject should not be told that a study has a probability of generating significant knowledge. Nor should potential subjects be led to believe that they will gain benefits that are possible but unlikely.

Although not usually covered in informed consent, administrators may wish to consider that risks and benefits may apply beyond individuals involved in a study. For example, families may be affected by the time a family member spends participating in a study, his reactions after sharing information, or the monetary benefit of a study on training intended to increase job skills. The jury study example suggested a project that would neither harm nor benefit its participants but which could have had a serious negative effect on an important social institution.

Selection of Subjects

Selection of research subjects should be unbiased and take into account who will benefit from the study. Examples such as the Tuskegee study and the Jewish Chronic Disease Hospital raised questions about studying vulnerable populations. Subsequently researchers and IRBs are expected to be especially diligent in reviewing work that relies on subjects from vulnerable populations and to make sure that appropriate measures are taken to solicit informed consent. The perspective on selection of subjects has shifted from exclusion to inclusion. Excluding certain groups from studies is also viewed as ethically questionable, however, because the group excluded does not benefit from the research.

Prior to the 1990s, women were routinely excluded from clinical trials. Much more was known about men's health and how males reacted to various therapies than was known about women's health.[29] A major consequence was that women and men with heart disease were treated the same. Subsequent research has found gender differences in women's symptoms, response to treatment, and outcomes.[30]

Recruiting Subjects

Recruiting subjects is the first step in ensuring informed consent. Announcements should state that participants are sought for a research project. They should avoid terms that oversell potential benefits or emphasize inducements for participating. The words appearing on flyers, media advertisements, Internet sites, and letters to potential subjects may act as a questionable inducement. Consider the attractiveness of being asked to test out a "new" or "exciting" treatment, to participate in a "free" program, or to receive "$1,000 for a weekend stay in our research facility."

If participants are recruited through personal contact, the investigator must be particularly sensitive to not pressuring participation.[31]

The recruitment process may also raise privacy concerns. Later in this chapter, we write about privacy in the context of confidentiality, that is, the ability to control disclosure and dissemination of information about oneself. Another dimension of privacy is physical privacy, that is, not having to endure unwanted intrusions.[32] Ethical concerns about intrusion vary. If a person is likely to wonder, "How did they get my name?" there may be an ethical problem. For example, before contacting recipients of social services to evaluate a program, someone in the agency who is familiar with the clients should ask their permission to be contacted. Along the same lines, agency clients should be assured that whether they choose to participate will not affect services they normally receive.

Ethical Selection

The key principle in the ethical selection of a study population is distributive justice: equitably distributing research benefits and risks, treating like situations the same way, and offering equal access to participate in research.[33] Currently, the ethical debate on subject selection is confined to biomedical research, where both the individual risks and benefits can be quite high. Because each population is different, a general rule will not cover different racial and ethnic groups, institutionalized people, and international studies. Rather, researchers should be aware of why they are studying a particular group and the ethical dimensions of their choice.

Selection of individual subjects should be fair. For example, in medical experiments, people assigned to a placebo group may feel cheated and lobby to receive the experimental treatment. Placing a favored subject in a more beneficial treatment group or placing a disliked subject in a risky treatment group is unethical. Note that good research practice coincides with ethical practice.

Sample size is not included as a component of informed consent, nor is it discussed in behavioral and social science texts on research ethics. Still, the number of desired subjects may affect how vigorously subjects are recruited and open the door for subtle coercion. The American Statistical Association includes sample size in its ethical guidelines. Ethical statisticians should "avoid the use of excessive or inadequate numbers of research subjects—and excessive risk to research subjects (in terms of health, welfare, privacy, and ownership of their own data)—by making informed recommendations for study size."[34] An inadequate sample can affect the quality of the statistical analysis, for example, decisions about whether findings occurred by chance. Conversely, an overly large sample may squander resources, including participants' time, and should not be gathered just to alleviate distributive justice concerns.

Protecting Privacy and Confidentiality

Understanding the terms *privacy, confidentiality, anonymity*, and *research records* provides a basis for ethical research practice (see Table 8.1).[35]

Confidentiality and Intrusive Questions

The requirements of voluntary participation and informed consent uphold the individual's right to have control over information about himself. While a researcher may promise anonymity or confidentiality, a potential subject may not necessarily trust her to follow through. Guarantees of confidentiality neither ensure candor nor increase propensity to participate in research; rather, limited research has found that respondents tend to view promises of confidentiality skeptically.[36] Nevertheless, researchers must respect participants' privacy and maintain confidences as part of their professional responsibilities to subjects.

Table 8.1 Foundations of Ethical Research Practice

Terms	Definitions
Privacy	An individual's ability to control the access of other people to information about himself
Confidentiality	The protection of information, so that researchers cannot or will not disclose records with individual identifiers
Anonymity	Collecting information using strategies whereby researchers cannot link any piece of data to a specific, named individual
*Research records**	Records gathered and maintained for the purpose of describing or making generalizations about groups of persons
Informed consent	Consent given with full knowledge of the possible risks and consequences of participation

* Unlike administrative or clinical records, research records are not meant to make judgments about an individual or to support decisions that directly and personally affect an individual.

Some research questions may seem unduly intrusive. Questions may stir up unpleasant recollections or painful feelings. Such topics include research on sexual behaviors, victimization, or discrimination. People who hold controversial opinions may prefer to keep this information to themselves. Disclosure of behaviors such as drug use, child abuse, or criminal activity may cause a respondent to fear that she will be "found out." For a study of a sensitive topic to be ethical:

1. The psychological and social risks must have been identified.
2. The benefits of answering the research question must offset the potential risks.
3. The prospective subjects must be informed of the risks.
4. Promises of confidentiality must be maintained.[37]

Anonymity

To avoid the problems associated with confidentiality, the researcher may gather anonymous information. Anonymous participation occurs if no records are kept on the identity of subjects and data cannot be traced back to a specific individual. If researchers are conducting a study for an agency, an approach that approximates anonymity is for the agency to collect information and delete any information that directly identifies the respondents. This approach may be taken if revealing a client's identity to outsiders compromises the client's privacy. An agency may select the sample, distribute questionnaires, or collect data from agency files. Alternatively, an agency may ask clients' permission to give their names to the researchers.

Often anonymity is impossible. Researchers must know and record subject names to:

- Follow up on nonrespondents or to compare respondents and nonrespondents.
- Combine information from a subject with information from agency records.
- Carry out panel studies and collect information from an individual at different points in time.
- Conduct a study audit to verify that the research was done and that accurate information was collected and reported. Auditors may need access to identifiable records. Auditing can present problems of confidentiality, but without the possibility of conducting an audit, incompetence or malfeasance may go undetected and potentially cause even greater harm.

Exactly how much care to take in protecting the identity of respondents varies from topic to topic. Some topics and some information require stringent safeguards. For the most part, social science researchers record individual information only to keep track of respondents and their data. They should take adequate safeguards to prevent disclosing identifiable information about an individual. Normally, at a minimum, disclosure may be disquieting. Removing identifying information from the data and strictly limiting access to lists with subject names are sufficient for studies requiring stringent safeguards to protect confidentiality.

Reporting Information

Confidentiality may be breached by carelessness, legal demands, or statistical disclosure. To avoid accidental disclosure of personal information, a researcher should separate identifying information from an individual's data. A code can be assigned to each record and a list of names and matching codes kept separately in a secure place. Lists with names of subjects may be kept separate from information collected on them. For longitudinal studies or other studies that combine information from more than one source, each person may be assigned an alias so that the information can be combined. A respondent may choose her own alias, that is, information that others could not easily obtain, such as her mother's birth date. The success of having respondents choose their alias depends on their ability to remember it each time they are asked. The list of names and aliases should be kept separately from the collected data. Theoretically, researchers can have their records subpoenaed. Federal policies offer some protections to participants and researchers. The Confidential Information Protection and Statistical Efficiency Act (CIPSEA), passed in 2002, requires that federal statistical agencies inform respondents and get their informed consent if the information is to be used for nonstatistical purposes.[38] Congress is currently revisiting this policy, but updated legislation has not yet been passed. Certificates of confidentiality by and large protect researchers investigating sensitive subjects, for example, mental illness or drug abuse, from having to provide identifying information. The U.S. Department of Health and Human Services offers certificates to researchers whose subjects might be harmed if they were identified. The Department of Justice offers certificates to researchers conducting criminal justice research. A certificate cannot be obtained after data collection is completed.[39]

When reporting data, one should be sensitive to inadvertent disclosures. For example, if a state has very few female city managers, tables distinguishing between the responses of male and female city managers may compromise the confidentiality of the female managers. Similarly, case studies may require more elaborate procedures to protect identities, including pseudonyms, and alteration of some personal information, such as occupation.

Sharing research records constitutes special problems for confidentiality. As we discuss in the following chapter, data sharing and secondary analysis may reduce research costs and increase research quality. Nevertheless, once records leave the control of the researcher, issues of informed consent and confidentiality become a particular concern. With regard to informed consent, the subject may have agreed to participate in research for a specific purpose without giving blanket authorization for other uses. Researchers should inform potential subjects of anticipated future uses of the data, including their availability for independent verification of the study's implementation and replication of its analysis.[40] Before releasing data, the researcher should remove identifiers such as names, addresses, and telephone numbers.

Deductive Disclosure

A more formidable problem is that of *deductive disclosure*. If the names of participants in a research project are known, someone may be able to sort through the data to identify a specific person. Imagine a study of employee satisfaction in a state agency. With a list of respondents, someone may sort the data by age, race, sex, and position and deduce a respondent's identity. One way to protect against such abuses is to not disclose the list of respondents.[41] Deductive disclosure is a concern with publicly available electronic databases. For example, if census data were released as reported, one could learn detailed information about the three Hispanic families in a county or a state's six female-owned utilities companies. To prevent such abuses, the Census Bureau has developed procedures to release as much information as possible without violating respondents' privacy. One procedure is to suppress the data if the number of cases falls below a minimum threshold.[42] As is true with many of the Census Bureau's statistical practices, these procedures can serve as a model for other researchers.

Social Media Research

Research is now frequently conducted using data from social media platforms (like Twitter and Facebook). There are disagreements about what online communications should be considered public as opposed to private and how researchers should treat the information. In the absence of specific requirements from institutional review boards, common sense should lead to ethical decisions. Information found on social media may be treated the same as other textual material, and one would not normally seek informed consent or be concerned about privacy. On the other hand, participants on social media platforms may not expect to read their comments in a research article. Although it may be naïve to assume that any communications sent out into cyberspace are private, we do not agree that one should forgo obtaining informed consent.

To grapple with this problem, social media platforms such as Twitter have adopted privacy policies[43] for users. For example, Twitter states in its privacy policy, as of 2019, that "Twitter is public and Tweets are immediately viewable and searchable by anyone around the world." It also asserts that it gives you other options and that "give you non-public ways to communicate on Twitter too, through protected Tweets and Direct Messages . . . or use a pseudonym if you prefer not to use your name."

Federal Policy on Protection of Human Subjects and Institutional Review Boards

In 1991, a uniform federal policy, the Common Rule for the protection of human subjects, was published.[44] Its hallmark was a requirement that every institution receiving federal money for research involving human subjects create an *institutional review board* and appoint its members. The Common Rule was recently revisited (2011) and revised (2017). This final rule is "intended to better protect human subjects involved in research, while facilitating valuable research and reducing burden, delay, and ambiguity for investigators. These revisions are an effort to modernize, simplify, and enhance the current system of oversight."[45] The new requirements are currently in the process of being implemented in 2019.[46] While most of the same standards apply, two changes are of note. First, the new rule allows for multicenter funded studies to use a single review process. Prior to this change, each institution involved in a research project had to receive separate institutional review board approval. Second, the rule broadens the types of research that qualify for exemptions.

In addition to this, in 2010, the federal government, as a part of the America COMPETES Act, requires education for responsible conduct of research (RCR) for postdocs, graduate students, and undergraduate students if applying for National Science Foundation grants.[47] The National Institutes of Health (2011)[48] and the U.S. Department of Agriculture (2013)[49] also have adopted similar policies required for all applicants.

Institutional Review Board Requirements

An IRB determines if a proposed project meets the following requirements:

1. Risks to subjects are minimized.
2. Risks are reasonable in relation to anticipated benefits.

3. Selection of subjects is equitable.
4. Informed consent will be sought and appropriately documented.
5. Appropriate data will be monitored to ensure the safety of subjects.
6. Adequate provisions exist for ensuring privacy of subjects and confidentiality of data.[50]

Two of these criteria merit further mention. First, the long-range effects of the knowledge gained from the research are explicitly excluded in determining the risks and benefits of participation. Second, the policy reflects concern with possible abuses of vulnerable populations. IRBs are reminded to consider whether research on a specific population is consistent with an equitable selection of subjects and to make sure that these populations' vulnerability "to coercion or undue influence" has not been exploited.

To perform these tasks, an IRB as a whole should be professionally competent to adequately review the proposals it commonly receives and sensitive to general ethical issues, especially issues affecting "vulnerable" populations, for example, prisoners, children, pregnant women, people with mental or physical disabilities, and otherwise disadvantaged people. An institution should consider each appointee's training, race, gender, cultural background, and sensitivity to community attitudes.

Failure to comply with the Common Rule can result in termination or suspension of federal support, which is a powerful incentive to ensure that the IRBs are created and follow through on their responsibilities.

Understanding What an Institutional Review Board Does

An IRB reviews all research involving human subjects under the purview of the institution. To review only publicly or privately supported research implies that only funded projects have to conform to ethical practices. To understand what an IRB does requires knowledge of what constitutes research and what constitutes a human subject. Covered research is "a systematic investigation, including research development, testing and evaluation, designed to develop or contribute to generalizable knowledge."[51] A *human subject* is a "living individual about whom an investigator (whether professional or student) conducting research obtains (a) data through intervention or interaction with the individual, or (b) identifiable private information."[52]

The IRB chair or the chair's designee first determines if a project is exempt from further review, appropriate for an expedited review, or must be subject to full review. Exempt and expedited projects receive less close scrutiny. These categories allow research involving minimal risk to avoid the long delays associated with a full IRB review. Minimal risk applies to those projects where the risks of participating in the research are similar to the risks of daily life.

Common Rule

The Common Rule exempts demonstration projects that study or evaluate public benefit or service programs, procedures for obtaining benefits or services under these programs, changes in or alternatives to the programs or procedures, and changes in the level or method of payment.[53] The U.S. Department of Health and Human Services wanted demonstration projects exempt because needed studies could not be conducted if subjects could withdraw at any time or if they could receive entitled benefits if they refused to participate. A 1972 study required an experimental group of employable welfare recipients to accept public-sector employment; if they refused to work, their benefits were discontinued. Clearly, such a study would have problems with internal validity if subjects could change their experimental-group status at will. The study ended up in federal court, which ruled that the experiment was acceptable because it was consistent with the Social Security Act's objective to remove people from the welfare rolls and move them into gainful employment.[54]

The Common Rule implicitly allows IRBs to waive written documentation of informed consent for surveys and observation of public behaviors. Waivers are not permitted if responses could be traced to a specific individual and if disclosure could result in civil or criminal liability or damage subjects' financial standing, employability, or reputation.

Importance of the Institutional Review Board System

The federal regulations are more detailed than we have presented here. Furthermore, practices are still evolving, and the regulations receive frequent scrutiny.[55] The 1999 death of an 18-year-old participant in a federally funded gene therapy study led to an examination of the IRB system. He suffered from a disorder that was controlled by diet and drugs; he was not terminally ill. One of the immediate reactions was to pay more attention to informed consent forms and to stress that having people sign informed consent forms was not sufficient evidence that they understood the information, especially the risks.

Federal agencies, advisory commissions, and research organizations have drafted and reviewed policies that may culminate in "a unified, comprehensive federal policy embodied in a single set of regulations and guidance."[56] The policies have been drafted by people in the fields of bio-ethics and biomedical research. Nevertheless, these proposed policies will affect social science research, and social science organizations are monitoring and evaluating them.

Federal policies protecting human subjects require compliance by federal agencies, institutions, and individual researchers. If you are a student or an employee of a university, a medical facility, or other institution that receives federal research funds, you should consult with your IRB before proceeding to conduct research that has you interacting with people, manipulating them, or using identifiable private information. Public administration students and professionals are most likely to conduct "research on individual or group characteristics or behavior . . . or research employing survey, interview, oral history, focus group, program evaluation, human factors evaluation or quality assurance."[57] All of these may be eligible for an expedited review.

Issues of Interest to Administrators

As an administrator, you may not work in settings that have an IRB. Nevertheless, you should adhere to standard practices as presented in the Belmont Report, discussed earlier in this chapter. Recall that these are respect for persons, beneficence and justice or getting informed consent, having a favorable cost–benefit ratio, and fair subject selection. Even if you never conduct your own research on human subjects, you may be approached by researchers who want to study your clients or employees. Administrators who work in educational, penal, or health care institutions are especially likely to be approached by researchers. In considering the ethical aspects of research, the administrator's major interest is to make sure that the researcher implements standard procedures to reduce risks.

Determine Research Context

First, an administrator should understand the context of the research and get an overall picture of the planned study. Why does the researcher want to study her agency's clients or employees? How will the research subjects be chosen? Will they be anonymous or confidential? What steps will be taken to protect their identities? What will they be asked to do? How much time will it take? What are other potential risks? She will want to make sure that selection of research subjects or assignment into different experimental groups is fair and will be seen as equitable to both participants and nonparticipants.[58]

To protect the identity and privacy of clients and employees, an agency representative should contact the affected people, explain the nature of the research, and seek their permission to give their names to the researchers. The researchers should not initiate contact with clients, and the agency representative should take care in using non-threatening language in approaching potential subjects. Public announcements, such as posters, may be used to recruit employees or clients. Voluntary participation is less likely to be compromised with posted announcements than by personal solicitation.[59]

Eligibility and Consent

Second, the administrator should make certain that contacted clients understand that their eligibility for agency programs or services will not be affected by a decision to participate. Research with employees may be problematic. For employee participation to be voluntary, whether employees decide to participate should not affect performance ratings or pay decisions, and this should be clearly communicated to them.[60]

Third, the administrator should review the informed consent procedures. If subjects will be at risk, the administrator should remember that signing an informed consent form is not sufficient evidence that the subjects understood what was asked of them and the risks involved. The administrator may want to learn whether the researcher has submitted the research for review to an IRB. The administrator may want to determine if the proposal was reviewed to ensure that the researchers are not blinded by their hopes for the research. Input from representatives of the participant community is especially valuable in identifying potential concerns. For example, prisoners, ex-convicts, and prison guards might review proposals involving prisoners.

The administrator should always engage the researcher in a discussion of potential risks. This lets the administrator decide whether the researcher has considered the ethical implications of the research. The administrator may learn more about research risks and improve his or her ability to judge the risks involved in later projects that come to the agency. Similarly, the administrator should check to see that subjects will be debriefed, if appropriate, at the end of their participation. For example, subjects who perform some task may be disappointed or frustrated about their performance. A debriefing offers the researcher an opportunity to observe any negative effects of the research and to answer questions or concerns that a subject may have.

After the Study

Fourth, if the study involves creating a program, the administrator may want to find out what the researcher plans to do when the research is finished. In prisons or psychiatric hospitals, the subjects may participate in a research project, such as a therapy group, that seems to end prematurely because the researcher's data collection has ended. Employees or students may participate in an experimental course only to find that their new skills are not wanted by their employer or do not fit into their school's curriculum.

A sticky issue is what will become of the findings. Does the researcher plan to put the data into an archive? If so, the subjects should be told where the data will be stored and how individual identities will be protected. If the agency expects to receive the research data, it should work out the details beforehand. This must be done if the researcher is going to act ethically and inform subjects of what will become of the information. If the agency, whether it employs the subjects or provides them with benefits, is going to get individual information, then the subjects must be told this as part of giving their informed consent. What information will the subjects or the agency receive? Will they receive the final report? What about interim reports? What will be done with the completed questionnaires, recordings, or other research documents?[61]

Fifth, the administrator should consider how the proposed research may affect the agency's reputation. No matter what the agency's role, its reputation may be enhanced by research that others consider valuable or harmed by research that they consider worthless or intrusive. An administrator may question whether a planned study will unduly infringe on respondents' privacy or abuse their time. He should decide if a study requiring agency resources, including time, represents a good use of the public's money. He should be convinced that a study will yield valued information. Unreliable items and items that serve no clear purpose cannot yield valued information. Studies that assume others will act on the findings should be reviewed by relevant actors, or at least the investigators should try to anticipate their reactions realistically. Unreliable items, unwanted items, or unwanted studies waste respondents' time. Items that are not operationally valid may abuse respondents' goodwill, insofar as their responses contribute to incorrect or misleading conclusions. To ensure that a proposed study provides usable and useful information, administrators should review research instruments and make sure that the researchers pilot-tested all instruments, establishing their reliability and operational validity.

The detrimental effects of an unwanted or poorly designed study go beyond the respondents. Future studies also are affected. Seeking too much information or seeking it too often can build resistance to future requests for information. Consider the complaints of businesses and state and local governments that must churn out data only to meet federal information requests. A similar outcome may occur if respondents perceive that the data are merely collected but not used.

Summary

Whether they are conducting research or permitting researchers to gather data on their organization, employees, or clients, administrators must make sure that all research participants are treated ethically. They should adhere to the dictates of the Belmont Report and ensure that all participants are treated with respect, beneficence, and justice (see Figure 8.1).

Obtaining informed consent is central to most research on humans. Informed consent means that a subject agrees to participate after receiving information on the purpose of the study, the potential risks associated with participating, and the probable benefits associated with participating. The information should be clear; the investigators should not oversell the benefits of participating nor discount social or psychological discomfort. Subjects must be told what will be done to them and what will be expected of them. Although informed consent may need to be

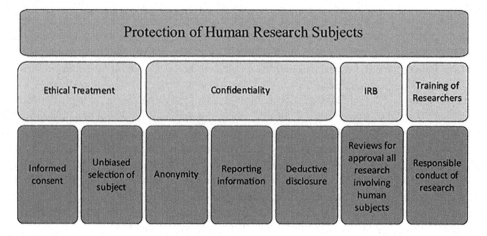

Figure 8.1 Protection of Human Research Subjects

documented, this should not take precedence over making sure that subjects understand what is being asked of them.

Selection of research subjects should be unbiased and take into account who will benefit from the study. The participation of subjects must be voluntary. Participants may receive incentives, but they should not be so large or attractive that they act as an "undue inducement." Researchers who maintain a professional or personal relationship with potential subjects must be aware that the relationship may make a request to participate in research exploitative and preclude the possibility of voluntary participation. Potential subjects must be told that their eligibility for benefits, other than benefits associated directly with participation, does not depend on participation. Similarly, their continued eligibility for benefits cannot be affected if they decide to withdraw from a study. As part of the informed consent process, the potential subject must be informed that he or she can withdraw from the research at any time.

Potential subjects should be told what will be done with the information after the study is completed. The research records should not be made available to others, including a subject's parents, physician, or employer, without the subject's explicit permission. This prohibition does not preclude allowing authorized auditors to examine records to verify the integrity of the research; however, the procedures for an audit must take proper precautions to maintain confidentiality, and disclosure to auditors should be part of the informed consent. If data are to be sent to a databank, the researcher should remove identifying information and take precautions to prevent deductive disclosure of a person's identity.

Agencies that receive federal money for research involving human subjects must have an institutional review board that reviews and approves all research involving human subjects. The IRB may expedite review of research involving minimal risk to subjects and waive certain aspects of informed consent for research involving surveys, observations of human behavior, and social experiments sponsored by a government agency or government officials. Whether IRB approval is required or not, researchers and project sponsors should satisfy themselves that a study adequately protects human research subjects.

The administrator who agrees to have research conducted with an agency's cooperation should interview the researcher and ascertain that the researcher will employ standard ethical practices and use appropriate procedures to get informed consent. Procedures for selecting subjects and assigning them to experimental groups must be perceived as fair by both participants and nonparticipants. Care must be taken that clients recognize that participation will not affect services they receive from the agency, and employees must know that their participation does not affect their working conditions. The administrator will want to learn, as appropriate, whether subjects will be debriefed and how research programs, such as therapy groups, will be terminated. If the agency expects to have access to the research data, this must be worked out beforehand and included in the information given to obtain a potential subject's informed consent.

More recently, the National Institutes of Health and the National Science Foundation have required universities receiving federal funds to support certain disciplines to provide training in the responsible conduct of research. Some universities are creating special courses to do this, while others are able to use existing courses.

Notes

1. "Training Graduate Students on the Responsible Conduct of Research (RCR) at UNC Charlotte," UNC Charlotte, The Graduate School, 2014.
2. L. K. Altman, "Volunteer in Asthma Study Dies After Inhaling Drug," *New York Times*, June 15, 2001.
3. K. Sucato, "Education: Student Survey's Unexpected Lessons," *New York Times*, February 5, 2001.
4. N. Hershey and R. D. Miller, *Human Experimentation and the Law* (Germantown, ND: Aspen Systems Corporation, 1976), 8–10, summarizes the bureaucratic history of the Tuskegee study. Hershey and

Miller observed that the Tuskegee study continued during the time when protection of human subjects was receiving so much attention.

5. The text of the Nuremberg Code can be found at https://history.nih.gov/research/downloads/nuremberg.pdf. The code is included in many research ethics texts, including R. Amdur and E. Bankert, *Institutional Review Board Management and Function*, 2nd ed. (Sudbury, MA: Jones and Bartlett Publishers, 2006).

6. Jay Katz, *Experimentation with Human Beings* (New York: Russell Sage Foundation, 1972), 10–65, reproduced a slightly edited version of the record from the investigation. Hershey and Miller, *Human Experimentation*, on 6–7, summarize the issues.

7. This summary of the critics' points of view is based on information from Tom L. Beauchamp, Ruth R. Faden, R. Jay Wallace, Jr., and LeRoy Walters, *Ethical Issues in Social Research* (Baltimore: Johns Hopkins University Press, 1982), 11–15. Edited versions of key criticisms also are found in Katz, *Experimentation*, 325–329.

8. Joan Sieber and Martin Tolich, *Planning Ethically Responsible Research*, 2nd ed. (Los Angeles: Sage Publications, 2013), 67–75. The authors discuss Humphreys' work and other controversial research.

9. Katz, *Experimentation*, 358–365, reprints Milgram's description of the study; on pages 403–405, Katz has reprinted some of the reactions to the Milgram study.

10. Richard Friedman, "Cases: Long-Term Questions Linger in Halted Breast Cancer Trial," *New York Times*, October 21, 2003.

11. Sam P. K. Collins, "How Can We Get More Black People the Organ Donations They Need?" *Think Progress*, available at https://thinkprogress.org/how-can-we-get-more-black-people-the-organ-donations-they-need-91ca8b53039/.

12. J. H. Jones, *Bad Blood: The Tuskegee Syphilis Experiment* (New York: Free Press, 1993), 221; Cynthia Santos, "A Legacy of Mistrust: Research and People of Color," *The BodyPro* (Spring 2009), available at http://img.thebody.com/cria/2009/achieve_spring_2009.pdf#page=16.

13. J. B. Braunstein, N. S. Sherber, S. P. Schulman, E. L. Ding, and N. R. Powe, "Race, Medical Researcher Distrust, Perceived Harm, and Willingness to Participate in Cardiovascular Prevention Trials," *Medicine* 87, no. 1 (January 2008): 1–9.

14. Santos, "A Legacy of Mistrust: Research and People of Color."

15. Katz, *Experimentation*, 68–103, reproduced material from the hearings. The subcommittee chairman was Senator James O. Eastland. His quote is found on page 80 of Katz's book.

16. See *The Belmont Report: Ethical Principles and Guidelines for the Protection of Human Subjects of Research*, available at www.hhs.gov/ohrp/regulations-and-policy/belmont-report/index.html.

17. Ibid.

18. Ibid.

19. Ibid.

20. Ibid.

21. Ibid.

22. See American Psychological Association's *Ethical Principles of Psychologists and Code of Conduct* (Washington, DC, 2008) for the use of deceptive research.

23. This suggestion is made by Hershey and Miller, *Human Experimentation*, 33.

24. E. D. Prentice et al., "Research Involving Prisoners," in *Institutional Review Board Management and Function*, 2nd ed., eds. R. Amdur and E. Bankert (Sudbury, MA: Jones and Bartlett Publishers, 2006), 394–398.

25. L. K. Altman, "Fatal Drug Trial Raises Questions About 'Informed Consent'," *New York Times*, October 5, 1993, B7.

26. Robert M. Groves, Floyd J. Fowler, Jr., Mick P. Couper, James M. Lepkowski, Eleanor Singer, and Roger Tourangeau, *Survey Methodology* (Hoboken, NJ: John Wiley & Sons, Inc., 2004), 352–355; a summary of research on informed consent is found on pages 364–366. See also J. M. Oakes, "Survey Research," in *Institutional Review Board Management and Function*, 2nd ed., eds. R. Amdur and E. Bankert (Sudbury, MA: Jones and Bartlett Publishers, 2006), 418.

27. For example, National Institutes of Health, "Update on the Requirement for Instruction in the Responsible Conduct of Research," NOT-OD-10-019 (2009). Additional information can be located at https://grants.nih.gov/grants/guide/notice-files/not-od-10-019.html.

28. See B. G. Gordon, J. S. Brown, C. Kratochvil, and E. Prentice, "Paying Research Subjects," in *Institutional Review Board Management and Function*, 2nd ed., eds. R. Amdur and E. Bankert (Sudbury, MA: Jones and Bartlett Publishers, 2006), 183–190. This article covers the arguments for and against reimbursing subjects and models for reimbursing subjects.

29. N. Kass, "Gender and Research," in *Beyond Consent: Seeking Justice in Research*, eds. J. P. Kahn, A. C. Mastroianni, and J. Sugarman (New York: Oxford University Press, 1998), 67–87.

30. See "Cardiovascular Disease and Other Chronic Conditions in Women," Agency for Healthcare Research and Quality, available at www.ahrq.gov/sites/default/files/publications/files/womheart.pdf.

31. For further discussion of ethics and recruitment procedures, see the following articles in Amdur and Bankert, *Institutional Review Board*: M. Khin-Maung-Gyi and F. Whalen, "Recruitment of Research Subjects," *Reference Work Entry*, 147–148, doi:10.1007/978-1-4419-1005-9_1062; R. Homer, R. Krebs, and L. Medwar, "Advertisements for Research," *Tutorialspoint*, 149–153, available at www.tutorialspoint.com.

32. For further discussion, including references, see H. Cho and R. LaRose, "Privacy Issues in Internet Surveys," in *Social Surveys*, vol. II, ed. D. de Vaus (Thousand Oaks, CA: Sage Publications, 2002), 206–222; Oakes, "Survey Research," 430, also discusses privacy and recruitment.

33. C. R. McCarthy, "The Evolving Story of Justice in Federal Research Policy," in *Beyond Consent: Seeking Justice in Research*, eds. J. P. Kahn, A. C. Mastroianni, and J. Sugarman (New York: Oxford University Press, 1998), 11. See also *The Belmont Report*.

34. American Statistical Association, "Ethical Guidelines for Statistical Practice," (2018): 4. The guidelines are available on the association's web site www.amstat.org/ASA/Your-Career/Ethical-Guidelines-for-Statistical-Practice.aspx.

35. R. F. Boruch and J. S. Cecil, *Assuring the Confidentiality of Social Research Data* (Philadelphia: The University of Pennsylvania Press, 1979), 23–27.

36. A. G. Turner, "What Subjects of Survey Research Believe About Confidentiality," in *The Ethics of Social Research: Surveys and Experiments*, ed. J. E. Sieber (New York: Springer-Verlag, 1982), 151–165. For a summary of more recent research on the public's attitude about the confidentiality of data, see *Expanding Access to Research Data: Reconciling Risks and Opportunities* (Washington, DC: National Academies Press, 2005), 52–54.

37. National Human Genome Research Institute, "NHGRI GDS Policy Information & Resources," 2019, available at www.genome.gov/about-nhgri/Policies-Guidance. Researchers interested in strategies for identifying sensitive subjects and asking sensitive questions should see D. Fahie, "Doing Sensitive Research Sensitively: Ethical and Methodological Issues in Researching Workplace Bullying," *International Journal of Qualitative Methods* 13, no. 1 (2014): 19–36.

38. National Research Council, "Expanding Access to Research Data: Reconciling Risks and Opportunities," Division of Behavioral and Social Sciences and Education, Committee on National Statistics, & Panel on Data Access for Research Purposes, 2005.

39. Ibid., 56.

40. T. E. Hedrick, "Justifications and Obstacles to Data Sharing," in *Sharing Research Data*, eds. S. E. Fienberg, M. E. Martin, and M. L. Straf (Washington, DC: National Academy Press, 1985), 136. Hedrick cites sources that discuss this issue in more depth. See also the *Ethical Guidelines in Statistical Practice, D4* (Alexandria, VA: American Statistical Association, 1989).

41. For strategies to prevent deductive disclosure, see J. Steinberg, "Social Research Use of Archival Records: Procedural Solutions to Privacy Problems," in *Solutions to Ethical and Legal Problems in Social Research*, eds. R. F. Boruch and J. S. Cecil (New York: Academic Press, 1983), 249–261; Boruch and Cecil, *Assuring the Confidentiality of Social Research Data*, Chapter 7.

42. For information on specific strategies and when they are applied, see U.S. Census Bureau, "A Monograph on Confidentiality and Privacy in the U.S. Census," July 2001, available at www.census.gov/history/pdf/ConfidentialityMonograph.pdf.

43. https://twitter.com/en/privacy (accessed September 4, 2019).

44. U.S. Science and Technology Policy Office, "45 Code of Federal Regulations 46 (45 CFR 46) Federal Policy for the Protection of Human Subjects: Notices and Rules," *Federal Register* 56 (June 18, 1991): 28002–28018.

45. Federal Register/Vol.82, No.12/Thursday, January 19, 2017 Rules and Regulations, available at www.govinfo.gov/content/pkg/FR-2017-01-19/pdf/2017-01058.pdf.

46. J. Menikoff, J. Kaneshiro, and I. Pritchard, "The Common Rule, Updated," *New England Journal of Medicine* 376, no. 7 (2017): 613–615.

47. Effective January 4, 2010, the National Science Foundation requires education in the responsible conduct of research for all undergraduate students, graduate students, and postdocs funded by its grants (see Section 7009 of the 2007 America COMPETES Act [42 U.S.C. 1862o-1]; Federal Register of 08/20/2009). This requires that "each institution that applies for financial assistance from the Foundation for science and engineering research or education describe in its grant proposal a plan to provide appropriate training and oversight in the responsible and ethical conduct of research to undergraduate students, graduate students, and postdoctoral researchers participating in the proposed research project."

48. https://grants.nih.gov/grants/guide/notice-files/NOT-OD-10-019.html.

49. https://grad.ncsu.edu/wp-content/uploads/2016/03/nifa-213.pdf.
50. 45 CFR 46 Section 46.111.
51. 45 CFR 46 Section 46.102(d).
52. 46 CFR 46 Section 46.102(f).
53. 45 CFR 46 Section 46.110.
54. M. J. Breger, "Randomized Social Experiments and the Law," in *Solutions to Ethical and Legal Problems in Social Research*, eds. Robert F. Boruch and Joe S. Cecil (New York: Academic Press, 1983), 104–105, citing material from *Aguayo v. Richardson*, 352 F. Supp. 462 (S.D.N.Y. 1972).
55. An excellent source of current activity is the Office of Human Research Protections web page (www.hhs.gov/ohrp/). For social science concerns, see "Protecting Participants in Social, Behavioral, and Economic Science Research: Issues, Current Problems, and Potential Solutions," *Responsible Research: A Systems Approach to Protecting Research Participants* (Washington, DC: National Academies Press, 2003), Appendix B.
56. Quote from page 290 of Publications & Reports. *Health Affairs* 20, no. 5 (2001): 290–293. See also www.bioethics.gov/ for further work on this matter.
57. U.S. Food and Drug Administration, "Protection of Human Subjects: Categories of Research that may be Reviewed by the Institutional Review Board through an Expedited Review Procedure," available at https://www.fda.gov/science-research/clinical-trials-and-human-subject-protection/categories-research-may-be-reviewed-institutional-review-board-irb-through-expedited-review.
58. For a more detailed discussion on recruiting volunteers, employees, and vulnerable populations, see "Protecting Human Subjects and Other Ethical Considerations, section on Selection of Subjects," National Institutes of Health, Office for Human Research Protection, 1993 Institutional Review Board Guidebook. Chapter 6, available at www.hhs.gov/ohrp/education-and-outreach-/archived-materials-index.htm; A. J. Kimmel, *Ethical Issues in Behavioral Research* (Cambridge: Blackwell Publishers, 1996), 215–235.
59. "Protecting Human Research Subjects," 53, Chapter 6.
60. Ibid., 55.
61. Paul Oliver, *The Student's Guide to Research Ethics*, 2nd ed. (Maidenhead, PA: Open University Press, 2011), Chapter 4, has a useful discussion related to these concerns.

Terms for Review

subjects at risk
Common Rule
informed consent
voluntary research participation
privacy
confidentiality
anonymity
deductive disclosure
research records
institutional review board
human subject
responsible conduct of research

Questions for Review

The following questions should indicate whether you have a basic competency in this chapter's material.

1. What four main pieces of information must a potential research subject have to give informed consent? Explain why each is needed.
2. How are informed consent and voluntary participation related?
3. To learn if traumatic events are associated with drug abuse, researchers propose to interview drug abusers about their childhood and adolescence. The researchers know that the

interviews may be painful and cause participants emotional distress. What should an IRB consider before allowing the research to be conducted?

4. What types of benefits should a potential participant be told about? How expansive should an investigator be when informing potential participants about the benefits of participation?

5. What types of risks should a potential participant be told about? How expansive should an investigator be in informing potential participants about the risks of participation?

6. Why is debriefing subjects important? What is the role of debriefing in lessening the risks to the subject? In teaching the researcher more about the risks of participation?

7. Explain why the following may be considered inappropriate:

   a. A principal asking teachers in her school to participate in a study she is doing for a graduate class.

   b. Giving each return envelope in an "anonymous" survey a different box number so that each participant can be identified.

   c. Explaining in English the planned research to potential participants who recently immigrated from a non-English speaking country.

   d. A social agency turning over a list of clients to researchers who are studying how clients view agency services.

   e. Conducting research on developmentally disabled adults based on the consent of their guardians and not the subjects.

8. Explain the difference between collecting confidential and anonymous information. Under what circumstances will a researcher prefer confidential information? Under what circumstances will a researcher prefer anonymous information?

Problems for Homework and Discussion

1. Imagine you are the head of a large agency, such as a school district. Outline a protocol for approving research involving agency employees and/or clients (students or students' parents, for example).

2. How does a researcher identify the risks associated with their study?

3. A researcher proposes testing an accelerated math program on talented third-graders in a school system. The program will last for one year, although data collection on participants may continue through elementary school. What risks do you foresee for children who participate in the program?

4. A young researcher contends that never putting subjects at risk is the only ethical way to conduct research. Comment on this position.

5. The American Statistical Association ethical guidelines imply that having too many subjects or too few are ethical concerns. Why do you think that this is so?

6. Does your college or university have an IRB? If so, obtain a copy of its requirements for informed consent. For what types of studies is informed consent required?

7. Find an actual consent form and identify the components critical to giving informed consent. Are any pieces of critical information missing?

8. A researcher is studying the effectiveness of educational programs offered to prisoners. In his interviews with prisoners, he learns that several of them smuggle drugs. Should he inform the prison authorities? Why or why not?

9. A researcher conducts genetic tests on subjects. One genetic test shows that Mary Doe carries the gene for Huntington's disease. What should the researcher tell Mary Doe?

10. Investigators are conducting a 24-month demonstration project to study the effectiveness of a job-training project. The investigators propose to randomly assign persons eligible for

job training to either an experimental or control group. Control group members will neither receive training nor will project staff inform them of other training opportunities. Comment on the ethics of this proposal.

11. Determine if your university has any courses to train students in the responsible conduct of research. What components are covered in the courses?

Working With Data

Review the IRB requirements for your university. Then review the General Social Science Survey and the Focus Group Report files. (1) Explain what requirements a researcher using the GSS Survey would need to meet. Under what circumstances would the research be exempt? (2) Would the participants in the Focus Group study need to sign a consent form? Assuming they did, draft one that could be used for the study.

Recommended for Further Reading

For up-to-date information on federal policies affecting human subject research, the Office of Human Subjects Research, National Institutes of Health, home page is recommended (https://irbo.nih.gov/confluence/).

A valuable resource is Amdur, R., and E. Bankert, *Institutional Review Board: Management and Function*, 2nd ed. (Sudbury, MA: Jones and Bartlett, 2006). It includes copies of major policies and short presentations on the IRB process, components of informed consent, vulnerable populations, and issues associated with various research approaches.

The following study recommended changes in research involving prisoners. It provides an excellent review of the Belmont principles, concerns with confidentiality and privacy, and the application of federal policy. Gosten, L. O., C. Vanchicri, and A. Pope, eds., *Ethical Considerations for Research Involving Prisoners* (Washington, DC: National Academies Press, 2006).

Boruch, R. F., and J. S. Cecil, *Assuring the Confidentiality of Social Research Data* (Philadelphia: The University of Pennsylvania Press, 1979); Lune, H., and B. L. Berg, *Qualitative Research Methods for the Social Sciences* (New York: Pearson Higher Education, 2016). Covers issues associated with confidentiality, strategies for maintaining confidentiality, and legal issues.

Division of Behavioral and Social Sciences and Education, *Expanding Access to Research Data: Reconciling Risks and Opportunities* (Washington, DC: National Academies Press, 2005), focuses on competing demands for data access to microdata and maintaining confidentiality.

Fisher, Celia B., *Decoding the Ethics Code: A Practical Guide for Psychologists* (Thousand Oaks, CA: Sage Publications, 2013), covers both research ethics and other ethical concerns of non-research psychologists. Other professional associations, including those of historians, anthropologists, statisticians, sociologists, and program evaluators, have ethical guidelines (available on their professional societies' home pages), which are worth consulting to get a better understanding of the ethical issues.

National Institutes of Health (NIH), "Update on the Requirement for Instruction in the Responsible Conduct of Research," NOT-OD-10-019 (2009).

Oliver, Paul, *The Students' Guide to Research Ethics*, 2nd ed. (Maidenhead, PA: Open University Press, 2011), is a good introduction to ethical concerns.

9   Finding and Analyzing Existing Data

In this chapter, you will learn:

1. Advantages and disadvantages of secondary data analysis.
2. Strategies for identifying, accessing, and evaluating the quality of secondary data.
3. About data included in the major U.S. Census Bureau population surveys and vital records.
4. About using census data.
5. About data warehouses and big data.

In our eagerness to get a study underway, we may overlook the possibility that appropriate data may already be available. Organizations collect and store data for many purposes. For this chapter, the take-home point is that data originally collected for one purpose may often be retrieved at a later time, even by others, and used to answer a range of questions beyond those originally asked.

Data used in this manner have a specific name—*secondary data*. These are existing data that investigators collected for a purpose other than the given research study. Secondary data may result from the research efforts of an individual researcher, a research team, an agency division, or a research organization. The data may have been collected for a specific study or as part of a database. The data may have been collected as part of routine record keeping or to monitor agency performance. Investigators with differing backgrounds, needs, and questions regularly consult and use secondary data. Statistical organizations, including the U.S. Census Bureau, state offices of vital statistics, public opinion firms, and university research groups, such as the Inter University Consortium for Political and Social Research, based at the University of Michigan, exist to collect, compile, and interpret data. Professional associations, such as the International City/County Management Association (ICMA), and public interest groups routinely collect and publish survey data on topics of interest to their members. For example, each year ICMA publishes *The Municipal Year Book*. Investigators with differing backgrounds, needs, and questions consult and use such data regularly.

In this chapter, we survey the benefits and costs of using secondary data and how to identify appropriate databases, gain access to them, and evaluate the quality and applicability of a particular secondary dataset for use in a new study. We introduce you to a few important data sources and provide an overview of some of the major public sources of secondary data available. The sources covered are certainly not an exhaustive review of the sources from which secondary data may be obtained. The chapter also gives an overview of the contents of the Census of Governments and two population surveys conducted by the U.S. Census Bureau.

Working With Secondary Data

Secondary data analysis allows a researcher to reduce her costs markedly. She can dispense with the costs of instrument design, data collection, and compilation. It has other benefits as well but also has some costs.

The Benefits and Costs of Secondary Data Analysis

Secondary data analysis enables researchers to conduct studies that are otherwise not feasible. It also opens up research to public scrutiny, allowing for verification, refinement, or refutation of the original results. Over time, it may improve research quality.

Secondary data analysis may also simply be necessary if investigators want comparative or longitudinal data. For example, Raj Chetty and his colleagues at the Equality of Opportunity Project looked at the effect of race, gender, and even the quality of neighborhoods that kids grow up in to see if they have an impact on intergenerational economic mobility. To do this, Chetty and colleagues analyzed the tax records of 40 million children and their parents between 1996 and 2012. To do this, they had to gain access to data from the Internal Revenue Service and then map the deidentified data onto the counties in the United States. Their descriptive work with data over time allows for further causal analysis of the determinants of income mobility in the United States.[1]

The investigators could not have conducted their study without government statistics. They would have had to dig through records with hundreds of millions of observations. This would have been too time consuming and costly to do by hand. They would have had to locate the appropriate records and get permission to review them. This would have required a veritable army of graduate students. If records did not exist, a cross-sectional study might have been tried, but logistic requirements would have restricted the study population. Most importantly, they would not have been able to answer their research question of the determinants of economic mobility over time.

Limitations of Government Statistics

Relying on government statistics has limitations. The researchers have to determine that the data are sufficiently reliable and comparable to those collected at different times or in different locations. The design may have to be modified if data for certain years or for some countries are missing or unattainable. Such modifications may introduce bias. For example, we assume that those countries that share their data differ in many ways from those that do not.

The ability of various countries to gather and disseminate data varies dramatically. One of our colleagues once worked on a study of international income inequality. Wage data reported by some countries were more reliable than data on income. To increase the number of countries included in the study, the research team decided to use the wage data rather than income. Even by using the proxy measure of wages to expand the scope of the study, data were still missing from large areas of the world, such as the Middle East and most of Africa.

In a similar vein, different levels of government or organizations have much more data than other levels. More data are available on a national level than on the state level and on the state level than on the local level. For example, comparing recent data between two states on the average manufacturing wages paid may be possible; comparing the same data between two cities in these same two states may be impossible. It would be unwise to assume that the wages in the cities reflected the same relationship as between the states.

Secondary Data Analysis

Secondary analysis is a necessary component for open science; it allows others to scrutinize a researcher's work. In the field of public policy, scrutiny can be especially important. Secondary analysis may be undertaken if the investigator's veracity is questioned; however, it is more likely to occur within the context of challenging and improving research that affects policy decisions. Some academic journals, notably in health fields, require public access to the data used in a study as a condition of publishing the results. For example, before it even reviews an article, the journal *Molecular Systems Biology* requires that all data be submitted to a public database. This is a common requirement for certain research studies published in prominent journals such as *Science* and *Nature*.

Analysts working with existing data may reexamine the findings by including supplemental data, adapting the measures, or applying new or different statistical models. Their efforts may confirm and extend the research findings. If their work contradicts the original research, their findings may contribute to greater insight into the complexity of the policy issue and its solution.

Ideally, secondary analysis should improve the quality of data collection and documentation. The researcher who knows that her work will be scrutinized may pay more attention to documenting the research process:

- the decisions that were made about which variables to include and how to measure them
- the results of pretests
- the sampling design and possible nonsampling errors
- the details of how data were compiled and analyzed

An investigator working with secondary data may have his mind jogged. He may think creatively about the research design and have insights on better ways to measure the concepts or to analyze the data. His experience is analogous to most researchers' reactions when they analyze their own data and find themselves rethinking the research design, including their model and how they measured the variables.

The ability of researchers to share data electronically has made outside scrutiny easier, with more cases of fraud being exposed as a result. Between 1995 and 2000, Karen Ruggiero of Harvard University became a national expert on the psychological effects of race and sex discrimination. Ruggiero was exposed as a fraud after she refused to share her raw data, which turned out to be made up. In 2006, for the first time in U.S. history, a university researcher was sentenced to prison for using false data to obtain federal research grants. Since then, there have been a string of cases that have shown that researchers have manipulated data. One of the most notable was the case of a British medical doctor, Andrew Wakefield, that was paid to falsify data to show a link between the measles, mumps, and rubella vaccine and autism. This has sparked a new wave of anti-vaccination efforts that have led to at least three people dying in Ireland and likely many more in the United States and the rest of Great Britain.[2]

While secondary data reduce the costs of the secondary analyst, they can increase the costs to the initial investigators.[3] They may incur costs in increased data preparation and documentation. In modest studies, many of the details are carried in the investigator's memory. Modest studies have their place, but investigators conducting them should not absolve themselves of proper documentation and retention of research records. Later analysis of existing data can prevent needless requests for information from subjects or replication of earlier mistakes. We know of a colleague who kept the original surveys from her master's thesis for 15 years. Two months after they were destroyed, another researcher requested the surveys in order to update the analysis. Increasingly, researchers are being asked by journals and funders to make their data available through online platforms like ICPSR at the University of Michigan (www.icpsr.umich.edu/icpsrweb/) or Harvard's Dataverse (https://dataverse.harvard.edu/). These online data repositories facilitate replication of the original study and often provide the data that students need to conduct their own studies.

Misuse of Secondary Data

Conclusions drawn from secondary data are not without controversy. Publicly available government statistics may be accepted as being accurate and reliable. And, while they may be, they can be misused and misinterpreted. Complete information about how they are collected and variable definitions may not be available.

In their book *The Data Game*, Mark Maier and Jennifer Imazeki demonstrate the use of secondary analysis to investigate policy questions and the value of scrutinizing reported findings. The book includes thumbnail sketches of policy controversies fueled by analysis of publicly available data; for example, "Does capital punishment deter murder?" "Are the rich getting richer?" "Is the workplace safe?" The authors discuss the Crime Index and how reporting methods can cause it to be biased. Maier and Imazeki conclude that the misuse of statistical data (for example, by treating a short-term change as a long-term trend) and different conceptual definitions of such phenomena as white-collar crime, illiteracy, and a nation's economic health contribute to policy debates and questionable claims about social conditions.[4]

Finding Secondary Data

If you want to find out whether data exist to meet your needs, where do you begin? Technology is rapidly changing search techniques, and the number of databases available electronically is constantly expanding. Official statistics, national surveys, and other data can be accessed through the Internet. Internet search engines may help identify potential databases. Nevertheless, any given search engine may "miss" obvious web sites. A search engine can direct you to an out-of-date, biased, or poorly maintained database. As we write, no surefire, efficient method exists to locate needed data. At a university, a reference librarian may be the best source of suggestions on how to conduct an online search. Depending on the databases to which your current institution subscribes, you may have access to organizations such as the Inter University Consortium for Political and Social Research (ICPSR) or the Roper Center that archive a large number of datasets. Many state and local government agencies have datasets that are available to the public, as do agencies of the U.S. national government. Many survey research and polling organization archive datasets and make them available to the public.[5]

The reader will find it instructive to access each of these sites to get an idea of the range of datasets, topics, and resources they make available. The ICPSR site is www.icpsr.umich.edu/icpsrweb/ICPSR. The Roper Center site is www.ropercenter.uconn.edu.[6] The ICPSR will have a wide range of types of datasets covering numerous topics. The Roper Center specializes in survey and opinion poll datasets. Although accessing each site directly can be instructive, virtually all institutions that are members of the ICPSR or the Roper Center will have a specialist to help users.

Published works usually indicate where their data came from, alerting readers to databases that they may wish to examine. Some works cite privately held data, that is, data owned by the researcher or a private organization. You may infer that a database contains variables of interest from the description of its sample, purpose, or general content or from the identity of the agency sponsor(s). You may or may not have access to these data, depending on the willingness of the researcher or her sponsors to make them available or to perform requested analyses for you. Even your access to public data may be limited by confidentiality guarantees to the respondents. In spite of these practicalities and because of increasing concerns about the reproducibility of research results, many journals in political science and public administration now either require data to be posted in a repository (i.e., *American Journal of Political Science*, https://dataverse.harvard.edu/dataverse/ajps) or strongly encourage it, such as the new *Journal of Behavioral Public Administration* (https://dataverse.harvard.edu/dataverse/JBPA).

Reference librarians can alert you to other resources for locating databases. Some databases are located almost by luck. Throughout agencies and universities, there are individuals who hold data. Administrators or analysts may conduct a survey, analyze the data, and write up the findings. The data may have been entered into a computer file with little or no formal documentation of their existence, or paper questionnaires may be stored on an office shelf. Public organizations may have inventories of the data files they have kept. If so, investigators at a later date may find new results by analyzing the data and avoid conducting a new study. From the organization's perspective, however, the cost of documenting and storing the data and protecting confidentiality may outweigh the benefits.

Database Access

For some research questions, investigators may be able to confine themselves to aggregated and published data, obviating the need to collect original data in full. After identifying a database, however, a researcher must determine if he can access it and must be able to review its documentation. If adequate documentation is not available, the data may not be worth using. The researcher must also be able to access the data with current technology.

Using secondary data is not limited to using single datasets. Many researchers combine data from more than one. Example 9.1 shows how researchers compiled information from four published sources to study factors associated with public-employee work stoppages. The researchers studied variations in work stoppages among the states because local government statistics were not published at the time the research was conducted. Another example, the County Data File accompanying this text, was assembled from several data files kept by agencies of a state government.

Example 9.1 Combining Aggregated Databases

Problem: Academic researchers wanted to identify the factors associated with the amount budgeted at the local level for election administration.

Strategy:

1. Collect actual cost data on elections administration for the years 1994–2014 from a database of county annual financial reports.
2. Identify sets of variables measuring county social and economic characteristics, political characteristics of the counties, and the different types of voting equipment used to conduct elections.
3. Identify data sources measuring variables.
4. Conduct analysis: Use statistical programs to create aggregate spending amounts over the years between elections, analyze the distributions of variables to make sure that transformations were not needed, then run statistical analysis using both bivariate and multivariate statistical models.

Unit of analysis: Counties in North Carolina
Sample: All 100 counties

Data in the analysis

1. Total expenditures per registered voter
2. Votes for Republican presidential candidate as a percentage of all votes
3. County commission partisanship
4. Total population
5. Voting equipment (optical scan machine, punch card machine, lever machine, paper ballot, direct recording electronic [DRE] machine)

6. Switch in voting equipment model
7. Covered by Section 5 of the Voting Rights Act of 1965
8. Average registered voters per precinct
9. Total property value in 1000s per capita
10. Nonwhite voting age population as a percentage of the total population
11. Percentage of voting age population under 25
12. Percentage of voting age population over 64
13. Whether the county is in an urban area
14. Public high school graduation (pseudo-graduation rate)

Data sources for variables in study:

1. North Carolina Annual Financial Information Reports; downloaded from the State Treasurer web site, https://lgreports.nctreasurer.com/lgcfinancial/
2. Log into North Carolina (NC LINC) database of historical and demographic information, www.osbm.nc.gov/facts-figures/linc
3. North Carolina State Board of Elections and authors collected from individual county websites, https://pages.uncc.edu/zachary-mohr/election-voting-equipment-data/
4. U.S. Census Bureau, www.census.gov/
5. U.S. Department of Justice, www.justice.gov/crt/jurisdictions-previously-covered-section-5

Illustration of data compilation procedures for election cost data:

1. From the Annual Financial Reports, the authors first had to collect all expenditures related to election administration for a single year (i.e. operating, capital, etc.) and add them up for individual years.
2. The unit of analysis was election years. So, the expenditures for the election year and the previous year had to be added together to get election spending for both the election year and the non-election year prior.
3. Once the election administration costs were calculated, they then had to be divided by number of registered voters of the county so that cost would be on a consistent basis.

Discussion: First, the actual study collected data on 20 variables. Second, all data were collected or aggregated at the county level. Election precincts would have been the preferred unit of analysis. The researchers did not sample election precincts, however, because statistics at this level are not published. Third, some variables of interest like the high school graduation rate are not calculated at the county level every year, and appropriate measures had to be created. See the online technical appendix of the article and the data source for more information.

Source: Mohr, Z., J. V. Pope, M. E. Kropf, and M. J. Shepherd, "Strategic Spending: Does Politics Influence Election Administration Expenditure?" *American Journal of Political Science* 63, no. 2 (2019): 427–438.

Data Source: Pope, J. V., M. E. Kropf, M. J. Shepherd, and Zachary Mohr, "Replication Data for: Strategic Spending: Does Politics Influence Election Administration Expenditure?" (2019). https://doi.org/10.7910/DVN/8KNF3I, Harvard Dataverse.

Flexibility

However, a researcher who relies on aggregated and summarized data is limited in the analysis he can perform. For maximum flexibility, investigators require access to individual records for their analysis, such as would be in a spreadsheet. They may get access in one of three ways:

1. through extracted files containing a portion of the data, such as

   a. a sample of cases
   b. a subset of variables on all cases

2. through direct access to the dataset via

   a. the original records, which may even be hand recorded
   b. the purchase of electronic dataset via the Internet or on a disk by mail
   c. a link allowing the investigator to access the database or download the data

3. through an agreement for the database holder to perform the requested data extractions and manipulations

While getting a direct link to the dataset is the most ideal, the investigator has no guarantee that he can access the data at all, especially if it is held by a private organization. Agency policy and the researcher's affiliation with the agency affect whether he or she can access data stored on an agency information system. The goodwill of the original researcher largely determines whether he or she can access individually held data. Contractual agreements may either guarantee or prohibit public access. Furthermore, confidentiality considerations may limit access to the database or the ability to perform certain tabulations.[7]

Locating and accessing public data is becoming easier all the time. The University of Michigan's Document Center (www.lib.umich.edu/govdocs/stats.html) has a comprehensive list of publicly available data and databases organized by subject matter. FedStats (http://fedstats.sites. usa.gov/data-releases) is similar to the Michigan site. FedStats directs users to publicly available federal data. Users can track trends and access official statistics. At either of these sites, a user can easily locate needed data and receive information on how to access them. Some databases can be analyzed or downloaded online. Other databases are available for purchase. In addition, each state has a state data center that provides users with technical assistance in accessing and using census and other public data.[8]

Researchers should always document the date and online source from which they accessed a database. Many databases are like living things—they are updated on a regular basis, and data previously listed are modified. Sometimes online data are also removed and no longer available.

Verifying a Database

After a database is downloaded, its content and scaling should be verified. Analysts need to understand how the original data were collected, cleaned, and prepared for analysis. For this reason, one should be sure to retrieve the codebook or other data documentation along with the actual data. It is a common mistake to download data without the appropriate coding key to variable names, dates data were gathered, original sources, and data definitions.

We suspect some problems in using secondary data are most likely to occur in working with older databases, databases where information was transferred from one medium to another, or databases developed or used by smaller organizations. We know of instances where investigators put a dataset aside until they were ready to do the analysis, only to discover the dataset was seriously flawed. To verify a database's content, an analyst confirms that

1. the number of cases or records conforms to the number indicated in the documentation
2. variables listed in the documentation actually appear in the dataset
3. summary statistics reported in the documentation can be confirmed

An error may have been made in downloading the data, or the wrong data may have been sent. Furthermore, the possibility that errors can be traced and resolved decreases as time passes, and people involved in the collection, documentation, and compilation of the original database can no longer be reached.

Next, the scaling should be verified. This may have been done as part of documenting the content, but, if not, the researcher should review the variables of interest or a sample of variables. He should check that codes used in the database conform to the codes found in the documentation. For example, if the documentation indicates that males are coded as "1," females as "2," and missing data as "9," no other codes should appear in the database. We have found discrepancies occur because of errors in transferring the data into the computer file or typographical errors in the documentation. In working with a database, variables can be dropped or lines deleted with the unintended (and perhaps unnoticed) keystroke. After one author's analysis of thousands of lines of federal budget data, she discovered data from a small federal agency were missing. It had been skipped in the data download. The error was not discovered for weeks, and the analysis had to be redone. Typographical errors may be especially serious, since an analyst may be manipulating the wrong variable or assuming the wrong values for a variable.

If an analyst is unfamiliar with the database or unsure about the quality of the original survey or the accuracy of the documentation, he may wish to verify the accuracy of the sample. The analyst may apply techniques to answer the question, "Could the data here have been produced by the methodology described in the documentation?" The analyst may want to note the frequency of nonresponses and missing data. Sample verification may be especially important when databases are being merged and the study will be used to estimate parameters.

Special care should be taken in merging data series from different databases. Even something as simple as unseen formatting can cause different series to be misaligned. In merging databases and eliminating cases where observations are missing for some of the variables, the number of cases may be reduced, altering the representativeness of the original samples. The techniques for sample verification and data missingness can become quite complex. The reader interested in learning more about determining data quality should consult the article by Smith and colleagues and the references they cite.[9]

Evaluating the Database

Quantitative researchers are continually reminded about their vulnerability to having "the tail wag the dog." They may unintentionally define a problem so that they can solve it using their methodological skills. Similarly, they may radically alter a problem to make it conform to available secondary data. In working with secondary data, some shifts in the original research question may be necessary. Nevertheless, a researcher should determine the impact of the shift and whether the change will undermine his ability to accomplish the study's original purpose.

In working with secondary data, a researcher must first remember that secondary data can be no better than the research that produced them. Analytical sophistication cannot compensate for poorly conceived measures and sloppy data collection. Investigators may be wary of working with data accompanied by haphazard documentation. To decide whether the secondary data meet his needs, the researcher needs to know the following:

1. What constituted the sample?

    a. What was the population?
    b. What was the sampling frame?
    c. What sampling strategy was used?
    d. What was the response rate?

2. When were the data collected?
3. How were the data collected?
4. How were the data coded and edited?
5. What were the operational definitions of measures?
6. Who collected the data and for what purpose?

Sample Population

Information on the sample population lets the researcher know whether the data represent the population of interest. If the population of interest and the database's population do not coincide, the researcher must decide the impact of the difference. This is especially important to research-ers working in public administration or studying nonprofits. For example, data for smaller units of government are often less readily available than national-level data. If they are not, and a researcher is interested in a local government issue, she must decide if regional or state data can be used with any confidence, although even state-level data can be difficult to obtain. In the research on work stoppages, the researchers decided to settle on state data rather than collecting local data; they reasoned that state data would uncover patterns of association that could be investigated later. Similarly, nonprofits vary by size and type. A database representing national nonprofits may not tell the researcher about what goes on in grassroots organizations at state and local levels.

Information on the sampling frame, sampling strategy, and response rate all affect the quality of the sample. Furthermore, a low response rate may render the data inadequate for the researcher's purpose. He also may decide that a low response rate indicates low research quality.

When Were the Data Collected?

Knowing when the data were collected can be extremely important in correctly interpreting some study findings. Consider school data. Achievement scores gathered on fourth-graders in Octo-ber should be interpreted differently from achievement scores gathered in May. Time can affect public opinion data. Investigators examining public concern about environmental warming may hypothesize that the public's opinion is linked to weather patterns. Consequently, investigators would want to know when surveys were conducted, so they could learn if, and how, extreme temperature or weather-related disasters were associated with variations in the public's opinions. Another issue to consider with public opinion data is the target population. Surveys based on nationally representative samples usually cannot be used to draw conclusions about the opinions of residents within a particular city, county, or even state. This usually is true even if a dataset provides information on where respondents live.

How Were the Data Collected?

Information on how the data were collected helps an investigator make inferences about data quality. He wants to know if respondents were surveyed by mail, over the Internet, by telephone, or in person and how interviewers were trained and supervised. Information on data coding and editing procedures also relates to data quality. Specifically, the investigator wants to know whether someone checked for errors in coding data or entering them into a computer. He also wants to learn how atypical responses, missing responses, and atypical answers were handled. In evaluating the information, the investigator uses his judgment to decide whether the evidence suggests sound research procedures.

Knowing the operational definitions of measures allows the investigator to determine the reliability, operational validity, and sensitivity of the measures. Ideally, the documentation explains and evaluates the measures used. Studies generated using the database also may provide evidence of the quality of the measures. Finally, a copy of the instrument used to collect the data helps the investigator to reach his own conclusions.

An Example of Collecting and Documenting Data

Data published by the Bureau of the Census are accompanied by fairly detailed documentation. In Example 9.2, we have quoted from an issue of *Current Population Reports*. The purpose of the example is to show you what to look for in reviewing documentation and to provide you with a model of what to include when you write up your own research.

Example 9.2 Documenting Secondary Data: An Example from the Annual Demographic Survey

> *Situation:* Each March, the Current Population Survey (CPS) conducts an Annual Demographic Survey, which supplements its monthly data on labor force participation with data on money received the previous calendar year, education, household, and family characteristics.
>
> *Population and sample:* The Annual Demographic sample consists of the CPS sample of 60,000 housing units, 2,500 additional units with at least one Hispanic member, and members of the Armed Forces, which are excluded from the CPS labor force survey.
>
> *Sampling frame and strategy:* The sampling frame was selected from 2010 Census files and updated to reflect new construction. The current CPS sample is located in 824 areas; an area consists of a county, several contiguous counties, or minor civil divisions in New England and Hawaii. The sample consists of independent samples in each state and the District of Columbia, and each state sample is specifically tailored to the demographic and labor market conditions that prevail in that particular state.
>
> *Response rate:* About 50,000 occupied households were eligible for interview. Interviews were not obtained from about 3,200 occupied units, because the occupants were not found at home after repeated calls or were unavailable for some other reason. Between 92 and 93 percent of households provide basic labor force information; between 80 and 82 percent complete the Annual Demographic Survey supplement.
>
> *When data were collected:* March 12–19, 2015.
>
> *How data were collected:* Face-to-face interviews, with the interviewer using a laptop computer, were conducted with households that had not been interviewed previously, with poor English language skills, or without telephone access. Telephone interviews were conducted with other households. The CPS questionnaire is a completely computerized document that is administered by Census Bureau field representatives across the country through both personal and telephone interviews.
>
> Conceptual definition of involuntary part-time: Individuals who give an economic reason for working 1 to 34 hours during a week. Economic reasons include

slack work or unfavorable business conditions, inability to find full-time work, and seasonal declines in demand. Those who usually work part-time must also indicate that they want and are available to work full-time to be classified as part-time for economic reasons.

Operational definition (partial): In the weeks that (name/you) worked, how many weeks did (name/you) work less than 35 hours in 1995? What was the main reason (name/you) worked less than 35 hours per week? (1) Could not find a full-time job, (2) Wanted to work part-time or only able to work part-time, (3) Slack work or material shortage, (4) Other reason.

Source: Information drawn from Current Population Survey web site www.census.gov/programs-surveys/cps/technical-documentation/methodology.html.

Changes can occur to data sources that reflect new uses for the original data. In 2001, the Census Bureau expanded the sample used for the Annual Demographic Supplement, a part of the Current Population Survey. It did so primarily to improve state estimates of the number of children without health insurance. Why was this necessary? Federal funds for the State Children's Health Insurance Program (SCHIP) are based on these estimates. States wanted better estimates so funding needs would be more accurately identified. Thus, the original data source was expanded. If a researcher using SCHIP funding data noted an increase in the estimate over time, she might think there was a rise in the number of children eligible for the program or that benefits were increased. In reality, the increase might just be due to the change in how the sample used for the estimates was selected.[10]

Who Collected the Data?

Knowing who collected the data and why can be valuable. Imagine you find data supporting the health benefits of eating chocolate. Would you be skeptical if the report was from a trade association of chocolate manufacturers? What information would convince you that the data were not biased or misleading? The answers to questions about the sample, the operational definitions, and how the data were coded and analyzed might lessen your suspicions.

Learning to Use Secondary Data

Although this chapter includes information on secondary data and its uses, some readers may wish to have more specific guidelines and examples related to their disciplines. They may also want information on particular databases for conducting studies in their disciplines. An Internet search will help locate numerous such books and articles. Several are available for psychology. One focusing on social work is particularly helpful to users in working through the process of using secondary data. Another provides detailed information on several longitudinal datasets on health and aging. Additional information on these is provided in the section "Recommended for Further Reading" at the end of this chapter. As noted earlier, the reference staff of most university libraries will be able to guide you to other appropriate data sources.

U.S. Census Data

Official statistics, statistics collected by governments, are a major source of secondary data. In the United States, most federal cabinet-level agencies regularly collect data needed for policy making. Important agency producers of statistical series include the Departments of Education,

Health and Human Services, Agriculture, Justice, Labor, and Commerce. Major federal databases can be accessed at three main web sites:

- https://data.census.gov/cedsci/
- www.usa.gov/statistics
- www.govinfo.gov/

The U.S. Census Bureau conducts periodic and special studies to describe the characteristics of the American people, their governments, and their businesses. The periodic studies include:

- Census of Governments
- Decennial Census of Population and Housing
- Current Population Survey
- American Community Survey[11]
- Survey of Business Owners
- Economic Census

We limit this chapter to U.S. Census Bureau surveys because of the extensive use of census data, especially in public administration, and the bureau's longstanding efforts to reduce nonsampling errors. The data are critical to political and policy decisions. Census data users include federal agencies, state and local governments, nonprofit associations and businesses, and private citizens pursuing business and personal interests.

This chapter discusses the Census of Governments and the bureau's two major population surveys. We summarize how the bureau decides on questionnaire content and documents data. The methodological discussion should improve your knowledge of data collection and your appreciation of Census Bureau procedures. Note that the Census Bureau and other federal agencies do not do public opinion polling. Instead, official statistics provide factual information about governments, other organizations, and individuals.

The U.S. Bureau of the Census

The U.S. Bureau of the Census traces its origins back to the Constitutional requirement that the U.S. population be counted every 10 years. The head count forms the basis for reapportioning the number of seats a state holds in the U.S. House of Representatives. By law, within nine months of "Census Day," the Census Bureau must give the president a count of state populations so that congressional delegations can be apportioned. Within a year, state legislatures must receive population totals for specified political subdivisions so they can draw legislative districts.

A hallmark of the Census Bureau has been its record for protecting the confidentiality of the information it collects. The principle of confidentiality has contributed to its success in getting people and businesses to answer questions about themselves. Current law on confidentiality states:

- Census data may be used only for statistical purposes.
- Publication of census data must not enable a user to identify individuals or establishments.
- Every person with access to individual data is sworn for life to protect the information.

The bureau releases its information on an individual only with that person's specific consent. For example, some individuals have asked for census records to corroborate their ages and demonstrate their eligibility for Social Security benefits. Census officials must anticipate how computer users can manipulate data to uncover the identity of and private information on specific individuals or establishments.[12] To respond to this threat, the bureau uses statistical methods to avoid unanticipated disclosures, including suppressing or modifying some data, to screen out any that might identify a specific business or individual. It also has an internal Disclosure Review Board to determine if any risks to confidentiality exist before data are released.

The Census Bureau is authorized to release and make available to the public the individual census records 72 years after Census Day. Because of that authorization, the records from the 1940 Census were released in April of 2012. Those from the 1950 Census will be released in 2022.[13]

The Census Bureau is not restricted to collecting data on and from individuals. It also collects data on the nation's economic activity, including manufacturing, agriculture, transportation, and government. The bureau protects the confidentiality of business information. Data on governments are based on public records and are not confidential.

After census data are compiled, the immediate interest is in the snapshots they supply of the nation and its economic activity. Nevertheless, the real importance of the census statistics is their role as time-series data. Members of the public can compare conditions from 1940 to today by accessing the Census Bureau web site. Census data from 1790 through 1940 are maintained by the National Archives and Records Administration and can be accessed at http://1940census.archives.gov.[14] Thus, the bureau is concerned with producing high-quality data and ensuring their comparability over time.

Censuses of Governments

In years ending with two or seven, the Census Bureau's Government Programs Division conducts a Census of Governments. In addition, it conducts annual sample surveys to gather data on government finances and public employment. Data from the censuses are used to estimate the government-sector economic activity in the United States. These estimates are incorporated into the National Income and Product Accounts, the account that measures the gross domestic product.[15] The census collects information on four major areas:

1. governmental organization
2. taxable property values
3. governmental employment
4. governmental finance

Public officials and scholars may cite these data when comparing governments or looking for patterns of fiscal health. Regularly reported data can answer questions such as:

*   How much revenue does California realize from state-run lotteries? What percentage goes to prizes? How does its profitability compare with New York's state-run lotteries?
*   In how many counties in Iowa can the public do business with the county government via the Internet? What percent of municipalities offer the same service? How does this differ between municipalities of different sizes?
*   What percentage of state and local government employees work in corrections? What percentage of payroll goes to corrections? How does this differ by state?

Rules governing statistics collected by the Government Division differ from typical census operations in two respects. First, as mentioned, governmental information is not confidential; the bureau regularly publishes statistics for individual governments. Second, the sovereignty of the American states exempts them and their subdivisions from the U.S. code (Title 13) requirement to supply the bureau with requested information. Therefore, this is a voluntary survey. In the 2012 Census of Governments, 25 percent of governmental units did not respond.[16]

A major challenge for the bureau's Government Division is to procure comparable data. Achieving high response rates and receiving comparable information necessitate developing data collection procedures that do not burden respondents. The bureau collects some data from reports submitted to other federal agencies. The bureau receives some financial data for individual local governments from state governments. Some information is based on responses to earlier surveys or other census databases. All these strategies avoid having a state or local government receive more than one request for the same information.[17] To further decrease respondent burden, staff review existing documents and rely on information available via the Internet. If necessary, they will use traditional mail surveys.

Staff manually review responses for completeness and consistency. A computer check is performed to identify improbable data and to show differences from previous years. Staff may conduct extensive follow-up contacts with the governments to fill in incomplete data, to clarify ambiguities, or to verify questionable responses.

Publication of the data is relatively quick. The Census Bureau constantly updates contact and legal status information in preparing for the coming government census. For the 2012 Census of Governments, the official survey period was October 2011 to April 2012. Preliminary results were released in August of 2012.[18] More detailed results were published periodically in the years following. For government finances, tax information and pension system information are typically the first forms of information available. For employment, preliminary data are released in the same year, with updates, more details, or revisions taking place periodically. Because the data are not confidential and largely consist of publicly available information, errors are constantly being identified and the data refined. A city manager or mayor who sees her city's revenue has been misreported is unlikely to let it pass without comment.

The Census of Population and Housing

The mainstay of the Census Bureau is the *Decennial Census of Population and Housing*, the constitutional reason for its existence. As we have suggested, counting the population is neither an easy task nor one that can be done with complete accuracy. From the first census, government officials added to the work of census takers and asked them to collect additional information about the population. Thus, the Census Bureau has had to develop procedures to accurately count the population and to determine what additional information to gather and from whom.

Cost

The Census Bureau, government officials, and researchers have given considerable attention to how the census can obtain an accurate population count. In 1990, 75 percent of occupied housing units returned census forms. This compares to an 83 percent response rate in 1980. In preparation for the 2000 Census, the Census Bureau focused on containing costs and increasing the response rate. The result was a slight reverse (or at least stabilization) of the trend of declining return rates. For the 2000 Census, 78.4 percent of occupied housing units mailed back a response (including late responses).[19] This was seen as a major success of the 2000 Census. However, the return rate for the 2010 Census was 74 percent (Figure 9.1).[20]

Often state and local governments are encouraged to increase the response rate to the Census. The following statement from the State of Ohio's Facebook appeal from the 2020 census illustrates the appeals made to citizens and some of the official uses for the 2020 census.

Why the Census is important.

Every community, county, and state uses Census data for planning and funding of schools, businesses, and other things that benefit citizens. Funding is often allocated based on the population of a certain area.

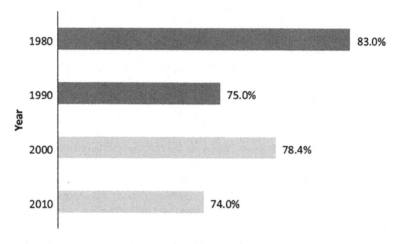

Figure 9.1 Census Response Rate

The State of Ohio also made the data on the 2020 census available to state and local partners such as teachers and college students in various formats ranging from basic flyers to social media charts and graphics. It will be interesting to see if the 2020 Census will reverse the long-term trend of declining response rates.

Accuracy

The phenomenon of the Census missing people completely is called the *undercount*. The problem of undercounting goes beyond underestimating the total size of the U.S. population. Certain groups, particularly urban minorities, are more likely than others to be missed, and their need or demand for services may be seriously underestimated. Urban areas with a sizeable uncounted population may end up shortchanged in the number of legislative representatives or in the size of their allotment of government program funds. To address the problem of the undercount and to keep costs in line, the bureau recommended sampling a portion of the households that did not return their forms. The plan to sample, however, ran into congressional resistance. Court challenges resulted in a Supreme Court decision in 1999 stating that the Census could not use sampling for congressional seat apportionment, and in 2000, the bureau was stuck with the costs of attempting to contact every nonresponding household.[21] In the end, the 2000 Census had an estimated undercount of 1.18 percent, or slightly over 3 million people. More than $100 billion in federal programs alone are allocated by using census data. A report to Congress from the U.S. Census Monitoring Board estimated that the 2000 Census undercount would lead to a net loss of almost $.5 billion to the states between 2002 and 2012.[22]

The 2020 Census is likely to lead to one of the worst undercounts in many decades. Lack of funding for the census, undertested process changes, and demographic changes could lead to a serious undercount. The Census will *not* include a citizenship question, as had been requested by the Trump administration late in the planning cycle for the Census. Federal judges blocked this change that experts suggested could lead to an undercount of as much as 3.68% for African Americans and 3.57% for Latinos. This may not sound like too large of a percentage, but that is 1.7 million African Americans and 2.2 million Latinos.[23]

Census Length

What to include in the census requires negotiation and evaluation. Other federal agencies, state and local government users, interest groups, and individual citizens suggest questions. To decide what additional questions to ask, census officials assess whether the proposed information will serve a broad public interest. Census officials approach the issue of the public interest by deciding whether the information justifies the expenditure of public monies. The criterion of public interest commonly leads to the elimination of questions primarily of interest to businesses, such as information on the number of pet owners in the United States.

One way the Census Bureau avoids overburdening respondents is to limit the number and content of questions so a respondent can complete the form within a reasonable amount of time. The Decennial Census used to have two forms, a short and long form. The short form asks all households to answer no more than two facing pages of questions. A sample of about 1 in 6 households used to receive a longer form. The longer form asked the same questions as the short form, plus one and a half pages of housing questions and two additional pages of information on each household member.

In an effort to address some problems of earlier censuses, the 2010 Census introduced a major change. It discontinued use of the long form and relied only on the short form. The 2010 Census asked only each household member's name, relationship to the respondent, sex, race, ethnicity, and age. Information previously gained through the long form is now gathered through the American Community Survey, a more frequent sample survey. The Census Bureau plans to use more technology to reduce the cost of the Decennial Census, such as using handheld computers to record information. Even with all the changes, however, the final cost of the 2010 Census was $13 billion.[24] For 2020, the Census will continue to rely on the short form and use even more technology, such as using more handheld computers and smart phones to collect data.

Pretests

Approximately five years before "Census Day," a series of census pretests begin. The pretests gather data on diverse aspects of the census process. Analysts evaluate the questionnaire, any computer-assisted interviewing techniques, and response rates. For the 2010 Census, dress rehearsals took place in 2008 in Fayetteville, North Carolina, and San Joaquin County, California. These rehearsals are major affairs—each generated thousands of local temporary jobs. Other tests had previously taken place throughout the country.[25]

The pretest for the 2020 Census began in June 2014. The Census Bureau conducted the tests in parts of Washington, D.C., and Montgomery County, Maryland, to experiment with less costly and less paper-intensive ways of counting the population. For instance, it had test respondents respond to the questionnaire through the Internet or, for some households, through a smart phone application.[26]

In asking its questions, the bureau considers historical comparability. If a question is reworded or response categories are altered, the change affects the answers. The bureau must weigh whether the benefit of a change balances the loss of comparability of information from one census to another.

Race and Ethnicity

The history of the race and ethnicity questions, as shown in Example 9.3, illustrates the trade-off between consistency and the need to make question changes that reflect current social conditions.

Example 9.3 Comparability Among Censuses

Summary of the Recent History of Racial and Ethnic Questions

1920: Census taker decided appropriate category. Categories: White, Negro, Mulatto (Black-White mixture), Chinese, Japanese, Indian, Other.

1930: Mulatto category dropped; a person identified as Mulatto counted as Negro.

1940: Mexican (Mexican birth or Mexican parents) added; people who qualified as Mexican but who were identifiable as Negro, Chinese, Japanese, or Indian were no longer counted as Mexican.

1960: Self-identification of race on mailed census forms. If data collected by a census taker, the census taker observed and filled in the racial data.

1970: Combination of self-identification and census-taker observation continued. Categories: White, Black or Negro, Japanese, Chinese, Filipino, Korean, Vietnamese, Indian (Amer.), Asian Indian, Hawaiian, Guamanian, Samoan, Eskimo, Aleut, Other.

1980: Question added to 100 percent count, "Is this person of Spanish/Hispanic origin or descent? Categories: No (not Spanish/Hispanic); Yes, Mexican-Amer.; Yes, Chicano; Yes, Puerto Rican; Yes, Cuban; Yes, Other.

1990: Changed the "Other" response category in both the race and Hispanic questions to allow the respondent to name the specific Asian or Pacific Islander or Spanish/Hispanic group.

2000: Allowed respondent to check more than one racial group; allowed respondent to select "some other race" with space to write in an answer.

2010: Hispanic origin: Three changes were: (1) The question was changed to ask "Is this person of Hispanic, Latino, or Spanish origin?" (2) The question provided no instructions for non-Hispanic respondents—they did not have to mark the "No" box. (3) The "Yes, another Hispanic, Latino, or Spanish" category "provided examples of six Hispanic origin groups . . . and instructed respondents to 'print origin.'" In contrast, no Hispanic examples were given in 2000.[27] Race: Two changes were: (1) change to "What is this person's race? Mark one or more boxes," and (2) examples were added to some race categories. In 2000, the question said "What is this person's race? Mark one or more races to indicate what this person considers himself/herself to be."[28]

2020: "Significant changes from the 2010 Census questions regarding race and ethnicity include: collecting multiple Hispanic ethnicities such as Mexican and Puerto Rican; adding a write-in area and examples for the White racial category and for the Black racial category; removing the term 'Negro'; and adding examples for the American Indian or Alaska Native racial category. The Census Bureau will not use a combined question format for collecting race and ethnicity; or a separate 'Middle Eastern or North African' category on the census form."

Source: Citro, C. F., and M. L. Cohen, eds., *The Bicentennial Census: New Directions for Methodology in 1990* (Washington, DC: National Academy Press, 1985), 205–214; 1990 Census Questionnaire. 2000 Information from Census 2000 form available on Census Bureau web site. 2010 information from Census 2010 available at Census Bureau web site and "Census Briefs."; 2020 from Albert E. Fontenot "2020 Census Program Memorandum Series 2018.02 (January 26, 2018).

This abbreviated summary suggests the social conditions that gave rise to the particular data-gathering strategy and the problems in comparing racial or ethnic changes from one census to the next. For example, note that until 1960, racial classification was not based on self-identification; rather, by asking specific questions on parentage or by observation, census takers decided a household's race. In 1960 and 1970, households that received a mailed form identified their own race, but if a census taker collected data, he or she decided, based on observation, the appropriate racial category. By 1980, the race question included 15 categories, mixing traditional concepts of race with ethnic or geographic identities. Also in 1980, respondents in all parts of the country could identify themselves as Eskimos or Aleuts; previously, these groups were listed only on census forms distributed in Alaska. The 2000 form simplified the directions on how to identify the applicable Hispanic group, for example, Colombian or Peruvian. An analysis of the New York City data found a sharp increase in the number of "other Hispanics" since 1990. At the same time, city agencies had overestimated the size of specific Hispanic groups. Explanations for the change other than the wording could not be ruled out. One critic speculated that younger, U.S.-born Hispanics might not identify with a specific nationality.[29]

Deciding on racial categories was a major controversy in planning for the 2000 Census.[30] People who had more than one racial background expressed their resentment at having to choose one racial group over another. Adding "bi-racial" and "multi-racial" as categories was considered. This strategy was opposed by those who feared that many African Americans would choose these categories and diminish the ability of analysts to identify patterns of discrimination. The final decision was to allow respondents to check more than one racial group.[31] The respondent could also have marked "some other race" and filled in an answer. The controversy over wording of this question continued after the 2000 Census. The Census Bureau found that in 2000, when people responded "some other race," they did not always identify what that race was. So the Bureau investigated dropping the option, effectively forcing people to choose one of the stated options or skip the question. However, the decision was taken out of its hands. In an example of the direct intervention Congress can make, the 2005 Consolidated Appropriations Act included language directing the Census Bureau to use the "some other race" option with all future censuses. In 2020, the suggestion of adding a citizenship question was blocked by federal judges. Given that the Census is used for electoral representation, federal funding, and many other public and private purposes, it is not surprising that all three branches of government often get involved with the Census.

Postcensus

The bureau conducts postcensus evaluations of the census coverage and content. The evaluation findings indicate data quality and suggest future changes. Postcensus evaluations largely consist of information supplied by respondents who are re-interviewed. The bureau also checks administrative records. For example, public utilities records have been compared with respondents' reports of utility expenditures. Medicare, income tax returns, and similar governmental records have been used to improve the accuracy of the population count.[32] The Census Bureau conducted 87 evaluations in the wake of the 2000 Census, covering everything from the impact of the paid advertising campaign to focus groups to explain respondent behavior. For the 2010 Census, the Bureau kept up the emphasis on postenumeration work, but this time it distinguished between auditing data for coverage and veracity on the one hand and outcome-focused evaluations of Census programs on the other.

The Census Bureau does not escape data-collection problems, but in general, it does as well as or better than other organizations that collect data. The bureau's resources, including its reputation, add to its advantage. In general, people are more willing to respond to government requests for information; consequently, census surveys have higher response rates than similar surveys conducted by nongovernment agencies and researchers.

Current Population Survey

The CPS is a monthly household survey to gather current population and labor force data. The U.S. Bureau of Labor Statistics releases the CPS data each month in its report on the nation's employment and unemployment rates. The Census Bureau analyzes the population data and reports them in *Current Population Reports*.

The data describe the personal characteristics of the labor force, including the age distribution, race, and sex of American workers. Data on who works, who works full-time, who works only part-time, and who is unemployed give us a picture of who gets ahead or falls behind in the labor force. For instance, the CPS reports separately the employment patterns for whites and African Americans, for men and women, for teenagers and adults, and for rural residents and urban dwellers. What will be done with this information depends on one's responsibilities. Interest groups, journalists, and legislators cite data to document social problems or to advocate policy changes. Program managers, especially in education and job-training programs, consult the data to structure programs to meet their clients' needs or simply to give their clients accurate information and advice.

Planners in programs delivering services to specific age groups, such as schoolchildren or the elderly, need current data on the population's age distribution so that they can estimate demand for services. Business analysts examine the data to identify population trends that may change the demand for products and services; administrators can undertake similar studies to improve their program planning or implementation.

Census analysts construct the CPS sample, consisting of approximately 60,000 households, so that it is representative of the nation's population. The sample is large enough that one can estimate parameters for individual states and large cities such as Los Angeles, New York City, and the District of Columbia.[33] Example 9.4 reviews sampling statistics and illustrates why you may be able to use a given database to estimate parameters for a state but not for its communities.

Example 9.4 Estimating Parameters from the CPS: A Hypothetical Example

Problem: Can an analyst use CPS data to estimate the proportion of a city's households with a certain characteristic, a nominal variable? What will be the size of the standard error if the characteristic is split 50–50? 75–25?

Table 9.1 Sample Size and Standard Errors

Sample	N	SE 50–50	95% Confidence Interval	SE 75–25	95% Confidence Interval
National	50,000	.0022	49.55–50.45%	.0019	74.63–75.37%
Regional	11,000	.0048	49.06–50.93%	.0041	74.19–75.81%
State	1,700	.012	47.65–52.35%	.0105	72.94–77.06%
City	85	.054	39.37–60.63%	.047	65.79–84.20%

Discussion: Recall that to establish a 95 percent confidence interval, the SE is multiplied by 1.96. The standard error will be larger for interval-level data because they are more variable. Note that as the sample sizes become considerably smaller, the standard error becomes larger. And the standard error becomes even larger when one tries to estimate the distribution of a characteristic among a specific group, for example, income of Hispanic families for a specific jurisdiction.

In reviewing Example 9.4, you may wish to note what a small standard error means at the national level. In 2012, the size of the civilian labor force was 163.45 million. If the unemployment rate was estimated to be 8.1 percent, then 13,239,450 people were believed to be unemployed. If an analyst assumed a 91.9–8.1 split between employed and unemployed, the standard error would be .0011 (0.11 percent). This means that at the 95 percent confidence level the estimate of the percent unemployed would be off by plus or minus 0.22 percent (0.11 times 1.96). The actual number of unemployed could be as high as 8.32 percent, or 13,599,040, or as low as 7.88 percent, or 12,879,860. Thus a very low standard error, 0.22 percent, can yield an estimate off by as many as 359,590 persons.

Using Census Data

The Census Bureau aggregates and releases data from the Decennial Census by political and statistical areas. The political areas are states, counties, minor civil divisions (such as townships and New England towns), and incorporated places. Statistical areas have been created to describe functionally integrated areas that are relatively homogeneous in population characteristics, economic status, and living conditions. Most important of these are the *census tract* and the *block*. Census tracts are large neighborhoods, averaging 4,000 in population and generally with a population between 1,500 and 6,000. Census tracts are not to cross county lines. As of the 2000 Census, all counties had census tracts, even rural areas. Blocks are sub-units of tracts and are the smallest geographical unit for which the Census Bureau collects 100 percent data—that is, data on all the households, not a sample. Over 8 million blocks were used in the 2000 Census.[34]

Macrodata and Microdata

Traditionally, census data have been publicly available as *macrodata*, that is, data aggregated by political or statistical area. Beginning with the 1940 Census, samples of individual records, called *microdata*, have been made available for public use. These used to be the systematic, stratified samples of the long questionnaires filled out by individuals. The American Community Survey is now the source for microdata information through its Public Use Microdata Sample.[35] A microdata record includes the responses reported to the Census Bureau with all personal identifiers removed. Information on geographic locations and data on very small and visible subgroups also may be eliminated to protect the confidentiality of respondents.

The advantages of the microdata are that the user can access individual records and design cross-tabulations of the variables in these records. Users can analyze variables in any combination they choose, something not possible with the summary data files. Since the microdata represent a sample, users must deal with the issues associated with making inferences from samples to populations. Users can analyze any state in the nation, any of the nation's over 3,000 counties, or any city or neighborhood. Data are in a standardized format for every place, so the same analysis can be easily repeated for any jurisdiction.

An Example of the Advantages of Microdata

These files also are large, and users need adequate computer support to do the analysis. These days, virtually all census data are available online and can be downloaded electronically. Menu-driven applications make it simple to access information on almost any geographic level from almost any survey through www.census.gov. Example 9.5 illustrates how a smaller community can use census data to select an area for a pilot project serving impoverished preschoolers.

Example 9.5 Using Census Data on Census Tracts to Select a Site for a Program

Situation: A local human services program plans to launch a pilot program for impoverished preschoolers. The program will combine a full-day nursery school program with nutritional and health services. A human services analyst has reviewed data on city census tracts to identify communities where the program could be located; see Table 9.2.

Table 9.2 Households Below Poverty Level in Selected Census Tract

	Census Tract				
	A	B	C	D	E
Population below poverty level	4,390	3,458	2,670	5,262	3,417
Number of related children less than six years	151	114	91	173	41
Number of related children less than six years old in a single-parent household	77	51	56	63	12
Percentage population below poverty level	28	25	29	21	16

Among the census tract, which sites are more suitable for the program and which one is inappropriate?

Decision: Look for a site in Census Tracts A or D. Indicators for both communities show a high level of need: number of young children living in poverty, number of young children living in a single-parent household, large percentages of population below poverty level. Census Tract E is an inappropriate site for the community project because of the small number of children in the target population and a distinctly lower level of need evidenced by indicators.

In addition, researchers are continuously constructing new databases and sharing information. For example, the Minnesota Population Center at the University of Minnesota has developed the Integrated Public Use Microdata series, which has pulled together a variety of high-precision samples over time for social science and economic research.[36]

Census Data Uses

Census data are important components of demographic analysis, such as studies of patterns of fertility, mortality, and the population's age distribution. Public services rely on demographic data to help with planning. School planners attempt to provide advance notice that school capacity is greater or less than needed for the anticipated number of children that the community must educate. In recent years, communities have paid more attention to the needs of their aging populations. Communities with an increasing proportion of elderly persons may experience a marked change in demand for certain services—for example, an increased need for nursing homes and specialized medical care.

The most immediate and obvious uses of the decennial data are for legislative reapportionment, funding allocations, and policy decisions. The national, state, and local government data give politicians, administrators, and journalists a snapshot of the nation's population on one day. The population can be viewed in cross-section or longitudinally. Cross-sectional studies look at the patterns among variables in the census dataset; longitudinal studies measure changes from one census to another.

Despite these numerous advantages, an obvious problem with decennial data is its timeliness. Over the course of a decade, the accuracy of the counts declines substantially. Population changes at the block level can be rapid and dramatic. Within a matter of months, vacant lots, fields, and wooded areas may be replaced by residential housing. Conversely, housing of marginal quality may be condemned and disappear. The switch from the long form previously used in the Decennial Census to the American Community Survey (ACS) may alleviate this problem. About 3.5 million households are surveyed each year with the ACS.

In 2012, data for areas with a population of 65,000 or more, including all 50 states, the District of Columbia, and Puerto Rico, were available for one-year estimates; data for areas with a population of 20,000 and above, including the 50 states, the District of Columbia, Puerto Rico, most of the metropolitan statistical/micropolitan statistical areas (98 percent), and 59 percent of the counties or equivalent were available for three-year estimates; five-year estimate data were available for all areas, including the small geographical units such as census tracts and even block groups.[37]

Dissemination

Each state has a state data center that helps disseminate census data to other state agencies and local governments and aids those wishing to use census data. State data centers are the main contact between the Census Bureau and state and local governments. Many of these centers also coordinate the collection, documentation, and dissemination of data by all state government agencies. They can update the ACS population count for the state and major subdivisions, such as counties and cities. States may also project or estimate the state's future population and that of its subdivisions, particularly counties. If you have good reason to believe that an agency in your state collects certain data, chances are that the state data center can tell you if it is available and how to obtain access to it.[38]

We have only scratched the surface in our description of census data and their uses. We have only touched on data gathered in the census of governments and the economic censuses conducted every five years. And we have not mentioned many of the products derived from the Decennial Census. Similarly, we have ignored many other important statistical collections and agencies. Federal agencies regularly gather statistics describing the nation's health, education, agriculture, crime, and criminal justice systems.

Census Bureau Webpage

A variety of resources are available at the Census Bureau webpage for users and other members of the general public. An interactive mapping tool allows users to investigate demographic values for states, counties, and municipal areas. The tool is easy to use and engaging. One is available at: https://data.census.gov. Simply type in the name of the jurisdiction for which you want to find data on the Explore Census Data menu, and you will get basic Census data. From there, you can connect with other data sources to get information on a variety of social topics (i.e. education, employment, etc.).

We had three reasons for writing at length about census data. First, we wanted to alert you to their availability and potential. Second, the Bureau of the Census continually appraises its data collection and compilation procedures; thus, it is an important source of information on current developments in survey research methodology. Third, the information accompanying Census Bureau data serves as a model for documenting primary data; researchers need similar information to decide if a dataset is suited to their needs. The Census Bureau continually strives to make the census data more available and easier to use. With the Internet and other advances in technology, more people can access this important source of secondary data.[39]

Vital Statistics

Vital records are another important secondary data source used by investigators from various disciplines and with different interests. Vital records and the resulting *vital statistics* give information on births, deaths, marriages, divorces, abortions, communicable diseases, and hospitalizations. Most countries have a system for recording vital statistics. These data were among the first collected by governments and nongovernmental organizations. Before national censuses were conducted, these records were used to estimate populations.

In the United States, federal, state, and local government agencies cooperate to collect, compile, and report vital statistics. Individual counties and states collect the data. The state data are forwarded to the National Center for Health Statistics, which compiles the data. The center publishes U.S. vital statistics reports and provides technical assistance to state agencies and other data users. As is standard with much secondary data, much of these data are online.[40]

Investigators use vital statistics to assess the state of a community's mental and physical health. Policy makers can examine vital statistics to evaluate the effectiveness of current programs, change policies or programs to better meet existing needs, and forecast future needs.

Typically, a county collects data required by state statute and reports them to the state. Hospital administrators, physicians, funeral directors, and medical examiners may collect the actual data. In most states, the state health department maintains vital records and releases them to the public in printed form or on computer media after removing information that identifies individuals. State offices of vital statistics also issue periodic statistical reports describing the health of the states and their communities.

The information gathered on a live birth illustrates the extensive information included in vital records:

- where birth occurred
- institution of delivery
- mother's residence
- mother's marital status
- mother's race
- mother's total pregnancies
- mother's previous number of live births
- mother's previous number of fetal deaths
- date of mother's last live birth
- date of mother's last fetal death
- outcome of mother's last delivery
- mother's number of previous children still living
- prenatal care
- baby's Apgar score (a medical rating scale done at birth)
- complications with this pregnancy
- congenital malformations

Accuracy

At first, vital records may appear to be objective, and a user may not doubt the quality of the data. Actually, the accuracy of any one vital record is subject to many possible errors. Think of distortions that can occur in the information on a live birth. For example, information on previous pregnancies may be misreported. Fear of censure may cause a woman not to tell her physician or midwife about previous pregnancies that resulted in an abortion, miscarriage, or adoption. Many women may not have recognized miscarriages early in a pregnancy. Added to the problem of misreporting are possible errors in recording the data and different standards in diagnoses; considering these factors, you can begin to appreciate the difficulties of maintaining data quality.

Similarly, social values affect death reports. Vital statistics on causes of death are obtained from death certificates and are coded according to an international code. Consider the early problems in getting accurate data on deaths from AIDS. AIDS patients and their families may have feared the social stigma attached to the disease. Physicians sensitive to the feelings of patients and their families have admitted to indicating cancer or another related disease on the death certificate rather than AIDS. Similarly, accurate reporting of suicide varies widely. Societies that consider suicide a shameful act are likely to underreport its occurrence. Depending on community mores, then, a physician may decide to attribute a death to diseases or events that cause a family less embarrassment.

Vital statistics and some other health records are based on complete counts of all relevant individuals. The Center for Health Statistics attempts to count all births and deaths, and the Centers for Disease Control and Prevention attempts to collect a complete count of many diseases and causes of death. However, a great deal of useful health-related information is collected with sample surveys. The National Center for Health Statistics, several other U.S. government agencies, and many private organizations conduct extensive surveys on health-related topics. Large amounts of health survey data are available.[41]

Databases, Data Warehouses, and Big Data

Most data available for statistical analysis exist in a single data file called a dataset. A dataset is a single data file that consists of information structured in rows and columns. Although some datasets can be very large, often organizations and researchers need to pull data from different datasets. However, with more research and management functions conducted electronically, organizations have an increased ability to link datasets, create data repositories, and use unstructured or big data. We discuss several types of data arrangements in this section of the chapter.

Relational Database

Researchers often combine datasets from several sources into a relational database, which is often referred to simply as a database. Organizations need to link separate files with data on the unit of analysis, such as departments, employees, customers, and so on. The unit of analysis is connected across datasets based upon the database key, which connects the rows in the separate datasets.

Often the database is managed by a relational *database management system* (DBMS), which are programs that link separate files with common subjects together based upon a common database key. A DBMS provides managers and analysts the ability to generate reports conveniently. They can extract information from different files to combine in a single report. DBMSs are often written in Structured Query Language, or SQL, but others, like Access databases, are also effectively DBMSs. While a DBMS is great for creating a single dataset from several datasets and running basic queries, typically DBMSs do not provide much ability to conduct statistical analysis. Often the researcher has to get the data from the DBMS, and then the work of running statistical analysis can begin.

Data Warehouses and Data Portals

A data warehouse is a data repository that aggregates information from multiple databases for the purpose of analysis and reporting. For instance, it may contain organizational data from individual transactions, such as paying a fine or a fee for a zoning or code enforcement violation.

This information may then be useful to public safety and sanitation, which can then merge the data on zoning with their databases in the data warehouse to get a comprehensive picture of public health and safety in the city.[42]

To enable queries and analysis, the data warehouse is designed to be "subject oriented."[43] That subject is usually an individual but might be something else, such as a service, an organization, or a court case. The company Oracle notes that data warehouses "must put data from [several] disparate sources into a consistent format."[44] The goal is to "resolve problems" such as different names for the same things and the use of different measurement tools in different time periods. This can be very difficult and time consuming. When these items are consistent across files, the files are said to be "integrated."[45]

The health care industry uses data warehouses extensively. People in that industry emphasize the importance of analytics for increasing the quality of services and reducing costs. Without the ability to store and analyze large amounts of data to provide answers to questions, managers will have trouble achieving these goals. As one expert put it, "Improving quality and [reducing] cost requires analytics. And analytics requires a data warehouse. . . If you get data into . . . [a relational database], you can report on it. If you get it into a data warehouse, you can analyze it."[46]

For an example of a data warehouse, we consider one created for a county criminal justice system. This warehouse includes data files from the following:

1. Arrest Processing Center
2. Administrative Office of the Courts
3. County Jail

The data warehouse allows managers to produce comprehensive reports on any aspect of the criminal justice system in the county quickly and accurately. They can also produce various statistical analyses using data from all of the files. The system also records data from all sources daily, so it is current every day. This allows managers to develop reports based on any previous time period.[47]

A relational database would have allowed the county to assemble reports on various subjects such as characteristics of those arrested, court processing, or jail sentences. With the files and data in a data warehouse, they can conduct various statistical analyses as well as generating reports on the entire criminal justice system. They can also obtain information from all files on individuals meeting specified criteria. For example, the analyst could identify all males between 20 and 40 years old arrested for simple assault in 2014 and review the court outcome for each. Although a relational database might enable them to link separate files, analysts and managers in one part of the system, such as the county jail, may not be able to identify trends or conduct statistical analysis of all parts of the system without the data warehouse.[48]

An emerging area for governments is to make their data warehouses open to the public through a data portal. A data portal uses the publicly available data in a data warehouse and makes them open to the public for purposes of exploring, visualizing, and downloading the data. For example, the City of Charlotte, NC, says on its data portal that it encourages users to "search for data, learn about the City's projects and resources, download data for your own use or create brand new maps and apps" (data.charlottenc.gov/). More information about open data portals around the world can be found at dataportals.org.

Big Data

The decreasing cost of computer capacity and data storage and the increasing ease of collecting data electronically have resulted in organizations assembling and holding immense amounts of data. Hence, some databases have become very large, and many computer systems are not able to manage them. *Big data* refers to these huge amounts of data, immensely larger than the datasets of secondary data analysis or even data warehouses. It is now quite common for information technology researchers to have data on millions of people using Facebook or Twitter that are just a subsample of users. Moreover, the data are not usually stored in consistent form or organized in related files.

The Basics of Big Data

The term has become popular to describe massive structured or unstructured datasets. Such data can be too large to process using traditional database and analysis software. Many business leaders believe that more data leads to more accurate analyses and better decisions and therefore greater efficiency, lower costs, and reduced risk.[49] Big data, however, is not limited to businesses; many other organizations and professions use big data. It has become increasingly important in health care and infectious disease control, where combining disease statistics with maps and social media has proven effective.[50]

Structured data means that numbers or other symbols are used to identify values for a characteristic, and the same characteristics are recorded for each case. A file of specific dimensions can be created for structured data, and it is relatively easy to analyze. This is the kind of data in traditional databases. Unstructured data has much less form. The information is usually qualitative, it tends to be textual, and specific characteristics are not necessarily identified or measured. The amount of information for each individual case may be different. The information is analyzed by looking for common themes and ideas. Much of the data from social media sources is unstructured or partly structured.

Challenges of Using Big Data

The volume of data of all kinds has increased. Businesses store information on transactions, customers, and vendors. Nonprofit organizations store information on services provided and on recipients. Governments have numerous files of statistical data, administrative records, and maps. Almost all of this is structured. Unstructured data is collected from social media such as Facebook and Twitter, along with information from business applications or from e-mails, video and audio files, or financial transactions.[51] Until recently, it was almost impossible to store such vast amounts of data.

With larger and cheaper computer capacity, organizations have been tempted to collect and keep data, hoping to find them useful. This creates problems such as deciding what data are relevant and how to use existing tools to analyze them usefully. Some organizations have been tempted to store everything and then search through it to discover meaningful information. However, this may have extreme costs for public organizations. Take, for example, police body cameras that are relatively cheap, easy to install, and issue to officers. However, that is only a fraction of the cost as the hundreds and thousands of hours of police video then has to be stored, usually on a rented server. Often smaller towns cannot afford this cost, but even larger cities like Arlington, Virginia, decided not to implement police body cameras when it was found that it would need to spend $300,000 per year to maintain its police body camera program.[52]

Big data is defined less by volume, however, than by its increasing variety and velocity. Volume refers to the vast amount of data, velocity refers to the speed at which data arrive, and variety refers to the different kinds of data and forms in which they appear.[53] Data stream in at increasingly high speed. As the company Statistical Analysis Systems (SAS) notes, "[r]eacting quickly enough to deal with data velocity is a challenge for most organizations." Data also come "in all types of formats."[54] "Managing, merging and . . . [analyzing] different varieties of data" is difficult.[55] For some organizations, up to 85 percent of an organization's data are unstructured, but they still must be included along with quantitative analysis for decision making.[56]

Another challenge, especially for government and smaller organizations, is recruiting enough skilled, well-trained data scientists who understand and can use the new big data analytics. A recent article noted that U.S. government agencies cannot use all of their data, as they do not have enough qualified people who understand and can use the new technologies.[57]

The scope of big data is constantly changing. Relational database management systems and standard statistical packages are usually not adequate for dealing with big data. Many software vendors have programs, typically called analytics, available for using with big data. Two well-known companies with big data analytics are SAS and Google (Google Analytics).

The tendency of organizations to mine big data turns the standard research model upside down. In the traditional model, based on a deductive approach, the analyst uses a theory to generate research questions or to formulate hypotheses and then uses the appropriate data to answer the questions or to test hypotheses. In contrast, much of the research using big data analytics is based on an inductive model. The analyst searches through the data for relationships and patterns.[58] Because of the possibility of storing large amounts of useless data, many analysts recommend analyzing some data as they are collected to determine their value and decide if they are worth keeping.

The Future of Big Data

It remains to be seen just how useful big data will prove to be. We can find examples of success stories using big data.[59] It also has its critics.[60] As noted earlier, big data presents challenges to analysis and access. One of the perils of big data is said to be that greater amounts of data collected does not necessarily lead to more usable knowledge. One researcher noted that analysts could fall into the trap of continually exploring the possibilities in the data. This researcher repeated the concept that sometimes "less is more."[61] This comment highlights two related concerns:

1. Evaluating the quality of the data.
2. The need to identify and select those analysis projects that return the most value for the user. Hyoun Park recommended focusing on the following strategies with regard to the future of big data.[62] Note that these do not include "make it bigger."

   Make it better. Pay more attention to the quality of big data and do not assume that the amount will overcome lack of quality. Also improve access to it so that more users— employees and customers—can comment on the data and add context.

   Make it stronger. Organizations should work on guaranteeing error-free storage of all data. Big data needs to be protected. It has to be kept uncorrupted, and procedures for recovery need to be in place should something happen.

   Make it faster. In addition to being able to collect and store big data more quickly, organizations need to be able to make the results of analyzing it available more quickly.

Comparing Data File Structures

Dataset. Also called a data table. Structured data will typically be in one file with many cases and variables that fit on rows and columns.

Relational database. A database that includes several files, all linked with one or more common items. Users can generate reports using data from several files.

Data warehouse. A relational database of many databases that provides the user the ability to generate reports, conduct statistical analysis, and extract individual cases meeting specific criteria for an entire organization.

Big data. A massive amount of data, both structured and unstructured, "so large that it's difficult to process using traditional database and . . . analysis software."[63]

Metadata

With so many different types of data coming from many sources, it is often difficult to keep track of where they all come from, how the data were collected, and what the data are supposed to represent. These higher-level data are increasingly useful as data proliferate, often just for analysts to know what is available. They provide structure to the many sources of data and what is available for the analysts. Therefore, metadata is simply "data that provides information about other data."[64]

While this definition may sound simple, the level of complexity of data in the modern world is not. Therefore, metadata is often broken out into different types of data that include descriptive metadata, structural metadata, and administrative metadata, among others. Descriptive metadata is information about the resources available. It includes items of information such as the title, a brief abstract, the authors of the data, and any keywords that may be of interest to researchers. Structural metadata is information about how compound objects are put together. For example, if a variable is located in a dataset that pulls from two other datasets, it is important to know the source location of the dataset and any transformations that may have been done (i.e., revenue contained in one dataset divided by population contained in another dataset that is saved in a third dataset). Finally, administrative metadata is important information that is used to manage data such as data types, permissions, when the data were last updated, and other related items.

Metadata is clearly important in administrative practice so that managers and analysts know what data are available and who has the permissions to use the data. Metadata is also increasingly important in the research process. Often you need to catalog your data sources and the type of information that is available in each. While the use of metadata in a well-designed research flow is outside of the scope of this text, interested readers are encouraged to consult J. Scott Long "The Workflow of Data Analysis Using Stata."[65] While it is written for users of Stata, the principles that he describes are good practices for anyone doing complex data analysis.

Summary

Secondary data are existing data that investigators collected for a purpose other than the given research study. Secondary data can be inexpensive, high-quality data adequate to define or solve a problem. Analyzing an existing database requires fewer resources than collecting original data. Some databases have higher-quality data than a researcher can hope to gather. Organizations specializing in collecting data typically have well-trained, professional staff to check the reliability and operational validity of measures; to design, implement, and document a sound sampling procedure; and to collect and compile data (see Figure 9.2).

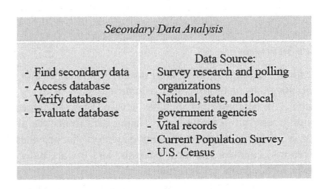

Secondary Data Analysis	

- Find secondary data
- Access database
- Verify database
- Evaluate database
	Data Source:
	- Survey research and polling
	  organizations
	- National, state, and local
	  government agencies
	- Vital records
	- Current Population Survey
	- U.S. Census

Figure 9.2 Secondary Data Analysis

Secondary analysis can contribute to the quality of primary databases. First, secondary analysis requires that the original researchers fully document a database. Second, secondary analysis enables researchers to see whether they can replicate the original researcher's findings. The need to document and the ability to check findings will encourage researchers to attend to research quality.

As documentation becomes routine, data archives are kept, and access to secondary data becomes easier, investigators may increasingly turn to existing data to conduct preliminary research and to hone their research questions. Whether secondary data are appropriate for the final research question depends on the question and the data. As investigators work on a research problem, they begin to understand what population and what measures are needed to answer their questions. Sometimes they may modify the question so that it is consistent with data in an existing database. Such adjustments should be made only after the investigators fully consider what is lost in making such a shift.

The U.S. Bureau of the Census is a major source of data on the country's population, governments, and businesses. Investigators find census data valuable because of their content and the quality of the data collection. Census data may be analyzed by geography, by demographic characteristics, and over time. Maintaining comparability and reducing respondent burdens have been especially challenging in a changing society.

Other federal bureaus, state offices of vital statistics, survey organizations, and professional associations routinely collect data that others study. Virtually any organization is a potential source of data, as are individual researchers. Internet sites may help you locate existing data. Occasionally, by asking agency personnel, individual investigators may uncover a fugitive or unknown database.

Once investigators locate a database, they have to find a way to access it. Some questions can be answered by working with published statistics, but often one needs access to the database. Depending on who holds the database and the contractual provisions for releasing data, a researcher may either access the database or a portion of it directly or have the database holder perform the analysis. Nevertheless, researchers cannot assume that access to a database is guaranteed. Organizational policies, contractual guarantees, and researcher inclination may become important factors in any agreement to allow someone to access data.

If access is obtained, the investigators need to verify the content of the database. Occasionally, the wrong database is accessed or information about a variable and its coding is incorrect. The researchers also need to review information on the sample, the measures, when the data were collected, how they were collected, and coding procedures to infer data quality.

Many organizations are collecting and keeping larger amounts of data as more administrative and research functions are done electronically. In addition to traditional data files, researchers and administrators seeking secondary data are likely to encounter relational databases, data warehouses, and increasingly public organizations are making their data available to the public through data portals. Big data refers to large amounts of data, in some cases too large for the existing computer systems and software of many organizations to handle.

Notes

1. Raj Chetty, Nathaniel Hendren, Patrick Kline, and Emmanual Saez, "Where Is the Land of Opportunity? The Geography of Intergenerational Mobility in the United States," *The Quarterly Journal of Economics* 129 (2014): 1553–1623. For more information on the papers that have come out of this paper and to get the data, see https://opportunityinsights.org/.
2. "Former UVM Researcher Sentenced for Falsifying Work," *Boston Globe*, June 28, 2006, available at www.boston.com/news/local/vermont/articles/2006/06/28/former_uvm_researcher_sentenced_for_falsifying_work/.
3. Gregory A. Poland and Robert M Jacobson, "The Age-Old Struggle Against the Antivaccinationists," *The New England Journal of Medicine* 364, no. 2 (2011): 97–99.

4. For more discussion of the costs of secondary analysis, see Committee on National Statistics, *Commission on Behavioral and Social Sciences and Education, National Research Council, Sharing Research Data*, eds. S. E. Fienberg, M. E. Martin, and M. I. Straf (Washington, DC: National Academy Press, 1985), 15–18.

5. Mark H. Meier and Jennifer Imazeki, *The Data Game: Controversies in Social Science Statistics*, 4th ed. (Armonk, NY: M. E. Sharpe, 2013).

6. Check with your university to see if it participates in or operates a survey research program. Also consult the web sites for major polling and research organizations such as Gallup, Harris, and Pew. For an example of a way to access state government agency data, see www.osbm.state.nc.us.

7. These addresses are accurate as of April 1, 2015. The Roper Center may move at a later date.

8. J. S. Cecil and E. Griffin, "The Role of Legal Policies in Data Sharing," in *Sharing Research Data*, discuss laws that affect researchers' access to data.

9. See www.census.gov/sdc/.

10. For information on SCHIP and its history, see: www.SCHIP-info.org.

11. Since 2010, the American Community Survey has provided current data on an annual basis for all levels of geography, including the smallest for which aggregate data are generally available—census tract and block groups. It replaces the traditional "long form" of the Decennial Census of Population and Housing.

12. For a further discussion of confidentiality, see C. P. Kaplan and T. L. Van Valey, *Census '80: Continuing the Factfinder Tradition* (Washington, DC: U.S. Bureau of the Census, 1980), 65–79; C. P. Kaplan and T. L. Van Valey, "Plenary Session V: Confidentiality Issues in Federal Statistics," in *First Annual Research Conference Proceedings* (Washington, DC: U.S. Bureau of the Census, 1985), 199–233.

13. See www.census.gov/history/www/genealogy/decennial_census_records/census_records_2. html.

14. www.census.gov/history/www/genealogy/decennial_census_records/census_records_2.html.

15. For information on the National Income and Product Accounts, see Jean Stratford and Juri Stratford, *Major U.S. Statistical Series* (Chicago, IL: American Library Association, 1992).

16. http://www2.census.gov/govs/cog/g12_org.pdf.

17. Each major Government Division report includes a section on data collection, including any sampling, in its introduction. The fiscal surveys are more complicated and tend to use the full array of strategies described here.

18. www.census.gov/govs/go/information_collection.html.

19. See page 17 of Herbert F. Stackhouse and Sarah Brady, "Census 2000 Mail Response Rates Final Report," January 30, 2003, available at www.census.gov/pred/www/rpts/A.7.a.pdf.

20. Jennifer Williams, "The 2010 Census: Background and Issues," *Congressional Research Service*, October 18, 2012, available at www.fas.org/sgp/misc/R40551.pdf.

21. To learn more about the politics surrounding the Census Bureau's plans to sample, see M. J. Anderson and S. E. Fienberg, *Who Counts? The Politics of Census-Taking in Contemporary America* (New York: Russell Sage Foundation, 1999); I. I. Mitroff, R. O. Mason, and V. P. Barabba, *The 1980 Census: Policymaking Amid Turbulence* (Lexington, MA: Lexington Books, 1983), detail the political, legal, and statistical aspects of the undercount.

22. See final report by PriceWaterhouseCoopers, available at http://govinfo.library.unt.edu/cmb/cmbp/reports/080601.pricewaterhouse/downloads.asp.htm.

23. *National Public Radio*, "2020 Census Could Lead to Worst Undercount of Black, Latin People in 30 Years," June 4, 2019.

24. *The Economist*, "Censuses: Costing the Count," June 2, 2011.

25. For more information, see www.census.gov/2010census/news/releases/dress-rehearsal/.

26. U.S. Bureau of the Census, www.census.gov/newsroom/press-releases/2014/cb14–104.html. Also, interesting information related to the change in procedures due to the coronavirus as the Census was being administered: www.census.gov/library/fact-sheets/2020/dec/2020-operational-adjustments-covid-19.html.

27. www.census.gov/prod/cen2010/briefs/c2010br-02.pdf.

28. www.census.gov/prod/cen2010/briefs/c2010br-02.pdf.

29. J. Scott, "A Census Query Is Said to Skew Data on Latinos," *New York Times,* June 27, 2001.

30. For an excellent discussion of the issues involved in the racial categories, see L. Wright, "One Drop of Blood," *New Yorker,* July 25, 1994, 46–55.

31. A March 14, 2001, press release, "Questions and Answers for Census 2000 Data on Race," discusses issues associated with categorizing racial data and comparing the races over time. The press release is available at the Bureau's web site, www.census.gov/Press-Release/2001/raceqandas.html.

32. N. L. Stevens, *Census Reform: Major Expansion in Use of Administrative Records for 2000 Is Doubtful* (Washington, DC: General Accounting Office, GAO/T-GGD-92–54, 1992), 4–6.

33. *Design and Methodology: Technical Paper 63* (Washington, DC: Current Population Survey. March 2000), H-1.

34. Dowell Myers, *Analysis with Local Census Data: Portraits of Change* (San Diego: Academic Press, 1992), 16–17. Those interested in doing research with census data, particularly for the investigation of local areas, should see this book. Beginners as well as those with considerable experience working with census data will find it useful.

35. See www.census.gov/geo/reference/puma.html.

36. See https://ipums.org/.

37. www.census.gov/acs/www/data_documentation/areas_published/. Also see the table in the following link showing when to use one-year, three-year, or five-year estimates: www.census.gov/acs/www/guidance_for_data_users/estimates/.

38. For example, see North Carolina State Data Center, available at http://sdc.state.nc.us/.

39. The American Community Survey has developed a series of handbooks for data users. These can be downloaded in PDF form. See www.census.gov, and from the American Community Survey page, select "Handbooks for Data Users." Users can also investigate several interactive tools by selecting "Which Data Tool Should I Use?"

40. See www.cdc.gov/nchs/nvss.htm.

41. ICPSR, the Inter-University Consortium on Political and Social Research, is another good source for computer-accessible secondary data. Its web address is www.icpsr.umich.edu.

42. This paragraph draws heavily from information provided in "Oracle9i Data Warehousing Guide Release 2 (9.2)," available at http://docs.oracle.com/cd/B10500_01/server.920/a96520/concept.htm.

43. See http://docs.oracle.com/cd/B10500_01/server.920/a96520/concept.htm.

44. Quoted from http://docs.oracle.com/cd/B10500_01/server.920/a96520/concept.htm.

45. See http://docs.oracle.com/cd/B10500_01/server.920/a96520/concept.htm. See also Drew Cardon, "Database vs. Data Warehouse: A Comparative Review," *Health Catalyst Insights*, available at www.healthcatalyst. com/database-vs-data-warehouse-a-comparative-review.

46. Cardon, "Database vs. Data Warehouse: A Comparative Review," 4, available at www.health catalyst. com/wp-content/uploads/2014/05/Database-vs-Data-Warehouse-A-Comparative-Review.pdf.

47. Michael Griswold, *MPA Mecklenburg County Government, Criminal Justice Services* (Charlotte, NC: Personal Correspondence, April 2015).

48. *Griswold,* Mecklenburg County Government, Criminal Justice Services*, Mecklenburg County, NC.*

49. For one discussion, see Theos Evgeniou, Vibha Gaba, and Joerg Niessing, "Does Bigger Data Lead to Better Decisions?" *Harvard Business Review*, available at https://hbr.org/2013/10/does-bigger-data-lead-to-better-decisions/.

50. Emily Carlson and Carolyn Beans, "Forecasting Infectious Disease Spread with Web Data," *Live Science*, December 10, 2014, available at www.livescience.com; Karen Biala, "The Quest for Big Data—Does It Ever End?" *The Disease Daily,* June 18, 2014, available at www.healthmap.org/site/diseasedaily/article/quest-big-data-%E2%80%93-does-it-ever-end-61814.

51. www.aventurineinc.com/big-data-analytics/.

52. P. R. Lockhart, "Why Some Police Departments Are Dropping Their Body Camera Programs," *VOX*, January 25, 2019, available at www.vox.com/2019/1/24/18196097/police-body-cameras-storage-cost-washington-post.

53. See SAS, "Big Data: What It Is & Why It Matters," See also Doug Laney, "3D Data Management: Controlling Data Volume, Velocity, and Variety," *META Group*, February 2001, available at http://blogs.gartner.com/doug-laney/files/2012/01/ad949-3D-Data-Management-Controlling-Data-Volume-Velocity-and-Variety.pdf.

54. SAS, "Big Data: What It Is & Why It Matters."

55. Ibid.

56. Larissa Moss, "Managing Unstructured Data," *EIMIArchives* 3, no. 9 (October 2009), available at www.eiminstitute.org/library/eimi-archives/volume-3-issue-9-october-2009/Managing%20Unstructured%20Data.

57. "Making Big Data Work for Governments," *Industry Insights* (November 2012), available at http://assets1.csc.com/public_sector/downloads/BigDataIssueBrief_11.2012_final.pdf.

58. Biala, "The Quest for Big Data—Does It Ever End?"

59. Ibid; Carlson and Beans, "Forecasting Infectious Disease Spread with Web Data."

60. Ernst Davise, "Recent Critiques of Big Data: Small Bibliography," (n.d.), available at www. cs.nyu.edu/faculty/papers/BigDataBib.html.

61. Biala, "The Quest for Big Data—Does It Ever End?"

62. Hyoun Park, "Bigger, Better, Faster, Stronger: The Future of Big Data," *CMS Wire* (October 2014), available at www.cmswire.com/cms/big-data/bigger-better-faster-stronger-the-future-of-big-data-027026.php#null.

63. Brian Sherman, "Channel Trends: Is Big Data a Real Opportunity for the Typical Provider?" (January 16, 2013), available at www.comptia.org/about-us/newsroom/blog/13–01–16/channel-trends_is_big_data_a_real_opportunity_for_the_typical_provider.aspx.
64. "Merriam Webster,>" October 17, 2019, available at www.merriam-webster.com/.
65. J. Scott Long, *The Workflow of Data Analysis Using Stata* (College Station Texas: Stata Press, 2009), 138–142.

Terms for Review

secondary data
replication data
Current Population Surveys
Decennial Census of Population and Housing
American Community Survey
census undercount
census tract
census block
macrodata
microdata
vital statistics
dataset
database
data warehouse
data portal
big data
metadata

Questions for Review

The following questions should indicate whether you have a basic competency in this chapter's material.

1. What is the value of secondary data? What are their main weaknesses? When would you recommend collecting original data instead of relying on secondary data?
2. A regional agency plans to study the relationship between highway features and accident rates.

    a. Briefly describe how you would go about locating existing databases.
    b. Briefly describe how you would decide whether existing databases were adequate for the planned study.
    c. Assume that you have obtained a copy of a dataset. What information would you need to be able to use and interpret the data?

3. You examine a dataset and note that it reports 70 managers and 200 nonmanagers. The documentation indicates the data represent 60 managers and 210 nonmanagers. What would you do?
4. Why is an undercount of the population during the Decennial Census treated as a serious problem?
5. In conducting a survey regularly, such as the Decennial Census, what are the trade-offs between changing a question's wording or its responses and keeping the wording the same?
6. Why did the Census Bureau switch from using a long-form survey every 10 years as part of the Decennial Census to the American Community Survey, conducted every year?

7. How can local government planners use American Community Survey data?
8. What is the difference between a dataset, a database, and a data warehouse? What is a data portal?

Problems for Homework and Discussion

1. Refer to Example 9.5. In Census Tract A:

   53 percent of the adults are high school graduates
   100 percent of the population over 5 years of age speak English
   1,105 rent housing
   Median family income of female householder with no husband present and children under 18, $8,774
   Median family income in census tract, $12,580

   a. A program in Census Tract A compiles information on community services. If you are an advisor to the governor on education policy, on which of the following would you place highest priority—information on tenants' rights, high school equivalency requirements and resources, or English-language resources? Justify your recommendation.
   b. Based on these data, what types of eligibility requirements would you initially suggest adopting to make sure that those needing the services the most will receive them? (Your recommendations will be tentative, and they will be refined as you become more familiar with the community.)

2. Obtain census data on the following for a state and a city of your choice:

   a. number of people over age 65
   b. number of people under age 5
   c. number of households with incomes between $35,000 and $85,000
   d. median value of owner-occupied homes
   e. percent of labor force employed in public administration

3. Go to the Census Bureau web site and use Census Explorer to obtain information for a county on household income and percent of adults with a bachelor's degree.
4. Using the Census Bureau Census Explorer, report median household income for five states and the value of one other variable you think may be related to income. Explain why you think the variables may be related.
5. Select a state and access its state data center. Obtain information on three vital statistics measures.
6. Assume that a hospital located in your community is assessing the need to open a long-term care facility (nursing home). Gather census data on no more than five variables. Based on these census data, what would be your initial recommendation? What other data would you want?
7. Go to dataportal.org and find a data portal for a local government of your choice. What type of information were you able to find in this data portal? Were you able to find metadata about the data in the data portal, and what type of metadata was presented?

Working With Data

The state legislature has funded a limited project to aid the state's poorest areas in economic development. You work for the agency that will implement and oversee the project monies. You have been asked to select no more than six counties to receive project funds. The primary criterion for the selection must be based on income.

1. Decide on an operational definition to measure a county's need for this program. The dataset, County Data File, provided with this book contains several possible variables: mean family income, median household income, the percent of persons in poverty, median family income, and number of families in poverty. A second file, County Data Notes, also provided, includes descriptions of the variables and the sources of the data.

   a. Of those, identify which definition of county poverty you would choose to use. Make a preliminary argument for its operational validity.

   b. Use your operational definition to select the project counties.

   c. By examining the data, identify at least three other variables that would support your choice of these particular six counties.

   d. Write a short memo (no more than one page, single-spaced) for the state legislature identifying your county choices and providing evidence to justify them.

   e. Compare your recommendations with those of your classmates. Can the class as a whole reach a consensus on which counties should receive funding?

2. We have found that students make some common errors when working with secondary data. To illustrate these errors, examine the following relationships in counties using the dataset, County Data File, provided with this book:

   a. The relationship between median family money income and number of families in poverty. Are these two indicators measuring the same concept? Why or why not?

   b. The number of families in poverty and the number of violent crimes. Are these two variables related? According to these data, what can you say about the hypothesis that poverty is the root of violent crime?

   c. The crime index and median family income. Are these two variables related? Is this a better test of the hypothesis that poverty is the root of violent crime? Why or why not?

   d. Median family income and the number of child abuse and neglect cases. Can one conclude from the evidence that poorer parents are more likely to abuse or neglect their children? Justify your answer.

3. Identify three vital statistics from the County Dataset. Calculate the highest and lowest values for each and which counties have them.

4. Identify two variables from the GSS dataset that could be indicators of risk-taking behavior or risk aversion.

5. Using one or more of the Census Bureau sites, select a county in your home state and find values for five of the indicators in the County Data file. Compare these values with those for a county in the County Data file.

Recommended for Further Reading

For information on secondary analysis, see *Secondary Analysis of Available Data Bases*, ed. D. J. Bowering (San Francisco: Jossey-Bass, New Directions for Program Evaluation, 1984). This collection includes J. C. Fortune and J. K. McBee's detailed essay on merging and verifying databases.

The Committee on National Statistics, *Commission on Behavioral and Social Sciences and Education, National Research Council, Sharing Research Data*, eds. S. E. Fienberg, M. E. Martin, and M. I. Straf (Washington, DC: National Academy Press, 1985), includes the committee's report and essays. Its essays discuss data sharing in the social sciences, with a list of data-sharing facilities and legal policies covering data sharing.>

Meier, Mark H., and Jennifer Imazeki, *The Data Game: Controversies in Social Science Statistics*, 4th ed. (Armonk, NY: M. E. Sharpe, 2013). This book includes an excellent compendium of some commonly used databases and their limitations. It is especially strong in identifying common problems of misinterpreting data.>

Vartanian, Thomas, *Secondary Data Analysis* (New York: Oxford University Press, 2011), provides examples of studies using secondary data and lays out the procedures for conducting a study. He also looks at the 29 most widely used social science datasets. Although the author focuses on the field of social work, the information is generally applicable.>

The following article describes a process for conducting research using secondary data. The authors provide an example and list numerous sources of data of interest to different disciplines. Andersen, Judith P., Jo Ann Prause, and Roxanne Silver, "A Step-by-Step Guide to Using Secondary Data for Psychological Research," *Social and Personality Psychology, Compass* 5, no. 1 (2011): 56–75, doi:10.1111/j.1751-9004.2010.00329.x.>

The following work provides extensive information on several datasets and where to locate them. Huguet, Nathalie, Shayna D. Cunningham, and Jason T. Newsom, "Existing Longitudinal Data Sets for the Study of Health and Social Aspects of Aging," in *Longitudinal Data Analysis: A Practical Guide for Researchers in Aging, Health, and Social Sciences*, eds. J. Newsom, R. Jones, and S. Hofer> (New York: Routledge, 2012), Chapter 1.>

Stratford, J. S., and J. Stratford, *Major U.S. Statistical Series: Definitions, Publications, Limitations* (Chicago: American Library Association, 1992), is another work that identifies major databases and measures.>

Descriptions of Census Bureau products, data, and methodology may be found at its web site (www. census. gov). The reports prepared by the General Government Division of GAO are a good source of various details (www.gao.gov). Census staff also report on census research activities at professional conferences, including the American Statistical Association's sections on survey research methods. Fienberg, Stephen E., and Margo J. Anderson, *Who Counts? The Politics of Census-Taking in Contemporary America* (New York: Russell Sage Foundation, 1999), have a discussion of the controversy surrounding the undercount and sampling and racial classifications.>

On understanding the issues surrounding the changes to the Decennial Census, see Reengineering the 2010 Census: Risks and Challenges, Cork, Daniel L., Michael L. Cohen, and Benjamin F. King, eds., *Panel on Research on Future Census Methods, Committee on National Statistics Division of Behavioral and Social Sciences and Education National Research Council* (Washington, DC: National Academies Press, 2004). For a broad understanding of issues surrounding the 2020 Census, see the Urban Institute's data http:// apps.urban.org/features/2020-census/.

10 Combining Indicators and Constructing Indices

In this chapter, you will learn:

1. About combining measures and using indices.
2. The difference between an index and a scale.
3. Issues involved in developing indices.
4. Some common methods of developing indices.
5. How to standardize measures for use in indices.
6. How factor analysis is used to combine indicators.
7. About index numbers and their uses.

Administrators often find it useful to combine several indicators to form a single measure. For example, one *indicator* is rarely adequate to measure a community's need for health services, the amount of crime in a neighborhood, or the level of sanitation. To measure the health status of a community, an administrator would want to use more than one indicator of a disease or condition and would combine several indicators into a composite measure called an index. *Index* is a term for a set of variables used as a measure for a more abstract concept. Each variable is called an item or indicator; an index will be composed of two or more items or indicators.

Readers interested in specific scales or in the various types of scaling procedures should consult the books listed at the end of the chapter. We focus on index construction in this chapter and will not delve further into scaling as such. We do, however give an example of a commonly used scale.

Some indices have been validated through long and varied experience and testing. Before constructing a new index, analysts should attempt to locate and review indices that have been developed by others doing similar kinds of research. We will discuss four important issues in developing indices. These are:

1. defining the concept to be measured
2. selecting the items to be included in the index
3. combining the items to form an index
4. weighting the separate items

Indices and Scales

Researchers often do not distinguish between indices and scales and may use the terms interchangeably. An index usually refers to any combination of indicators. A *scale* combines indicators according to rules that are designed to reflect only a single dimension of a concept.[1] This chapter focuses on indices and has a brief discussion of scales.

Indices

We are all familiar with a number of commonly used indices, such as the Consumer Price Index (CPI); the National Association of Security Dealers Automatic Quotations (NASDAQ), which tracks stock market activity; and students' grade point averages (GPAs). The CPI is used to measure the prices of goods that consumers purchase. Obviously, the price of more than one good must be included to obtain an accurate and valid measure. Analysts combine the prices of a basket of representative goods to arrive at a value for the CPI. The NASDAQ provides an indicator of the performance of the stock market by combining a measure of the performance of a set of selected stocks. And we are all familiar with the grade point average illustrated in Table 10.1. Grades are assigned a numerical value and combined to provide a measure of academic performance.

The value of the index is computed using an equation that includes all the items that are part of the index. The result is a composite of two or more numbers, each representing an item in the index. We can usually obtain a more operationally valid and reliable measure by using several components of an attribute rather than only one. A measure of neighborhood crime that combines the rates of several different crimes, including burglaries, assaults, homicides, and robberies, will be more valid and reliable than one that uses only one crime, such as the rate of home burglaries. Indices are often constructed so that seemingly different things can be combined and described by a single number. For example, an index of environmental health includes measures of child mortality, water quality, adequate sanitation, and indoor air pollution.[2]

Scales

Scales are constructed by assigning scores to patterns of responses arrayed so that some items reflect a weaker degree of the variable, while others reflect a stronger degree.[3] Strictly speaking, a scale assigns a number to a case based on how that case fits into a previously identified pattern of indicators. True scales can be complex and difficult to construct and their operational validity and reliability difficult to establish. Numerous published scales are available for the study of attitudes, opinions, beliefs, and other mental states. They range from those developed for specific projects to standardized (and copyrighted) scales for the measurement of common concepts such as managerial stress and job satisfaction.

A Guttman cumulative scale, often used in social science research, illustrates important characteristics of a scale. This scale, named for its developer, Louis Guttman, uses a continuum to measure one dimension of a concept. The researcher constructs it by selecting a set of items that form a scale of intensity from low to high, so if respondents agree with an item on the scale, they would also agree with all items coming before it. A respondent's score is assigned based on that highest item. Guttman scales have been used to measure respondents' attitudes toward outsiders such as members of ethnic groups, a religious organization, or immigrants.

Table 10.1 Example of an Index: The Grade Point Average (GPA)

Course	Letter Grade	Numerical Grade	Credit Hours	Points (grade × hours)
Math	C	2.0	3.0	6.0
History	A	4.0	3.0	12.0
Economics	B	3.0	4.0	12.0
Total			10.0	30.0

GPA = Total points divided by total credit hours
    = 30/10
    = 3.00

Table 10.2 Indices Versus Scales

Indices	Scales
An *index* usually refers to any combination of indicators.	A *scale* combines indicators according to rules that are designed to reflect only a single dimension of a concept.
The value of the index is computed using an equation that includes all the items that are part of the index. The result is a composite of two or more numbers, each representing an item in the index.	Scales are constructed by assigning scores to patterns of responses arrayed so that some items reflect a weaker degree of the variable, while others reflect a stronger degree.
Commonly used indices • The Consumer Price Index • The National Association of Security Dealers Automatic Quotations • Grade point averages	Numerous published scales are available for the study of attitudes, opinions, beliefs, and other mental states.*

*For information on some of these, see the citations in "Recommended for Further Reading" section for this chapter: D.C. Miller and N. Salkind, *Handbook of Research Design and Social Measurement*; J. P. Robinson, R. Anthansiou and K. Head, *Measures of Occupational Attitudes and Occupational Characteristics*; J. P. Robinson and P. P. Shaver, *Measures of Social Psychological Attitudes*.

The Research Methods Knowledge Base has an example of a Guttman scale to measure attitudes toward immigrants. The details of how such a scale is developed are discussed there.[4] Assume that researchers develop a set of 10 statements, each describing a level of acceptance of an outsider. The first statement might be: "I believe that our country should allow more immigrants to move here." Each statement would indicate a higher level of acceptance. Statement five might say: "It would be fine with me if an immigrant family moved down the street from me." Statement 10 could be: "I would be fine if one of my children married an immigrant." After a set of items forming a scale is developed, they are presented to respondents, although not necessarily in order. Respondents are usually instructed to check Yes if they agree with a statement.

If the scale is developed properly, someone who marks Yes to number 4 would also agree with statements 1, 2, and 3. If the person did not agree with statement 5, they would not agree with statements 6 through 10. Their scale score would be 4. Developing a reliable and valid scale can be very time consuming. Table 10.2 provides a quick overview of the differences between indices and scales.

Defining the Concept

Developing an index requires a clear understanding and definition of the concept being measured. The analyst should be able to explain what the concept means and, theoretically, what its components are. Therefore, the first step in developing a measure is to define and describe the concept to be measured. In doing so, we review the literature to find definitions used by others. We should define a concept the same as others have unless there is good reason for not doing so.

An operational definition of a concept details how the concept is to be measured; this is done in the remaining steps. However, the conceptual definition will determine the appropriateness of each possible measure included in the operational definition. We want a close correspondence between the conceptual and operational definitions. When selecting items for a measure, the analyst should be able to discuss how each item measures the concept as defined.

Selecting the Items

Choosing items carefully for a composite measure is important. The analyst should be concerned about the following:

1. Using the right items—those that represent the dimension of interest and no other.
2. Including enough items to distinguish among all the important gradations of a dimension and to improve reliability. The emphasis here is on thinking of lots of items. Items that do not result in any variation when applied are not useful and should not be included. Such items would classify all cases as the same even if there are differences among them.
3. Deciding whether every item is supposed to represent the entire dimension of interest or whether each is to represent just part of the dimension.
4. Keeping costs down and reducing unnecessary complexity by excluding items that provide no additional information.

Use the Right Items

Recall that an operationally valid measure should be relevant and representative. The items selected for the index should cover the characteristic being measured in a balanced way. An analyst constructing an index to measure management performance for an agency would include those aspects he believed to be important. For example, if personnel turnover, absenteeism, and the number of grievances were all believed to be important, a measure leaving out any of these may be biased.

Include Enough Items

A developer's knowledge of the subject matter related to a topic being measured can be very important in selecting items. Items should be theoretically or conceptually related to what the analyst wants to measure. Those that logically relate to the concept being measured should receive top priority. Developers should consider evidence from several sources that indicate an item is related to the concept being measured. For each selected item, the analyst should be able to write at least a few sentences explaining why it should be included.

Measure One Dimension

Analysts may use several methods to choose items for an index. They may rely only on their own judgment. Or they may talk to people knowledgeable in the area and to potential users of the index to determine what they think is important. One danger in talking to a large number of users, however, is that they may suggest including any item that interests them. You must guard against including unneeded items.

One step in the process is to assess how well a set of items "hangs together," that is, the extent to which each item in a set contributes significantly to the reliable measurement of a well-defined concept. Items hang together when each reflects the underlying dimension in roughly the same way. They also hang together when several measures are all associated—correlated—with each other. When this happens, the separate measures may all be tapping the same concept, and the analyst may be able to drop some items. One method for identifying items that measure the same concept, for reducing a large number of associated items and combining them into fewer indices, is factor analysis. Later in the chapter, we discuss factor analysis and how it can be used to select among variables that could be included in an index. For an index, analysts want measures that are related to each other but not too strongly. If two measures are perfectly related—that is, correlated—only one of them needs to be included.

Keep Costs Down

Analysts should be careful to include only relevant information. As noted in other parts of this text, researchers are often inclined to capture lots of information and determine how to use it later. This is unnecessary and wasteful, both for the participant and the researcher in terms of the time the participant takes to respond and the researcher takes to analyze the results. Therefore, only items that contribute to the understanding of the concept should be included.

The validity of a measure is greatly dependent upon the items selected for the index. The validity of an index's content can be evaluated by reviewing the items involved in measuring the concept. In developing an index to measure personnel performance, a supervisor would want to be sure that the items in the index represented the content of the performance being evaluated. That is, the content of the index should match what is thought to be important about the employee's performance. Example 10.1 illustrates the construction of an index to measure personnel performance and the content-based evidence of its operational validity.

Example 10.1 Index Construction and Operational Validity

Problem: A supervisor wants to develop an index to appraise the performance of her subordinates.

Procedure: After reviewing her agency's personnel policies and the job descriptions of those she supervises and discussing the task with other supervisors, she determines that the following characteristics are most important: accuracy in filling out standard forms, production (the number of forms completed and entered in the office database in a week's time), knowledge of her office's filing system, ability to use the filing system in her office, and facility with computer software. The supervisor then selects items for the index to measure the employee's performance in these areas.

Comment: The items that the supervisor selects must measure the employee's performance in these areas in order for the index as a measure to be relevant and representative. If one or more of these criteria were not included in the index, a critic could argue that it is not operationally valid based on its content. If evaluation of other characteristics were included, such as the employee's attitude or dress, this also would affect the validity of the index.

Combining the Items in an Index

An administrator can use a variety of methods to combine the values for separate measures into an index. The simplest and most common procedure is simply to add the values of the separate items.

A city stress index provides an example of simple adding, using rankings on four separate items. The authors of this index obtained data on four indicators of psychological stress for cities. These indicators were the rates for each of the following: alcoholism, divorce, suicide, and crime. They then ranked the 300 largest cities in the nation on each of the four indicators and added the ranks together for a composite ranking for each of the cities.[5] Another common procedure is to find an average of the values of the separate items. For example, in the stress index previously, the analysts could have averaged the rates, expressed as percents, of the four indicators. Items for an index also may be combined by a more complicated equation to compute a single value. The individual values may be multiplied or divided by each other, for example.

Ranges for indices include absolute ranges, such as items with a minimum and maximum value. Others use a relative standard, such as the degree to which each individual case has attained the group average on an item. Using the literacy rate for a country to measure development would be an example of the first. Literacy can range from zero to 100 percent. How close a country's average life span is to the average life span for all countries would be an example of the second.[6]

The analyst may need to transform the values of each separate item in order to combine them for an accurate measure. Often the separate items will be in different units and have different ranges of values. For example, in an index to determine grant eligibility for a community, one cannot meaningfully add the percent of unemployed adults to the per capita income of residents. Later in the chapter, we discuss methods of transforming measures so that they can be combined to accurately measure the characteristic of interest.

Weighting the Separate Items

Each item should contribute to the index in proportion to its importance. For some indices, we will want each item to have the same influence. For others, we may decide that some items are more important and want them to have greater influence on the final index values.

Adjusting the Influence of Separate Items

We can *weight* the separate items differently if we want them to have different influences on the index. You may create an index to evaluate the desirability of different job offers, for example, and decide that the following items are important: salary, location, and vacation time. Assume that you need to decide between two jobs and construct the following index.

Job Offer A	Job Offer B
Salary = 1	Salary = 2
Location = 2	Location = 1
Vacation = 1	Vacation = 2

The numbers are the ranks of each of the items for each job offer; the higher the number, the better the job. The number 2, for example, means that the job ranks higher on that item than the other job does. The salary for Job B is preferable to the salary for Job A, but the location for Job A seems better to you than the location for Job B, and so on. You also decide that salary is by far the most important thing that you are looking for in a new job and that the amount of vacation time is not nearly as important. So you decide to weight each item by how important it is and multiply the rank of each item by its weight.

The index provides a higher value for Job B than for Job A. If you have included the proper items and weighted them accurately, you would choose Job B:

Job Offer A	Weights	Job Offer B	Weights
Salary = 1	× 4 = 4	Salary = 2	× 4 = 8
Location = 2	× 2 = 4	Location = 1	× 2 = 2
Vacation = 1	× 1 = 1	Vacation = 2	× 1 = 2
Total Index Value	9		12

Table 10.3 The Crime Rate Index for Two Cities

City A		City B	
Crime	Rate	Crime	Rate
Homicide	5	Homicide	8
Manslaughter	7	Manslaughter	6
Rape	13	Rape	10
Burglary	35	Burglary	36
Total	60	Total	60

Rates for individual categories of crimes are usually given as the number of crimes per 100,000 of population.

The Uniform Crime Index

The Uniform Crime Index adds together the number of crimes per 100,000 people for each of a wide variety of crimes, including murder, manslaughter, rape, and burglary. Table 10.3 gives an example of the crime rate index for two cities:

> Note that the level of crime in the two cities is the same on the index, although the cities have different rates for each of the crimes. Each crime in the index is treated as being equally important. Since the Crime Index is dominated by property crimes, changes in the rates of other, more violent crimes impact it less. However, since 2004 the FBI groups these crimes differently and includes a caution with regard to using them to compare jurisdictions.[7]

Avoiding Undue Influences on the Index

Administrators may assign different weights to separate items when combining them into an index. The weight is determined by how important an item is to the attribute being measured. If an index is described as being "unweighted," it typically means that each item included in it is weighted the same. We need to be careful that we do not unintentionally allow some items to have undue influence on the index. This might happen if we were to combine items whose values had much different scales or ranges. An item with values that range from 1 to 50 differs considerably from an item whose values range from 1 to 10. The difference between 5 and 10 on the first item is probably not as important as the difference between 5 and 10 on the second. However, if we simply added them together, they both would have the same influence on the index.

An index to measure financial stress in communities, discussed in Example 10.2, gives an example of this situation. In Item 1 of the index, the range of values obtained went from 10 to 50, and in Item 2, the range was from 1 to 10.

Example 10.2 Community Financial Stress Index

Item 1. Percentage of population applying for any kind of public assistance. Lowest community = 10; highest = 50; Community C = 14; Community D = 20.

Item 2. Percent increase in the number of business failures from previous year. Lowest community = 1; highest = 10; Community C = 7; Community D = 1.

Table 10.4 Financial Stress Index Values

	Item 1	*Item 2*	*Total*
Community C =	14%	7%	= 21%
Community D =	20%	1%	= 21%

Both communities have the same total value on the index. However, the difference between the values for Item 1 is likely to be much more important than the difference between the values for Item 2. Although both items are measured in the same units—percent—they have greatly different ranges and simply adding them together gives greater weight to the first item.

Table 10.5 and Figure 10.1 present a brief overview of the issues involved in developing an index.

Defining the concept to be measured: Developing an index requires a clear understanding and definition of the concept and its components being measured. When selecting items for a measure, the analyst should be able to discuss how each item measures the concept as defined.

Selecting the items to be included in the index: The items selected for the index should cover the characteristic being measured in a balanced way. The validity of a measure is greatly dependent upon the items selected for the index. The validity of an index's content can be evaluated by reviewing the items involved in measuring the concept.

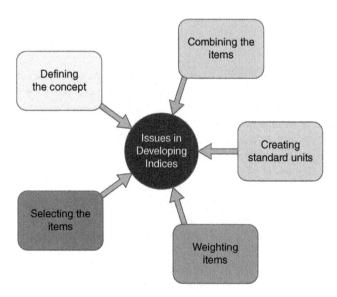

Figure 10.1 A Review of Issues in Developing Indices

Combining the items to form an index: A variety of methods can be used to combine the values for separate measures into an index. The simplest and most common procedure is simply to add the values of the separate items. The values of each separate item may need to be transformed to be combined. Often the separate items will be in different units and have different ranges of values.

Weighting the separate items: Each item should contribute to the index in proportion to its importance. We can *weight* the separate items differently if we want them to have different influences on the index. The weight is determined by how important an item is to the attribute being measured.

Standardizing the items in the index: If items are measured in different units and have different ranges, it may not be possible to combine them. For example, a measure of unemployment in percent cannot meaningfully be combined with total income measured in dollars. Some way of putting then into standard units is necessary so the items can be combined.

Creating Standardized Indicators for Indices

Another issue in preventing one indicator from having undue influence is to measure all items composing the index in comparable units. To achieve this often requires the analyst to *standardize* the individual measures so that each is based on the same scale.[8] As in the previous example, if this is not done, one item may contribute more to the final index than another simply because of different values or ranges of values. If we do want some items to have greater weight in the index than others, then after standardizing them, we can still weight the separate items differently. As in the job offer example previously, for some to have more weight, we can multiply the values of those items by a factor before combining them. In building an index, the analyst should address the two issues, transforming the measures and weighting the items, separately.

Inner-City Hardship

The indices discussed in the following sections were developed some years ago for use in urban research. We use them as examples because they illustrate several issues of index development. They are also important for policy makers, scholars, and students interested in the fields of urban poverty, health care, and human development, among others. We cite several recent examples of research in which these indices were used. Since its development, many cities have used the Hardship Index. It has been used to measure characteristics of neighborhoods and other government subunits. Recently, analysts used the Hardship Index to assess inequality among neighborhoods, to aid in planning, and for many other purposes.[9]

Nathan and Adams created an index of inner-city hardship by combining six measures, including unemployment rate and dependency, which is the percent of persons under 18 or over 64.[10] Although the dependency percent could vary more among cities, the unemployment rate may be more important, and a few points' difference in it may greatly affect the level of city financial stress. In order to measure each variable in the same units and to prevent the second measure from having undue weight, researchers standardized the variables. Example 10.3 presents a detailed example based on the research mentioned earlier.[11] While the data in the example are based on U.S. counties rather than on cities, indices like the one described earlier have been developed to measure city fiscal stress and community need for use in grant allocation formulas.[12] After the indicators in this index were standardized, all were weighted equally, although different weights could have been assigned to each.

Example 10.3 Application of a Hardship Index

Quantifiable Measures and Definitions

1. *Unemployment:* Percent of civilian labor force unemployed
2. *Dependency:* Percent of persons under 18 or over 64.
3. *Education:* Percent of persons 25 years of age or older with less than a 12th-grade education.
4. *Income level:* Per capita income.
5. *Crowded housing:* Percent of occupied housing units with more than one person per room.
6. *Poverty:* Percent of families below 125 percent of low-income level.

Note that the range of each of these variables could differ considerably. Also, the income level variable is not a percent, as are the other five. In order to standardize them, each of the variables was treated in the following manner: The lowest value of the variable was assigned the value 0; the highest was assigned a value of 100.
The values for counties in between were determined by the following formula:

$$x = ((y - y_a)/(y_b - y_a)) \times (100)$$

where:
x = standardized value to be created for each county
y = value on a specific measure of need for each county
y_a = value of y indicating least need
y_b = value of y indicating greatest need

Thus, for each measure, the county with the greatest need (the highest unemployment, the lowest per capita income, and so on) was assigned a score of 100, and the county with the lowest need was assigned a score of 0.[13] Consider unemployment and per capita income as examples.

> *Unemployment:* The county with the lowest unemployment is William, with 0.9 percent; the highest rate occurs in Yuma, with 27.7 percent. Determine the standardized value for El Paso, with an unemployment rate of 8.8 percent.

> x = ((El Paso rate – lowest)/(highest – lowest)) × 100
> ((8.8% – 0.9%)/(27.7% – 0.9%)) × 100 = 29.47

> *Income:* Per capita income: highest = New York at $136,682; lowest = Hancock at $32,014. Lower per capita income indicates greater need. Determine standardized value for El Paso County, with per capita income at $54,980.

> Standardized measure of income for El Paso =
> (($54,980 – $136,682)/($136,682) – $32,014)) × 100 = 78

Table 10.5 Developing Standardized Values for the Items in an Index for Counties (2012 Data)

El Paso County

Variable	High	Low	Unstandardized	Standardized
1	27.7%	0.9%	8.8%	29.47
2	52.5%	5.9%	39.8%	72.46
3	55.1%	2.5%	26.7%	46.0
4	$32,014[a]	$136,682[b]	$54,980	78.0
5	7.9%	0.0%	2%	25.32
6	40.2%	0.6%	20.8%	51.0
				Total 302.25

[a]greatest need
[b]lowest need

Since a higher income level is considered more desirable than a lower one, the lowest value for income is given the value of 100. This county, then, has the highest value of need on the per capita income measure. Both variables are now standardized to a range of 0 to 100 and are measured in the same units. (The units, of course, are percentages. They are not the amounts of unemployment or of income.) The standardized measures for each variable can be added together to obtain an index value for each city.

Each separate variable in the index is standardized in the fashion described and summed for each county. Table 10.5 shows the low and high values for each variable and the standardized value for the example county, El Paso.

Note that the highest possible value for this index is 600. A county, city, or other organization would have this value only if it were the highest on each of the separate variables.

The separate item values in an index such as this can be averaged. That is, the total value for each case would be divided by the number of items in the index. For this index, the total for each county would be divided by 6. For El Paso County, the six-item average would be 50.38. The maximum average for the index would be 100.

Data and Data Sources for Example 10.3

1. Unemployment (as defined by the Bureau of Labor Statistics)

   County Level (2013 Annual Average—Bureau of Labor Statistics: www.bls.gov/lau/#cntyaa)
   Highest: Yuma County, Arizona—27.7%
   Lowest: William County, North Dakota—0.9%
   El Paso County, Texas—8.8%

2. Dependency (percent of individuals under 18 or over 64 in the general population)

   County Level (ACS 2012 5-Year Estimates S0101)
   Highest: Sumter County, Florida—52.5%
   Lowest: Loving County, Texas—5.9%
   El Paso County, Texas—39.8%

3. Education (percent of individuals age 25 and older that do not have a high school diploma or equivalent)

County Level (ACS 2012 5-Year Estimates S1501)
Highest: Starr County, Texas—55.1%
Lowest: Ouray County, Colorado—2.5%
El Paso County, Texas—26.7%

4. Income Level (individual mean earnings)

   County Level (ACS 2012 5-Year Estimates DP03)
   Highest: New York County, New York—$136,682
   Lowest: Hancock County, Georgia—$32,014
   El Paso County, Texas—$54,980

5. Crowded Housing (1.51 or more occupants per room)

   County Level (ACS 2012 5-Year Estimates DP04)
   Highest: Passaic County, New Jersey—7.9%
   Lowest: Platte County, Wyoming—0.0% (several are at zero percent)
   El Paso County, Texas—2%

6. Poverty (percentage of family and people whose income in the past 12 months is below the poverty level).

   County Level (ACS 2012 5-Year Estimates DP03)
   Highest: Shannon County, South Dakota—40.2%
   Lowest: Glasscock County, Texas—0.6%
   El Paso County, Texas—20.8%

   ACS: American Community Survey

Using Z-Scores to Standardize Measures for Indices

Each of the indicators in the index in Example 10.3 was standardized by the procedure shown and weighted equally. A different procedure for standardizing measures could have been used, however, and other weights could have been assigned. Another method of creating standardized measures is to calculate the z-scores of cases on each of a series of separate measures and then add the z-scores. Remember that a z-score is also known as a *standard score*. It is the number of standard deviations that a raw score, or original value, is above or below the arithmetic average.[14]

The z-score is calculated by the following formula:

$$z = (x - \bar{x})/s$$

((x = value of a case for a specific measure) – ($\bar{x}$ = average value of the measure for all cases))/ (s = standard deviation of the values)

In an index developed to measure the fiscal strain of cities in the 1970s, analysts used five quantifiable variables and calculated z-scores for each city on all measures. The variables were:

- population change, 1972 to 1976
- per capita income change, 1969 to 1974
- own-source revenue burden change, 1969 to 1974
- long-term debt burden change, 1972 to 1976
- change in local full market property value, 1971 to 1976

Table 10.6 Variables and Weights

Variable	Weight
Population change	.37
Per capita income change	.27
Revenue burden change	.12
Long-term debt burden change	.12
Change in full market value	.12

The differences in measurement units and range for the different measures required that a standardized measure be used. Statistical z-scores were developed for each city for each indicator. The five z-scores were summed for each city to obtain a Total Fiscal Strain value.[15]

Any of the standardized measures can be given greater weight than others if desired. In the example used here, some of the z-scores were weighted more heavily, as the analysts thought that these indicators were more important than others. The weights in Table 10.7 were assigned and multiplied by the z-scores for each city before being added together for an index value. Note that the weights total 1.00.

Population change was weighted more highly than any other indicator in the index.[16] Analysts thought it created more fiscal stress than the other items. Another example of using z-scores in an index is shown later in Example 10.7.

Likert Scaling and Other Indices

A common method of building indices is called *Likert scaling*.[17] Likert scales also are called summated rating scales. However, they are not scales according to the definition given earlier. A Likert method produces an additive index. The procedure is relatively easy to use. To develop an index using Likert scaling, the analyst selects a set of statements, each of which reflects favorably or unfavorably on a concept that he wants to measure. A rating form is provided for each item, with several ranked responses. The dimension is then rated on each item according to the responses provided. A numerical value is assigned to each response, and the values are summed to obtain a single numerical value. An alternative is to sum the separate ratings and divide by the number of items to provide an average of the responses to each item.

Measuring Opinions and Attitudes

Likert scaling is often used to measure opinions or attitudes of individuals. If it is used in an interview or survey, respondents are asked to indicate on the rating scale the degree to which they agree or disagree with each statement. The agreement scale may have only two choices (Agree–Disagree), or it may have more choices, permitting a respondent to indicate a level of agreement or disagreement. Five categories are commonly used: Strongly Agree, Agree, Neutral or No Opinion, Disagree, and Strongly Disagree. Some forms omit the neutral category, and some add even more categories to permit finer distinctions. An equal number of positive and negative statements is recommended. An unequal number of positive or negative response categories may bias responses. For positive statements, the categories are scored 1, 2, 3, 4, 5, with 1 indicating "Strongly Disagree" and 5 indicating "Strongly Agree." If the statement is unfavorable toward the subject, the scoring is reversed, as in Example 10.4. The respondent's index value is the sum of the values on the separate items. Example 10.4 shows a typical Likert scale used as an index to assess clients' satisfaction with agency service.

Example 10.4 An Example of a Likert Scale Index

The following was developed to assess client satisfaction with a public agency.

Item 1. The staff of this agency always treat me with respect.

| (5) | (4) | (3) | (2) | (1) |
| —Strongly Agree | —Agree | —Neither Agree nor Disagree | —Disagree | —Strongly Disagree |

Item 2. I never have to wait longer than 20 minutes for service at this agency.

| (5) | (4) | (3) | (2) | (1) |
| —Strongly Agree | —Agree | —Neither Agree nor Disagree | —Disagree | —Strongly Disagree |

Item 3. I am often told to come on the wrong day by staff of this agency.

| (5) | (4) | (3) | (2) | (1) |
| —Strongly Agree | —Agree | —Neither Agree nor Disagree | —Disagree | —Strongly Disagree |

Item 4. I am not satisfied with the services provided by this agency.

| (5) | (4) | (3) | (2) | (1) |
| —Strongly Agree | —Agree | —Neither Agree nor Disagree | —Disagree | —Strongly Disagree |

The numbers under the responses are the values for each item. They can be added together to obtain a value for the index. Of course, more items would usually be included. The numbers representing the values of each response category would not necessarily appear on the questionnaire. Notice in Table 10.7 that the direction of Items 3 and 4 is reversed from that of Items 1 and 2. Items 1 and 2 are positive statements, and if the client agrees with them, a higher value is given. Items 3 and 4 are negative statements, and if the client agrees with them, a lower value is given. People often get into the habit when answering questions of this nature of either agreeing or disagreeing without giving the item adequate thought. It is therefore wise to change the direction of some of the items so that the client is more likely to think through an answer before responding. Note also the "Neither Agree nor Disagree" category. This response is often included in Likert scaling.

Two alternatives are used to combine the responses to these items into an index. The first is to add the numerical value of the response to each item. Assume that a client gave the following responses to the questionnaire in this example.

The total would be the index value. If one of the items did not apply to this agency, or if the client did not respond to one or more of the items, adding the total responses would be inappropriate. The alternative is to add the total values of each response and divide by the number of items answered.

Table 10.7 Items, Response, Value

Item	Response	Numerical Value of Response
1.	Agree	4
2.	Agree	4
3.	Strongly Disagree	5
4.	Strongly Disagree	5
Total		18

Table 10.8 Items, Response, Value

Item	Response	Numerical Value of Response
1.	Agree	4
2.	Agree	4
3.	(No answer. Clients come in at their convenience.)	NA
4.	Strongly Disagree	5
Total		13

This would give an average value for each item and would be more accurate if some items were not answered. Using this approach for the previous client would result in an index value of 18/4 = 4.5.

If we used the index in another agency where Item 3 did not apply, we might find a client giving the responses shown in Table 10.8.

To compare this total to the total for the first agency would be inappropriate and inaccurate. Dividing the total by the number of items answered gives a more accurate value and one that can be compared to the average of the first agency.

Total/number of items answered = 13/3 = 4.33.

Constructing a Likert Scale

The first stage of constructing a Likert scale is the selection of the items. Although no definitive set of procedures exists, we offer the following general suggestions. First, understand the concept to be measured. The analyst should be able to write a few paragraphs about the concept and why each item is chosen and explain how she thinks each item relates to the concept being measured. Much of what applies to writing good questions for questionnaires applies to the construction of items for Likert scales. After statements have been selected, they should be pretested. The relationship of each item to the total score should be examined. In general, one wants items that are highly associated with the total scores. The items in a Likert scale need not be weighted equally. However, in practice, most Likert scales treat items with parity. One criticism of Likert-type indices is that the same total score can be obtained in a variety of ways. One of the main advantages is that they are easily constructed.

Indices based on the Likert scaling technique do not always use the "Strongly Agree" to "Strongly Disagree" continuum. And although each item has the same number of response categories, every item may not have the same responses. Example 10.5 shows such an index.

Example 10.5 Example Index

A city's public works director wishes to evaluate the cleanliness of neighborhoods surrounding the city's solid-waste transfer and materials recycling stations. He prepares an index for inspectors to use once a month. Inspectors are to check one rating for each item.

Neighborhood Cleanliness Index

1. The entire area looks like a pleasant residential neighborhood.

    _____Very much (A)
    _____Somewhat (B)
    _____Not at all (C)

2. Loose trash such as paper, bottles, and cans are visible.

    _____Not at all (A)
    _____Somewhat (B)
    _____Very much (C)

3. Grass is neatly trimmed and weeds are kept out.

    _____Very much (A)
    _____Somewhat (B)
    _____Not at all (C)

4. Streets and sidewalks are dusty or dirty.

    _____ Never or almost never (A)
    _____ Occasionally (B)
    _____ Very often or usually (C)

5. Numerical values are assigned to each response in the following manner:

    A = 3; B = 2; C = 1

If an inspector rated a neighborhood as shown subsequently, the corresponding numerical values would be assigned:

The total would be 10. The inspector could also divide this by 4—the number of items—to obtain an average cleanliness index value of 2.5 for the neighborhood. This number could be used to compare the cleanliness of the neighborhood with other neighborhoods or with the same neighborhood at some previous time.

Note that the higher the value of the index, the cleaner the neighborhood. If the value of the index falls below a certain level, the director would probably wish to address the problem and take corrective action.

The operational validity and reliability of the index also should be determined. If the items constituting the index were known to be the factors that most concerned nearby residents, then it would probably be a content-valid measure. However, the director would need to verify that each inspector used the index in the same way and that the index allowed the inspectors to assess cleanliness accurately.

A Likert-type index represents an ordinal level of measurement. The items do not really measure the quantity of a characteristic, but we can use the items to rank the cases. However, by adding together the numbers assigned to the response categories for each item or by calculating the mean for the scale, we are treating the measurement as if it were interval. This practice allows us to use more statistical techniques for analysis.

Table 10.9 Neighborhood Cleanliness Values

Item	Rating	Numerical Value
1.	B	2
2.	A	3
3.	A	3
4.	B	2

Opportunity Index Example

The Opportunity Index, first published in 2011, illustrates many of the concepts involved in constructing an index. It is interactive so readers can access it on the Internet and explore values for states and counties. The index was developed to measure the level of opportunity available to residents of all U.S. states and 2,900 counties. The index was originally composed of three dimensions and 16 themes and indicators but has been updated and modified since then. For the 2018 version, the index was expanded to include four dimensions and 20 indicators for states and 17 indicators for counties. The Themes component has been dropped. A Health dimension was added to the three original dimensions. The four dimensions are now : (1) Economy, (2) Education, (3) Community, and (4) Health.[18] Indicators are linked to the dimensions. Example 10.6 lists the indicators for the dimension Economy and the description for each indicator.

A web site provides more details on the Opportunity Index and how it was developed. These include definitions for each of the indicators and sources of data for each. The reader is encouraged to spend some time reviewing these. The discussion of the methodology of developing the index includes descriptions of the concepts discussed in this chapter. We recommend reading that discussion.[19] Another part of the web site allows users to find the overall opportunity index score for any state or county—including their own. The site also provides the score for each individual indicator for any jurisdiction. The user can compare states or counties to each other and to the national average. Because of changes in the index over time, the reader should be very careful when making comparisons between time periods.

Example 10.6 Economy Dimension of the Opportunity Index

Indicators are often measured in different units. For example, some are in percent, while median household income is in dollars. Therefore, the indicators are adjusted to a common scale, and each indicator is weighted equally. The values for indicators within each dimension are averaged. The scores from the four dimensions are also weighted equally—each counting one-fourth toward the total—and averaged to determine the Opportunity Index for any jurisdiction.

    http://opportunityindex.org/methods-sources/

Table 10.10 Indicators for Economy Dimension of the Opportunity Index

Indicator	Description
Jobs	Unemployment rate (% of the population ages 16 and older who are not working but available for and seeking work).
Wages	Median household income (in 2010 dollars).
Poverty	Percent of population below federal poverty level (the amount of pretax cash income considered adequate for an individual or family to meet basic needs).
Income Inequality	80/20 ratio; the ratio of the income of households at the 80th percentile to the income of the households at the 20th percentile (ratio; for example, 4 to 1).
Access to Banking Services	Number of banking institutions commercial banks, savings institutions, and credit unions (per 10,000 residents).
Affordable Housing	Percent of households spending less than 30 percent of income on housing-related costs.
Broadband Internet Subscription	Percent of households with subscriptions to broadband Internet service.

Factor Analysis

Factor analysis is used to determine the different dimensions captured by individual items used to measure a concept. It can be used to help select the best items from a larger number of possible measures for an index and to develop a standard measuring unit. For example, one recent survey had 21 questions regarding the use of performance information in public organizations and the instruments used to measure it.[20] Researchers used factor analysis to determine that there were four dimensions of the performance management: performance information to use for internal purposes, performance information for engaging with external stakeholders, general performance instruments, and performance instruments for decentralization.

Factor analysis is often used to investigate the relationship between theoretical concepts and empirical indicators and to reduce a large number of items to a smaller, more manageable number of indices. Analysts use it to select items and determine how important they are to an index. One method of factor analysis transforms a set of variables into a new, smaller set of composite variables. It can show which items should be used in an index and how they should be weighted. The technique often involves a large number of calculations; however, computer programs will quickly do the computations and provide the necessary information. You are likely to see the application of factor analysis in reports using or developing indices.[21]

Factor analysis was developed to help construct indicators of abstract concepts. The basic ideas were developed in the context of using examination grades to measure general intelligence. Suppose we have a large number of students' exam grades on six courses taken during a semester. We would expect that a student's exam grades in the different courses would be related to each other. The original investigators suggested that the link among the exam grades was the individual's general level of intelligence and was not directly observable. The investigators suggested that grades in all subject areas would depend to some extent on this factor. General intelligence could be considered a factor common to the performance on all of the tests. The several exam grades can then be used as a measure of this factor.

When factor analysis is used to condense a large number of items into fewer indices, the composite should have greater reliability and operational validity than the items taken separately. Factor analysis calculates how closely a large number of variables, assumed to be associated with each other, are related to a common dimension or factor. The analyst can pick variables that seem to be most closely associated with that factor and then use them in the index. As noted earlier in the chapter, factor analysis can also be used to identify the number of factors represented by a larger set of variables. An example later in the chapter also illustrates this.

For example, assume an analyst has data on several characteristics of counties in his state. These might include utilization of public services such as hospitals and parks; measures of conditions such as prevalence of diseases, the number of high school graduates and dropouts, and the number of handicapped; and measures of per capita income, employment, types of jobs, assessed evaluation of county real property, and so on. The analyst wants to develop indices of human service needs and separately of financial condition. A factor analysis could help determine which of the separate indicators best measure these two dimensions. It would show the analyst how the individual measures are related to the two conceptual dimensions of human service needs and financial condition. The analysis can help him pick out the indicators to use in the two indices and in weighting these indicators.

Factor Analysis and Association

Factor analysis is based on the concept of association. Analysts assume that associations among variables reflect the extent to which the variables measure the same trait or factor. Factor analysis begins by calculating association measures between each pair of variables of interest. Then a

Table 10.11 Tourist Area Property Owners' Perceptions of How Important Certain Actions Are to Sustainable Development in Their Community

Items	Factor Loadings
Operational-Oriented Sustainability (Factor 1)	
Reducing and managing greenhouse gas emissions	0.85
Managing, reducing and recycling solid waste	0.795
Reducing consumption of fresh water	0.79
Managing waste water	0.62
Being energy efficient	0.80
Purchasing from companies with certified green practices	0.85
Training and educating employees on sustainability practices	0.81
Normative-Oriented Sustainability (Factor 2)	
Conserving the natural environment	−0.88
Protecting our community's natural environment for future generations	−0.895
Protecting air quality	−0.855
Protecting water quality	−0.87
Community-Oriented sustainability (Factor 3)	
Reducing noise	0.65
Preserving culture and heritage	0.73
Providing economic benefits from tourism to locals	0.78
Full access for everyone in the community to participate in tourism development decisions	0.69

The authors developed a measure which averaged the respondents' scores for each dimension.

factor that maximizes the associations among the variables in the set is created mathematically. Next, coefficients measuring the association between each variable and the factor are calculated. These coefficients, called *factor loadings*, vary between zero and 1; the closer the loading is to 1, the more closely the variable is associated with the factor. Variables showing high loadings with a factor are considered to be measures of that factor. The analyst using factor analysis to help build an index would use as items in that index those variables that had the highest loadings with the factor.

The following example shows using factor analysis to determine which underlying dimensions are measured by a group of separate indicators. Hao, Derrick, and Long surveyed a sample of property owners in tourist areas to identify their perceptions of how important a set of 15 sustainable actions were to developing their county's tourism industry.[22] These authors applied factor analysis to the responses to identify the factors underlying the series of variables measuring sustainable actions. The factor analysis showed three themes representing three factors or dimensions of sustainability: (1) operational-oriented, (2) normative-oriented, and (3) community-oriented sustainability. Table 10.12 shows the factors, the items strongly related to each, and the factor loadings for each item.[23]

Factor Score Coefficients and Factor Scores

Analysts can also use the results of factor analysis to determine a value for each case for the index. The factor-analysis program provides a second coefficient, called a *factor score coefficient*, for each variable. These are the weights to be used. The analyst multiplies each by the original value of the case on that variable, producing a *factor score*. The results for all variables are added to obtain the index value for each case. This value is also called a *factor score*. Example 10.7 shows an index developed with factor analysis and the factor score coefficients. (The details on calculating a factor score follow Example 10.7.)

In the 1970s, researchers sought to develop indices of health, social service, recreational, environmental, public safety, and transportation needs for cities. They analyzed a large number of separate measures of community characteristics and used factor analysis to select items for each index.[24] The factor analysis showed how the individual variables were associated with five different factors. Five indices were developed from the loadings on these five factors. Example 10.7 shows the items selected to form an index of health needs for cities because of the high loading of each of these items with the factor named "Health Needs." The factor score coefficients also are shown. Computer programs that perform factor analysis will also produce the factor score coefficients and can calculate the composite value, the factor score, for each city.

Factor analysis requires interval-level measurement, although analysts sometimes use it when the variables are measured at ordinal and even nominal levels. A large number of cases is also needed to develop reliable factors.[25]

Example 10.7 Need Index

Table 10.12 Calculating Factor Scale Scores

Factor	Composite Variables	Factor Score Coefficients
Health needs	1. Local suicide rate	.73
	2. Infant mortality rate	.64
	3. Birth rate	.55
	4. Death rate	.51
	5. Families with income below $7,500	.44

The composite value computed for each city on the index, the factor score, is obtained by multiplying the coefficient for each variable by the value that the city has for that variable (Table 10.12).

The formula that is applied in each case is:

factor score = .73 × (value on variable 1) + .64 × (value on variable 2)
$\qquad$ + 55 × (value on variable 3) + .51 × (value on variable 4)
$\qquad$ + 44 × (value on variable 5).

In combining the values for each separate item, it was necessary to use the cities' z-scores for each variable rather than the cities' actual value. The z-score is used to account for different units of measurement and different ranges of the separate items. As noted earlier in the chapter, the z-score is calculated:

z value for each city = ((city's value for a variable) × (average value of the variable for all cities))/(standard deviation of the variable)

The mean and standard deviation for each variable are provided by the factor analysis program, and most programs also will calculate the z-scores. Most factor analysis programs also have routines that will provide the factor scores for each of the cases so that the investigator does not have to do this separately.

Table 10.13 shows the factor score calculations for one city.

Table 10.13 Factor Score Calculations

Variable	City A's Values	z-Scores for City A	z-Score × Coefficient
1. Local suicide rate	3 per 1,000	.5	.5 ×.73 = .365
2. Infant mortality rate	5 per 1,000	−.2	−.2 ×.64 = .128
3. Birth rate	9 per 1,000	1.0	1.0 ×.55 = .550
4. Death rate	2%	.5	.5 ×.51 = .255
5. Families with income less than $7,500	5,000	2.0	2.0 ×.44 = .880
Total			1.922

City A's value on the index is 1.922.

Index Numbers

Index numbers express the relationship between two figures, one of which is the base. Index numbers are used to describe changes over time in such things as prices, production, wages, and unemployment. Indices such as the CPI are used extensively in the analysis of economic conditions. The CPI combines prices for a large number of separate items to measure the cost of living and to describe changes in economic variables over time. This index produces a summary statistic called an *index number*.[26]

The CPI, as an index number, provides information on

1. a composite price of a standard set of goods and services
2. the percent change that has occurred in the composite over a given time

To facilitate comparison, the index is expressed as a percentage, a proportion, or a ratio.

The CPI, probably the most commonly used and well-known index number, has many subcomponents. It groups the prices of more than 400 commodities and services into eight categories.[27] The index measures price changes from a designated reference date, which is set equal to 100. An increase of 50 percent, for example, is shown by the index as 150. This change also can be expressed in dollars as follows: The price of a set of goods and services in the CPI has risen from $10 in the base year to $15. One can treat the index number as a percent to determine by what percent the price of goods and services has increased since the base year. Currently, the period 1982–1984 is used as the base for the components of the CPI. Table 10.14 illustrates how index numbers show the relative increase in prices. The index number for year 5, for example, shows that the prices of goods and services have increased by 11.2 percent since the base year.

Administrators often wish to assess growth in the amount of money spent by the organizations they direct. However, they usually like to separate increases in expenditures due to increased services and activities from growth due to inflation. An index number helps them to do that. For example, the administrator of a county hospital can use a component of the CPI to compare

Table 10.14 Changes in the CPI Over Time

Year	1 (Base)	2	3	4	5	6	7	8
Index	100	101.9	105.5	109.6	111.2	115.7	121.1	127.4

Source: U.S. Department of Labor, *CPI Detailed Report for March 1993* (Washington, DC: Bureau of Labor Statistics).

increases in the cost of hospital and health care goods and services to increases in the hospital's budget.

Summary

When a single indicator is inadequate to measure a concept, investigators often use multi-item measures called indices.

Four important related issues in constructing indices are:

1. *selecting items* to be included in the measure, which involves being very clear about what you want to measure and not omitting any important aspect;
2. *standardizing items* so that each is in comparable units so that they can be combined reasonably and to help avoid any one item having undue influence;
3. *weighting the separate items* so the index satisfactorily represents the concept you had in mind when you started; and
4. *combining items* to form an index, which requires you to think about strengths of indices, categories of the index, and type of index.

Procedures and methods for accomplishing each of these were discussed. These include Likert scaling, factor analysis, and procedures for standardizing variables.

Likert scaling is most likely to be used when the cases can be rated on several items. It is easy to do and can be done even if only a few items are available. Although each item represents an ordinal scale, by assigning a numerical value to each category and adding the values for each, analysts sometimes treat the index as an interval-level measurement. The resulting index value is often treated as an interval value as well. This has the advantage of allowing the analyst to apply statistics that assume interval-level measurement.

If the separate items measure the case's attributes on an interval scale, you may prefer one of the other methods. Factor analysis is used to reduce a large number of measures to a few underlying common factors. The variables associated with each factor can be combined in a composite index. The factor analysis provides information to help in choosing the variables for the index and weights to use for each variable. Factor analysis requires a large number of cases and requires that the measures be interval level.

If separate items in the index are measured in different units, the measures must be standardized. Even if the items are measured in the same units, if the range of values for the items differs

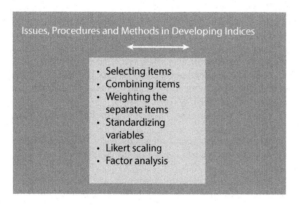

Figure 10.2 Constructing Indices

greatly, the scale or range should be standardized. If not, the greatly differing ranges may cause some of the items to have a greater impact than others, contrary to the intent of the analyst.

In constructing indices (Figure 10.2), one should determine whether each item measures the variable appropriately and then decide how to weight each item. If the ranges or units used to measure items are different, the analyst should first standardize them and then separately apply a weighting procedure.

The CPI is an example of an index number. Index numbers typically express the change in some factor relative to a designated date called the base year. The CPI is set to 100 for the base year, and the number for each succeeding year indicates the percent change in prices since that year.

Notes

1. R. A. Singleton, B. C. Straits, and M. M. Straits, *Approaches to Social Research*, 2nd ed. (New York: Oxford University Press, 1993), 397; Earl Babbie, *The Basics of Social Research*, 6th ed. (Belmont, CA: Thomson-Wadsworth, 2014), Chapter 6 discusses scales and indexes and the differences between them.
2. Daniel C. Esty, Christine Kim, Tanja Srebotnjak, Marc A. Levy, Alex de Sherbinin, and Valentina Mara, *Pilot 2006 Environmental Performance Index* (New Haven: Yale Center for Environmental Law and Policy, 2006).
3. Earl Babbie, *The Practice of Social Research*, 13th ed. (Belmont, CA: Cengage Learning, 2012), Chapter 6.
4. Research Methods Knowledge Base, available at www.socialresearchmethods.net/kb/scalgutt.php. Gutmann scaling is also discussed in R. L. Gordon, *Unidimensional Scaling of Social Variables* (New York: Palgrave Macmillan, 1977).
5. Robert Levine, "City Stress Index: 25 Best, 25 Worst," *Psychology Today* (November 22, 1988): 52–58.
6. Allen C. Kelley discussed these issues in a critique of a human development index. See Kelley, "The Human Development Index: 'Handle with Care'," *Population and Development Review* 17, no. 2 (June 1991): 315–324. For current information and country rankings on this index, see United Nations Development Program Human Development Reports, data.undp.org.
7. The Uniform Crime Reports are described in the Federal Bureau of Investigation, U.S. Department of Justice, Crime in the United States 2012 and About UCR, The Uniform Crime Reporting Handbook, July 1992, available at www.fbi.gov. Also see Mark H. Maier and Jennifer Imazeki, *The Data Game*, 4th ed. (Armonk, NY: M. E. Sharpe, 2013), Chapter 6. Maier and Imazeki discuss the Crime Index and how reporting methods can cause it to be biased in Chapter 6. They also report on other sources of crime statistics and their shortcomings.
8. See opportunityindex.org/methods.
9. David J. Wright and Lisa M. Montiel, "Divided They Fall: Hardship in American Cities and Suburbs," (Rockefeller Institute of Government, November 2007). "Map 19: Hardship Index- Los Angeles Department of City Planning (planning.losangelescity.org/cwd/framework/healthwellness/maps), accessed September 1, 2019, available in PDF. "Economic Hardship Index Shows Stark Inequality Across Chicago," September 19, 2016. Great Cities Institute, University of Illinois, Chicago Circle. Accessed at greatcities.uic.edu on September 1, 2019.
10. See Robert W. Burchell, David Listokin, George Sternlieb, James W. Hughes, and Stephen C. Casey, "Measuring Urban Distress: A Summary of the Major Urban Hardship Indices and Resource Allocation Systems," in *Cities Under Stress: The Fiscal Crises of Urban America*, eds. R. Burchell and D. Listokin (Rutgers: The State University of New Jersey, 1981), 159–229. The specific index on which Example 10.3 is based was developed by Richard P. Nathan and Charles F. Adams, Jr. They measured city hardship using 1970 Census data in "Understanding Central City Hardship," *Political Science Quarterly* 91, no. 1 (1976), and updated the study with 1980 Census data in "Four Perspectives on Urban Hardship," *Political Science Quarterly* 104 (Fall 1989): 483–508.
11. Nathan and Adams, "Four Perspectives," 504–506; Burchell et al., "Measuring Urban Distress," 159–229.
12. Data for some indicators used in the Nathan and Adams index were not available for several cities when the authors updated this example. Therefore, we based the example on counties for which a complete set of data was available.
13. For another application and discussion of this technique, see opportunityindex.org/.
14. Chava Frankfort-Nachmias, *Social Statistics for a Diverse Society* (Thousand Oaks, CA: Pine Forge Press, 1997), 447–448; Lawrence Giventer, *Statistical Analysis for Public Administration*, 2nd ed. (Sudbury, MA: Jones and Bartlett Publishers, 2008), 81.

15. Burchell et al., "Measuring Urban Distress," 159–229.
16. Ibid.
17. Although other common scaling methods, such as Thurstone, Guttman, and the semantic differential, are used in basic research, administrators are seldom likely to use them. Readers interested in pursuing these topics will find them discussed in A. B. Anderson, A. Basilevsky, and Derek Hum, "Measurement: Theory and Techniques," in *Handbook of Survey Research*, eds. Peter Rossi, James D. Wright, and Andy B. Anderson (New York: Academic Press, 1983), 231–287. For a discussion of the distinction between indices and scales, how they are constructed, and major types of scales, see Babbie, *The Practice of Social Research*, Chapter 6. Other sources for these topics are listed in "Recommended for Further Reading" at the end of this chapter.
18. http://opportunityindex.org/methods-sources/.
19. Ibid.
20. Zachary T. Mohr, Ringa Raudla, and James W. Douglas, "Comparing Cost Accounting Use Across European Countries: The Role of Administrative Traditions, NPM Instruments, and Fiscal Stress," *Public Administration Review* (forthcoming). See the appendix for a succinct overview.
21. Huili Hao, Derek H. Alderman, and Patrick Long, "Homeowner Attitudes Toward Tourism in a Mountain Resort Community," *Tourism Geographies: An International Journal of Tourism, Space, Place, and Environment* (2013), doi:10.108/1416688.2013.823233.
22. Huili Hao, Patrick Long, and J. Kleckly, "Factors Predicting Homeowners' Attitudes Toward Tourism: A Case of a Coastal Resort Community," *Journal of Travel Research* 50 (2011): 627–640.
23. Ibid.
24. Gregory Schmid, Hubert Lipinski, and Michael Palmer, *An Alternative Approach to General Revenue Sharing: A Needs Based Allocation Formula* (Menlo Park, CA: Institute for the Future, 1975).
25. Hao, Alderman, and Long, "Homeowner Attitudes Toward Tourism in a Mountain Resort Community."
26. Wayne Daniel and James Terrell, *Business Statistics*, 7th ed. (Boston, MA: Houghton Mifflin, 1995), 69–81. See David S. Moore and William Notz, *Statistics: Concepts and Controversies*, 7th ed. (New York: W. H. Freeman and Company, 2009), 308–327, for a thorough and accessible discussion of the CPI, trends, and how to convert the purchasing power of a dollar at one time to another. The reader should note that the CPI received a great deal of criticism in 1997, and a major revision of the methodology for calculating it was discussed and researched. CPI data can also be found at www.bls.gov/cpi.
27. For a more detailed discussion of the Consumer Price Index, see U.S. Department of Labor, Bureau of Labor Statistics (BLS), *Handbook of Methods*, vol. II, "The Consumer Price Index," *Bulletin 2134–2*, April 1984; "The Consumer Price Index: 1987 Revision," *BLS Report* 736, January 1987; David Ammons, *Tools for Decision Making* (Washington, DC: CQ Press, 2002), 108–114, for a discussion of the Implicit Price Deflator (IPD), an index number appropriate for use by local governments. Ammons has clear instructions for using the IPD.

Terms for Review

indicator
index
scale
standardizing measures
weighting
Likert scaling
factor analysis
factor score coefficient
factor score
index number

Questions for Review

The following questions should indicate whether you have a basic competency in this chapter's material.

1. What benefits does using an index to measure a characteristic provide to an investigator?

2. How does operational validity apply to index construction?
3. Explain why an index should provide a more valid and reliable measure than a single indicator.
4. Give an example of an index for which using unweighted items is not likely to cause any problem. (Unweighted items means that all items are weighted the same.)
5. Give an example of a second index for which unweighted items are likely to result in an invalid measure.
6. Explain why transforming variables so that they are in comparable units and weighting the items are usually accomplished separately.
7. Consult a library to obtain a recent copy of the United States Department of Labor's CPI: Detailed Report or access the department's web site at www.bls.gov/cpi. Determine how much goods and services costing $50,000 in 1982 would cost in the current year or in the year of the report. What is the CPI for the year used?

Problems for Homework and Discussion

1. Develop a Likert index to rate the quality of a public transportation system. List 10 items that could be included. Use the index to evaluate the bus system in a town or city with which you are familiar.
2. Refer to Example 10.4 to answer the following: Assume that a client gave the following responses to the items in the index: Agree, Disagree, Strongly Disagree, No Opinion.

   1. Calculate the index value for this client.
   2. What other item(s) would you consider adding?

3. In Example 10.3, work out the calculations for Items 2, 3, 5, and 6. (Items 1 and 4 are done in the text.) Check your answers against those listed in the example.
4. Pick a sample of five cities or counties in your state or surrounding states, and apply the index discussed in Example 10.3. Remember that the lowest and highest levels of need will be based on the values from your sample, not on those in the example.
5. Name five indicators of health needs in addition to those listed that may have been included in the set of measures factor-analyzed in the research reported in Example 10.7. Which of these indicators do you think would be most closely associated with a factor called "Community Health Needs"?
6. Find a research article or discussion for an example of each of following indices:

   a. Likert scaling
   b. Factor analysis
   c. Different weights for at least some of the items
   d. A method of standardizing items that measure characteristics in different units such as dollars, years of education, and percent unemployed.
   e. Describe how each was used and what it was intended to measure.

7. Consult a recent edition of the County and City Data Book and develop an index of "city hardship" using three of the following indicators: percent unemployed, average level of education, income level, amount of crowded housing, percent of residents under age 18 and over age 64. Copies of the Data Book can be found on the Internet at http://fisher.lib. Virginia.edu/ccdb. Most academic libraries also have printed copies.
8. Access the following for the Opportunity Index discussed in the chapter and respond to the following items. http://opportunityindex.org/methods-sources/

   a. Select an indicator for one of the dimensions and read over the definition of that indicator. Would you say that the definition is (i) conceptual, (ii) operational, (iii) both? Explain.

b. Do you think the indicator selected is a valid measure of the dimension? Explain. Do you think it is likely to be reliable? Explain.

c. Select an indicator that you think may not be a valid or reliable measure of the dimension. Explain why you think this.

d. Select a state to compare to the national average and comment on the differences.

e. Compare the county where you spent your youth to the county where you are now living. Comment on the differences.

Working With Data

1. Refer to Example 10.3 and how the Hardship Index was constructed. Using the following four variables in the County Data file, follow the procedure in that example to create a hardship index for the counties: percent of families in poverty, median household income, percent of adults with a college degree, unemployment rate. Before carrying out this exercise, you should have worked out Problem 3 in Problems for Homework and Discussion. Select 10 counties at random and calculate their scores on your Hardship Index. What is the average score for the five counties with the highest score? The lowest?

2. For each of the five counties with the greatest hardship and the five counties with the least hardship, identify the percent of the labor force employed in: manufacturing, retail trade, and service industries. Put your findings in a table similar to Table 9.1. Do these data suggest what types of economic development might be most beneficial? Write a brief memo (no more than one page, single spaced) suggesting types of economic development activity which should be explored in more depth to benefit less-well-off counties.

3. Using three of the variables in the County Data file, create your own index to measure one of the following: county health status, social well-being, level of economic development.

4. Do the following in developing your index: (1) provide a conceptual definition for the concept you want to measure, and (2) write a few sentences justifying the three variables you choose for the index. Explain why you think these variables should be in the index. Then apply your new index to the 10 counties selected for Problem 1. (If you did not do Problem 1, select 10 counties at random from the dataset.)

5. You have been tasked with using the General Social Survey dataset to come up with a proposed index of support for government intervention.

    a. Select four to six variables from the GSS and justify how they could help measure support for government intervention.

    b. Then, propose a series of weighting for this concept.

    c. What steps could you take to determine whether this weighting is necessary?

    d. The variables that you choose may not have the same number of responses. Suggest how you might combine responses so that each variable has the same number of responses.

    e. What would be the lowest and the highest values for the index if the values for each response were simply added?

    f. Write a few sentences critiquing your index.

Recommended for Further Reading

Miller, D. C., and N. Salkind, *Handbook of Research Design and Social Measurement: A Text and Reference Book for the Social and Behavioral Sciences*, 6th ed. (Thousand Oaks, CA: Sage Publications, 2002). Miller and Salkind review many existing indices covering a variety of subjects. Although the majority of these are used in basic research, many others are likely to be of interest to managers and administrators.

These include measures of organizational structure, organizational effectiveness, community services, leadership in the work organization, morale, and job satisfaction.

For an extensive discussion of job-satisfaction scales and similar indices, see Robinson, J. P., R. Athansiou, and K. Head, *Measures of Occupational Attitudes and Occupational Characteristics* (Ann Arbor: University of Michigan, Institute of Social Research, 1969). A publication similar to Robinson et al. discusses a large number of scales and indices measuring social, psychological, and related attitudes. See Robinson, J. P., and P. P. Shaver, *Measures of Social Psychological Attitudes* (Ann Arbor: University of Michigan, Institute for Social Research, 1973).

McDowell, I., and C. Newell, *Measuring Health: A Guide to Rating Scales and Questionnaires* (New York: Oxford University Press, 1987), review and evaluate several indices of health and well-being.

The following publications include discussions of factor analysis:

Hamburg, M., *Basic Statistics*, 3rd ed. (New York: Harcourt, Brace, Jovanovich, 1985), Chapter 14, discusses index numbers and procedures for combining and weighting items.

Jackson, B. B., *Multivariate Data Analysis: An Introduction* (Homewood, IL: Richard D. Irwin, 1983).

Kim, J., and C. W. Mueller, *Factor Analysis: Statistical Methods and Practical Issues,* Sage University Paper Series on Quantitative Applications in the Social Sciences, 07–014 (Beverly Hills, CA and London: Sage Publications, 1978).

Mertler, Craig, and Rachel A. Vannatta, *Advanced and Multivariate Statistical Methods*, 5th ed. (Glendale, CA: Pyrczak Publishing, 2013), Chapter 9 has an easy-to-follow discussion of factor analysis with examples of computer output.

For more information on the Consumer Price Index, see the Bureau of Labor Statistics web site: www.stats.bls.gov/cpihome.htm; Moore, David S., and William Noltz, *Statistics: Concepts and Controversies*, 7th ed. (New York: W. H. Freeman and Company, 2009).

11 Univariate Analysis

In this chapter, you will learn:

1. How to prepare and organize data for analysis.
2. About computer programs for managing, analyzing, and mapping data.
3. About displaying data in graphs, tables, and charts.
4. About statistics for analyzing one variable at a time.
5. To calculate and interpret measures of relative frequency.
6. To calculate and interpret measures of central tendency and measures of dispersion.
7. About new techniques for presenting and visualizing data.

After collecting and assembling a set of data, analysts want to get a sense of its nature: What are the values of individual variables? How similar are the values for different cases? How different are the cases? Over what range are values distributed? Even if the analyst's purpose is to evaluate associations between variables, he also should describe and summarize the values for individual variables. Often the main purpose of a study is to obtain data on individual variables, for example, when the fire chief or city manager wants to find out the dollar amount of fire damages in the city for each month of the past year.

Many of the questions we ask as administrators, analysts, and researchers require us to describe the *distribution* of a single variable: How many individual programs were funded last year by the county budget? What percent of welfare recipients have lived in the county for less than two years? How much do the per capita expenditures for police differ among cities in the state? How does the tax rate in my county compare to the most common rate? Interest in these issues lead to the following question: What is a good way to characterize a set of data with statistics?

Two important distinctions in statistics are (1) between descriptive and inferential statistics and (2) between statistics for univariate and those for multivariate distributions. *Descriptive* statistics are used to summarize and describe the data on cases included in a study. *Inferential* statistics are used to make inferences to larger populations, to use data from the cases studied to conclude something about the cases not studied. *Univariate* statistics tell us about the distribution of the values of one variable. *Multivariate* statistics measure the joint distribution of two or more variables and are used to assess the relationships between and among variables. *Bivariate* distributions are special cases of multivariate distributions; they are the joint distributions of two variables.

In this chapter, we discuss descriptive statistics for univariate distributions and several other quantitative measures for describing single variables and their distributions. We also discuss tables and graphs—important tools for analyzing and illustrating the nature of variable distributions—and computer programs for managing, analyzing, and presenting data.

Basic computer programs have eased the burden of calculating statistical measures. However, people may not fully understand the statistics they ask a computer program to provide or may not use those statistics correctly. We discuss several basic statistics—such as the arithmetic mean, the median, and the mode—useful to administrators and analysts. We also discuss common visual presentations of data.

We do not intend for this or the other chapters on statistics in this book to substitute for courses in statistics. We provide formulas for the statistics and, in some cases, worked-out examples. The reasons for providing these examples, however, are to clarify the nature and purpose of statistical measures and to illustrate their applications. A separate section of the chapter includes additional information and worked-out examples of some of the statistics. Another section includes an example and discussion of preparing data for computer analysis. These sections can be read separately from the rest of the chapter.

Computer Software for Data Management, Mapping, and Analysis

A variety of computer programs are available for managing and analyzing data. These include spreadsheets, statistical analysis programs, database managers, and geographic information systems. Administrators are more likely to have spreadsheets and database managers available on their desktop computers than they are to have a statistical package. Spreadsheet programs are used extensively for managing numerical information. They can be used to calculate many of the statistics discussed in this and the following chapters. We discuss each of these after the next section.

Data Preparation

While some data can be entered directly into the programs used to manage and analyze them, analysts typically put data in a visual, sometimes printed, form first. This includes documents that allow other users to identify the variables and their values. The following discusses this process.

Preparing a data dictionary, also called a codebook, is an important step in organizing and preparing the data for analysis and reporting. Data are typically coded for storage and analysis by computers, and before collecting them, project staff should develop a plan for this. Coding means assigning a number to each value of a variable. In a *data dictionary*, the analyst lists the items, labels the valid codes for a variable's values, names each variable, and gives the location of the variable in the file. Given the data and the dictionary, an analyst should not need to consult the original data collection forms. Coding all information as numbers is useful for statistical analysis, although not necessary. However, data analysis with a statistical package will be easier with numerical coding.

Consider a data collection form, Injury Protocol Data Collection Form, used by a regional emergency medical service (EMS). Individuals suffering head injuries in a multicounty area were taken by ambulance to the emergency room of a regional hospital. An attendant recorded several pieces of information on each case, including the county where the injury occurred, the cause of the injury, the severity of the injury, and other information about the injured individual.

Data dictionaries typically contain the following information:

1. *Item number.*
2. *Item description.* This may include the actual question from a questionnaire or the instructions used in collecting a particular piece of data.
3. *Item codes.* These are the most important part of the dictionary and show the numbers, letters, or other codes assigned to each possible value of the item. Assigning a code for

unavailable data is very useful. Analysts usually try to use the same code, such as 9 or 99, for missing information for all items. Similarly, a code may be assigned to all "Don't Know" responses for all items. In a survey, for example, the number 8 might be used to code an answer of "Don't Know" and 9 to indicate missing information.

4. *Variable or field name.* Most statistical computer programs require that a name be assigned to each variable. The variable name is used in instructing the program to manipulate and analyze the data. Sometimes the variable name is determined early in the process and printed on the data collection form.

A dictionary for the Injury Protocol Data Collection Form is represented in Table 11.1. Note that the unit of analysis for this example is the injured individual.

An identifying number should be assigned to each case and placed on the data collection form for that case. This number should be included in the data record and the data file.

Spreadsheets

A *spreadsheet* is a matrix of rows and columns. On the computer screen, a spreadsheet program shows a form in which data can be entered in an array, usually with *each row as a case* and *each column as a variable*. The user can enter labels for columns and rows, change column widths to accommodate lengthier pieces of information, enter and manipulate data, and perform statistical and other mathematical analysis.[1] Numbers, text, formulas, and logical statements are used. A spreadsheet is useful for entering data and creating datasets, for recording budget and expenditure data, and for doing arithmetic and algebraic calculations. The data input form for some statistical programs also uses a spreadsheet format.

Spreadsheets are useful and often sufficient for managing small datasets, calculating univariate and some multivariate statistics, providing graphics, and entering data.[2] After a dataset has been entered into a spreadsheet program, a user can easily import it into a more powerful statistical

Table 11.1 Data Dictionary for Injury Protocol Data Collection Form

Item	Description	Codes	Variable Name
1.	Case identification number	001–total	ID
2.	County of injury	01 = Able 03 = Baker . . 99 = Unknown or not answered	County
3.	Cause of injury	01 = Car accident 02 = Fall 03 = Violence . . 99 = Unknown or not answered; include actual score: 1–10	Cause
4.	Severity scale	01 = Least severe 10 = Most severe 99 = Unknown or not answered	Severity
5.	Clinician's comments	Written description by clinicians.	Comments

analysis program for extensive analysis. Even fairly large data files are sometimes available as spreadsheets or comma-separated value (.csv) files that can be read by spreadsheet programs. Many datasets released by U.S. government agencies are made available in spreadsheet format. Others, including some census data files, are released in a database format, which can be transferred to a spreadsheet program and then to a statistical package.

Yet spreadsheet programs generally do not perform advanced statistical calculations. Most have built-in functions for basic and intermediate-level statistics, but a software program for statistical analysis is often needed to conduct the desired analysis of the data in a spreadsheet. Statistical software packages permit easy alteration of variables and their values, handle larger datasets than do many spreadsheet programs, and provide a wider range of statistical tools, allowing more sophisticated data analysis.

Statistical Software Packages

The most powerful and versatile types of software for statistical analysis are *statistical software packages*. These were developed to analyze large datasets and include numerous statistical routines. After the data are prepared and entered, the user can easily conduct many types of analysis. Statistical packages calculate various statistics and allow the user to subdivide the dataset, combine the values of different variables, and create new variables from earlier calculations. Three widely used statistical packages for administrative and social science data are SPSS, SAS, and Stata. SPSS, in particular, is relatively easy to use, is often taught in undergraduate and graduate courses, and performs most routines that practitioners are likely to need.[3]

Database Management Systems

In contrast to spreadsheet and statistical software programs, with a database management system, the user enters and stores large amounts of data under different headings. These programs are often used to link separate files that have related information. While statistical software programs are generally used mainly to store numerical values, a database management system (DBMS) can store textual information for each case. Forms containing the outline of a questionnaire or other data-collecting instruments can be shown on the screen. Information stored in a database can be edited and manipulated easily and specific information located quickly. DBMSs can sort and compile large amounts of data quickly for reports. If the database has more than one file that an administrator or researcher wishes to link, the files must be related to each other with some kind of *cross-walk*. The cross-walk links common elements in each file.[4] A database with two or more files with common variables, or fields, that allow information to be linked across files is called a *relational database*. The most powerful and useful database managers for personal computers are relational database managers. These help the user manage information in new ways and can be used to organize data from separate files for reports and statistical analysis.

Relational databases have several advantages. Old information does not have to be entered again when a new file is created. The first file does not have to save space for records in succeeding files that might have blank fields. Databases are often developed for multiple purposes and the data shared by many users. For example, a county government may have records on personnel, financial information (such as revenues and expenditures), and budget information in a large database accessible to all departments. Several different files may contain information about employees. One may contain employment history, another may contain benefit records, and a third may contain information on training and volunteer activities. Many include a data dictionary function that names and describes each piece of information and its location in the database. The dictionary aids users by letting them know what information is available and how it can be obtained.

Relational database managers can produce summary information, basic statistical analysis, and graphs. Database managers, however, are not powerful tools for statistical analysis. For this reason, analysts often transport the data to a statistical analysis package. If data from more than one file are used, transporting usually requires that the analyst create a new file containing those variables that he wishes to analyze further. The database manager program enables the analyst to do this easily. The resulting file is then transported (exported) to the analysis program. The major statistical analysis programs also provide for easy transfer of data from database programs, as they do with spreadsheets.[5]

Separate data dictionaries should be developed for each file in a database to explain the codes, labels, and location of each data item. The user should develop a plan for this before data entry. The quality of the data is paramount to a good database. The structure of the database makes little difference if the quality of the data or the documentation is poor. Throughout this book we have emphasized the importance of ensuring high-quality data. As data files are linked in databases and their use expands, the quality of the data becomes even more important.[6]

Geographic Information Systems

A geographic information system is a special type of relational database. Its unique and defining characteristic is that it records data that can be displayed geographically. Many data records stored by government agencies include important geographic location information. A few examples are property valuation and taxes; residence information; and locations of streets, water mains, buildings, utility lines, and crimes. A GIS has the ability to display this information on maps produced by the computer.

Examples of location information collected by governments and private organizations also include data on individuals: taxpayers, clients of various services, school-age children, homeowners, and users of public facilities. Such information may be kept in separate files in a relational database. With the addition of geographic location information, a GIS can show these data in various ways on maps and provide an important method for analyzing them.[7] The method of analysis provided by a GIS is called *analytic mapping*.[8] A GIS has the ability to map on a geographic display the values and distributions of variables. For example, a GIS can show the locations of users of services, existing and proposed school district lines, and numbers of school-age children. A GIS can present data so the analyst can literally see the relationship between location and values of a variable. An important and powerful use of GIS is in combining census data with mapping capability.[9]

Analytical Mapping

Analytic mapping using geographic databases has developed rapidly in recent years with advances in microcomputer technology, software, and improved collection and dissemination of geographic data. With this tool, the analyst can relate the values of variables to areas or locations. Although a GIS map shows the relationship of variables to locations rather than to each other, the distribution of several variables can be overlaid. For example, a city analyst working with a GIS could map crime data and unemployment data on the same map. The map could also show where the city government spends its money.[10] One of the more common functions of GIS mapping is showing the location of public facilities. Figure 11.1 is an example of GIS in use at a municipality.

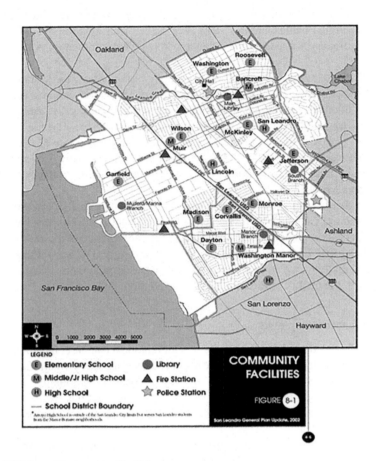

Figure 11.1 GIS Map of Community Facilities

Source: Downloaded from San Leandro County web site 4/29/2015, www.sanleandro.org/depts/cityhall/citymanager/it/gis/pdf.asp. Reprinted with permission of San Leandro County, CA.

Benefits of Geographic Information System Mapping

In 2011, the City of San Leandro, California, partnered with a local high-tech entrepreneur to build a high-speed fiber optic network. The city's geographic information system played a central role in the planning, implementation, and expansion of the system. A newer map displays multiple "layers," including the fiber route, city facilities, schools, libraries, parks, and their proximity to the high-speed fiber optic network. Produced with professional GIS software, it demonstrates the power of GIS to analyze and map discrete, non-associated datasets.[11] City management has continued to develop and use GIS mapping to bring and display important information visually to the citizens. A visit to their web site will show several interactive maps available to interested citizens.

A study conducted in 1990 without the benefit of GIS illustrates its benefits. An analyst mapped all stranger rapes reported in a city over a three-year period. He obtained police records and marked the location of each rape on a paper map of the city.[12] This took several days of full-time work. The resulting map identified clusters of rape cases and led to several conclusions for programs to prevent these crimes and apprehend offenders. Today, with the records in a GIS, this same map could have been prepared in a few hours or less.

Combining Statistical and Geographic Information System Approaches

The statistical capability of GIS programs has expanded along with their increased use and sophistication. Univariate and multivariate statistical techniques are available to analyze spatial relationships in new ways. Statistical analysis modules have been included in GIS, allowing more thorough analysis of variables.[13] These statistical routines allow the calculations of average distances between residents and facilities such as fire stations; the extent to which objects, such as houses with specified characteristics, are clustered; and locations of houses with different values.

In a 1992 book, G. David Garson and Robert Biggs discussed the uses of GIS in social science, public administration, and public policy and the relationship of statistical analysis to GIS and analytic mapping. They emphasized that statistical and GIS approaches are complementary. The substantial developments in GIS technology and statistically related tools have made these relationships even more important and useful today.[14] Combining statistical analysis and GIS approaches is easier and provides more powerful results than ever before. Given the importance and growth of this technology in the public sector, we agree with Garson and Biggs that a generalist knowledge of research methods should include information about GIS.[15] In the "Recommended for Further Reading" section at the end of the chapter, we cite several of the many books describing GIS now available, including the Garson and Biggs book. We also cite books on spatial statistics.

Learning More About Geographic Information Systems

GISs are used extensively by government and not-for-profit organizations. Governments record information on roads, streets, facilities, and real estate parcels and use GIS to produce maps showing these.

Those interested can find out more about government's applications of GIS by accessing state, local, and national sites. Your local city and county is a good place to start. Counties are usually responsible for keeping real estate records for both county and cities, so accessing a county site is recommended. For example, in your web browser, enter

[county name GIS]. Example: Olmsted County Minnesota GIS.

Or the same for a city or state, such as:

[Raleigh NC GIS] or [Maryland GIS]

Enter the name of your city, county, or state to find out more about its GIS use.

The most well-known uses of GIS by the national government are by the U.S. Bureau of the Census. Find out more at www.census.gov. Select Geography and then Interactive Maps or Maps and Data.

The YMCA—a national organization with state and local organizations and branches—has a GIS generated locator. Enter: ymca.net/find-your-y. Then entering a zip code will bring up a map showing the location of the nearest branch of the local YMCA and other branches nearby.

ArcGIS is a commonly used program. Students taking courses in GIS are likely to use it. The "Recommended for Further Reading" section at the end of the chapter lists sources of information on ArcGIS.

Organizing and Presenting the Data

Before analyzing data statistically, investigators prepare various displays to describe its nature and summarize the data. Numerical and visual displays are typically prepared for reports. We discuss several of the more common of these in the following sections.

Tabular Presentation

A straightforward way to present data is to report the measurements associated with each case. We do this in an *array*—a listing of each case along with the classification or value for each of the variables measured. An array provides useful information concerning the individual cases but does little to facilitate the analysis of the variables. For example, a researcher may be less concerned with the fire rates of individual cities than with how many cities have high rates and with an average rate for the group of cities. The frequency of high or low values or what constitutes a typical rate cannot be easily ascertained from a data array with a large number of cases. Table 11.2 is a data array for three variables and 20 cases from the Injury Protocol Data Collection Form (see Table 11.1 in the "Data Preparation" section). It would be unusual, by the way, to see a data array included in a report except as an appendix. Data arrays are often so large that they are seldom included in reports. Increasingly, if readers want to analyze the data further, the data array is available electronically.

Frequency Distribution

The first step in organizing data in an array for analysis is to group cases with the same or similar values and count them. This process is called *enumeration* and is the starting point for analysis. If the analyst's concern is with the variable distribution itself rather than with the value of individual cases, then developing a frequency distribution is the next step.

A *frequency distribution* lists the values or categories for each variable and the number of cases with each of the values. A *univariate distribution* is the distribution of a single variable. If the variable is the number of fires per 1,000 households per year and the population is the 340 towns and cities of a state, then the distribution consists of the 340 separate fire rates, as shown in Table 11.3. Remember, however, that one table may include several univariate distributions.

Table 11.2 Data Array for Three Variables: County, Cause, and Severity of Injury

Case	County	Cause	Severity
01	Baker	Fall	3
02	Charlie	Car	4
03	Charlie	Violence	6
04	Able	Car	4
05	Charlie	Violence	5
06	Baker	Fall	9
07	Charlie	Car	10
08	Baker	Fall	1
09	Able	Violence	5
10	Charlie	Violence	5
11	Charlie	Fall	7
12	Able	Car	4
13	Charlie	Car	7
14	Baker	Fall	6
15	Able	Fall	3
16	Baker	Car	5
17	Charlie	Car	5
18	Baker	Fall	6
19	Able	Car	4
20	Charlie	Violence	7

Table 11.3 Frequency Distribution of Fire Rates for Cities in Southeastern States 1998

Fire Rates (Number of fires per 1,000 buildings)	Number of Cities
0–1.99	49
2–3.99	87
4–5.99	112
6 and above	92
Total	340

Table 11.4 Frequency and Percent Distribution for County of Injury

County	Frequency	Percent
Able	5	25
Baker	6	30
Charlie	9	45
Total	20	100%

The categories in the frequency distribution must be exhaustive and mutually exclusive. The categories for any variable must be defined or set up in such a way that each case will be counted in only one category. Also, there must be a category into which each case will be counted.

Percent or relative frequency distributions are often included in the same table with frequency distributions and, as in Table 11.4, show what proportion of the total number of cases has a particular value or is within a given category. A frequency distribution should show the total number of cases. If a percent distribution is presented without the accompanying frequency distribution, enough information should be given so that the reader can determine the number of cases represented by each of the percents.[16]

Class Intervals

The letter f, for frequency, indicates the number of cases with each value or category of the variable. The letter N stands for the total number of cases in the distribution and equals the sum of the category frequencies. Developing a frequency distribution for a variable often requires the analyst to decide how many categories of the variable to present and how wide to make these categories, called *class intervals*.[17] If there are a large number of cases and many individual values, the values may be grouped into a smaller set of class intervals. A distribution for income, for example, could have so many values that presenting them all separately would not be very useful. Rather, an analyst would group, or collapse, the cases into a smaller number of income categories.

In the example of the fire rates for a number of cities, the investigator may have nearly 340 different values, because the rates include decimal values—for example, 5.75 fires per 1,000 buildings. The analyst needs to reduce the large number of separate rates presented in a data array to a much smaller number of categories for the frequency distribution. Table 11.3 illustrated this.

The intervals should allow the reader to make comparisons among them. If the variable being grouped is discrete, the highest value of one category is clearly distinct from the lowest of the next. If the variable is continuous, the values of the categories could seem to overlap. For continuous variables, class intervals include the left endpoint but not the right. In Table 11.3, the first interval contains 0 and up to, but not including, 2; a city with a fire rate of 1.99 would be in the first group, whereas a city with a fire rate of 2.0 would be in the second group. The intervals should not be so broad that important differences are overlooked. Nor should they be so narrow that too many intervals are required. The number of cases in the data array and the range of values in the data suggest to the researcher how to set up the class intervals.[18] The analyst may need to experiment with different widths and choose according to how well the shape of the distribution is best communicated.

Table 11.4 is a frequency and a percent distribution for the variable "county of injury" developed from the array in Table 11.2. The frequency distribution lists each value or category of the variable in one column and the number of cases with that value—the frequency—in another. Note how this table is fully labeled.

Cumulative Frequency Distribution

A cumulative frequency distribution shows how many cases are below a certain value of the variable. Table 11.5 shows a frequency and cumulative frequency distribution for the ages of a number of employees.

The entries in the Cumulative Frequency column are obtained by adding the number of observations in an interval to the total number of observations from the first interval through the preceding one. For example, Table 11.5 shows that 39 employees are under 50 years of age.

Tables are commonly used to present data arrays and frequency and percent distributions. Students have produced and interpreted tables since grade school. However, tables can be easily misinterpreted and are sometimes confusing. Analysts must take care that their tables can be easily and correctly understood. Administrators must interpret correctly the various tables they are presented.

Conventions important in constructing tables are:

1. Provide a descriptive title for each table. This applies as well to charts, graphs, and figures.
2. Label variables and variable categories. All rows and columns should also be fully labeled.
3. All terms open to interpretation should be defined in the footnote section below the table.
4. The source of the data should be indicated in the footnote section.

Visual Presentation of Data

Visual presentations of data often can illustrate points more clearly than do verbal descriptions. In addition to tables, charts and graphs help the analyst see what is in the data as well as presenting them to an audience. The analyst should prepare a picture of the data for his own use. Visual presentations can be more precise and revealing than conventional statistical computations. Maps showing the distribution of variable values and frequencies are also very useful and easily produced with GIS software. The injury data from Table 11.2, for example, could be shown on a map of the counties involved. Visual displays should:

Table 11.5 Frequency and Cumulative Frequency Distribution of Ages of Employees

Age in Years	Frequency	Cumulative Frequency
20–29	9	9
30–39	14	23
40–49	16	39
50–59	21	60
Total	60	

- show the data
- entice the reader to think about the substance of the information
- avoid distorting what the data have to say
- make large datasets coherent
- encourage the eye to compare different pieces of data
- serve a clear purpose—to describe, explore, tabulate, or elaborate
- enhance the statistical and verbal descriptions of the data[19]

Develop the habit of fully labeling each table, graph, and other display. Although each graph, table, and figure in a report should be discussed, if only to identify it and its purpose, readers should be able to comprehend and interpret each without reference to the discussion.

Several conventions have been developed for tables showing the joint distribution of two variables. Usually the independent variable is the column variable, with its label and categories along the top of the table. The dependent is the row variable, with its categories forming the rows. The frequencies in each column of the independent variable are then totaled at the bottom. The frequencies in each row of the dependent variable are added across and shown at the right side of the table.

Bar Graphs

Bar graphs are a particularly effective and simple way to present data. A *bar graph* shows the variable along one axis and the frequency of cases along the other. The length of the bar indicates the number of cases possessing each value of the variable. Figure 11.2 is a bar graph for the variable "county of injury" from Tables 11.2 and 11.6. It is a graphical representation of the frequency distribution for that variable.

The bars in a bar graph should be of the same width for all categories of the variable. It can be misleading to show bars of different widths for different categories. In fact, that technique is used at times to intentionally mislead the reader.[20] Bar graphs also are used to present percentage distributions for variables.

Bar graphs are sometimes developed with the bars displayed horizontally so that the frequency is indicated by the length of the bar. In this case, the value labels for the variables would be placed along the side of the vertical axis, as in Figure 11.3.

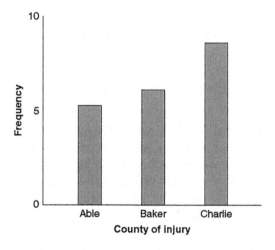

Figure 11.2 Bar Graph for County of Injury

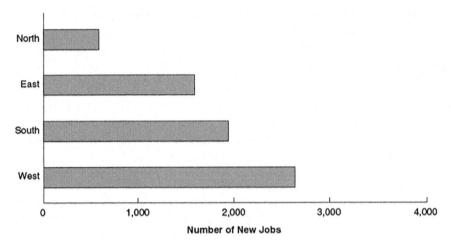

Figure 11.3 Number of New Jobs by Sector, 1984–1990

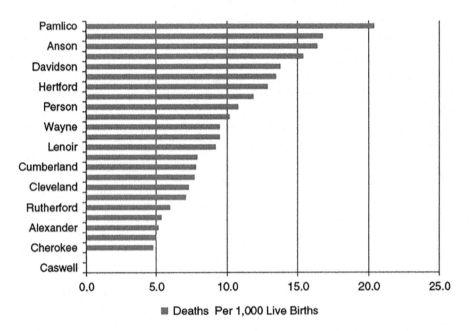

Figure 11.4 Infant Death Rate in a Sample of NC Counties, 2012

Many computer programs generate bar graphs with horizontal rather than vertical bars, and some authors prefer this. Whether the bars are presented vertically or horizontally depends on several factors, including the number of categories of the variable, the length (or height) of the bars, and the amount of variation in frequency that must be displayed. The horizontal format is useful when there are many categories, as it provides more room for the category labels. The choice of horizontal or vertical bars depends on which display communicates more effectively and clearly.

Figure 11.4 is a horizontal bar chart of the infant mortality rate in a sample of North Carolina counties. The bars show the mortality rate in descending order, making it easy for administrators and elected officials to note the rates and to compare counties.

Histograms

A *histogram* is a type of bar graph made by plotting frequency of occurrence against the values obtained for interval- and ratio-level variables. It is the common way of representing frequency distributions in statistics and also is used for relative frequency distributions, that is, those showing percents. With the histogram, the width of the bar also portrays information, so that the width of the bar and its length or height are related.[21] Each column of the histogram represents a range of values, and the columns adjoin one another because the range of values is continuous. The variable and its values are displayed along the horizontal axis, and the frequency or percentage is displayed along the vertical axis of a histogram. Figure 11.5 is an example of a histogram. A histogram represents numbers by area, not height. Histograms differ from bar graphs in that the width of the bars represents specific quantitative values or ranges of values. Although histograms are used for percentage distributions, the percents are not plotted unless the class intervals are equal.[22]

Sometimes a frequency polygon is used as an alternative to a histogram, although both can display the same information. The two are closely related in the type of information and the way that information is presented. A frequency polygon has as its horizontal axis a scale showing the values of an interval- or ratio-level variable. Frequencies or percentages are placed on the vertical axis. A line is formed by marking a position, a "dot" over the value of the variable at the height corresponding to the number of cases—frequency—possessing that value and connecting these dots. Alternatively, the height of the line could indicate the percentage of cases possessing any value of the variable. For a distribution with many individual values, the line will appear continuous and smooth. The area under the line—the polygon—is important.

The polygon in Figure 11.6 is an example of an important type of distribution, the normal curve. It can be described by a mathematical formula that allows analysts to make precise observations about the population whose values are displayed.[23] The normal curve will be discussed in more detail near the end of this chapter. Figure 11.7 is a histogram showing the same information with the addition of more values. Note how it also forms the shape of a normal curve. As more values are represented by bars, the histogram becomes closer to forming a smooth curve.

The polygon in Figure 11.6 shows the number of cases with IQ scores in increments of 10. The histogram in Figure 11.7 presents IQ scores in increments of 5. The resulting graph looks more like a smooth normal curve.

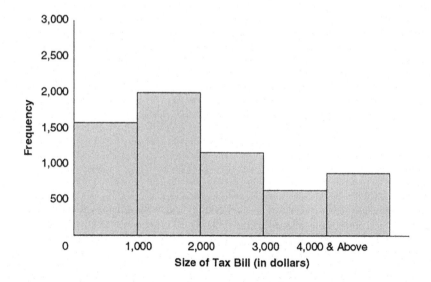

Figure 11.5 Number of Families by Size of Tax Bill in Southwood

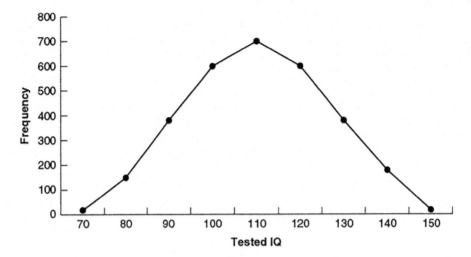

Figure 11.6 Frequency Polygon: Tested IQ of a General Population

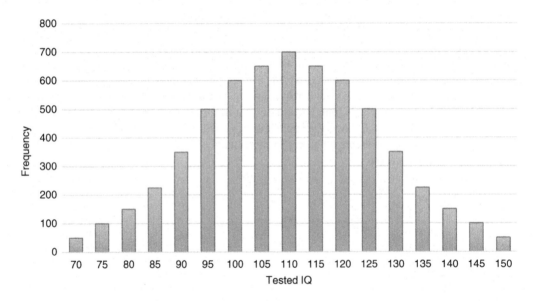

Figure 11.7 Histogram: Tested IQ of a General Population

Pie Charts

A *pie chart* represents a complete circle—indicating a quantity—that is sliced into a number of wedges. This graph conveys what proportion of the whole is accounted for by each component and facilitates visual comparisons among parts of the whole. The portions usually are expressed as percentages of the whole. Financial information, such as source of revenue, category of expenditure, and so on, is often displayed in this manner. The circle represents 100 percent of the quantity of the resource or other factor displayed. The size of each wedge or "slice of the pie" corresponds to a percent of this total. Consult the budget document for your city or county, and you are likely to see information displayed in one or more pie charts. Figure 11.8 is a pie chart showing the sources of revenue for a city government. Some authors advise that, when possible, the largest slice of the pie should begin at the 12 o'clock position and the other slices should follow in a clockwise direction according to size, with the smallest slice last. Analysts may violate this rule when two pie charts are used for comparison.[24] Although no formal rule exists for the proper number of slices in a pie chart, analysts should take care not to have "too many slices." A very large number of slices can confuse the reader and even mask the information presented.

As a contrast, the same information can also be shown by a bar graph, as in Figure 11.9; however, the pie chart more effectively illustrates the relationship of parts to the whole and allows

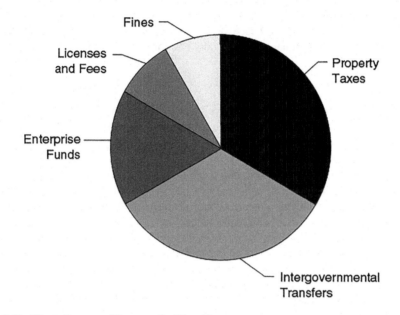

Figure 11.8 Pie Chart: Sources of Revenue for West City

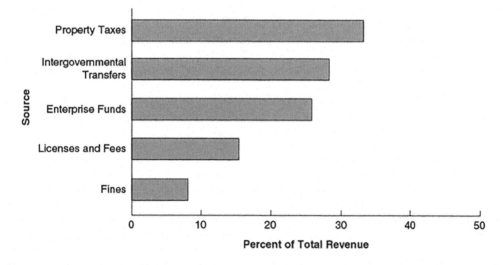

Figure 11.9 Bar Graph: Sources of Revenue for West City

the reader to more readily see the relative sizes of various categories. Pie charts are commonly used to enhance presentations. Many computer programs, such as spreadsheets and data analysis programs, can generate them.[25]

In Figure 11.9, note how the bars are arranged in declining order of magnitude. The longest bar is at the top, indicating the largest source or frequency, and the other bars become shorter as we look down the chart. This makes it easy to spot the largest and smallest revenue sources. Arranging bars this way is particularly effective in charts with numerous bars, as in Figure 11.4.[26]

Line Graphs

Line graphs used for displaying frequencies are primarily of two types: the frequency polygon, described earlier, and the time series, which has been addressed in Chapter 2. The frequency polygon displays the number of cases for each value of a variable, whereas the time series shows the value of a variable over subsequent time periods. The following sections discuss the time series.

Time Series

In a time series, the units of time are displayed along the horizontal axis, and the frequency of some occurrence or the values of a variable are scaled along the vertical axis. The time scale marks the passage of time in units such as days, months, or years. Figure 11.10 shows a line graph of a time series—county property tax levy by year.

For many time series, the occurrences being enumerated are given in *rates*. Time-series graphs of accident rates, crime rates, unemployment rates, and so forth allow the analyst to compare these factors from year to year without worrying about changes in population, because these changes are accounted for in the process of calculating rates.

Quantitative Measures

Numerous measures are available for summarizing and comparing datasets, as well as for measures of individual quantities. These include traditional statistical measures such as arithmetic means as well as newer measures used for specific purposes.

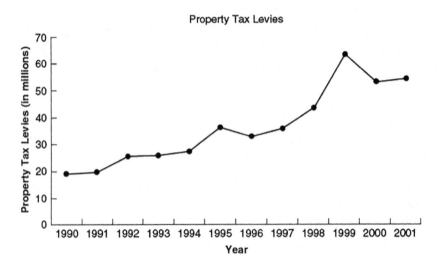

Figure 11.10 County Property Tax Levy by Year

Percentage or Relative Frequency Distribution

The most well-known relative frequency is the percent or percentage. Almost everyone is familiar with percents and can quickly understand presentations based on them. A *percent* reports the number of units as a proportion of 100. It is calculated by dividing the frequency of one category of the variable by the total number of cases in the distribution and multiplying the result by 100. Percents are especially useful when the analyst wants to compare the frequency of the part to the whole or compare frequency distributions with different numbers of cases.

Proportion

Parts of a whole are sometimes given as proportions. A *proportion* is the same as a percentage, except we do not multiply it by 100.

The formula for a proportion is: proportion = f/N.

For example, the proportion of all injuries in the distribution in Table 11.5 that occurred in Able County is .25.

Percent Change

Analysts are often interested in the degree of change in the frequency or amount of something over time. The changes in budget amounts, tax collections, population, unemployment, and so forth are important to track over time. A relative frequency that provides a useful measure of change is the *percent change*. This converts to a percentage the amount of change in the value of a variable between two points in time for a case. The formula is:

Percent change = $((N_2 - N_1)/N_1) \times (100)$

The value of the variable at the earlier time (N_1) is subtracted from the value at the later time (N_2), and the difference is divided by the earlier value. The result is multiplied by 100 to produce a percent.

For example, over a 10-year period, 1990 to 2000, the population of Mecklenburg County increased from 511,433 to 695,454. By what percent did the population increase?

This is a 36 percent increase in population.

$N_1 = 511,433 : N_2 = 695,454$; Percent change = $((695,454 - 511,433)/511,433)$
$\times 100 = (184,021/511,433) \times 100 = 35.98 = 36\%$

Of course, the percent change could be negative, indicating a decrease in the frequency or value. Recently in a small city in Texas, property-tax collections declined from \$13,300,000 to \$12,700,000 from one year to the next. This is a 4.5 percent decrease.

Do not attempt to find an average yearly rate of change by dividing the percent change over a period of time by the number of years involved. Think of the percent change as an interest rate that is compounded at the end of each year. The base on which each succeeding year's growth is calculated changes from year to year.

In the previous example of population change in Mecklenburg County, for example, it would not be accurate to calculate an average yearly growth rate by dividing the total percent change for the decade by 10. The appropriate statistical measure to use in this case is the geometric mean. We will discuss this after discussing the arithmetic mean. Logarithms can also be used to determine the yearly average growth.

Ratio

A *ratio* compares the frequency of one variable category to the frequency of another variable category. Patient-to-physician ratios and student-to-teacher ratios are common. Often the denominator is reduced to 1; for example, the patient-to-physician ratio for a state may be 240 to 1, or the student-to-teacher ratio for a school system may be 30 to 1. Administrators can compare these ratios for different years or for different communities. We may also compare the number of cases in one category of a variable to that in another. For example, assume we found that an organization employed 45 men and 15 women; we would express this as a ratio of male to female employees of 3 to 1.

Rates

Rates are defined as the number of actual occurrences of an event divided by the number of possible occurrences of that event over some period of time. To compare the number of occurrences of an event, such as crimes or traffic accidents in a year for different cities or states, we evaluate the number of occurrences by standardizing the measure. Do this by dividing the frequency of an event in a jurisdiction by the total population of that jurisdiction.

For example, one year, Perth County had only 24 automobile accidents, whereas Maltiby County had 384. Which county had the greater accident problem? Directly comparing the number of accidents would not be useful, since Perth County had only 14,800 inhabitants, while Maltiby County had 307,700 inhabitants. Comparing the frequency of automobile accidents in each county relative to the population of that county is more useful.

Dividing the number of automobile accidents by the population for each county results in ratios of 24 to 14,800 or .001621 to 1 for Perth County and 394 to 307,700 or .00128 to 1 for Maltiby. The large imbalance between the numerator and the denominator makes it difficult to comprehend these figures. The decimal values are so small that readers and analysts find it hard to understand them. These problems can be corrected by multiplying by a base number that will convert the numerator from a decimal to a whole number. Multiplying by 1,000 as a base number produces 1.62 accidents per 1,000 of population for Perth County and 1.28 per 1,000 of population for Maltiby County. This is clearer and makes comparing these rates between counties over time much more meaningful. Table 11.6 illustrates this.

Rates and Denominators

A rate divides one frequency count N_1, by another frequency count, N_2, and multiplies the result by a base number. The numerator, N_1, in the rate formula contains the count for the variable of interest; for an accident rate, it is the number of accidents; for the murder rate, it is the number of murders. The denominator, N_2, is the population size or another indicator of the number of cases at risk. The formula for a rate is as follows:

$$\text{rate} = (N_1/N_2)(\text{base number})$$

Table 11.6 Automobile Accidents for Two Counties, 1990

County	Number of Accidents	Population	Rate
Perth	24	14,800	1.62 per 1,000 of population
Maltiby	394	307,700	1.28 per 1,000 of population

The selection of the denominator for the rate is very important but somewhat arbitrary. For many rates, the denominator is the size of the population. However, an analyst could compare the frequency of automobile accidents in a county relative to the number of automobiles in the county. Automobile death rates also are expressed in terms of millions of miles driven. Rates also are given for the size of the population at risk. The rate of a childhood disease, for example, means more if it is given relative to the number of children in the population rather than for the entire population.

Selecting the Correct Denominator for Rates

Example 11.1 illustrates how the choice of a denominator for the rate can influence our perspective on the subject variable.

Example 11.1 Selecting a Denominator for a Rate

Policy analysts in a large southeastern state investigated the potential impact of a proposed law requiring hunters to wear orange reflective clothing. Opponents of the proposal pointed out that the rate of hunting deaths in the state was very low and did not differ from states with a reflective-clothing law. They used the state's population as the denominator for the rate calculation. A more useful denominator might have been the number of hunting licenses issued. This would provide a more useful rate to compare with other states having a law requiring hunters to wear orange clothing.

Consider the following information from States A and B in Table 11.7.

Table 11.7 Different Base Information for Rate Calculations

State	A	B
"Wear orange" law	No	Yes
Number of hunting fatalities	9	5
State's population	5,500,000	3,200,000
Number of hunting licenses issued	18,100	16,900

Calculate the following rates for both states (use 100,000 as the base number):

1. Number of hunting fatalities per population
2. Number of hunting fatalities per licensed hunter

Rate calculation answers:

1. **A.** 16 per 100,000 population   **B.** 15 per 100,000 population
2. **A.** 49.7 per 100,000 hunters   **B.** 29.6 per 100,000 hunters

Do you think that the "wear orange" law is a good idea? Do you see how the choice of a denominator for the rate calculation influences the perspective?

Source: *The Charlotte Observer* (Charlotte, NC, December 28, 1986).

The following conventions apply in selecting an appropriate base number. Remember that these are guidelines, not rigid rules.

1. Be consistent with rates already in general use for given properties. Be consistent in your own use of rates for the same properties. Crime rates, for example, are usually given as crimes per 100,000 of the population. Birth and death rates, however, are usually given per 1,000 of the population.
2. The base number should produce a rate with a whole number of at least one digit and not more than four digits.
3. The base number is usually an exponent of 10 (10, 10^2, 10^3, and so forth).
4. Use the same base number when calculating rates of different units for comparison. For instance, in the case shown in Table 11.7, you would not use a base number of 1,000 for Maltiby County and a base number of 100 for Perth County.

Incidence and Prevalence

Rates are reported for birth, death, crime, accident, illness, unemployment, and countless other variables. Note that a rate is meaningful only if it is specified for a particular time period, usually a year. Many of the terms used in the discussion of rates come from the various fields of health care. Incidence and prevalence (see Table 11.8) are two of the most common rate measures developed by health care professionals.

We can apply these concepts to the occurrence of conditions other than disease, for example, being a victim of a crime, having an accident, being unemployed, and so on.

An incidence rate is calculated by dividing the number of people who get a disease or a condition over a specified period of time relative to the total number who are at risk of getting the disease or condition in the same period of time. In calculating incidence, only the new cases for the time period in question are included. If we investigate the incidence of a chronic disease, such as diabetes, for 2018, we would include in our rate calculations only the number of new cases occurring in 2018. Cases carrying over from 2017 or previous years would not be counted.

In measuring prevalence, cases carrying over from previous years are included in the count. Those individuals who developed diabetes prior to 2018 and in 2018 and still had the condition would be included in the 2018 prevalence count. Incidence and prevalence rates are usually calculated for a year's time. The number of people in the population of interest at midyear is then used as the denominator. Prevalence rates can be determined using a cross-sectional design; incidence rates typically require a longitudinal design. Usually a panel or a cohort design is used for incidence with a cohort design the more common. Ideally, to determine incidence, an analyst would follow a population over time.[27]

We are all aware of the discussion of AIDS. Public health and other officials are concerned with both the number of new cases each year and the total number of people with AIDS. The first of these concerns involves incidence, and the second involves prevalence. Another example is

Table 11.8 Incidence and Prevalence

Term	Definition
Incidence	Is defined as the number of people who get a disease or other condition during a specified period of time, usually a year.
Prevalence	Is defined as the total number of people who have a disease or condition at a given time.

the statistics on epilepsy in Olmsted County, Minnesota, where one year, the incidence was 30.8 cases per 100,000 of population, and the prevalence was 376 cases per 100,000 of population.[28]

Characteristics of a Distribution

In summarizing a distribution of values, analysts usually want to provide several kinds of information and typically use two types of statistics to describe the distribution of one variable:

1. measures of central tendency
2. measures of dispersion

These categories represent the two most important things an analyst usually wants to show about a group of values: how similar the individual values are and how different they are. Several measures are included in each of these categories. The choice of which to use depends on the level of measurement of the variable and the information the analyst wants to obtain. Other types of measures also are used by statisticians, but we do not discuss them here.

The measures of central tendency indicate what a typical value or case in the distribution is like. One might ask how well the measure of central tendency reflects the overall nature of the distribution. This leads to measures of the variability of the values in a distribution. Measures of variability include those of spread, variation, and dispersion. These, along with measures of central tendency, are useful for measuring and comparing groups.

Measures of Central Tendency

Measures of central tendency are used to indicate the value that is representative, most typical, or central in the distribution. These measures include mode, median, and arithmetic mean.

Mode

The simplest summary of a variable is to indicate which category is the most common. The *mode* is that value or category of the variable that occurs most often. In a frequency distribution, it is the value with the highest frequency. Table 11.9 shows the distribution of a nominal-level variable. The mode for this variable is "Council manager": more cities in the table have this form than any other. Remember that we find the mode by determining which category or value has the highest frequency. The mode can be determined for all levels of data:

- nominal
- ordinal
- interval
- ratio

Table 11.9 Frequency and Percent Distribution for Type of City Government

Type of Government	Frequency	Percent
Strong mayor	22	25
Council manager	45	51
Weak mayor	17	19
Commission	4	5
Total (*N*)	88	100

A mistake that students often make is to confuse the frequency of the modal category with the mode. For instance, the mode for Table 11.9 is "Council manager"; it is not 45. For nominal and many ordinal variables, the category that occurs most often will have a name; for interval and ratio variables, the value of the mode will be a number.

Median

The *median* is the value or category of the case that is in the center of the distribution. It is the value of the case that divides the distribution in two; one-half of the cases have values less than the median and one-half of them have values greater than the median. The median requires that variables be measured at the ordinal or interval level. It can be determined only if the values can be ordered. It makes no sense, by the way, to find the middle case if the cases have not been ordered according to their values on the variable of interest.

To find the median, we must locate the middle case in a distribution. To find the middle case, we add 1 to the number of cases and divide by 2, or: $(N + 1)/2$. If the number of cases (N) is odd, then the median will be the value of a specific case. If N is even, then the median is estimated as the value that is halfway between two cases. For example, if N is 21, then the median is the value of the 11th case. If N is 22, then the median would be halfway between the value of case number 11 and case number 12. Several examples follow.

Table 11.10 shows the distribution of an ordinal variable from a sample survey. The table also includes a column showing the cumulative percent, that is, the percent of cases with a particular value or less. The middle case is the case $(629 + 1)/2 = 315$. This case is in the rating category "Good." The median for this variable, then, is "Good." One-half of the cases rated county government as "Good" or better, and one-half rated it as "Good" or worse. Note that the cumulative percent reaches 50 in the "Good" category. Cumulative percent information allows us to quickly locate the category containing the middle case.

Table 11.11 shows the distribution of a small number of values—the amount of investment income for a year for a number of employees. We will discuss how to calculate the median income for this group. The first part of Table 11.11 shows the income data as they were collected; that is, the values are not in any particular order. To find the median, the cases must be ordered according to increasing or decreasing values. Sometimes, people unfamiliar with calculating the median overlook this. The second part of Table 11.11 orders the amounts from lowest to highest. The median is the amount of the middle case in the ordered set. Since the number of cases N in the distribution is 7, the middle case is $(N + 1)/2 = 8/2 = 4$. The value of the median is the investment income of that case, $24,000.

Table 11.10 Rating of County Government

Rating	Frequency	Percent	Cumulative Percent
Poor	36	5.7	5.7
Fair	206	32.8	38.5
Good	287	45.6	84.1
Excellent	100	15.9	100.0
Total (N)	629	100.0	

Table 11.11 Yearly Investment Incomes of Seven Employees (in Dollars)

(1) Unordered:	17,000	25,000	30,000	12,000	24,000	18,000	27,000
(2) Ordered:	12,000	17,000	18,000	24,000	25,000	27,000	30,000

Consider how the median would be affected if one more case were added.

In Table 11.12, N is 8, an even number, so the middle of the distribution is between case number 4 and case number 5. We would place the median value halfway between the values of these cases. This value would be $(24,000 + 25,000)/2 = 24,500$.

Notice how little the extreme value of the highest amount in the distribution affects the median. In Table 11.12, the one income amount of $58,000 is substantially higher than any of the others. However, the median is only $500 higher than the median of the distribution in Table 11.12. This is one of the most important characteristics of the median. An additional property is that when used with interval and ratio variables, the median is closer to all other values than any other point in the distribution.

The median should be used as the measure of central tendency for ordinal variables. It is also useful for interval variables if the distribution has a few extreme values, as in Table 11.12. A distribution such as this with a few extreme values is said to be skewed. The most serious problems of skewing seem to occur with variables measuring resources. These include income, in particular, but also the values of stock, land, and other real estate owned by individuals and families. Since the median is affected little by extreme numerical values, it gives an accurate picture of central tendency even for highly skewed distributions.

Computing the Median From Grouped Data

It is often necessary to compute the median from grouped data, that is, data already organized and presented in a frequency distribution with class intervals more than one unit wide. Table 11.13 shows such a distribution. Since we are not able to identify the individual values in the frequency distribution, we cannot find the exact value of the median. One method of estimating the median is to find the class interval containing the middle case and take the midpoint of that interval as the median. The middle case in Table 11.13 is in the third class interval, the interval with ages 40–49. The midpoint of this interval is 44.5 years; thus, 44.5 is the estimated median.[29] A more exact procedure for estimating the median from grouped data is explained in most introductory statistics textbooks.[30]

Arithmetic Mean

A third way of measuring central tendency is with the arithmetic average or arithmetic mean. This is probably the most commonly known measure of central tendency. The *arithmetic mean* (mean or $\bar{x}$) is the arithmetic balance point or arithmetic center of the distribution. By that, we mean

Table 11.12 Yearly Investment Incomes of Eight Employees (in Dollars)

12,000	17,000	18,000	24,000	25,000	27,000	30,000	58,000

Table 11.13 Ages of Employees in Central Agency

Age (in years) (X)	Frequency (f)	Midpoint of Interval (m)	(f) × (m)
20–29	9	24.5	220.5
30–39	14	34.5	483.0
40–49	16	44.5	712.0
50–59	21	54.5	1,144.5
Total	60		2,560.0

that the sum of the differences from the mean for values above the mean is equal to differences from the mean of the values below the mean. Although the mean is sometimes the middle of the distribution and has the same value as the median, we cannot always expect this to be so. To use the mean requires variables to be measured at the interval or ratio level.

The mean is created by mathematical calculations involving the values of each case. It is useful by itself as a measure of central tendency but is also an important component of many other statistical formulas. To calculate the arithmetic mean, add the value of the variable for each case and divide by the number of cases. The following is the formula for the arithmetic mean.

$$\text{arithmetic mean} = \bar{X} = \Sigma X_i / N$$

The symbol means to sum the value of the variable for each individual case. X_i is the value of the variable for each case. Calculate the arithmetic mean for the distribution in Table 11.12. The distribution in Table 11.12 would have the arithmetic mean as follows:

$\Sigma X_i = 12{,}000 + 17{,}000 + 18{,}000 + 24{,}000 + 25{,}000 + 27{,}000 + 30{,}000 + 58{,}000 = 211{,}000$
Arithmetic mean = 211,000/8 = \$26,375

For a frequency distribution, the following formula is more useful:

$$\bar{x} = \left(\Sigma f_i x_i \right) / N$$

This directs you to multiply each value of the variable by the frequency of that value (f), then add these values and divide by the total number of cases (N).

Calculate the mean for the distribution in Table 11.11. You should obtain \$21,857. This illustrates how a few extremely high or low values will greatly affect the value of the mean. The reason that the mean for the distribution in Table 11.12 is so much higher than the mean for the distribution in Table 11.11 is the one very high earnings figure of \$58,000. If a distribution has a few extreme values, the mean is not as useful a measure of central tendency, and the median is the preferred measure. Providing both the mean and the median for distributions of interval and ratio variables is often useful, and we recommend it.

Computing the Mean From Grouped Data

As with the median, analysts often want to have the mean for data grouped in class intervals and presented in tables. In this situation, the analyst multiplies the value of the midpoint of each interval of the variable by the frequency of that interval.[31] The resultant values are added and divided by the total number of cases to obtain the mean. The procedure is demonstrated in Table 11.14, which reproduces the data from Table 11.13. The arithmetic mean for this distribution is 42.7 years. The reader should compare this to the value of the estimated median.

Table 11.14 Age of Employees in Central Agency Calculation Table

Age (in years) (X)	Frequency (f)	Midpoint of Interval (m)	(f) × (m)
20–29	9	24.5	220.5
30–39	14	34.5	483.0
40–49	16	44.5	712.0
50–59	21	54.5	1,144.5
Total	60		2,560.0

Geometric Mean

The *geometric mean* is used to average ratios and rates of change. Assume that an agency's budget increased from $100,000 to $150,000 over a period of five years. The change is 50 percent during this time. But how much has the budget grown on a yearly basis? If we used the arithmetic mean and divided the total percentage change by the number of years—five—the average yearly change would be 10 percent. However, this would be too high. If you start out with the beginning budget of $100,000 and increase it by 10 percent each year, the ending budget will be larger than $150,000. (It is a useful exercise to do this.) The actual average yearly growth is less than 10 percent. The geometric mean is used to calculate this. Several sources are available for those who wish to pursue this topic.[32] Our concern here is that the reader understand that the geometric mean is used with ratios and rates of change, especially in calculating the mean percent change over time, giving the annual average growth rate.

Guidelines for Selecting an Appropriate Central Tendency Measure

Spindle County was applying for a community improvement grant for one of its low-income neighborhoods. "But the average income in that neighborhood is over $37,000," said County Commissioner Crane. "Yes," agreed Hap, the county's grants manager, "but one-half of the families earn less than $25,000." Charlie, a member of the neighborhood, commented, "The most common income level here is around $30,000." The grant application required the reporting of an "average income figure." Which average or measure of central tendency should the grants manager include?

The analyst needs to be aware of and understand the different measures of central tendency and their characteristics to know which to use. Since each of the measures provides somewhat different information, she must take care to choose the proper one. The appropriate measure of central tendency depends on the level of measurement, the nature of the resulting distribution, and the information the analyst wants to present. The following guidelines should be considered:[33]

1. The median and the mean require ordinal and interval levels of measurement, respectively. If the variable is nominal, only the mode can be used. Nevertheless, the mode is appropriate when the purpose is to report an actual typical figure. In Spindle County, Charlie is reporting that $20,000 is a typical income. If the distribution has two or more modes, all of these should be reported.
2. In a distribution of an interval or ratio variable, if the distribution is unimodal and symmetrical, or nearly so, the mean is the preferred measure of central tendency. In such a distribution, the mode, median, and mean will be the same or nearly so. If, however, the distribution has a few extreme values, either high or low, the mean will be distorted, and the median should be used. Such a distribution is said to be *skewed*. Figure 11.11(a) shows a distribution positively skewed or skewed to the right. The mean is higher than the median and the mode. The distribution in Figure 11.11(b) is negatively skewed or skewed to the left. The mean is lower than the median and the mode.

Distributions of the measures of income tend to be skewed. For example, in any community, most people will have relatively low or moderate incomes. But a few people or families will have very high incomes. These high values will distort the mean. This has probably happened in the neighborhood in Spindle County. The grants manager would want to report the median, as it will be affected less by any extreme income values.

3. The geometric mean has an explicit application. It is used primarily to average rates of change and does not apply to the Spindle County example.

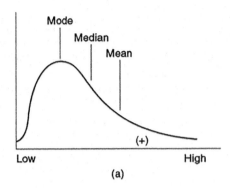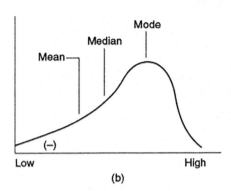

Figure 11.11 Examples of Skewed Distributions

Unimodal distributions that are not symmetrical are skewed. Extremely high or low values in a skewed distribution pull the mean away from the median toward the extreme value. The more the degree of skew, the greater the difference between the values of the mean and the median. The mean may give a misleading figure for a skewed distribution. The median should also be provided, and the analyst should compare the median and the mean to judge if the distribution is skewed.

Measures of Variation and Dispersion

Two distributions may have similar means yet very different values overall. So in addition to measures of central tendency, analysts present measures of dispersion and variation that indicate the extent to which individual values in a distribution are different from each other. *Measures of dispersion* describe the uniformity of the data. Relatively smaller values on the measure of dispersion for a variable imply more uniformity, whereas relatively larger values imply more diversity or variation. Some measures indicate only the difference between two observations in an ordered set of values. Other measures consider all observations in a distribution.

In comparing distributions of income for two groups of employees, measures of central tendency would be used. The average incomes of the two groups shown in Table 11.15 are nearly equal, but, as a visual inspection of the data will confirm, the individual incomes are very different. Measures of dispersion must be calculated to provide complete information.

Maximum variation for ordinal, interval, and ratio variables is defined as occurring when all cases are equally divided between two extreme categories or values. Maximum variation for nominal variables occurs when cases are evenly distributed across all categories. The more the cases are clustered in one category, the less the variation. If all cases were in one category, then variation would be zero.[34]

Common measures of dispersion for interval and ratio variables are the range, midranges (such as the inter-quartile range), standard deviation, and variance. Somewhat less common but useful as quantitative measures are the average deviation and median absolute deviation.

Table 11.15 Data and Arithmetic Means for Investment Incomes of Two Groups of Employees (In Dollars)

Group A:	12,000;	17,000;	18,000;	24,000;	25,000;	27,000;	30,000;	58,000	$\bar{x} = 26,375$
Group B:	22,500;	24,500;	25,000;	26,000;	27,000;	28,000;	28,500;	29,000	$\bar{x} = 26,312.50$

Range

The simplest measure of dispersion is the *range*. It is the difference between the highest value and the lowest value in a distribution and indicates the range over which the values are spread. However, it has many disadvantages, since one extreme value will greatly affect its size. Note the ranges for the investment income distributions from Employee Groups A and B from Table 11.15.

Group A Range = Highest value − Lowest value = $58,000 − $12,000 = $46,000
Group B Range = Highest value − Lowest value = $29,000 − $22,500 = $6,500

These range figures make it obvious that although the two distributions have similar arithmetic means, they are different in other respects.

Because the extreme values in a distribution affect the size of the range, statisticians have developed a number of midrange measures that eliminate some portion of the low and high ends of a distribution. Whereas the range is very sensitive to extreme values, the midrange measures are not. The most common of these is the *inter-quartile range* (IQr). It is also known as the midspread.[35]

Inter-Quartile Range (IQr)

This specifies the range of values within which the middle 50 percent of the observations are found. The *first quartile* is that value below which 25 percent of the cases are found. The *third quartile* is that value below which 75 percent of the cases are found. The *second quartile*, of course, is the median. In determining the inter-quartile range, the lowest 25 percent of the observations and the highest 25 percent are omitted. The resulting range is less affected by extreme values than is the range (see Table 11.16).

Note the inter-quartile ranges for the incomes of the two groups of employees shown in Table 11.15. IQr (distribution A): Range of the middle 50 percent of values, for example, the four values between $18,000 and $27,000. To find the first quartile, drop the lowest 25 percent of the cases, that is, the two cases with the lowest values. To find the third quartile, drop the upper 25 percent of the cases, that is, the two cases with the highest values. For this distribution, the inter-quartile range = $27,000 − $18,000 = $9,000. The spread of the middle 50 percent of the cases is $9,000. Compare this to the IQr for distribution B. IQr (distribution B): $28,000 − $25,000 = $3,000. The middle 50 percent of the cases are spread over a range of $3,000.

Calculating Quartiles and the Inter-Quartile Range

The inter-quartile range indicates the extent of variation among individual values. It is useful as a means of comparing the variation or spread of two or more distributions measuring the same variable. Many analysts use it to compare distributions over time or to compare the distributions for two or more groups. For example, in a quality-improvement program, we might record the number of errors made by accounting clerks. To show success, we would expect that the inter-quartile range would become smaller over time and that those involved in the program would show

Table 11.16 Quartile

Quartile	Value
First quartile	Value below which 25 percent of the cases are found
Second quartile	Median = value below which 50 percent of the cases are found
Third quartile	Value below which 75 percent of the cases are found

a smaller variation than those not involved. By itself, however, without a context or perspective, the inter-quartile range has little meaning. A more precise method for calculating quartiles and the inter-quartile range is shown next.

Finding Quartiles

The following is a more precise method for determining the value of quartiles of a distribution. For ungrouped data, such as in the example array from Group A in Table 11.15, the first and third quartiles may be found as follows:

First Quartile (Q1)

Find the value of the ordered observation number: $(N + 1)/4$

For this example, it would be $(8 + 1)/4 = 2.25$. The value of the quartile would be between two observations. Case number 2 has the value 17,000, and case number 3 has the value 18,000. The quartile would be 25 percent of the way between 17,000 and 18,000, or 17,250.

Third Quartile (Q3)

Find the value of the ordered observation number: $3(N + 1)/4$

For this example, it would be $3(8 + 1)/4 = 6.75$. This value also comes between two observations: case numbers 6 and 7. The third quartile would be 75 percent of the way between the values of these two cases, 27,000 and 30,000. This would make it 29,250.

The inter-quartile range is the distance between the third and the first quartiles and would be $29,250 - 17,250 = 12,000$.

From this, it follows that the formula for the second quartile, the median, also may be written:

Median $= Q2 =$ value of ordered case number $2(N + 1)/4 = (N + 1)/2$.

This case is one-half the way between case 4 and case 5, putting the value of the median at $24,500.

Group A: 12,000, 17,000, 18000, 24,000, 25,000, 27,000, 30,000, 58,000

In reporting on frequency distributions with a large number of cases, investigators often quote percentiles of the distribution. A *percentile* is a value below which a certain percent of the ordered observations in a distribution are located. The 90th percentile of a distribution of applicants' job-registry ratings is the rating value below which 90 percent of the applicants fall. Only 10 percent of the applicants equal or exceed the 90th percentile. The 50th percentile, of course, is the same as the median: 50 percent of the values are lower, and 50 percent are higher. The 25th percentile and the 75th percentile are the first and third *quartiles*, respectively, and mark the endpoints of the inter-quartile range. Students are usually familiar with instructors "curving" the exam scores from large classes to assign grades. Those students at the 90th percentile or above get As; those between the 80th and 89th get Bs, and so forth. *Quartiles, percentiles*, and similar measures are called location parameters because they help locate a distribution on the axis showing the variables' values when the distribution is graphed.[36]

Standard Deviation and Variance

Analysts also need a measure that includes all values in a distribution. Two such measures for quantitative variables are the standard deviation and the variance. The *standard deviation* is a measure of the average distance of values in a distribution from the arithmetic mean of the distribution. The *variance* is the square of the standard deviation. Both of these measures are used in higher-level statistical tests; they are two of the most important of all statistical measures. Unfortunately, it is difficult to obtain an intuitive understanding or interpretation of them. The following calculations are presented to help explain them.

Formulas for Variance and Standard Deviation

The formula for the variance is given first, as we usually calculate the variance and then take its square root to find the standard deviation:

$$S^2 = \text{variance} = \frac{\Sigma(X_i - \bar{X})^2}{N}$$

The formula for the standard deviation is the square root of the variance:

$$s = \sqrt{\Sigma(X_i - \bar{X})^2 \, N}$$

The steps for calculating the variance and standard deviation are these:

1. Subtract the mean from each individual value $(X_i - \bar{X})$. This shows how much the value of each case deviates from the mean of the distribution.

2. Square each deviation value $(X_i - \bar{X})^2$.

3. Add the squared deviations $\Sigma(X_i - \bar{X})^2$.

4. Divide the total of the squared deviations by the number of cases $\Sigma(X_i - \bar{X})^2 / N$. This is the variance.

5. Take the square root $\sqrt{\Sigma(X_i - \bar{X})^2 / N}$. This is the standard deviation.

The variance and standard deviation of the salaries in Group A, Table 11.15, are $173.40 and $13.16, respectively.

The formulas presented previously are to be used with a population. If you are working with a sample, the denominator, N, is replaced with $n - 1$. n is the number of cases in the sample, whereas N denotes the number in the population. Many calculators and computer packages have $n - 1$ programmed in the equations used for the variance and standard deviation.[37] If the means of two distributions are about the same and the standard deviations differ, the larger standard deviation indicates greater dispersion in the distribution.

Alternative formulas for calculating the standard deviation and variance are shown and illustrated in the following section. Although these formulas do not demonstrate the principles underlying the measures as well as the earlier discussion, the alternative formulas—next—are easier to use with calculators.

Variance and Standard Deviation Calculation

The following illustrates the calculation of the variance and standard deviation. For both calculations, Formula 1 represents the formula given in the previous section and is the definitional formula. Formula 2 is an alternative that is more efficient if you have a large number of cases and a calculator. Many calculators have a standard deviation routine built in.

Find the variance and standard deviation of the distribution in Table 11.17, a useful calculation setup table. (Note: Some values are rounded.)

Table 11.17 Calculation Table for Variance and Standard Deviation

X *(in thousands)*	X²	$\left(X_i - \bar{X}\right)$	$\left(X_i - \bar{X}\right)^2$
12	144	−14.37	206.5
17	289	−9.4	87.8
18	324	−8.4	70.5
24	576	−2.9	8.3
25	625	−1.4	1.92
27	729	.62	.38
30	900	2.62	13.1
58	3,364	31.6	999.6
Σ 211	6951		1,387
(211)² = 44,521			

Variance: $S^2 = 173.4$

Standard deviation: $S = \sqrt{S^2} = 13.2$

Formula 1.

$$S^2 = \frac{\Sigma\left(X - \bar{X}\right)^2}{N} = \frac{1387}{8} = 173.4$$

$$S = \sqrt{S^2} = \sqrt{173.4} = 13.2$$

Formula 2.

$$S^2 = \frac{\Sigma X^2 - \frac{\left(\Sigma X\right)^2}{N}}{N}$$

$$S = \sqrt{S^2}$$

$$S^2 = \frac{6951 - \frac{44,521}{8}}{8} = \frac{6951 - 5565}{8} = 173.4$$

$$s = \sqrt{173.4} = 13.2$$

These formulae apply to a population. If the data are from a sample, then $n - 1$, the number of cases in the sample, replaces N in the denominator of the formula. In the example here, the denominators would be 7 instead of 8.

Other Deviation Measures

The *average deviation* and the median absolute deviation are two other useful quantitative measures of dispersion, although they lack the mathematical properties necessary for inclusion in other statistics. Unlike the standard deviation, which is used as a component in the formula for many other statistical measures, the average and median absolute deviations are not used in other formulas.

Average Deviation

The *average deviation* measures how far, on average, the cases deviate from the arithmetic mean. To calculate it, add the absolute value of the deviation of each case from the mean of the distribution and divide by the number of cases. The formula is:

$$Average\,Deviation = \frac{\Sigma\left|X_i - \bar{X}\right|}{N}$$

The value for the average deviation for the salaries in Table 11.15, Group A, is: 8.98. (The reader is encouraged to calculate this measure.) For most people, the average deviation is more intuitively appealing than is the standard deviation. Forecasters often use it to estimate the accuracy of models. In a normal distribution, the average deviation is about 20 percent smaller than the standard deviation.[38] The average deviation also is less sensitive to extreme values than is the standard deviation.

Median Absolute Deviation

This measure calculates the average deviation of a set of cases from the median of the distribution. A property of the median is that the total deviations and the average deviation from the median are smaller than from any other point in the distribution. This fact makes the median absolute deviation useful in some financial applications, such as measuring the uniformity of property tax assessments of local governments.[39]

The Standard Deviation and the Normal Curve

The standard deviation is used extensively in inferential statistics. It allows us to estimate population parameters, such as means and variances. From our knowledge of the distribution of sample means, we can use the standard deviation to estimate how close any particular sample characteristic is to the corresponding population parameter.

The standard deviation is a key part of an important type of distribution in statistical analysis. This distribution is the *normal curve* and has the following characteristics:

1. It is bell shaped and symmetrical.
2. The mode, median, and arithmetic mean have the same value at the center of the distribution.
3. A fixed proportion of the observations lie between the mean and any other point.[40]

This last characteristic is very useful. If a distribution is normal, or approximately so, we know what proportion of observations found in it lie between any two values of the variable measured. Typically, these values are measured in standard deviation units above and below the mean. This is illustrated in Figure 11.12.

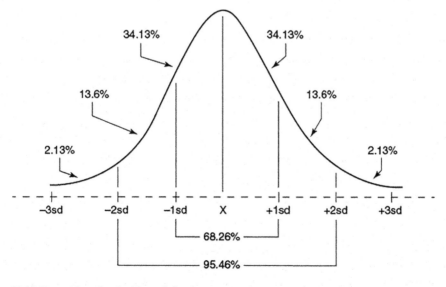

Figure 11.12 Proportions for the Normal Curve

The mean of the distribution divides it exactly in half: 50 percent of the observations are above it, and 50 percent are below. Between the mean and one standard deviation above it lie 34.13 percent of the observations. The same proportion is between the mean and one standard deviation below the mean. Exactly 68.26 percent of all observations in the distribution are between one standard deviation below the main and one standard deviation above the mean.

Standard scores, also called z-scores, are used to measure the values of the individual observations in a normal curve. Standard scores express the values in terms of units of the standard deviation. A z-score of 1.5, for example, indicates that a case's value is 1.5 standard deviations above the mean; a z-score of 2.0 shows that the value is two standard deviations below the mean, and so on. An observation with a z-score of 2.0 would be an unusually low observation relative to the rest of the observations. Standard scores are calculated by the following formula:

$$z = X_1 - \bar{X}/s$$

where:
z = standard or z-score = number of standard deviation units
X_1 = an observation
$\bar{X}$ = arithmetic mean of the distribution
S = standard deviation of the distribution

For example, if the mean of the distribution in Figure 11.12 is 100; the standard deviation is 20; and an individual observation, x, has the value 115, the z-score for that individual would be:

$$Z = (115 - 100)/20 = .75$$

If the observations in the distribution are the means of all possible samples, we can determine how likely it is that any particular mean is within a certain distance of the center of the distribution. The center of such a distribution is the mean of all sample means; it also is the population mean. This information is very useful in inferential statistics, as discussed later in the book. And, as discussed in Chapter 5, it allows us to calculate the size of the standard error and to determine how likely it is that the confidence interval around any particular sample estimate encompasses the population parameter.

Exploratory Data Analysis

The concepts of central tendency and variability are among the oldest in statistics. In addition to traditional ways of treating them, newer techniques have been developed in recent years. Based on the work of statistician John Tukey and others, the field of applied statistics has developed many techniques of exploratory data analysis, often abbreviated as EDA.[41] EDA emphasizes becoming thoroughly familiar with a set of data rather than just computing a set of summary statistics. The adherents of this approach have introduced new terms and devised new measures of center and spread. New diagrammatic and graphical means of displaying distributions also have been developed and adapted for personal computers.

Trimmed Means

New measures of central tendency include order-based measures and some variations of the arithmetic mean, such as *trimmed means*.[42] These measures are helpful in understanding the central part of the distribution of a variable. Variations of the arithmetic mean, such as trimmed means, are not influenced by cases with extreme values, or outliers, as these means are calculated after adjusting or removing such cases. The 5 percent trimmed mean, for example, drops the top 5 percent and the bottom 5 percent of cases before calculating the mean. The midmean uses only the middle 50 percent of the cases. Although these means are useful in comparing the centers of distributions without influence from outliers, some authorities feel that they pay too little attention to extreme values. Also, remember that these measures do not enter into the calculation of other statistics.

Box Plots

Several new graphic procedures for data exploration are now available. They appear to have been motivated both by developments in EDA and by the enhanced graphics capabilities of computers. These techniques and computers provide opportunities for visual analysis of distributions in ways not generally used before.[43]

One popular technique is the *box plot*, also called a box and whisker diagram. This shows, in the same graphic, the median of a variable, its minimum and maximum values, quartile locations, inter-quartile range, and, in some cases, outliers. Box plots give a quick and clear view of both central tendency and spread.

Figure 11.13 illustrates a box plot for the 1999 murder rates from 23 western and southern states in the United States.

The horizontal axis gives the possible values of the variable. The extreme left part of the plot, the end of the whisker, shows the minimum value in the distribution. The maximum value is marked by the extreme right of the plot. The box enclosed by two vertical bars and two horizontal bars marks the middle 50 percent of the cases in the distribution. This, of course, is the inter-quartile range, with the left vertical bar marking the first quartile and the right vertical bar marking the third quartile. The median is marked within the box by an asterisk. Some graphics programs will show the median as another vertical bar.

The minimum, maximum, median, first quartile, and third quartile values are called the *Tukey five-number summary*, after the statistician who developed several EDA measures. These are readily shown on the box plot and often are included in the computer output. For the set of data illustrated by the plot in Figure 11.13, these values are:

minimum = 2.0: maximum = 10.7: median = 6.0: first quartile = 3.4: third quartile = 7.8.

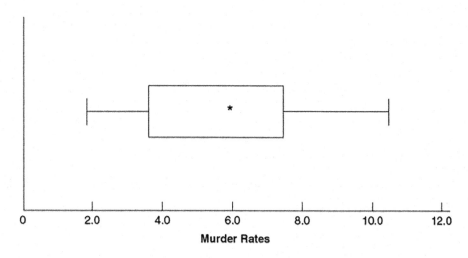

Figure 11.13 Box Plot of 1999 Murder Rates in Western and Southern States

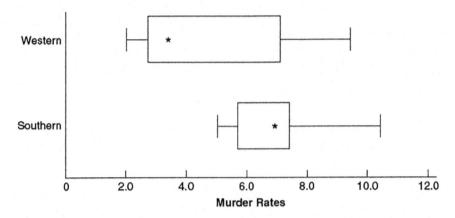

Figure 11.14 Two Box Plots of 1999 Murder Rates in Western and Southern States

Figure 11.14 shows two box plots on the same diagram—one for western states and the other for southern states. Visual comparison of the two reveals much about murder rates in the two groups of states. The data for these box plots are included in Problem 9 at the end of the chapter.[44]

Data Visualization

The presentation of data in pictorial and graphical format is a long-standing and common practice. Visually representing data in charts and maps helps readers to understand information more quickly and easily.[45] As noted earlier in this chapter, with the increased capacity of individual computers, the vast increase in data, and the increased recognition of the need to use data for improved decision making, analysts and decision makers place a greater emphasis on presenting information so that it is best perceived and understood by those reading it. A newer, general term for this is "data visualization."

Data visualization has been explained as: "presenting data to an observer in a way that yields insight and understanding." It is done following "[the study] of human perception and cognition."[46]

Many organizations are using the capability of computer technology and knowledge of how humans view and process visual information to present data in effective ways. Universities and computer vendors have courses on data visualization.[47] Governments have created visualizations to better transmit information to users and other members of the public.[48]

Principles discussed in older works illustrating the impact and usefulness of appropriately displaying quantitative data can now be implemented very quickly.[49]

Summary

After data are collected, appropriate ways of managing and analyzing them must be found. This chapter discusses preparing and organizing data for analysis. Data are usually coded and entered into a computer file.

A data dictionary enables the user to interpret the meaning of the data codes without having to go back to the original collection forms.

Computer Programs

With the increased availability, versatility, and use of personal computers, hundreds of programs to manage, analyze, and store data are available. Three types of computer software available for the administrator and the analyst to manage and analyze data are statistical packages, spreadsheets, and database managers. Statistical packages are perhaps the most useful for thorough statistical analysis of large datasets. Spreadsheets can assist in the entry of small datasets, as a medium of data transfer between application programs, and for some simple statistical calculations. Geographic information systems use geographic data and allow users to show the distribution of other variables on maps produced by the computer. Newer GIS programs are capable of calculating numerous statistics, and new statistical techniques have been developed to take advantage of the spatial characteristics of geographic data.

Frequency Distributions

Investigators typically organize data in frequency distributions for further analysis and presentation. In frequency distributions, cases are grouped by their variable value, and the number of cases with each value is shown. Relative frequencies, especially percents, also are included in frequency distributions. Percents, rates, rates of change, ratios, and proportions communicate information about a variable quickly and effectively. Most people can easily visualize and interpret percentage data. Ratios facilitate comparisons between groups or jurisdictions, whereas percent changes and rates of change indicate how the values of a variable change over time.

Analyzing and Presenting Data

Various graphs and tables are used in analyzing and presenting data. Bar graphs and histograms show the frequency of cases for each variable value with a bar whose height or length indicates the number of cases. Histograms are limited to interval and ratio variables. Line graphs also are used

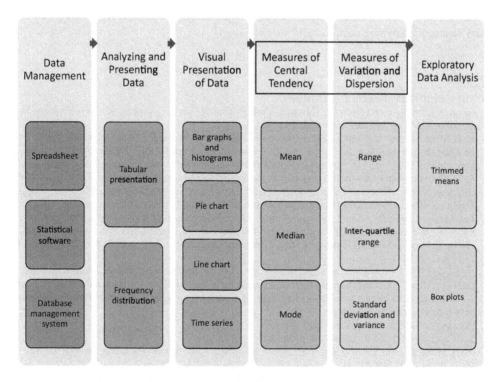

Figure 11.15 Univariate Analysis

to summarize and display interval- and ratio-level variables. Time-series and frequency polygons are two types of line graphs. Frequency polygons show the frequency of cases with each value of a variable. Time-series graphs show how the value of a variable changes over time.

Central Tendency

Measures of central tendency describe one aspect of a frequency distribution. The mode, a nominal measure, is the most frequently occurring value in a distribution. The median, an ordinal measure, is the value at the center of a distribution's cases. It should be used with interval and ratio variables when the distribution is skewed, as its value is not affected by extreme values. The arithmetic mean, an interval measure, is preferred to summarize those interval variables that are neither bimodal nor skewed. The mean has properties that make it an important measure for many statistical procedures.

Dispersion

Measures of dispersion indicate how well the measure of central tendency describes a variable. A number of measures of spread, such as the range and inter-quartile range, help assess the configuration of a distribution. The variance and standard deviation describe the degree of variation in the distribution of interval and ratio variables. The variance and standard deviation are important components of other statistical measures and are useful in estimating population parameters.

Exploratory Data Analysis

Recent work in statistics has produced measures for exploratory data analysis. This approach emphasizes the benefit of becoming thoroughly familiar with a set of data. Graphic techniques, such as box plots, based on this approach are readily produced by computers, providing additional visual means of analyzing data.

Notes

1. The County Data File is in a spreadsheet.
2. Gregory J. Privitera and Darryl J. Mayeaux, *Core Statistical Concepts with Excel: An Interactive Modular Approach* (Los Angeles: Sage Publications, 2018).
3. L. B. Bourque and V. Clark, *Data Processing: The Survey Example*, Sage University Paper Series on Quantitative Applications in the Social Sciences, Series no. 07–085 (Newbury Park, CA: Sage Publications, 1992), 45–48. Several introductory manuals for SPSS are available. Some focus on using SPSS software; others are for the purpose of teaching SPSS and statistical analysis. For examples of the latter, see Stephen Sweet and Karen Grace-Martin, *Data Analysis with SPSS: A First Course in Applied Statistics*, 4th ed. (Boston, MA: Longman-Pearson, 2010); Samuel Green and Neil Salkind, *Using SPSS for Windows and Macintosh: Analyzing and Understanding Data*, 7th ed. (Upper Saddle River, NJ: Pearson Education, Inc., 2014). Stata has also become very popular. For an introduction to Stata, see Phillip Pollock and Barry Edwards, *A Stata Companion to Political Analysis*, 4th ed. (Washington, DC: CQ Press, 2018.).
4. A cross-walk is a way of linking separate files in a relational database. Each file has a common element, such as a client's identification number. Cross-walking is the linking of the files in the database: http://searchsqlserver.techtarget.com/definition/database-management-system. Also see http://the-difference-between.com/crosswalk/database.
5. Bourque and Clark, *Data Processing*, 46–47.
6. See Chapter 9, this book, section on metadata.
7. Andy Mitchell, *Zeroing in: Geographic Information Systems at Work in the Community* (Redlands, CA: ESRI Press, 1997–1998).

8. G. David Garson and R. S. Biggs, *Analytic Mapping and Geographic Databases*, Sage University Paper Series on Quantitative Applications in the Social Sciences, Series no. 07–087 (Newbury Park, CA: Sage Publications, 1992), 2–3.

9. For a brief overview of GIS basics, how files are created, common types of maps and examples of GIS applications, see Irvin Vann, "Working with Geographic Information Systems," in *Practical Research Methods for Nonprofit and Public Administrators*, eds. E. O'Sullivan, G. Rassel, and J. D. Taliaferro (Boston, MA: Longman-Pearson, 2011), Chapter 13. Also see D. Smith, S. Moore, C. Harder et al., *Understanding GIS: An ArcGIS Pro Project Workbook*, 4th ed. (Redlands, CA: ESRI Press, 2018).

10. Garson and Biggs, *Analytic Mapping and Geographic Databases*, 2.

11. *Fiber and Facilities Map*, by M. Hamer. Copyright 2014 by City of San Leandro. Reprinted with permission. The authors are grateful to the City of San Leandro, CA, for permission to use the map in Figure 11.1 and to Michael Hamer and Tony Battala, Department of Information Technology, City of San Leandro, for their help and additional information. For more information, visit www. sanleandro.org.

12. See Roger S. Guy, *Selected Patterns of Stranger Rape in the City of Charlotte, 1986–1989* (Charlotte, NC: University of North Carolina at Charlotte, 1990).

13. Andy Mitchell, *The ESRI Guide to GIS Analysis, Vol. 2: Spatial Measurements & Statistics* (Redlands, CA: ESRI Press, 2005); David Wong and Jay Lee, *Statistical Analysis of Geographic Information With ArcViewGIS and ArcGIS* (Hoboken, NJ: John Wiley & Sons, Inc., 2005). The Wong and Lee book is the more advanced of these two.

14. For research using the mapping capability of GIS to compliment other data collection and statistical techniques, see: Huili Hao, Derek H. Alderman, and Patrick Long, "Homeowner Attitudes Toward Tourism in a Mountain Resort Community," *Tourism Geographies: An International Journal of Tourism, Space, Place, and Environment* (2013), doi:10.108/1416688.2013.823233.

15. Garson and Biggs, *Analytic Mapping and Geographic Databases*, 76.

16. Remember that the percent is calculated by dividing a part by the whole and multiplying by 100. For example, the percent of injuries in Table 11.4 occurring in Able County is the number of injuries in that county divided by the total, then multiplied by 100: $(5/20)100 = 25\%$. The reader is encouraged to calculate percents for the frequencies in Table 11.5.

17. Wayne Daniel and James Terrell, *Business Statistics for Management and Economics*, 7th ed. (Boston, MA: Houghton Mifflin, 1995), 19.

18. Ibid., 18–21, for discussion of a formula, Sturges's rule, used to guide analysts in deciding on the number of intervals. An alternative is the Rice rule, which puts the number of intervals as twice the cube root of the number of observations. See David Lane for a discussion of Sturges's and Rice's rules: *On Line Statistical Education: A Multimedia Course of Study*, available at http://onlinestatsbook.com.

19. Edward R. Tufte, *The Visual Display of Quantitative Information* (Cheshire, CT: Graphics Press, 2001), 13. Edward Tufte has several books, all published by Graphics Press, LLC, all dealing with presenting and interpreting information presented visually. These include *Envisioning Information* (1990); *Visual and Statistical Thinking: Displays of Evidence for Making Decisions* (1997); *Visual Explanations: Images and Quantities, Evidence and Narrative* (1997).

20. Ibid., 54–58; Darrel Huff, *How to Lie with Statistics* (New York: W. W. Norton & Company, 1954) is an older yet entertaining, relevant, and informative treatment of techniques used intentionally and unintentionally to mislead and misrepresent data.

21. David Freedman, Robert Pisani, and Roger Purvis, *Statistics*, 4th ed. (New York: W. W. Norton & Company, 2007), 25–31.

22. Ibid., 30–31.

23. Frequency polygons of a particular shape are called normal curves or normal distributions. These have characteristics that are extremely useful in statistics. See Freedman, Pisani, and Purvis, *Statistics*, 69–87; Joseph F. Healey, *Statistics: A Tool for Social Research*, 3rd ed. (Belmont, CA: Thomson-Wadsworth, 1993).

24. Freedman, Pisani, and Purvis, *Statistics*, 39–40.

25. Misty Vermaat, R. Enger, S. Freund, and S. Sebok, *Microsoft Office 2013: Introductory (Shelley Cashman Series)* (Boston, MA: Cengage Learning, 2014).

26. Another effective example is a bar graph of the murder rates of the 50 states in descending order, with horizontal bars indicating the murder rate. See "The Curse of the South," *The Charlotte Observer*, August 2, 1998, 1C, 4C.

27. For information on measures of incidence and prevalence and a discussion of formulas, see a basic text in epidemiology, for example, R. Bonita, R. Beaglehole, and J. Kiellsrom, *Basic Epidemiology*, 2nd ed. (Geneva: World Health Organization, 2006) or Charles H. Hennekens and Julie E. Buring, *Epidemiology*

in Medicine (Boston, MA: Little-Brown and Company, 1987). For a comprehensive list of concepts and definitions used in epidemiology, see John Last, ed., *A Dictionary of Epidemiology*, 3rd ed. (New York: Oxford University Press, 1995).

28. Rochester, Minnesota, the home of the Mayo Clinic, is in Olmsted County. The medical conditions of its population are among the most thoroughly studied of any county in the nation.

29. H. F. Weisberg, *Central Tendency and Variability* (Newbury Park, CA: Sage Publications, 1992), 26–27.

30. For example, see K. J. Meier, J. L. Brudney, and J. Bohte, *Applied Statistics for Public and Nonprofit Administration*, 6th ed. (Belmont, CA: Thomson-Wadsworth, 2006), Chapter 5. Later editions do not include the detailed instructions. Also see B. P. Macfie and P. M. Nufrio, *Applied Statistics for Public Policy* (Armonk, NY: M. E. Sharpe, 2006), Chapter 4. Weisberg also demonstrates this on page 11. The reader also may wish to visit the following for a demonstration: www.mathisfun.com/data/frequency-grouped-mean-median-mode.html.

31. In the calculation of the median, we assumed that the lower limit of the class interval was the whole number ending in 0. We make that assumption here as well. In this case, the midpoint of the interval is then halfway between two numbers. If the true lower limit of the interval 20–29 is 20, for example, then the midpoint of the interval is 24.5.

32. See Weisberg, *Central Tendency*, 41–43, or one of the other "Recommended for Further Reading" suggestions for a formula to calculate the geometric mean. Logarithms also can be used, and readers familiar with the mathematics of finance will have used present-value and interest-rate tables and formulae that can be used to determine the geometric mean.

33. These criteria consider the major properties of measures of central tendency but are not the only ones. See Weisberg, *Central Tendency*, 35, for a more extensive list.

34. Healey, *Statistics*, 93–97, discusses the calculation of IQV, a measure of variation for nominal and ordinal variables.

35. Weisberg, *Central Tendency*, 63–64.

36. Daniel and Terrel, *Business Statistics*, 65–66.

37. For a more detailed discussion of the difference between N and $n - 1$ in the formulas for computing the variance and standard deviation, consult one of the statistics texts listed under "Recommended for Further Reading"; see especially Hubert Blalock, *Social Statistics* (New York: McGraw-Hill, 1972), 81, 185–193.

38. Richard Chase and Nicholas Aquilano, *Production and Operations Management: A Life Cycle Approach*, 5th ed. (Homewood, IL: Irwin and McGraw-Hill>, 1989), 250–254.

39. See Weisberg, *Central Tendency*, 66, for more information about this measure and its calculation. For application to property tax procedures, see John Mikesell, *Fiscal Administration: Analysis and Applications for the Public Sector*, 7th ed. (Belmont, CA: Thomson-Wadsworth, 2007), 439–443.

40. For more discussion of the normal curve, see "Recommended for Further Reading," especially Freedman, Pisani, and Purvis, *Statistics*, 69–87; Healey, *Statistics*, 120–130.

41. See the following for discussions of exploratory data analysis: John Tukey, *Exploratory Data Analysis* (Reading, MA: Addison-Wesley, 1977); D. Hoaglin, F. Mosteller, and J. W. Tukey, eds., *Understanding Robust and Exploratory Data Analysis* (New York: Wiley Interscience, 1983); Frederick Hartwig and Brian E. Dearing, *Exploratory Data Analysis* (Newbury Park, CA: Sage Publications, 1979); Weisberg, *Central Tendency and Variability*.

42. Tukey, *Exploratory Data Analysis*; Hartwig and Dearing, *Exploratory Data Analysis*; Weisberg, *Central Tendency*.

43. See Tufte, *The Visual Display of Quantitative Information*. Two books by William S. Cleveland are also recommended: *Visualizing Data* (Summit, NJ: Hobart Press, 1994); *The Elements of Graphing Data* (Summit, NJ: Hobart Press, 1994). Given the relative ease with which visual displays can now be produced with computer technology, analysts have increased the emphasis on properly and clearly presenting information visually. Edward Tufte has produced excellent books illustrating in dramatic fashion the impact of data visualization. His work draws from extensive research on the topic. Stephen Few also has several works on clearly presenting data visually. His books are inexpensive. We cite works by Tufte and Few in the "Recommended for Further Reading" section. The reader will find it instructive to visit Edward Tufte's web site: www.edwardtufte.com.

44. Although some statisticians distinguish between box plots and box and whisker diagrams, others use the terms interchangeably. We do so here as well. The statistical terms in this chapter are used regularly, although some texts use variations. For early scholarly work on the box plot, see R. McGill, J. W. Tukey, and W. A. Larson, "Variations of Box Plots," *American Statistician* 32, no. 1 (1978): 12–16. For a visual demonstration of creating a box plot, see: www.khanacademy.org/math/probability/descriptive-statistics/box-and-whisker-plots.

45. "SAS: Data Visualization: What It Is and Why It Is Important," available at www.sas.com/en_us/big-data/data-visualization.html.
46. University Courses, "The University of Illinois at Urbana," available at www.coursera.org/course/datavisualization.html.
47. University of Illinois at Urbana; other; SAS—Statistical Analysis System.
48. "FEMA's New Data Visualization Tool Maps Where Disasters Hit," *Government Technology*, available at www.govtech.com; U.S. Bureau of the Census, "Data Visualization Gallery-Census," available at www.census.gov>Data.
49. See, for example, sources cited in Endnotes 19, 43, 45, 48.

Terms for Review

descriptive statistics
inferential statistics
univariate statistics
multivariate statistics
bivariate statistics
statistical software packages
spreadsheet
database management system
relational database
GIS
data array
frequency distribution
univariate distribution
class interval
bar graph
histogram
pie chart
rate
incidence
prevalence
ratio
percentage change
mode
median
arithmetic mean
measures of dispersion
range
inter-quartile range
percentile
quartile
standard deviation
variance
average deviation
normal curve
box plot
data dictionary
data visualization
big data

Questions for Review

The following questions should indicate whether you have a basic competency in this chapter's material.

1. How does a bar graph differ from a histogram?
2. Describe the differences between a frequency polygon and a graph of a time series.
3. Why are pie charts so often used in budget documents?
4. Compare percentages to rates. How are they similar? How are they different?
5. When would the median be preferred to the arithmetic mean as a measure of central tendency? In what situations would you recommend using both?
6. "The standard deviation can be used as a measure of how adequate the arithmetic mean is as a measure of central tendency." Explain.
7. Compare the uses of the standard deviation and midrange measures of variation or dispersion. What are the advantages and disadvantages of each?
8. When would an analyst use the inter-quartile range for interval or ratio-level data instead of the standard deviation?
9. When is the average deviation best used?
10. Discuss the criteria to use for choosing a measure of central tendency.
11. What statistics are used to construct a box plot? When would you use box plots to present your data?

Problems for Homework and Discussion

1. Develop a data dictionary for the following data collection instrument. You will need to anticipate the responses to Part E, an open-ended question.

   Library Users' Study

   a. Do you currently have a County Public Library card?

      Yes 1
      No 2

   b. How often have you visited the Public Library within the past 12 months?

      At least once a week 1
      Less than once a week but at least once a month 2
      Less than once a month but at least once every three months 3
      Less than once every three months but at least once a year 4

   c. Which of the following keeps you from using the Public Library more frequently?

      Hard to find parking 1
      Hours are inconvenient 2
      Too far away from home 3
      I use another library 4
      Lack of time 5
      No need for library services 6
      Library doesn't have what I need 7
      I'm not a reader 8
      Other (please specify) 9

d. Do you usually find the materials you need?

Yes 1
No 2

e. What subject area or types of materials would you like to see the library add to its collection? (please specify)

f. What is your age?

16–20 years 1
21–30 years 2
31–50 years 3
51–65 years 4
Over 65 years 5

g. What is your gender?

Male 1
Female 2

2. Write the data record for a case having the following answers to the Library Users' Study survey.

Item Response

a. Yes
b. At least once a week
c. Hours inconvenient
d. Yes
e. Architectural history
f. 31–50 years
g. Female

3. Obtain a copy of a recent budget for a city, town, or county and answer the following questions concerning it. Which types of graphical presentations are used? For what purposes? Which type is most effective? Would you change any of them in order to improve them? How?

4. Change one of the graphical displays from the budget you obtained by using a different format. That is, change a pie chart to a bar graph. Or change a bar graph to a pie chart to show the same information.

5. A users' survey to determine the number of personal computers owned by government agencies in a five-county planning region generated the data shown in Table 11.18. Construct a frequency polygon using these data. Write a short paragraph describing the results of the survey.

6. Use the data array given in Table 11.19, which is a slightly revised version of Table 11.2, to work the following:

a. Calculate injury rates for Able and Charlie counties. (The population of these counties is 15,300 for Able and 79,200 for Charlie.) Which is the most dangerous county in which to live?

b. What is the ratio of injuries due to violence to those due to falls in Charlie County?

c. What is the mode for "cause of injury"?

d. For "severity index," calculate the following: mode, median, arithmetic average, range, 90th percentile, variance, and standard deviation. Also calculate the z-score for case 07.

Table 11.18 Data Array for Number of Personal Computers

Agency ID Number	Number of Computers	Agency ID Number	Number of Computers
01	1	16	3
02	4	17	4
03	9	18	2
04	7	19	5
05	7	20	8
06	7	21	11
07	10	22	6
08	6	23	6
09	4	24	4
10	5	25	8
11	9	26	5
12	10	27	7
13	5	28	3
14	1	29	5
15	2	30	3

Table 11.19 Data Array for Three Variables: County, Cause, and Severity of Injury

Case ID	County	Cause	Severity Index
01	Baker	Fall	3
02	Charlie	Car accident	4
03	Charlie	Violence	6
04	Able	Car accident	4
05	Charlie	Violence	5
06	Baker	Fall	9
07	Charlie	Car accident	10
08	Baker	Fall	1
09	Able	Violence	5
10	Charlie	Violence	5
11	Charlie	Fall	7
12	Able	Car accident	4
13	Charlie	Car accident	7
14	Baker	Fall	6
15	Able	Fall	3
16	Baker	Car accident	5
17	Charlie	Car accident	5
18	Baker	Fall	6
19	Able	Car accident	4
20	Charlie	Violence	7

7. The population of Mecklenburg County in 2000 was 695,453. In 2010, it was 919,628. (censusviewer.com/county/NC/Mecklenburg). Calculate the percent change in population during this time. Compare the change in population in this period to the change worked as an example in the text. (The 1990–2000 population percent change was used in the text example in the "Quantitative Measures" section under "Percent Change".)

8. Draw box plots illustrating how the distributions from the following figures in the chapter might look: Figure 11.11(a) and (b), Figure 11.12.

   You will not be able to get precise values from the figures. The purpose is to approximate the distribution by drawing a box plot to show the various locational measures of center and spread.

9. Table 11.20 shows murder rate data for 12 southern and 11 western states for 1999 and 2009. These include murder and non-negligent manslaughter and show the number of these crimes per 100,000 of population.

Table 11.20 Murder Rate Data for Selected States, 1999 and 2009

State	1999 Murder Rate	2009 Murder Rate	State	1999 Murder Rate	2009 Murder Rate
Alabama(s)	7.9	6.8	Montana(w)	2.6	3.3
Arizona(w)	8.0	5.8	Nevada(w)	9.1	5.9
Arkansas(s)	5.6	6.2	New Mexico(w)	9.8	9.9
California(w)	6.0	5.3	North Carolina(s)	7.2	5.2
Colorado(w)	4.6	3.2	Oregon(w)	2.7	2.3
Florida(s)	5.7	5.5	South Carolina(s)	6.6	6.7
Georgia(s)	7.5	5.8	Tennessee(s)	7.1	7.4
Idaho(w)	2.0	1.6	Utah(w)	2.1	1.4
Kentucky(s)	5.4	4.3	Virginia(s)	5.7	4.7
Louisiana(s)	10.7	11.8	Washington(w)	3.0	2.9
Maryland(s)	9.0	7.7	Wyoming(w)	2.3	2.0
Mississippi(s)	7.7	6.6			

(s) = southern (w) = western

a. Calculate the arithmetic average for the data for each year. Also determine the following for each year: minimum, maximum, median, first and third quartiles, and interquartile range. Draw a box plot for the dataset from each year and label the components corresponding to the Tukey five-number summary.

b. Construct a frequency distribution of the rates for 2009. Group the values into intervals of five, that is, 0–4.99, 5.0–9.99, and so on. Draw a histogram from the frequency distribution.

c. Review the box plots in Figure 11.14 and write a page describing the differences between the two groups of states in their murder rates.

d. Draw separate box plots for southern and western states for the 2009 rates. (State region is indicated by (s) or (w) in Table 11.20.)

e. Write a memo comparing the rates in southern and western states. Discuss how these rates have changed from 1999 to 2009 (see Figures 11.13 and 11.14).

Working With Data

These exercises require you to become familiar with statistical software and its outputs. Begin by using the County Data File accompanying this textbook.

1. This exercise requires you to use a statistical software package or spreadsheet to learn about variation in counties' net migration.

a. Get the following information on net migration across the counties: frequency distribution, relative frequency distribution, and cumulative percents.

b. Produce a bar graph (for both number of counties and percent of counties) and a pie chart describing counties' net migration. Which would you include in a report? Justify your choice.

c. If you were including the information on migration in a report to the state's governor, what class intervals would you use? Justify your choice.

d. Get information on the mode, median, and mean for counties' net migration. Use the procedure "Computing the Mean From Grouped Data" described in the chapter to calculate the estimated mean. Compare this with the mean on the output. Can you conclude that either mean is accurate? Why or why not?

2. This exercise asks you to analyze variations in the demographics in counties, using them to make decisions about allocating health care resources.

   a. With data on the counties' population, use spreadsheet formula functions to calculate the number of active primary care physicians and social security beneficiaries per 1,000 population in each county. Produce a histogram and frequency polygon for these variables. Experiment with creating additional visuals communicating the distribution of each variable. Which would you be most likely to include in a report? Justify your choice. Would you consider using bar graphs or pie charts to present these data? Explain your choice.

   b. For the variables Average SAT total and Average public school expenditure per 1,000 population (which you must calculate), produce a frequency distribution and obtain the following statistical information: mode, median, mean, minimum, maximum, range, first quartile (25th percentile), third quartile (75th percentile), standard deviation. What do these statistics tell you about the similarities and differences between the two distributions?

3. Do counties with high fiscal stress tend to spend in a different fashion than those with low fiscal stress? Explore this relationship by creating the variable of county debt service per person to represent county fiscal stress. Create a new variable that codes each county as high or low fiscal stress.

   a. Create two sets of box plots for high and low fiscal stress counties. Have one set of box plots compare cultural and recreation spending per person. Have the other set of box plots compare human services expenditures. Can you comment on the initial question? Do county spending patterns differ?

   b. Write a one-page single-spaced memo for the state association of counties, reporting your findings and their implications.

4. Use the General Social Survey data accompanying this textbook for the following.

   a. Prepare a frequency distribution, cumulative percent distribution, and bar chart for INCOME16. Describe this distribution.

   b. Prepare a frequency polygon for INCOME16. Comment on the difference between the frequency polygon and the bar chart. Does the information given by each differ?

   c. Obtain and report the median for INCOME16. What other statistics, if any, would you recommend reporting?

5. Using the General Social Survey data, prepare frequency and percent distributions and a bar chart for WRKSTAT. Comment on the distribution. What statistics would be appropriate to report for these data?

Recommended for Further Reading

Blalock, H. M. Jr., *Social Statistics*, 2nd ed. (New York: McGraw-Hill, 1972), is the standard source of details on statistics commonly used by social scientists.

Healey, J. F., *Statistics: A Tool for Social Research*, 3rd ed. (Belmont, CA: Thomson-Wadsworth, 1993), is an excellent introductory statistics text covering a wide range of topics.

Meier, K. J., J. L. Brudney, and J. Bohte, *Applied Statistics for Public and Nonprofit Administration*, 9th ed. (Stamford, CT: Cengage Learning, 2014), presents examples and problems suggesting how administrators can use statistics in decision making. The calculations for statistics are worked in detail and easily followed.

Several works discuss preparation of charts and graphs, theory of graphic design, visualization techniques, and the research supporting these principles. The following are recommended: Henry, G. T., *Graphing Data: Techniques for Display and Analysis* (Thousand Oaks, CA: Sage Publications, 1995); Tufte, E. R., *The Visual Display of Quantitative Information*, 2nd ed. (Cheshire, CT: Graphics Press, 2001) (also see the works by Edward Tufte cited in Endnote 19); Cleveland, William S., *The Elements of Graphing Data* (Summit, NJ: Hobart Press, 1994); Cleveland, William S., *Visualizing Data* (Summit, NJ: Hobart Press, 1994); Wainer, H., *Graphic Discovery: A Trout in the Milk and Other Visual Adventures* (Princeton, NJ: Princeton University Press, 2005). Tufte and Wainer provide interesting histories of the development of graphic and other visual techniques with examples from history and modern times.

The following, all by Stephen Few, are newer and more oriented to providing information for dashboard monitoring as well as for analysis. *Information Dashboard Design: The Effective Visual Communication of Data* (Sebastapol, CA: O'Reilly Media, Inc., 2006). *Now You See It: Simple Visualization Techniques for Quantitative Analysis* (Burlington, CA: Analytics Press, 2009). *Show Me the Numbers: Designing Tables and Graphs to Enlighten* (Burlington, CA: Analytics Press, 2012).

Tukey, J. W., *Exploratory Data Analysis* (Reading, MA: Addison-Wesley, 1977), discusses several techniques for exploratory data analysis, illustrates their use, and provides examples.

Weisberg, H. F., *Central Tendency and Variability* (Newbury Park, CA: Sage Publications, 1992), is a thorough and accessible discussion of these two topics.

For preparing survey data for analysis by a statistical package, see Bourque, L. B., and V. Clark, *Data Processing: The Survey Example*, Sage University Paper Series on Quantitative Applications in the Social Sciences, Series no. 07–085 (Newbury Park, CA: Sage Publications, 1992); Folz, D. H., *Survey Research for Public Administration* (Thousand Oaks, CA: Sage Publications, 1996), Chapters 5 and 6.

For GIS, see the following: Lang, Laura, *GIS for Health Organizations* (Redlands, CA: ESRI Press, 2000). This inexpensive book includes a tutorial compact disk for the beginner. ESRI (Environmental Systems Research Institute) has numerous publications describing GIS applications (www.esri.com). Bolstad, Paul, *GIS Fundamentals: A First Text on Geographic Information Systems*, 4th ed. (White Bear Lake, MN: Eider Press, 2012) is also recommended.

An excellent treatment of GIS and statistics is Wong, David, and Jay Lee, *Statistical Analysis of Geographic Information with ArcView GIS and ArcGIS*, 2nd ed. (Hoboken, NJ: John Wiley & Sons, Inc., 2005); Ripley, Brian D., *Spatial Statistics* (Hoboken, NJ: Wiley Interscience, 2004) is another recommended introductory text. For an older but readable text, see Isaaks, Edward, and R. Srivastavas, *An Introduction to Applied Geostatistics* (New York: Oxford University Press, 1990).

Huxhold, William E., *An Introduction to Urban Geographic Information Systems* (New York: Oxford University Press, 1991), provides an excellent introductory overview of GIS and is a good source for a discussion of the development of a GIS for urban government applications. He describes various types of computer files and the similarities and differences among them.

Universities have made available numerous YouTube videos on GIS. For example, see www.youtube.com, *Introduction to GIS*, uploaded by the University of Kentucky Libraries, March 6, 2017. An Internet search will lead the reader to others.

For a brief yet more detailed overview of the following types of applications programs—word processing, statistical packages, spreadsheets, and database managers—see Hall, L. D., and K. P. Marshall, *Computing for Social Research* (Belmont, CA: Thomson-Wadsworth, 1992).

Wright, D. B., *Understanding Statistics: An Introduction for the Social Sciences* (London: Sage Publications, 1997), is a very readable text in which the author integrates output from SPSS with his discussion of a variety of statistics.

12 Examining Relationships Between Variables With Tests of Statistical Significance

In this chapter, you will learn:

1. About testing hypotheses.
2. The meaning of the terms *null hypothesis, statistical significance, Type I* and *Type II errors,* and *power.*
3. How to determine the statistical significance of research results.
4. When to use chi-square, *t*-tests and analysis of variance.
5. How to apply and interpret tests of statistical significance.
6. The distinction between statistical significance and practical significance, and about assessing the importance of each.
7. About alternatives to tests of statistical significance.

Deciding how to analyze data involves many choices. Researchers choose from an array of analytical tools and statistics. The selection is based on a knowledge of specific tools and statistics, the questions the research is to answer, and the properties of the variables, data, and the sample. Investigators identify certain tools as appropriate for typical questions and data in their professions. Consequently, planners, psychologists, engineers, financial analysts, and administrators may prefer different approaches to analysis and work with different statistics. In carrying out management and policy-making tasks, public administrators rely on analytic approaches favored by social scientists. They use the tools of sociologists and political scientists in working with cross-sectional data, the tools of economists and statisticians in working with time series, and the tools of psychologists in working with experimental and quasi-experimental designs.

Public administrators and those studying administration select various methods of data analysis and statistics. Although methods of analysis are many, public and nonprofit administrators commonly encounter a few statistical procedures and tests. Among these are tests of statistical significance. We discuss three of the most common in this chapter. Other types of analysis examine the nature and strength of relationships among variables and are typically used in conjunction with tests of significance. We discuss these in the later chapters of this text. Our discussions are not intended to substitute for a course in statistics. Rather, we emphasize what practitioners and researchers need to consider in selecting a data analysis method or in interpreting statistical findings. We focus on questions that a statistic helps to answer, its data requirements, and its correct application and interpretation.

In examining data, the researcher usually has two general questions:

1. What is the probability that a particular finding, such as a relationship between variables, arose by chance?
2. What are the strength and direction of the relationship between independent and dependent variables? To answer the first question, the researcher works with tests of statistical

significance. To answer the second, he works with measures of association, comparison of means tests, analysis of variance, and regression.

In this chapter, we discuss the general process of testing for statistical significance. We show how hypotheses are set up for the test to provide information so the investigator can decide if the research hypothesis is supported. Three common statistical tests are discussed, presented, and illustrated with examples. We also discuss how to interpret the results of the tests. The formula, details of calculation, and tables of test values are in the final section of the chapter. The reader can learn about the tests and their application and separately read through the calculation details.

With careful reading, you should be able to follow the reasoning behind hypothesis testing and to interpret a finding of statistical significance. Rather than slowly reading, digesting, and thoroughly understanding each section of the following, you may find it more efficient to read through the entire chapter and then go back and review the specific sections.

Debate About Statistical Significance

Before reading further in this chapter, we want to alert you that respected social science research-ers argue that tests of statistical significance are overrated and often misunderstood. We agree. Nevertheless, we begin our discussion of statistical analysis with them. We do this because understanding the reasons for significance testing is important and because introductory sta-tistics courses typically follow an important tradition and focus on testing the null hypothesis. Consequently, you may be familiar with the steps in significance testing, and the chi-square and *t*-test statistics, which are often covered in a first course in statistics. Analysis of variance is more advanced and usually addressed in conjunction with analysis of data from experiments. All are discussed in this chapter. You may encounter these statistics in professional reports and research journals. Later in the chapter, we also discuss some more recent works on alternative approaches to assessing and reporting statistical significance.

Statistical significance can be difficult to understand. We have found in our classes that students feel frustrated if we rush through it during the last week of classes. Furthermore, the term *signifi-cance*, the time spent on hypothesis testing in statistics courses, and the frequent appearance of significance tests in research reports all suggest that tests of statistical significance are extremely important. To counter this perception, we point out the limitations of significance tests and sug-gest alternative statistical information and procedures that provide additional information about a hypothesis test.

Tests of statistical significance are a type of inferential statistic. *Inferential statistics* allow an investigator to infer population characteristics from sample data. To use inferential statis-tics correctly, the investigator must have a probability sample rather than a nonprobability one. (The reader may find it useful to review material on probability sampling before completing this chapter.)

Probability samples allow an investigator to estimate parameters using inferential statistics. A parameter can be estimated for a single variable or for a relationship between variables. With sample data, one can state the probability that a parameter falls within a specified range, called the *confidence interval*. From a sample, one cannot tell exactly what portion of the population has a certain characteristic, but one can say with 95 percent confidence that the parameter falls within ±1.96 standard errors of the sample's estimate. Remember that a 95 percent confidence level means that 95 times out of 100, the sample design will allow the investigator to accurately estimate a parameter.

Another use of inferential statistics is to learn the probability that the observed relationship between variables in a sample could have occurred if the two variables are *randomly related* in the population. "Randomly related" means that the two variables are independent of, that is, not related to, each other. Asking if variables are randomly related in the population is the same as asking, "Could this relationship have occurred by chance?" If we examined the entire population or another sample, would a relationship between the variables—the independent and dependent variables of a hypothesis—still be found? If a statistical test suggests that in the population, the relationship between the two variables is nonrandom, the relationship is said to be *statistically significant.*

Just because the relationship between two variables is statistically significant does not mean that the relationship is important or strong. Nor does it mean that all or any of the previous research steps were conducted correctly. An important, valuable, or useful research finding is said to have *practical significance.* Just because a finding is statistically significant does not mean that it is practically significant.

Statistical significance simply tells us something about the statistical relationship between variables. For example, with a large sample, many relationships will be statistically significant. Yet the independent variable may do a poor job of explaining the variation in the dependent variable; large samples often lend statistical support to trivial relationships.

Steps in Testing Statistical Significance

The traditional process of determining if two variables have a nonrandom relationship requires the researcher to carry out four steps, as shown in Figure 12.1.

You may have learned this four-step process or a similar one in a statistics course. We use it here to organize our discussion. Nevertheless, strict adherence to these steps may not be appropriate in social science research.[1] The term *significance* and the logical process in applying tests of statistical significance imply that these tests are more informative than they are. They provide evidence that may support a hypothesis, but they may be inadequate for making a final decision whether to accept or reject a hypothesis.

Before we explain each step, we want to assure you that many people are bewildered the first time they study hypothesis testing. It is probably one of the most complicated topics in statistical analysis. Furthermore, hypothesis testing has its own terminology. Hypothesis testing has developed from careful statistical and epistemological thinking, and this attention is reflected in the terminology. Some of the phrases may seem convoluted.

Stating Null and Research Hypotheses

A *hypothesis* states a relationship between two variables. The *research hypothesis* is the hypothesis that an investigator wants to study. It is a statement of the result the researcher expects to find. Typically, through model building, the investigator formulates a plausible research hypothesis; then he initiates research to see if it is supported. The *null hypothesis,* unlike the research hypothesis, postulates that no relationship exists; that is, the relationship between the same two variables is random.

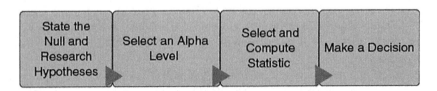

Figure 12.1 Review of Steps in Testing Statistical Significance

The following examples state three sets of research (H_1) and null (H_0) hypotheses:

Examples of Research and Null Hypotheses

H_1: Some job-training programs are more successful than other programs in placing participants in permanent employment.

H_0: All job-training programs are equally likely to place participants in permanent employment.

H_1: The more clearly written a research report, the more likely it will be used.

H_0: The clarity of a research report is not related to the probability that it will be used.

H_1: Male planners earn higher salaries than female planners.

H_0: Gender is not related to planners' salaries.

Confirming and Disconfirming Evidence

A researcher generally requires two types of evidence in order to claim support for the research hypothesis: confirming evidence, based on inductive reasoning, and disconfirming evidence, based on deductive reasoning. Confirming evidence provides support for the truth of the research hypothesis; disconfirming evidence indicates that the null hypothesis is false. Consider the hypothesis that participants in on-the-job training (OJT) programs are more likely to be placed in jobs than are participants in other programs. If one study is conducted, and it shows that OJT participants have more placements than those in other programs, can a researcher argue that her hypothesis is true? Evidence confirming her hypothesis would include a statistical relationship between variables and a demonstration that the study design satisfies the other criteria for establishing causality, including eliminating hypotheses about alternative causes. The subjects of one study represent one sample. If investigators consistently find higher placement rates among OJT trainees, the hypothesis receives stronger support.

When just one study is conducted, sampling error may have accounted for the higher placement rate of OJT trainees. The statistical test assesses whether the data are strong enough to reject the null hypothesis and provide disconfirming evidence. Another sample from the same population may not show a higher placement rate for OJT participants. Consequently, researchers generally do not (and should not) claim to have "proven" that a research hypothesis is true. Rather, they report on whether they are able or unable to reject the null hypothesis. When a researcher is able to reject the null hypothesis, he or she has support for the research hypothesis but has not proven it to be true.

To elaborate, hypothesis testing and tests of statistical significance rely on confirming and disconfirming evidence to support a research hypothesis. An investigator does not directly assert that her data support the research hypothesis. Rather, she states that her data show that the null hypothesis is probably false. She finds support for the research hypothesis when she is able to disconfirm the null hypothesis.[2]

The following outlines that process:

Stage 1: Investigator states research hypothesis that OJT has higher placement rates than other programs.

Stage 2: Investigator states null hypothesis of no relationship between type of training program and placement rate.

Stage 3: Data from sample support the research hypothesis providing confirming evidence; that is, the data show that the expected relationship exists.

Stage 4: Based on test of statistical significance, investigator rejects the null hypothesis as false. This is disconfirming evidence.

Stage 5: Investigator argues that because the null hypothesis is false, OJT is associated with a higher placement rate than other programs.

If her data did not show a difference in placement rates among programs, she would report failure to reject the null hypothesis. For the same reasons that she cannot prove the research hypothesis to be true, she cannot prove the null hypothesis true. She cannot use failure to reject her null hypothesis to argue categorically that OJT training does not affect placement rates more than other programs.

In conducting a hypothesis test, the researcher selects a statistical test to determine the probability that the hypothesized relationship in the population is random; this is the same as saying that the null hypothesis is true. The investigator uses data from a sample to make a guess about the population. In our example, the investigator found higher placement rates among OJT trainees; she selects a hypothesis test to determine the probability that the type of training program is not statistically related to placement rates in the population.

Type I and Type II Errors

In guessing about what is true in the population of interest, a researcher risks making one of two types of errors. She may decide to reject the null hypothesis when in reality the null hypothesis is true, or she may decide not to reject the null hypothesis when in reality the null hypothesis is false. Statisticians have named each of these errors. Rejecting a true null hypothesis is a Type I error. Failure to reject a false null hypothesis is a Type II error. Table 12.1 illustrates the linkage among the researcher's decision, the state of nature (what is "really" true), and the types of errors.

With a Type I error, we have concluded from sample data that the research hypothesis is true when in fact the research hypothesis is untrue. In general, researchers try to minimize the probability of a Type I error. This bias rests on their desire to minimize the risk of promoting research findings that may have occurred by chance. Conversely, researchers want to only "accept" findings that were unlikely to have occurred by chance. A Type I error may be thought of as a false alarm; in other words, it alerts one to the possibility of a relationship or situation, yet in reality, the relationship or situation does not exist. A Type II error may be thought of as a failure in signal detection; it fails to alert the researcher to an existing relationship or situation.

Prior to beginning the analysis, the investigator decides on a criterion for rejecting a null hypothesis. This criterion is a probability value and is called the alpha (α) level. The alpha level is a number between 0 and 1. Common alpha levels for hypothesis testing are .05, .01, and .001.[3]

The interpretation of the alpha levels is as follows:

$\alpha = .05$, 5% chance of committing a Type I error
$\alpha = .01$, 1% chance of committing a Type I error
$\alpha = .001$, 0.1% chance of committing a Type I error

Table 12.1 Type I and Type II Errors

Decision Based on Sample	State of Nature	
	H_0 *True*	H_0 *Untrue*
Reject H_0	Type I error	Correct decision
Do not reject H_0	Correct decision	Type II error

If alpha is set at .05, and an investigator tests 100 hypotheses where the null hypothesis is true, he can expect to make five Type I errors.

Power

Closely related to Type I and Type II errors is the concept of power. Just as we know that not all of the research hypotheses we accept are true, some of the null hypotheses we do not reject are untrue. *Power* refers to the probability that a test of significance results in the rejection of a false null hypothesis, which you may think of as the probability that an investigator will accept a true research hypothesis.

We debated whether to include a discussion of power in this text; most readers may rarely apply the concept of power directly in their work or encounter it in their professional reading. We decided to discuss power because it captures the relationship among the criteria for establishing significance (alpha-level), sample size, and the strength of a relationship.

The power of a test of significance is 1 minus the probability of a Type II error (β); if the probability of a Type II error is 20 percent, the power of the test of significance is 80 percent, or .80. If the power of the test is .80, an investigator knows that 80 times out of 100, she will correctly reject the null hypothesis, and 20 times out of 100 she will fail to reject an untrue null hypothesis.[4] Determining the probability that the null hypothesis is false requires calculating the probability of a Type II error, a tedious procedure. The hypothesis must be exact, that is, the hypothesis must state the strength of the anticipated relationship. In the absence of an exact hypothesis, an administrator can consider sample size and the anticipated magnitude of the effect to estimate the power of the test.

The magnitude of the effect might be measured by the size of the difference between or among groups or by the strength of a relationship. For example, if we hypothesize that OJT programs have higher placement rates than other training programs, we would consider how the size of the difference in rates is related to sample size. If we expect that the difference between OJT rates and other program rates is very small, we may need a relatively large sample to detect this; otherwise, we risk failing to reject a false null hypothesis. If we expect to find a large difference, a relatively small sample may be adequate. For example, a sample of 1,300 will detect a very slight effect 95 percent of the time at an alpha level of .05. On the other hand, a sample of 50 will detect a moderate effect only 46 percent of the time if alpha is set at .05.[5]

Selecting an Alpha Level

The specific alpha level selected depends on the practical consequences of committing a Type I error or a Type II error and on the anticipated strength of the relationship. Sorting out the consequences of Type I and Type II errors involves thinking through the research and the null hypotheses and the actions that will result from accepting or rejecting the null hypothesis.

Practical Consequences of Type I and Type II Errors

Example 12.1 describes how one might think through the practical consequences of committing a Type I or Type II error.

Example 12.1 Practical Consequences of Type I and Type II Errors

Case 1

H_1: Gamma cars are less safe than other cars.
H_0: Gamma cars are as safe as other cars.

Situation: You have an opportunity to buy a Gamma car for an excellent price. You collect accident data on a sample of Gamma cars.

To select an alpha level, you need to think through what will happen if you commit a Type I error or a Type II error. If you conclude from the data in your sample that Gamma cars are less safe than other cars when in fact Gamma cars are as safe as other cars, you will commit a Type I error. The practical significance? You miss out on a bargain.

If you conclude from the data in your sample that Gamma cars are as safe as other cars when, in fact, they are less safe, you will have committed a Type II error. The practical consequence? You may purchase and drive a car that is less safe than other cars. Eventually, this decision may lead to a serious personal injury that could have been avoided.

Your decision in this situation? In analyzing the data you collect on cars and safety, you want to minimize the probability of committing a Type II error. Specifically, you will select a higher alpha level, that is, a value closer to zero. Alternatively, you will select a test with high power, that is, a low probability of accepting a false null hypothesis.

Case 2

H_1: Some job-training programs are more successful than others.
H_0: All job-training programs are equally successful.

Situation: Data on job-training programs and placement outcomes are gathered and analyzed.

What happens if a Type I error is committed? Some job-training programs are erroneously assumed to be more successful than other programs. What are the practical consequences? Program continuation or cessation may depend on the findings. Programs that in reality are as successful as other programs may come to an end.

What happens if a Type II error is committed? Less effective programs will continue to offer training.

Your decision? It may depend on your point of view. If you represent training programs, you may want to minimize Type I error, which could result in effective programs closing. If you are a trainee, you may be concerned about being assigned to a less effective program. If you are a legislator, you may want evidence of the success of the programs you have supported.

The findings may reduce ambivalence about minimizing one type of error or the other. A small, but significant, difference in placement rates may not be worth the disruption of discontinuing programs.

The examples should lead you to question the wisdom of accepting a specific, rigid criterion, such as $\alpha = .05$, and applying it to a wide range of studies. The examples also illustrate that acceptable levels of Type I or Type II errors are affected by point of view and other practical concerns.

Other Parameters

The alpha level is one of four parameters that should be considered in testing statistical significance. The other parameters are the power of the test, the sample size, and the size of the effect. A small effect is unlikely to be detected with a small sample and the traditional $\alpha = .05$ cutoff.

Manipulating the alpha level can change the probability of Type I and Type II errors. If the alpha level is raised, the probability of Type I errors will increase, and the probability of Type II errors will decrease. If the alpha level is lowered, the probability of Type I errors will decrease, and the probability of Type II errors will increase. Type I and Type II errors both decrease with a larger number of cases in the sample.

How Do These Considerations Apply to Public Administrators?

While administrators may want to know the probability that an observed relationship is nonrandom in the population, the size of the effect is of far greater concern. For some situations, reacting to a "false alarm" or even responding to minor, nonrandom differences may be undesirable. Segments of the population may differ in their opinions about public policies or their experiences with public services. Nevertheless, administrators may prefer to ignore small differences. If older people are less satisfied with recreational facilities than other respondents, or if the police respond less quickly to calls from the south side of town than from other parts of town, administrators may not address these variations if they do not suggest definite patterns of discrimination or other problems. The political and economic costs may render resolving small differences impractical or undesirable.

In the context of program evaluation, we should not expect a test of significance alone to be adequate for judging a program's effectiveness. Evaluations based on small sample sizes and programs with modest effects may result in the erroneous conclusion that an effective program is ineffective. When social programs were first evaluated, researchers were disappointed in how little such programs accomplished. The disappointment was misplaced. It now seems naive to assume that a limited social program is going to markedly change behavior. Administrators and policy makers may have to be satisfied with smaller changes, albeit changes in the desired direction.

To limit research costs, evaluations of experimental programs may be limited to a relatively few subjects. If a study has a small number of subjects, and the program's effect is not strong, the probability of a Type II error is larger. If the outcome measures have poor reliability, more random error will be introduced, further increasing the probability of a Type II error.[6] If a Type II error occurs, a program that has a modest impact on participants may be judged to have made no difference. This illustrates the need for addressing Type II errors as well as Type I errors.

Selecting and Computing a Test Statistic

In this text, we discuss three common tests of statistical significance. If you need only a basic understanding of the research findings, you should find the information provided here adequate for your needs.

- The first test statistic we consider is *chi-square* (χ^2). Chi-square is a statistic for nominal-level data, although, as is true with other statistics for nominal data, it may be used to analyze ordinal- and interval-level data. It is often used to analyze data in contingency tables. (Contingency tables are also discussed in the chapter on analyzing relationships between nominal and ordinal variables.)
- The second test statistic is the *t-test*. The *t*-test examines the differences between the means of two groups, that is, the relationship between an interval dependent variable and a nominal independent variable.
- The third test is the *F*-test and is based on analysis of variance. Analysis of variance examines the differences among the means of three or more groups. It also examines the relationship between an interval dependent variable and a nominal or ordinal dependent variable.

Chi-Square

The chi-square test compares the observations contained in a set of sample data with the observations expected if the relationship between variables is random in the population. Chi-square only indicates the probability that in the population the variables are unrelated; it does not indicate the direction or strength of a relationship. Table 12.2 contains data gathered to test a hypothesis that different job-training programs achieved different outcomes. In the example, the dependent variable is the outcome of the training program, measured by the current status of the participants; the independent variable is the type of training program.

The units of analysis are the participants in the programs. Table 12.2 (top) compares three job-training programs and outcomes; the table contains the frequencies observed (f_o) in the collected data. Table 12.2 (bottom) shows what the data would look like if there were no relationship between a program and trainees' outcomes. The numbers in Table 12.2 (bottom) represent the frequencies expected (f_e) if the null hypothesis were true.

In the observed frequencies, we see that nearly 50 percent of all subjects were working (292 of 586). If the work status and program categories were not associated, the same percentage of participants in each program would be working. Hence, in Table 12.2 (bottom), nearly 50 percent of the clients of each program are assigned to the category "working." The ratios in the overall group for in-school and unemployed are similarly applied to the subgroups. A little less than 25 percent are assigned to the "in school" category, and somewhat more than 25 percent to the "unemployed" category. The footnote in Table 12.2 explains how the values for the table of expected frequencies are calculated. (The last section of the chapter shows the details of the calculation of expected frequencies for this example and in general.)

The next step is to calculate chi-square. Each cell from the table of frequencies observed is subtracted from its counterpart in the table of frequencies expected ($f_o - f_e$). The result is squared and divided by the frequency expected ($(f_o - f_e)^2/f_e$). For example, for the cell "Working" and

Table 12.2 Observed and Expected Frequencies— Current Status by Training Program Attended: Observed Frequencies (f_o)

Observed Frequencies

Program

	Vocational Education	On-the-Job Training	Work Skills Training	Total
Working	19	109	164	292
In school	19	82	31	132
Unemployed	26	82	54	162
Total	64	273	249	586

Expected Frequencies If Current Status and Training Program Attended Are Not Related.

Program

	Vocational Education	On-the-Job Training	Work Skills Training	Total
Working	31.9	136.0	124.1	292
In school	14.4	61.5	56.1	132
Unemployed	17.7	75.5	68.8	162
Total	64	273	249	586

"Vocational Education," the calculation would be $(19 - 31.9)^2/31.9 = 5.22$. Chi-square ($\chi^2$) equals the sum of these figures. Thus, the equation is

$$\chi^2 = \Sigma \left(f_0 - f_e\right)^2 / f_e$$

for the example $\chi^2 = 50.57$.

The "significance level" of a χ^2 value is typically reported by statistical software programs. The programs perform the calculations and report the χ^2 value and its "significance level." This significance level is the associated probability, that is, the probability that the specific value of χ^2 will occur if the null hypothesis is true in the population. (Later in this chapter, we show how to calculate chi-square and the relationship between its value and common alpha levels.) In interpreting computer program output, the investigator determines if the associated probability level is equal to or smaller than the alpha level he set. In this example, if $\alpha = .05$, $\chi^2 = 50.67$ indicates that the relationship is statistically significant. (The associated probability of a χ^2 value of 50.67 for this size table is less than .001.)

Three Characteristics of Chi-Square

Chi-square is a widely used and understood test of statistical significance. You should be aware of three of its characteristics.

- First, because chi-square is a statistic for variables measured at the nominal level, it does not provide information on the direction of any association. In our example, the chi-square evidence indicates that the relationship between type of training program attended and current status is probably nonrandom. It does not indicate which program is the most effective.[7]
- Second, the numerical value of chi-square tends to increase as the sample size increases. Thus, the chi-square value is partially a product of sample size. A larger chi-square value is more likely to be statistically significant, even if the practical effect of the independent variable is small. It does not directly measure the strength of the association between variables and should not be used as a measure of association. Some statistics to measure the strength of a relationship are based on chi-square, but they take into account and adjust for sample size and number of rows and columns in the table. One such statistic, Cramer's V, is discussed along with measures of association for contingency tables in another chapter.
- Third, chi-square does not reliably estimate the probability of a Type I error if the expected frequency of a cell in a table is less than 5. Research has found that violating the expected frequencies criterion causes relatively small errors; a researcher can easily ignore this problem if it occurs in relatively few cells. If most cells have few cases, the investigator may consider combining categories.[8]

Let's review with an example the steps taken thus far to test the hypothesis that the type of training program was related to program outcomes:

Step 1. State the Hypothesis and the Null Hypothesis

H[1]: Different job-training programs achieved different outcomes.
H[0]: The job-training programs did not differ in their achievement of outcomes.

Step 2. Select an Alpha Level

Set alpha at .05, a common default value.

Alternatively, a lower alpha level, for example, .01, may be selected because of the relatively large sample size and a desire to detect only a marked difference between program outcomes (making changes based on small differences would be costly).

Step 3. Select a Test Statistic

Chi-square is selected because data in the table are nominal.
Determine the value of test statistic: $X^2 = 50.57$.

Step 4. Determine if the Value of χ^2 Meets Alpha-Level Criterion

Locate the associated probability reported in the computer output. The associated probability must be less than or equal to .05. (If computer output is not available, consult a table of chi-square values. A condensed table is included at the end of the chapter. Note that you must also consider *degrees of freedom* or *df*. Degrees of freedom for the chi-square test is number of rows in the table minus 1 times number of columns minus 1.)

Step 5. Decide if Statistical Evidence Supports the Research Hypothesis

If associated probability is equal to or less than .05, the statistical evidence supports the research hypothesis, and the null hypothesis can be rejected.

Notes on Degrees of Freedom: In order to evaluate if the value of the statistical test is large enough to reject the null hypothesis, most tests of statistical significance require that this value be adjusted by the degrees of freedom (*df*). This is the number of values that can vary freely in the data and is important since the test assumes a probability sample. For some tests, *df* is based on the sample size; for others, on the size of the table, the number of groups, or number of variables involved.

t-Test

Assume that you want to learn whether men earn more than women in a particular occupation or whether children at School A obtain higher state achievement test scores than children at School B. One way to test these hypotheses is to rearrange the data into contingency tables and perform a chi-square test. Of course, to create contingency tables, you must change the dependent variable information from interval to ordinal and thus would lose information. And you might rightfully wonder whether the finding of significance or lack of significance was associated with how the values were combined. Alternatively, you could hypothesize that the average achievement test grades at School A are higher than at School B. You could test that hypothesis using a *t*-test.

The *t*-test is an interval statistic that can test hypotheses that two groups have different means. Each group is considered a sample, and the appropriate test is a two-sample *t*-test. The two-sample *t*-test can test a research hypothesis (1) that two groups have different means or (2) that one group's mean is higher than the other's. The first type of hypothesis specifies no direction; a two-tailed *t*-test is used to test nondirectional hypotheses. The second type of hypothesis has direction; a directional hypothesis is tested with a one-tailed *t*-test.

The degree of variability in each sample affects the choice of formula used to calculate the *t*-value.

- One formula is used if the variances of each population represented by the samples are assumed to be equal.
- Another formula is used if the populations are assumed to have unequal variances.
- A third formula is used if data for both groups come from the same subjects.

For example, a researcher may compare analysts' error rates when entering data on laptop computers as opposed to desktop computers. A single group of analysts would enter data into both laptop and desktop computers. The dependent variable would be error rate, and the independent variable would be type of computer. Other examples include measures on the same individuals taken before and after some event.

The *t*-test can also test a single sample hypothesis, for example, that a group's mean is greater or less than a specified value. The single sample *t*-test is useful for various situations such as staffing studies.[9] For example, a university library may assume that its reference desk handles an average of 25 inquiries per hour on weekends. The assumption determines the level of staffing at the reference desk on weekends. Periodically, data may be gathered and the average number of inquiries per weekend calculated. A single sample *t*-test is performed to evaluate the hypothesis that the average number of inquiries is greater or less than 25.

Illustrating a t-Test

To illustrate a *t*-test, we test the hypothesis that male planners earn higher salaries than female planners.[10] We assume unequal variances of the salaries of the two groups. Assuming unequal variances is less likely to result in a Type I error and consequently is more conservative than assuming equal variances; that is, it is less likely to result in "accepting" an untrue hypothesis.

Step 1. State the Research Hypothesis and the Null Hypothesis

H_1: The average salary of male planners is higher than the average salary of female planners.

H_0: The average salary of male planners is the same as or less than the average salary of female planners.[11]

This hypothesis requires a one-tailed test. In a one-tailed test, the null hypothesis is expanded to include a finding in the "wrong" direction. Thus, if the data showed that the average salary of female planners was more than that of male planners, the researcher would not reject the null hypothesis. Alternatively, with a two-tailed test, the hypothesis only indicates a difference in salaries. It does not specify which gender earns more. The null hypothesis only notes no difference in the average salaries of male and female planners. So a researcher could find that the average salary of female planners is greater than that of male planners and still reject the null hypothesis.

Step 2. Select an Alpha Level

$\alpha = .05$ often serves as a default alpha level. It simply suggests that the relationship probably did not occur by chance, directing the investigator to examine the size of the difference in salaries.

Step 3. Select a Test Statistic

A *t*-test is selected because the hypothesis postulates that the two samples have different means. A one-tailed test is selected because the hypothesis postulates which sample mean is greater than that of the other.

Step 4. Calculate the Test Statistic

This requires an assumption about the sample variances. The formula for unequal variances will produce slightly higher associated probabilities. Assuming unequal variances is a more conservative test and is reasonable. When the study on which this example is based was done, women had only recently entered the planning profession. Thus, they were less likely to have salaries consistent with extensive experience, and one can expect that their salaries would have less variance than the salaries of males. With this assumption that the variances are unequal, the formula to calculate t is:

$$t = \frac{\overline{X}_1 - \overline{X}_2}{\sqrt{\dfrac{s_1^2}{n_1 - 1} + \dfrac{s_2^2}{n_2 - 1}}}$$

where:

For the Male Sample   For the Female Sample

$n_1 = 403$ $n_2 = 132$

$\overline{X}_1 = 17,095$ $\overline{X}_2 = 14,885$

$s_1 = 6,329$ $s_2 = 4,676$

$s_1^2 = 40,056,241$ $s_2^2 = 21,864,976$

$$t = \frac{17,095 - 14,885}{\sqrt{\dfrac{40,056,241}{402} + \dfrac{21,864,976}{131}}}$$

$$= \frac{2210}{\sqrt{99,624 + 166,908}}$$

$$= \frac{2210}{\sqrt{266,550}} = 4.28$$

$$= 4.28, \ df = n_1 + n_2 - 2 = 533$$

Step 5. Determine if the Value of t Meets the Alpha-Level Criterion

From the associated probability in the computer output, associated probability must be less than or equal to .05. (If computer output is not available, consult a table of t-values.) Table 12.13, in a later section, shows the relationship between values of t and common alpha levels. Note also that degrees of freedom, *df*, must be considered. Degrees of freedom for a two-sample t-test is the total of the number of cases in the two groups minus two. The example here illustrates this.

Step 6. Decide if Statistical Evidence Supports the Research Hypothesis

If the associated probability is no more than .05, or if t is 1.645 or larger, the alpha-level criterion statistical evidence supports the research hypothesis. In this case, $t = 4.28$, which is much higher than the level we need, so we can reject the null hypothesis, and the researcher has support for saying that the average salary for male planners is higher than the average salary for female planners.

The Normal Distribution

Students who have studied statistics may wonder why we do not use the normal distribution (*z*-scores) to test hypotheses about sample means. To use *z*-scores properly, the population variance must be known. If the population variance is not known, it is estimated by the standard deviation, in which case *t*-tests as opposed to *z*-scores are appropriate. With larger samples, *t*-scores approximate *z*-scores. While using *z*-scores introduces little error with larger samples ($n \geq 60$), social scientists tend to prefer reporting *t*-tests.

Analysis of Variance

If the independent variable has more than two groups, the analyst must conduct several separate *t*-tests to determine if a statistically significant relationship exists. An alternative procedure is to use an *F*-test with analysis of variance to determine if the means of two or more groups differ significantly. Analysis of variance was developed early in the twentieth century by R. A. Fisher, a British statistician, who wanted a method for analyzing data collected in carefully structured experiments. He primarily studied agricultural practices. Hence, many early textbooks and experts treated analysis of variance in the context of agricultural research. The main statistical test used in ANOVA, the *F*-test, is named for Fisher.[12]

In an experiment, the subjects may be drawn from the same population and randomly assigned to an experimental or control group. The experimental and control groups constitute separate samples. Just as differences between a random sample's statistics and its parameters are due to chance, pretest differences between the means and variances of the experimental and control groups should be due to chance.

After the experimental intervention takes place—the independent variable of the experiment—the means of the groups may differ. Experimenters expect an experimental intervention to cause differences between the experimental and control groups. ANOVA is designed to determine whether the posttest differences between the experimental and control groups are large enough that they probably occurred for a reason other than chance.

Analysis of variance has the following assumptions:

1. dependent variables are interval or ratio level
2. variances and standard deviations for each group are equal or close to it
3. the dependent variable should be normally distributed for each group; this assumption is relaxed if the groups are large
4. the values of the dependent variables are independent of each other[13]

Correctly designed experiments fulfill these assumptions. However, in practice, some of these assumptions are relaxed.[14]

Two Types of Variances in Analysis of Variance

Analysis of variance examines two types of variances:

1. within-groups variances
2. between-groups variances

Within-groups variance refers to how each member within an experimental or control group varies from the group mean. *Between-groups variance* refers to how the means and variances of each group differ from those of other groups. In a pretest-posttest situation, if experimental and control group subjects are assigned correctly, that is, randomly, the between-group and within-group variances of pretest groups should be equivalent. If the experimental treatment had no effect, the statistics for the posttest groups also will be equivalent. If the treatment had an effect, the means of the groups and most likely the variances of posttest groups will differ.

ANOVA begins by establishing whether the differences between the experimental and control groups at the end of the study could have occurred by chance. Analysis of variance has the following basic research and null hypotheses:

H_1:The arithmetic means of two or more groups are unequal.
H_0:The arithmetic means of two or more groups are equal.

In experiments, the investigators compare the means of the posttest experimental and control groups to see whether the data support the research hypothesis. Analysis of variance determines whether differences between the groups (that is, between-group variance) are greater than differences within any of these groups (that is, within-group variance). The reader should note that ANOVA can be applied to situations in which only one set of groups is used. That is, a posttest-only design is used. If the members of the initial group are randomly assigned—or nearly so—then posttest differences should be due to the influence of the independent variable.

To judge whether the difference between two means is large enough to be statistically significant, we use the *F-test*, which is based on the following ratio:

Between-Groups Variance/Within-Groups Variance

If the value of the ratio is much greater than 1, the difference between group means is likely to be statistically significant. The ratio will be greater than 1 if the difference between group means is a relatively large value and the variability within groups is relatively small. If the researchers obtain a large enough value for the *F*-ratio, they reject the null hypothesis of the equality of the means and thus find support for the research hypothesis.[15]

ANOVA has widespread applications apart from experimental designs. It is used to assess the results of multiple regression, a technique applied to the joint relationship of several independent variables to a quantitative dependent variable.[16] It may be used to analyze cross-sectional data if the dependent variable is interval or ratio and the independent variable is nominal.[17] Remember that an ordinal variable can be treated as a nominal variable. To use ANOVA appropriately, the data must come from random samples. The assumption that variances in all groups in the population must be equal is often ignored if the number of cases in each group is similar. The analyst may wish to check this assumption, especially if the group sizes are markedly different.[18]

Illustrating Analysis of Variance

To illustrate the points about ANOVA made thus far, we review a study by an analyst with Majesty City's Division of Training and Job Placement. The analyst compared three programs sponsored by the agency:

1. Basic Skills
2. Classroom Training
3. On-the-Job Training

The agency assigned clients to one of the three programs. At the end of the training program, clients were placed in jobs.

The analyst wanted to determine if clients trained in the OJT program were earning more than clients trained in either of the other two programs. Previous research suggested that OJT participants were more successful. Furthermore, from her own observations, she believed that OJT participants were normally more "job ready" than other trainees and therefore were more likely to receive better job placements. The analyst collected data on trainees' monthly earnings during their first year of work. Her research and null hypotheses were:

H_1: The average earnings of OJT trainees will be greater after one year than the earnings of trainees in the Basic Skills and the Classroom Training programs.

H_0: The average earnings after one year will be equal for trainees in all three programs.

A sample of clients attending each program was selected from agency records. Interviewers telephoned the members of the samples and asked them for wage information. Table 12.3 shows the average monthly earnings for each group. The grand mean refers to the average monthly earnings for all 40 persons included in the study.[19]

Figure 12.2 supplements the table information and visually represents the data. The graph shows the average earnings and the range, a measure of variability, for each group. The asterisk indicates the average income, and the ranges are enclosed with the boxes.[20]

The data in Table 12.3 and Figure 12.2 support the research hypotheses. To supplement these findings, the analyst used statistics to answer two questions:

1. Are the differences among average salaries for the groups statistically significant? That is, are the differences larger than would be expected to occur by chance?
2. How strong is the relationship between type of program and average income?

The *F*-test answers the first question. A measure of association, *eta*, answers the second question. The *F*-test is discussed in Chapter 12; *eta* is discussed in detail in Chapter 13.

If all three groups were equal on the dependent variable (earnings), then the between-groups and within-groups variances would be equal and the value of the *F*-ratio would be 1. If the *F*-ratio is larger than 1, the investigator determines its statistical significance by consulting a table of *F*-values or obtaining the statistics from a computer program. To use a table of *F*-values, the analyst must know the value of the *F*-ratio and the degrees of freedom for the numerator and the denominator. In our example, the degree of freedom for the numerator is 2, the number of groups minus one, and for the denominator, it is 37, the total sample size minus the number of groups.

Table 12.3 Average Monthly Earnings by Program

		Program		
		Basic Skills	*Classroom Training*	*On-the-Job Training*
Average monthly income ($)	$\bar{X}$	433.5	528.7	765.1
Sample size	*N*	11	21	8

Note: Average for all participants = grand mean = 549.8; total number of participants = 40[1]

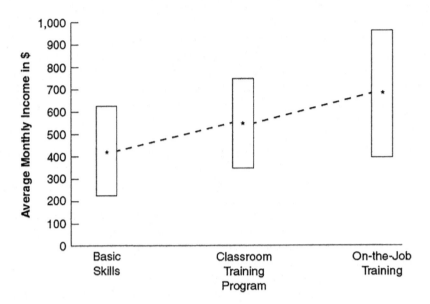

Figure 12.2 Average Monthly Earnings by Program

Applying Analysis of Variance With Statistical Package for Social Sciences

The data from the Majesty City study were analyzed by a computer program, Statistical Package for Social Sciences, using analysis of variance. Figure 12.3 shows the printout from this program. We will comment only on portions of the printout that help the analyst decide whether the data support the research hypothesis and allow you to reject the null hypothesis.[21]

D.F.: D.F. is the degrees of freedom for each group.

SUM OF SQUARES: The Sum of Squares measures the between-groups variation and the within-groups variation.

F-RATIO: The *F*-Ratio is between-groups mean square divided by the within-groups mean square.

F-PROB: The *F*-Prob. refers to the associated probability of *F*. In this case, the probability that an *F*-ratio as large as 5.2 would have occurred if the means of the three groups were equivalent is .01, or 1 percent.

The *F*-test, in this example, tells the investigator that at least two of the groups differ significantly from each other. To learn which pairs of means are significantly different, the analyst must perform separate *t*-tests. The *t*-tests of the example's data found a significant difference between the earnings of trainees who received Basic Skills training and those in the OJT group.

An *F*-test does not measure the strength of the relationship between variables. A measure of association, eta or *E*, based on analysis of variance, is used for this purpose. Eta can vary between 0.00 and 1.00. We discuss eta further in Chapter 13, along with other measures of association.

This discussion focused on the basic analysis of variance model. In one-way analysis of variance, the analysis involves one independent variable, and each case contributes one score to the analysis. In the school example mentioned at the beginning of this section, the independent

ONEWAY

By	Variable Variable	MWAGE PROGRAM	Monthly salary and wages Program participant enrolled in		

ANALYSIS OF VARIANCE

SOURCE	D.F.	SUM OF SQUARES	MEANS SQUARES	F-RATIO	F-PROB
BETWEEN GROUPS	2	528,965	264,482.5	5.2	.0100
WITHIN GROUPS	37	1,869,230	50,519.7		
TOTAL	39	2,398,195			

Figure 12.3 Analysis of Variance Table

Source: The "Source" column identifies the two types of variance, between groups and within groups.

variable was whether a student attended school in the neighborhood or elsewhere, and the piece of data contributed by each subject was the number of days missed from school. In the job-training example, the independent variable was type of job-training program, and the contributed score was each subject's monthly earnings.

The basic model can be adapted to include additional variables. Analysis of variance models including more than one independent variable are called multivariate analysis of variance (MANOVA) or, depending on the number of independent variables, two-way analysis of variance, three-way analysis of variance, and so on. Analysis of variance models also may include more than one measure on a subject; these are called repeated measures models.[22]

Making a Decision

Making a decision based on a test of statistical significance is relatively straightforward. The strategy presented thus far has the investigator (1) setting the alpha-level criterion to reject the null hypothesis, (2) selecting a test statistic, (3) comparing the associated probability to the alpha-level criterion, and (4) rejecting the null hypothesis if the calculated value yields an associated probability equal to or less than the alpha level—otherwise, the null hypothesis is not rejected. In our examples, an investigator would reject the null hypotheses that different job-training programs achieved the same outcomes, that male planners earned the same or less than female planners and that Majesty City's different programs' participants earned similar salaries. The associated probabilities for X^2, t, and F are less than the selected alpha level of .05.

An Alternative Decision Strategy

An alternative strategy begins with the associated probability. The investigator looks up the specific probability (p) associated with the calculated value of chi-square (χ^2), t, or F. If this associated probability is not larger than the alpha-level criterion, the researcher rejects the null hypothesis. Statistical software programs usually report associated probabilities. Example 12.2 reproduces computer output examining the relationship between respondent age and the grade given to a local police department. The output is from a popular statistical software program, SPSS, and identifies the relevant χ^2 (chi-square) information. Note that if an investigator had decided on an alpha level of .05 or .01, she would reject the null hypothesis. If she had decided on an alpha level of .001, she would not reject the null hypothesis.

Using Computer Output to Decide

Example 12.2 Interpreting χ^2 Information on SPSS Output

Situation: A probability sample of 830 citizens graded a city's police department. The investigators hypothesized: The younger the respondent, the lower the grade he or she gave the police department. (The null hypothesis was that age is not related to the grades given.)

Findings: Analysts run an SPSS program and Table 12.4, and statistics are reported. Note that relatively more of the oldest respondents gave the department an A or B, and that relatively more of the youngest respondents gave the department a D or F.

Table 12.4 Grade by Age

Age				
Grade		*Under 18 %*	*18–59 %*	*60 or older %*
	Codes	1.00	2.00	4.00 Row Total
A or B	1.0000	50.0	72.7	82.0 615
				74.1
C	2.0000	41.7	22.0	14.8 174
				21.0
D or F	4.0000	8.3	5.3	3.3 41
				4.9
	Column	24	623	183 830
	Total %	2.9	75.1	22.0 100.0
Chi-Square	**Value**		*df* **(Degrees of Freedom)**	**Significance**
Pearson	13.90036 (calculated value of χ^2)		4	.00762 (associated probability)

Notes:

"Pearson" refers to "Pearson's chi-square," which is the chi-square equation used in this text.

The X^2 information would be reported as follows: "$\chi^2 = 13.9$, $df = 4$, $p = .00762$." One may round p to .008.

The column labeled "significance" is actually reporting the associated probability. Some researchers do not report the specific associated probability; instead, they indicate if the hypothesis was supported based on the preset alpha level. For example, if the researcher set $\alpha = .01$, he would report "$\chi^2 = 13.9$, $df = 4$, $p < .01$."

If the researcher changes the number of rows or columns in the table, chi-square must be recalculated.

Evaluating Associated Probability

The associated probability does not indicate the probability of a Type I error, nor does it imply that a relationship is "more significant" or "stronger." Its contribution is more modest. It indicates the probability of a specific X^2 or t-value occurring if the null hypothesis is true.[23] Researchers combine information on the associated probability with the other evidence to make inferences about hypotheses. Such evidence may include evidence of internal validity or findings from previous studies.

Tests of statistical significance may appear to constitute a "pass/fail" test. Should they play this role? No, not in our examples or in similar situations. The investigator is only testing a statistical model. The data represent many decisions, including whom to sample, how to sample them, what information to gather, how to gather it, and when to gather it. The investigator applies statistical tests to a set of numbers; he obtains an answer even if he violates all the statistical assumptions and ignores sound methodological practice.

Significance Tests As Evidence

If the methodological decisions are sound and the statistics are applied correctly, the data still represent a single sample out of all the possible samples from the population. A test of significance alone should not bear the burden of demonstrating the value of a hypothesis. It is far more realistic and reasonable to consider each statistical finding as part of a body of evidence supporting the truth or error of a hypothesis. We doubt that any administrator would decide the effectiveness of a program or the fairness of salaries based on one piece of statistical evidence.

In some situations, investigators can appropriately use tests of statistical significance to make definite decisions. The tests establish inter-rater reliability, verify whether shifts in time series can be assumed to be random, and demonstrate the probability that respondents and nonrespondents are similar to one another.[24] Tests of statistical significance have been particularly valuable in quality control, where a sample of "products" is inspected to locate systematic problems. For example, a quality control unit may review a sample of probation cases. An analyst reads a case record and notes any errors. Working with small samples, she applies tests of statistical significance to determine the probability that the errors she finds are random. If it is highly likely that the errors are not random, the agency should take corrective action, such as retraining the probation officer. Table 12.5 provides a brief overview and summary of the steps in testing a hypothesis for statistical significance.

Reporting Tests of Statistical Significance

In most reports, tests of statistical significance may be inconspicuous. The reader unfamiliar with them may scarcely notice that they are mentioned. Readers with a serious interest in the research topic refer to them to evaluate research and its findings.

The researcher should report the test statistic used and its value, the *degrees of freedom*, and the associated probability. Citing the statistic allows readers trained in statistics to determine whether

Table 12.5 Review Steps in Hypothesis Testing

State the null and research hypotheses.	The *null hypothesis* postulates that no relationship exists; that is, the relationship between the same two variables in the research hypothesis is random.
Select an alpha level.	The specific alpha level selected depends on the practical consequences of committing a Type I error or a Type II error and on the anticipated strength of the relationship.
Select and compute a test statistic.	Chi-square, a statistic for nominal level data, is used to analyze contingency tables. The *t*-test examines the differences between the means of two groups, that is, the relationship between an interval dependent variable and a nominal independent variable. The *F*-test analyzes the difference among three means of three or more groups.
Make a decision.	If the calculated value yields an associated probability equal to or less than the alpha level, the null hypothesis is rejected.

the researcher used the appropriate statistical test. The statistic's value and degrees of freedom are needed to determine the associated probability. With this information, a reader can identify mathematical or recording errors and conclude how the data were analyzed. The associated probability allows the reader to make her own independent determination of the findings' "significance."

To report this information in the text, the researcher encloses the statistical details in parentheses. For example:

The type of job-training received is related to participants' current employment status ($\chi^2 = 50.57$, $df = 4$, $p < .001$).

The average salary of male planners is higher than that of female planners ($t = 4.28$, $df = 533$, $p < .001$).

If contingency tables are included in a report, the same information is reported immediately beneath the last row of table cells. Table 12.6 illustrates a common format for presenting a series of *t*-tests, which compares means for two values for each independent variable. The table reports the average (mean) values for male and female planners on nine variables. Underneath the mean values are the values of the standard deviations. The third column reports the values of the *t*-test. Note that marital status and race are dichotomous variables and the means represent the percent of married planners and the percent of white planners. The interested reader has all the information she needs to compute the values of *t* herself. The letters next to the *t*-values direct the reader to the table footnotes, where she can find the associated probability rounded to the nearest

Table 12.6 Tests Comparing Men and Women in Planning

	Men	Women	
	$\bar{X}$	$\bar{X}$	
	(s)	(s)	T
Individual characteristics Age	32.43	30.11	2.93[c]
	(8.28)	(7.72)	
Marital status (1 = married)	.75	.61	3.00[c]
	(.41)	(.49)	
Race (1 = white)	.93	.81	2.05[b]
	(.26)	(.34)	
Organization characteristics Agency size	4.78	5.29	−1.70[a]
	(2.88)	(2.99)	
Centralized management	5.01	5.25	−.85
	(2.61)	(2.73)	
Career behavior Job turnover	.51	.60	−2.19[c]
	(.37)	(.39)	
No. of roles	7.98	7.88	.53
	(1.67)	(1.79)	
Years of experience	3.67	3.42	7.26[c]
	(5.31)	(2.57)	
Career attainment Income	17,095	14,885	4.28[c]
	(6,329)	(4,676)	

[a] Significant at the .10 level.
[b] Significant at the .05 level.
[c] Significant at the .01 level.

Source: Mayo, M. Jr., "Job Attainment in Planning: Women Versus Men," *Work and Occupation* (May 1985): 152. (Copyright © 1985 by Sage Publications, Inc. Reprinted with permission: Sage Publications, Inc.).

commonly used alpha level. One asterisk represents $p = .10$, two asterisks represent $p = .05$, and three asterisks represent $p = .01$. The relationship between the number of asterisks and value of p varies from author to author. Usually, the more asterisks, the lower the value of p.

At the present time, no one convention for reporting associated probability has emerged. Using a preset alpha level to report the significance of all values is least common. In Table 12.6, the author distinguished among three levels of significance rather than reporting all the significant findings as significant at $p = .10$. Whether the specific associated probability is reported seems to depend on the researcher's ability to present the information in an uncluttered table. The asterisks in Table 12.6 would not work if the author wanted to report the exact associated probability. This author's strategy is the most common, that is, to round the associated probability up to .10, .05, .01, or .001 and report the rounded-up p value.

A researcher may simply note that a relationship was found to be "statistically significant." This observation merely means that the value of a specific statistical test met the researcher's alpha-level criterion. If the value did not meet the alpha-level criterion, the relationship is said to be "statistically insignificant." In order for such information to have any value, the researcher must indicate the statistical test and alpha level used.

Modifications and Alternatives to Tests of Statistical Significance

Here we reiterate some important misconceptions about tests of statistical significance.[25] They cannot remedy a flawed design. They only test a statistical hypothesis, not the theoretical hypothesis. The analyst applies an equation to a set of numbers, comes up with a statistical value, and makes a decision. Nothing prevents someone from dividing a set of random digits into groups, calculating a test statistic, and finding a statistically significant relationship. One researcher summed up the situation aptly with the phrase "the numbers don't remember where they came from."[26] The numbers do not know if the data represent a carefully designed study with reliable measures and a sample free of nonsampling errors or if the sample was biased and the measures unreliable.

The information produced by a test of statistical significance is modest. It simply indicates the probability that a relationship exists in the population. The relationship may be weaker, or stronger, than the one found in the sample data. Depending on the sample size and the p-level, statistically significant differences may be too slight to be of theoretical interest or practical use. Social scientists have recommended four alternatives to the traditional significance test.

Alternative Approaches

First, the null hypothesis can include a specific difference or relationship value. Consider an earlier example. Assume that administrators do not want to fund more on-the-job training programs unless they place at least 10 percent more of their trainees than other training programs.

To test this preference, the hypothesis and null hypothesis would be stated as:

H_1: On-the-job training programs place at least 10 percent more participants than other training programs.

H_0: On-the-job training programs do not place 10 percent more participants than other training programs.

Second, instead of reporting significance, researchers may report confidence intervals. (To become more familiar with the standard error and computation of confidence intervals, the reader should review material on probability sampling.) Proponents argue that confidence intervals are more informative. They provide information on the estimated size of parameters,

Table 12.7 Mean Age and Income of Male and Female Planners

	Mean	*95% Confidence Interval*
Age		
Males ($n = 300$)	32.43	31.49–33.37
Females ($n = 132$)	30.11	28.8–31.42
Income		
Males ($n = 300$)	17,095	16.379–17.811
Females ($n = 132$)	14,885	14.087–15.683

differences between them, and the direction of the differences.[27] To illustrate how this works, Table 12.7, based on the data comparing male and female planners, includes confidence intervals.[28]

Table 12.7 shows that male planners on average are older and earn more than female planners. These differences continue even when sampling error is taken into account. The confidence interval avoids implying that the sample means provide a precise estimate of the differences. Rather, the population means probably fall somewhere within the range indicated by the confidence intervals. We return to this issue later.

The third strategy to increase understanding about the nature of the relationship is to report a measure of association or other measure of effect size. Another set of statistical measures is used to quantify the effect of an independent variable on a dependent variable. These statistics, not tests of statistical significance, measure the strength of a relationship.

The fourth alternative is to replicate studies. One author states, "The results from an unreplicated study, no matter how statistically significant . . . are necessarily speculative. . . . Replications play a vital role in safeguarding the empirical literature from contamination from specious results."[29] Replications do not have to duplicate previous research exactly. Investigators may implement the research using a different population or another setting and see if the findings generalize to other populations or settings. Findings that go in the same direction, whether or not they attain a specified level, or that have overlapping confidence levels, provide more confirmation than a simple significance test.

Gary King and co-authors discussed another approach. They recommend that analysts and researchers, in reporting research results, present quantities of interest in ways that are accurate and informative but that the reader without special training will understand.[30] Such statements, according to these authors, would also include reasonable assessments of uncertainty about estimates. They note that this will require researchers to think more about which quantities are of interest and how to communicate to a wider audience.[31] Confidence intervals, according to King et al., could be a quantity of interest, as could a measure of association. They recommend doing more than reporting the quantity, however, and explain what it means in plain language. For example, in Table 12.7, an analyst might say that we can be 95 percent confident that, on average, male planners' annual income is $32,430 plus or minus $940. The actual procedure these authors presented and tested is complicated and beyond the scope of this text. However, their general ideas merit attention and consideration. The reader with the interest and skill to do so will enjoy reading this challenging article.

With the exception of measuring the size of the effect of the independent variable on the dependent variable, each alternative has constraints. A researcher may have inadequate knowledge or understanding of the question at hand to specify a relationship beyond anticipating some difference. Confidence levels work with interval data, but setting confidence levels for proportional data can become unwieldy. Replication requires the opportunity and resources to repeat a study.

Support for Using Tests of Statistical Significance

Tests of statistical significance can serve as a provisional test of a hypothesis. The term *provisional* avoids implying that any one hypothesis test is definitive. On the other hand, investigators may use them as an initial screen to filter out the weakest relationships and to identify relationships meriting closer examination. In using significance tests as a filter, an investigator should recall that with an alpha level of .05 and a small sample, a test of statistical significance will have low power and will probably identify only strong relationships as statistically significant.

One author surveyed the technical and theoretical arguments for and against using tests of statistical significance. While he found little support for the tests, he gave two justifications for their continued use. First, the tests allow a researcher to argue that he found a low probability of random error or chance. Second, since the tests yield "pure numbers," they standardize findings, allowing researchers to easily scan findings to identify results that are worthy of pursuing and those too weak for further study.[32]

Neither a test of statistical significance nor an alternative approach can indicate that a supported hypothesis is interesting or of any apparent importance. The findings may have no practical significance. Statistically significant findings may be uninteresting and unimportant. A practically significant finding is one that piques people's interest and leads to follow-up investigations, spurs policy decisions, or causes change in behavior. *Practical significance* is what most people mean by the term *significant* in everyday conversation.

Over the years, professionals have become more educated about statistics and more sophisticated in using them. Researchers typically report statistical significance along with measures of association. The readers are given sufficient information to decide whether they agree with the researcher's conclusions or to apply different statistical criteria and reach other conclusions. The researcher and the reader are both likely to understand that statistical significance is not the same as practical significance.

Calculating and Interpreting Chi-Square and *t*-Tests

This section reviews the equations for chi-square and *t*-tests, shows how to use the tables for their respective distributions, and discusses how to interpret and report the test results. Note that Tables 12.8a and 12.8b contain the same information as Table 12.2. This section details the calculations first to obtain expected frequencies and then to obtain chi-square using that information.

Chi-Square

The equation for chi-square is

$$\chi^2 = \Sigma(f_0 - f_e)^2 / f_e$$

The data contained in Table 12.8a show the values for the observed frequency (f_o), and the values in Table 12.8b are the expected frequency (f_e). The symbol f_e is the frequency expected if the independent and dependent variables are randomly related. To obtain its value for a cell, the frequency of the row total is multiplied by the frequency of the column total and divided by the total *n* represented in the table.

For the three cells under "Vocational Education," the calculations to derive f_e are:

f_e (voc. ed. & working) = (64 × 292)/586 = 31.9
f_e (voc. ed. & in school) = (64 × 132)/586 = 14.4
f_e (voc. ed. & unemployed) = (64 × 162)/586 = 17.7

Table 12.8 Observed and Expected Frequencies

A. *Current Status by Training Program Attended: Observed Frequencies (f_0)*

Program

	Vocational Education	On-the-Job Training	Work Skills Training	Total
Working	19	109	164	292
In school	19	82	31	132
Unemployed	26	82	54	162
Total	64	273	249	586

B. *Current Status by Training Program Attended: Expected Frequencies if Program and Status Are Unrelated (f_e)*

Program

	Vocational Education	On-the-Job Training	Work Skills Training
Working	31.9	136.0	124.1
In school	14.4	61.5	56.1
Unemployed	17.7	75.5	68.8

Table 12.9 Chi-Square Distribution

df	Probability			
	.10	.05	.01	.001
1.	2.706	3.841	6.635	10.827
2.	4.605	5.991	9.210	13.815
3.	6.251	7.815	11.345	16.266
4.	7.779	9.488	13.277	18.467
5.	9.236	11.070	15.086	20.515
6.	10.645	12.592	16.812	22.457

Source: Adapted from Walpole, R. E., and R. H. Myer, "Critical Values of the Chi-Square Distribution," in *Probability and Statistics for Engineers and Scientists*, 3rd ed. (New York: Macmillan, 1985), 577.

The next step is to subtract the value of f_e from the value of f_e for each cell, square this figure, and divide it by f_e. For the three cells under vocational education, these calculations are

$$\text{Cell 1,1 } \left(f_0 - f_e\right)^2 / f_e = (19 - 31.9)^2 / 31.9 = 5.22$$
$$\text{Cell 1,2 } \left(f_0 - f_e\right)^2 / f_e = (19 - 14.4)^2 / 14.4 = 1.47$$
$$\text{Cell 1,3 } \left(f_0 - f_e\right)^2 / f_e = (26 - 17.7)^2 / 17.7 = 3.89$$

To obtain the value of chi-square, all the values for $(f_0 - f_0)^2 / f_e$ are summed. For this example, $\chi^2 = 50.57$.

The degrees of freedom equal the number of columns minus 1 multiplied by the number of rows minus 1 $[(C - 1)(R - 1)]$. For the example, the degrees of freedom = $(3 - 1)(3 - 1) = 4$. Next, the investigator consults a chi-square distribution table, such as Table 12.9.

If the investigator has selected a specific alpha level, for example, $\alpha = .05$, he notes the value in the column headed ".05" and the row for 4 degrees of freedom, that is, 9.488. For $\alpha = .05$, he will reject the null hypothesis if the value of chi-square is equal to or greater than 9.488.

Alternatively, he can find the associated probability for $\chi^2 = 50.57$. He goes to the row for 4 degrees of freedom and looks for the value that is closest to, but not greater than, 50.57, that is, 18.467. He identifies which column 18.467 is in, .001; thus, .001 is the associated probability for $\chi^2 = 50.57$ with 4 degrees of freedom. In Table 12.9, $df =$ degrees of freedom. The level at which a value of chi-square is statistically significant is affected by the number of rows and columns in a table. The degrees of freedom for chi-square for a table is:

(number of rows $- 1$) $\times$ (number of columns $- 1$).

If he includes the table in the report, he puts the χ^2 value, the degrees of freedom, and the associated probability underneath the table. (For example, see Table 12.11 in "Problems for Homework and Discussion.") If the investigator combines two columns or makes other changes in the table, he must recalculate chi-square.

Two-Sample t-Tests

To test hypotheses that compare two means, an investigator chooses between two equations of t. If she assumes the variances for the two groups are unequal, she calculates t using the equation

$$s_1^2 = 40,056,241 \qquad s_2^2 = 21,864,976$$

$$t = \frac{17,095 - 14,885}{\sqrt{\dfrac{40,056,241}{402} + \dfrac{21,864,976}{131}}}$$

$$= \frac{2210}{\sqrt{99,624 + 166,908}}$$

$$= \frac{2210}{\sqrt{266,550}} = 4.28$$

If she assumes the variances of the two groups are equal, she calculates t using the equation

$$t = \frac{\overline{X}_1 - \overline{X}_2}{\sqrt{\dfrac{n_1 s_1^2 + n_2 s_1^2}{n_1 + n_2 - 2}} \cdot \sqrt{\dfrac{1}{n_1} + \dfrac{1}{n_2}}}$$

How does she decide between the two models? She may prefer to assume unequal variances, since the appropriate use of the equation does not require unequal population variances.[33] If she wishes, she can test the hypothesis that the variances of the two groups are unequal. If she fails to reject the null hypothesis (H_0: The variances are equal), she will calculate t with the second equation; if she rejects the null hypothesis, she will use the first equation.

For either equation, the degrees of freedom equals the sum of the number of observations in the two samples minus 2. Next the investigator consults a t-distribution, such as Table 12.10. If she has hypothesized a direction, she uses the columns reporting probabilities for a one-tailed test. Otherwise, she uses the columns reporting for a two-tailed test.

Let's consider the example of planners' salaries. We hypothesized that female planners would have lower salaries than male planners; thus, we would want to perform a one-tailed t-test. If we

Table 12.10 Distribution of *t*-Values

| df | Probability for One-Tailed Test | | Probability for Two-Tailed Test | |
	.05	.01	.05	.01
1	6.314	31.821	12.706	63.657
2	2.920	6.965	4.303	9.925
3	2.353	4.541	3.182	5.841
4	2.132	3.747	2.776	4.604
5	2.015	3.365	2.571	4.032
6	1.943	3.145	2.447	3.707
7	1.895	2.998	2.365	3.499
8	1.860	2.896	2.306	3.355
9	1.833	2.821	2.262	3.250
10	1.812	2.764	2.228	3.169
11	1.796	2.718	2.201	3.106
12	1.782	2.681	2.179	3.055
13	1.771	2.650	2.160	3.012
14	1.761	2.624	2.145	2.977
15	1.753	2.602	2.131	2.947
16	1.746	2.583	2.120	2.921
17	1.740	2.567	2.110	2.898
18	1.734	2.552	2.101	2.878
19	1.729	2.539	2.093	2.861
20	1.725	2.528	2.086	2.845
21	1.721	2.518	2.080	2.831
22	1.717	2.508	2.074	2.819
23	1.714	2.500	2.069	2.807
24	1.711	2.492	2.064	2.797
25	1.708	2.485	2.060	2.787
26	1.706	2.479	2.056	2.779
27	1.703	2.473	2.052	2.771
28	1.701	2.467	2.048	2.763
29	1.699	2.462	2.045	2.756
30	1.697	2.457	2.042	2.750
60	1.671	2.390	2.000	2.660
∞	1.645	2.326	1.960	2.567

Source: Adapted from Fisher, R. A., and F. Yates, *Statistical Tables for Biological, Agricultural, and Medical Research*, 6th ed. (New York: Hafner, 1968), Table III, 46.

select a specific alpha level, for example, $\alpha = .01$, we note the value in the column headed ".01" and the row for the appropriate degrees of freedom. In the example, the degrees of freedom was 533 ($n_1 + n_2 - 2$). (In general, the degrees of freedom for a *t*-test is the number of cases minus the number of groups in the study.) So we go to the row marked with ∞ (infinity) and find that if the value of *t* is equal to or greater than 2.326, we will reject the null hypothesis.

It is possible for *t*-values to be negative. The negative sign indicates which group has the larger mean. In reading and interpreting the *t*-distribution tables, the associated probability is the same whether *t* is negative *t* or positive *t*. In Table 12.9, *df* = degrees of freedom. The level at which a value of a *t*-test is statistically significant is related to the degrees of freedom. The *df* for a *t*-test is based on the number of cases sampled. For a two sample *t*-test, the *df* = $n_1 + n_2 - 2$.

Summary

Tests of statistical significance allow a researcher to determine the probability that variables related in a random sample are not related in the population. A test of statistical significance cannot indicate that a relationship is strong or important; it may not even indicate if the direction is as hypothesized. Nor does a finding of statistical significance imply that other parts of the research were carried out correctly. The test only makes a statistical statement about the nature of a relationship.

To carry out a test of statistical significance, a researcher

1. states the null and the research hypotheses
2. selects an alpha level
3. selects and computes a test statistic
4. makes a decision

Figure 12.4 illustrates these steps.

The null hypothesis postulates that an independent and a dependent variable are not related. For some statistical tests, the null hypothesis may state that the relationship goes in a certain direction or does not exceed a specific value.

In hypothesis testing, a researcher runs the risk of making two errors. First, he may reject a null hypothesis that is actually true. The researcher, then, has "accepted" an untrue research

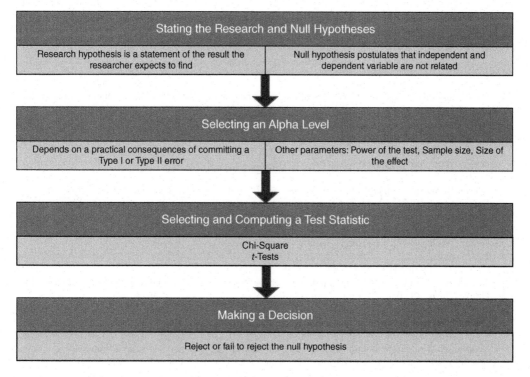

Figure 12.4 Testing Statistical Significance of a Relationship

hypothesis. This error is called a Type I error. Second, the researcher may accept a null hypothesis that is untrue. The researcher has failed to accept a true research hypothesis; that is called a Type II error.

In practice, researchers are more likely to make a judgment about a hypothesis based on the associated probability, the sample size, the magnitude of the effect, and the practical consequences of their decisions than they are to decide on a specific alpha level set a priori. The associated probability is the probability of a specific X^2 or t-value occurring if the relationship in the sample does not occur in the population, that is, if the null hypothesis is true. To detect small effects, the investigator will tend to set higher, that is, less strict, alpha levels, accept higher associated probabilities or use larger samples. With a larger sample and an interest in detecting moderate or strong effects, the investigator will want to work with lower alpha levels or prefer lower associated probabilities.

Depending on the nature of the data and the hypothesis, the researcher selects a test of significance. Chi-square and t-tests are two commonly used tests. The major piece of information needed is the probability that a given statistical value would occur if the null hypothesis were true, indicated by the ** value in the notation $p < **$.

Tests of statistical significance are easily misinterpreted and produce modest information. Researchers are urged to put aside the traditional process of focusing solely on testing a null hypothesis of no difference. Rather, they should provide alternative or additional information supporting the merits of a hypothesis. Researchers may include a specific difference in the null hypothesis, report confidence levels, and report measures of the strength of the relationship between variables. Replication of studies provides the best test of a hypothesis. Occasionally, administrative investigators base a decision on a single test of significance, for example, in deciding whether respondents and nonrespondents are similar, or if shifts in a time series are greater than shifts that can be attributed to long-term trends, cycles, seasonal effects, or random fluctuations.

Notes

1. For discussion on uses of tests of statistical significance in social sciences, see J. Cohen, "Things I Have Learned (So Far)," *American Psychologist* 45 (December 1990): 1304–1312; M. Oakes, *Statistical Inference: A Commentary for the Social and Behavioral Sciences* (New York: Wiley Interscience, 1986).
2. For clear explanations of the epistemological reasons for working with and rejecting null hypotheses, see A. Kaplan, *The Conduct of Inquiry* (San Francisco: Chandler, 1964); R. E. Henkel, *Tests of Significance*, Quantitative Applications in the Social Sciences, 07–004 (Beverly Hills, CA: Sage Publications, 1976), 34–40.
3. See D. B. Wright, *Understanding Statistics: An Introduction for the Social Sciences* (London: Sage Publications, 1997), 42–43, for a short but informative discussion of the nature and philosophy of probability or p values and the link to power analysis.
4. Wright, *Understanding Statistics*, 42–43, 77–83.
5. Cohen, "Things I Have Learned," 1309.
6. E. J. Posavac and R. G. Carey, *Program Evaluation: Methods and Case Studies*, 5th ed. (Englewood Cliffs, NJ: Prentice-Hall, 1997), 97.
7. The analyst would have calculated percents for the table and analyzed them to assess the relationship.
8. Henkel, *Tests*, 48–49.
9. T. H. Poister, *Public Program Analysis* (Baltimore: University Park Press, 1978), 213–214, illustrates the one sample t-test with a worked-out example. K. Meier, J. Brudney, and J. Bohte, *Applied Statistics for Public & Nonprofit Administration*, 9th ed. (Stamford, CT: Cengage Learning, 2014), discuss testing hypotheses with single sample t-tests. See Chapters 11 and 13. Also see Chava Frankfort Nachmias, *Social Statistics for a Diverse Society* (Thousand Oaks, CA: Pine Forge Press, 1997), Chapter 13.
10. J. M. Mayo, Jr., "Job Attainment in Planning: Women Versus Men," *Work and Occupation* (May 1985): 152.
11. An acceptable null hypothesis would also be: "The average salary of male planners is not higher than the average salary of female planners."

12. Wright, *Understanding Statistics,* 131–133; G. W. Bohrnstedt and D. Knoke, *Statistics for Social Data Analysis,* 2nd ed. (Itasca, IL: F. E. Peacock, 1988), 232–233. These assumptions are for between-subjects ANOVA. Other models require somewhat different assumptions; see Wright, *Understanding Statistics,* 141.

13. See Borhnstedt and Knoke, *Statistics for Social Data Analysis,* 240–244.

14. L. Meyers and N. Grossen, Behavioral Research: Theory, Procedure, and Design (San Francisco: W. H. Freeman and Company, 1974), 78–79. Also see Borhnstedt and Knoke, *Statistics for Social Data Analysis,* 232–233.

15. Meyers and Grossen, *Behavioral Research,* 78–79. Also see Bornstedt and Knoke, *Statistics for Social Data Analysis,* 232–233.

16. Wright, *Understanding Statistics,* 132, 183–184.

17. See Bornstedt and Knoke, *Statistics for Social Data Analysis,* 240–244.

18. J. F. Healey, *Statistics: A Tool for Social Research,* 9th ed. (Belmont, CA: Cengage Learning, 2011), 287–293; Wright, *Understanding Statistics,* 131–133, discusses using SPSS to test for equality of variances.

19. In this study, the members of the groups had not originally been randomly assigned. See H. Livengood, *JTPA: Follow Up and Evaluation of Selected Programs* (Charlotte, NC: University of North Carolina at Charlotte, 1993).

20. To help visualize the data and the model, you also could draw a box plot for each group and indicate in it the location of the mean instead of the median. Remember that ANOVA requires the calculation of the arithmetic mean. Wright, *Understanding Statistics,* 116–119.

21. Excel will also do analysis of variance. The output will look different, but a student familiar with the SPSS output should be able to follow Excel output. Also see "Real Statistics Using Excel" at www.real-statistics.com/one-way-analysis-of-variance-anova and Neil Salkind, *Statistics for People for People Who [Think They] Hate Statistics* (Los Angeles: Sage Publications, 2013).

22. Wright, *Understanding Statistics,* 131–133; Bohrnstedt and Knoke, *Statistics for Social Data Analysis,* 232–233. These assumptions are for between-subjects ANOVA. Other models require somewhat different assumptions; see Wright, *Understanding Statistics,* 141.

23. For a discussion on how to interpret the associated probability, see Oakes, *Statistical Inference,* 15–19.

24. Statistical tests in time-series analysis are key in model building, and the findings from a specific test determine subsequent decisions. See R. McCleary and R. A. Hay, Jr., *Applied Time Series Analysis for the Social Sciences* (Beverly Hills, CA: Sage Publications, 1980), 97–100. Homework Problem 5 in this chapter shows how tests of statistical significance are used to compare respondents and nonrespondents.

25. This section is based on the several works critiquing tests of statistical significance. The consulted works were: *What If There Were No Significance Tests?* Eds. L. L. Harlow, S. A. Mulaik, and J. H. Steiger (Mahwah, NJ: Erlbaum Associates, Publishers, 1997), especially "Significance Testing Introduction and Overview" by L. L. Harlow; M. N. Branch, "Statistical Inference in Behavior Analysis: Some Things Significance Testing Does and Does Not Do," *The Behavior Analyst* 22 (1999): 87–92; R. Hubbard and P. Ryan, "The Historical Growth of Statistical Significance Testing in Psychology—and Its Future Prospects," *Educational & Psychological Measurement* 60 (2000): 661–681. Geoff Cumming, *Understanding the New Statistics: Effect Sizes, Confidence Intervals, and Meta-Analysis* (New York: Routledge, 2012).

26. This phrase is cited by Oakes, *Statistical Inference,* 173; Cohen, "Things I Have Learned," 1310. Curiously, the citations differ slightly. Oakes quotes it as "don't remember" and Cohen as "don't know." The original is found in F. M. Lord, "On the Statistical Treatment of Football Numbers," *American Psychologist* 2 (1953): 750–751.

27. For information on using confidence intervals, see H. Rothstein and M. C. Tonges, "Beyond the Significance Test in Administrative Research and Policy Decisions," *Journal of Nursing Scholarship* 32 (2000): 66–70; Cumming, *Understanding the New Statistics,* Chapters 3 and 4. For a critique of using confidence levels, see J. M. Cortina and W. P. Dunlap, "On the Logic and Purpose of Significance Testing," *Psychological Methods* 2 (1997): 161–172; Cumming, *Understanding the New Statistics,* cited earlier, has proposed moving away from the traditional hypothesis testing model using significance and toward emphasizing effect sizes and confidence intervals. His work, although challenging at times, is thorough and interesting.

28. The standard error, used to compute the confidence level, was estimated by dividing the standard deviation by the square root of the size of the respective samples.

29. Hubbard and Ryan, "The Historical Growth of Statistical Significance Testing in Psychology."

30. Gary King, Michael Tomz, and Jason Wittenberg, "Making the Most of Statistical Analysis: Improving Interpretation and Presentation," *American Journal of Political Science* 44, no. 2 (April 2000): 341–355, available at gking.harvard.edu/files/making.pdf.

31. Ibid., 347. The authors give as an example a statement that satisfies the criteria for their proposal: "'Other things being equal, an additional year of education would increase your annual income by $1,500 on average, plus or minus about $500.' The sentence is substantively informative because it conveys a key quantity of interest in terms the reader wants to know. At the same time, the sentence indicates how uncertain the researcher is about the estimated quantity of interest."
32. Henkel, *Tests*, 87. On pages 78–87, he summarizes the literature on the utility of tests of statistical significance. He divides the literature into two categories: technical issues and philosophy of science issues.
33. For a discussion of these two equations, see H. M. Blalock, *Social Statistics* (New York: McGraw Hill, 1972), 220–228. For a discussion of the *t*-test and these two equations, also see Kenneth Meier, Jeffrey Brudney, and John Bohte, *Applied Statistics for Public and Nonprofit Administration*, 7th ed. (Belmont, CA: Thomson-Wadsworth, 2009), 217–228.

Terms for Review

randomly related
statistical significance
practical significance
hypothesis
null hypothesis
Type I error
Type II error
chi-square
t-test
ANOVA
F-test
degrees of freedom
associated probability
effect size
confidence interval
within groups variance
between groups variance

Questions for Review

The following questions should indicate whether you have a basic competency in this chapter's material.

1. What is meant by statistical significance? How is it different from practical significance?
2. Describe the steps in hypothesis testing.
3. Distinguish between Type I error and Type II error. What two strategies are available to reduce the probability of a Type I error? What two strategies are available to reduce the probability of a Type II error?
4. The common wisdom among social scientists is that tests of statistical significance are overrated. What is the value of tests of statistical significance? What are their limitations?
5. Evaluate the soundness of using an alpha level of .05 as a minimal cutoff point for rejecting a null hypothesis.
6. Discuss the adequacy of tests of statistical significance as evidence supporting a hypothesis.
7. Write a paragraph for each of the four alternatives to tests of statistical significance. Describe and evaluate each.

8. Why is the Majesty City Job Training study described as a posttest-only design?
9. When would an analyst use analysis of variance rather than a *t*-test?

Problems for Homework and Discussion

1. A parents' group has charged that minority students are more likely than other students to be suspended for breaking school rules. For example, minority students are more likely to be suspended for fighting. The school superintendent has a random sample of 200 cases drawn from the list of all students in grades 7 through 12 who have been cited for serious school infractions. For each sample, student data are gathered on the student's racial or ethnic group, nature of offense, and action taken.

   a. State a hypothesis that the superintendent may have tested.
   b. What would be the practical consequences of committing a Type I error? What would be the practical consequences of committing a Type II error?
   c. Should the superintendent seek to minimize a Type I or a Type II error? Justify your answer.
   d. What contribution would power make in deciding if the parents had a legitimate concern?

2. The data given in Table 12.11 are reported in a study of soup kitchen users. (Note: Soup kitchens serve free meals to homeless or very poor people. Table 12.11 is a duplicate of Table 12.9.)

   a. State the hypothesis and the null hypothesis that these data may have been testing.
   b. If you set $\alpha = .05$, what is the probability that you committed a Type I error? Would you reject the null hypothesis?

Table 12.11 Number of Meals Eaten Daily by Soup Kitchen Users

Age of Soup Kitchen User			
Number of Meals Eaten Daily	Less than 31 Years Old (n = 66) %	31–54 Years Old (n = 213) %	Over 54 Years Old (n = 144) %
1.	37.9	29.1	27.1
2.	47.0	49.3	40.3
3.	15.2	21.6	32.6

Note: $X^2 = 10.4$, $df = 4$, $p < .05$.

   c. If you set $\alpha = .01$, what is the probability that you committed a Type I error? Would you reject the null hypothesis?
   d. What is the value of knowing whether age is related to the number of meals a soup kitchen user eats daily? Assess the adequacy of using only the information provided by a χ^2 test to decide if these two variables are related.

3. The data given in Table 12.12 were reported in a study of children's knowledge of the effects of smoking. The column headed "χ^2 probability" reports the associated probabilities.

   a. How many hypotheses are being tested?
   b. Explain what information the associated probabilities reported in the chi-square analysis provide.

   c.   State the hypotheses that you would probably accept based on the associated probabilities results.

   d.   What other information would you include, if any, to support your hypotheses? Justify your answer.

Table 12.12 Percentage of Correct Answers by Grade Level

	6th	9th	12th	χ^2
	(265)	*(125)*	*(120)*	*Probability*
Lung cancer is higher among pipe smokers	10.2	5.6	10.8	.593
Smoking constricts blood vessels	38.2	70.8	57.7	.0001
Smoking helps circulation	43.3	65.8	61.3	.0001
Smokers live longer than nonsmokers	77.0	90.8	84.9	.0189

Source: Chen, T. T. L., and A. E. Winder, "When is the Critical Moment to Provide Smoking Education in Schools," *Journal of Drug Education* 16 (1986): 121–132.

4.   Consider the data in Table 12.13:

   a.   Based on information in the table, on which characteristics do male and female planners appear to differ most? On what did you base your answers?

   b.   What alpha level would you suggest for this dataset? Why did you select it?

   c.   Which null hypotheses would you fail to reject?

   d.   What information does a *t*-test give that chi-square doesn't?

   e.   What information would you like to have that is not provided by the statistical tests?

Table 12.13 Comparisons Between Male and Female Planners on Selected Characteristics

	Men $\bar{X}$	Women $\bar{X}$	
	(s)	(s)	*T*
Age	32.43	30.11	2.9[b]
	(8.28)	(7.72)	
Agency size	4.70	5.29	−1.7[a]
	(2.88)	(2.99)	
Job turnover rate	.51	.60	−2.9[b]
	(.37)	(.39)	
Number of planner roles	7.98	7.88	.53
	(1.67)	(1.79)	
Years of professional experience	3.67	3.42	7.26[b]
	(5.31)	(2.57)	
Income	17,095	14,885	4.28[b]
	(6,329)	(4,676)	

[a]Significant at the .1 level.
[b]Significant at the .01 level.

Source: Mayo, J. M. Jr., "Job Attainment in Planning: Women Versus Men," *Work and Occupation* (May 1985): 152. (Copyright © 1985 by Sage Publications, Inc. Reprinted with permission: Sage Publications, Inc.).

5.   A random sample of 407 adults was asked to participate in a study; 283 agreed to participate. Can the researchers conclude from the data in Table 12.14 that the participants were similar to the nonparticipants? Justify your answer.

Table 12.14 Comparison Between Participants and Nonparticipants on Selected Characteristics

	Participated %	*Did Not Participate %*
Race		
White	70 (186)	30 (81)
African American	88 (97)	12 (13)
$\chi^2 = 14.25$, $df = 1$, $p < .001$		
Gender		
Male	85 (115)	15 (21)
Female	70 (168)	30 (73)
$\chi^2 = 10.36$, $df = 1$, $p < .005$		
Education		
12 years or less	73 (101)	27 (37)
13–15 years	82 (80)	18 (17)
16 years or more	72 (102)	28 (39)
$\chi^2 = 3.52$, $df = 2$, $p < .25$		
Age		
18–39 years	80 (177)	20 (43)
40 years or older	67 (105)	33 (51)
$\chi^2 = 8.36$, $df = 1$, $p < .005$		

Source: Linz, D. et al., "Estimating Community Standards," *Public Opinion Quarterly* 55 (Spring 1991): 93.

6. Use $\alpha = .05$ to interpret the findings in Table 12.15.

Table 12.15 Differences Between Early and Late Adopters of Spreadsheet Software

Early Adopters			Late Adopters			
	Mean	S	Mean	S	t-value	Significance Level
Age	28.3	(7.3)	30.3	(10.1)	1.91	.059
Years of education	16.2	(1.7)	15.4	(1.8)	3.06	.002
Number in prof. assoc.	2.6	(3.5)	2.6	(4.9)	.05	.956
Number of professional journals read	8.2	(4.5)	6.7	(5.3)	2.21	.028
Size of work group	2.6	(1.2)	2.5	(1.1)	.65	.516

Source: Brancheau, J. C., and J. C. Wetherbe, "The Adoption of Spreadsheet Software: Testing Innovation Diffusion Theory in the Context of End-User Computing," *Information Systems Research* 1 (June 1990): 129.

Working With Data

Access the County Data File accompanying this textbook to complete the following exercises.

1. Compare the median household income of counties in the "plains" region of the state to that of counties in the "coastal" region.

   a. State a research hypothesis and null hypothesis.
   b. To test the hypothesis, would you use chi-square or a *t*-test? Justify your choice.
   c. Using the dataset, carry out the appropriate statistical procedure. Based on the analysis, would you reject your null hypothesis? Why or why not? What evidence did you use to make your decision? (Provide detail on the number of cases, the alpha level used, and the value of the statistic used.)

2. Analyze the data to see if county expenditures on cultural or recreation services vary by school enrollment. To conduct your analysis, categorize expenditures and school enrollment into three categories: high, medium, and low. Test two hypotheses.

    a. Report your findings in a three-by-three table with accompanying statistical results.
    b. State the hypotheses you tested.
    c. Were your hypotheses supported? Give evidence to support your decision.

3. Calculate the percentage of the total registered voters who are Hispanic. Use this percentage as a measure of political activity and consider whether minority political activity varies by region of the state. Use region as an independent variable and the percentage of registered voters that are Hispanic as a dependent variable.

    a. State a research hypothesis and null hypothesis.
    b. Using the appropriate test of significance, test the hypothesis with two regions of the state.
    c. What is the associated probability? What does it indicate about the relationship between region and politically active Hispanic population?
    d. Which statistical test did you use? If you included three regions of the state, which test would you use?

4. Review the General Social Survey variables and then access the GSS.save file. State research and null hypotheses about a relationship between the variables SEX and NATEDUC. Set up a table similar to that in Table 12.2. Use the variable SEX as the column variable and NATEDUC as the row variable. Using a computer program, obtain a value for chi-square. Select an alpha level. Is the value of your chi-square significant at less than the alpha level? What do you decide about your hypotheses? Explain.

5. This exercise gives you some experience in applying analysis of variance.
    Using data from the County Data File and a software program, analyze the relationship between SAT Grand Total Average Score and Region of the State.

    a. State research and null hypotheses and choose an alpha level.
    b. Create a graphic similar to Figure 12.2.
    c. Generate an analysis of variance table similar to Figure 12.3.
    d. What does your analysis tell you about the relationship between Region of the State and SAT scores?

Recommended for Further Reading

Theoretical orientation for this chapter was drawn from the insights of Michael, Oakes, *Statistical Inference: A Commentary for the Social and Behavioral Sciences* (New York: Wiley Interscience, 1986). This book will challenge and reward readers interested in statistical inference.

Cohen, J., "Things I Have Learned (So Far)," *American Psychologist* 45 (December 1990): 1304–1312, is an accessible work that identifies common misconceptions in interpreting and working with tests of statistical significance.

Cumming, Geoff, *Understanding the New Statistics: Effect Sizes, Confidence Intervals, and Meta-Analysis* (New York: Routledge, 2012).

Helena, Kraemer, and Sue Thieman, *How Many Subjects? Statistical Power Analysis in Research* (Newbury Park, CA: Sage Publications, 1987), discuss the relationship between sample size and the power of statistical tests. Also see the second edition of this text by Helena Kraemer and Christine Blasey, available in 2015.

Harlow, L. L., S. A. Mulaik, and J. H. Steiger, eds., *What If There Were No Significance Tests?* (Mahwah, NJ: Erlbaum Associates, Publishers, 1997). An excellent source on current thinking about tests of statistical significance. "Significance Testing Introduction and Overview" by L. L. Harlow is especially recommended.

13 Examining Relationships Between Nominal and Ordinal Variables

Contingency Tables, Measures of Association, and Control Variables

In this chapter, you will:

1. Learn how to construct and interpret contingency tables to study relationships between nominal and ordinal variables.
2. Learn how to select and interpret common measures of association for nominal and ordinal variables.
3. Learn about using nominal and ordinal variables as control variables in relationships.
4. Continue to learn about using analysis of variance to analyze relationships between nominal independent and interval dependent variables.

This chapter discusses a major approach to data analysis and some of its components: contingency tables, measures of association, and using control variables in contingency table analysis. The chapter also continues a discussion of using analysis of variance to address relationships. The measures of association generated from contingency tables provide evidence of the strength and nature of a bivariate relationship. Contingency tables are commonly used to investigate relationships between nominal and ordinal variables with a limited number of values. Such variables are typical of questions and other items used in survey research. Specific measures of association have been developed to apply depending on whether the variables are nominal or ordinal. Analysts typically examine contingency tables to learn about strength and direction of relationships between nominal and ordinal variables. They use percent and measures of strength and direction of association to aid them. They also use tests of significance to assess the statistical significance of data.

In Chapter 1, we discussed model building and the addition of control variables to a model. Contingency table analysis continues the use of models by applying them in the analysis of data. A two-variable contingency table presents the data to test a two-variable model expressed in a hypothesis. Analysts then extend the model testing and analysis by introducing a control variable and determining if that changes the original relationship. We discuss this procedure in Chapter 13.

Experimental research usually tests the relationship between a nominal or ordinal independent variable and an interval dependent variable. Researchers often analyze the data to assess these relationships with analysis of variance. Its primary output is a test of statistical significance, the F-test. The application of analysis of variance as a test of statistical significance is discussed in Chapter 12. Analysts can also obtain a measure of association, eta (η), from it with one additional calculation. We discuss eta in this chapter. Familiarity with ANOVA is also important for other reasons. It provides a procedure for assessing the relationship between a nominal or ordinal variable with more than two categories and an interval variable. Although analysts use it primarily to test statistical significance, they also can use it to obtain a measure of effect size. A researcher experienced in analysis of variance may seldom use contingency tables to analyze a dataset. The reverse also is generally true. Contingency-table analysis and regression, discussed in Chapter 14, dominate nonexperimental research by social scientists. Researchers often use these analyses when working with observational data. In comparison, analysis of variance dominates work by researchers using experimental designs. Familiarity with analysis of variance is also helpful in understanding some aspects of regression. In practice, analysts tend to work with one or the other approach, depending on their training, the nature of the data, and the type of research problem.

This chapter does not give a full, balanced coverage of both techniques. It should, however, give you a better idea of their separate roles.[1] If you have not done so, you may find it useful to read Chapter 12 on statistical significance before reading Chapter 13.

Constructing and Interpreting Contingency Tables

Analysts usually want to investigate the joint distribution of the values of two or more variables. They may want to know how the values of one variable differ for different levels or categories of a second variable and, to find out, they organize the data into a contingency table. A contingency table shows the frequency or relative frequency of each value of the dependent variable for each value of the independent variable. (Various software programs are available to do this. Microsoft Excel can create contingency tables using the pivot table routine.)[2] An analyst may use measures of association to summarize the relationship depicted in the contingency table. Measures of association are statistics that indicate the strength of the relationship between a dependent and an independent variable.

Constructing Contingency Tables

Contingency tables and measures of association help test hypotheses. To "accept" a research hypothesis requires empirical evidence that the relationship between variables was

1. nonrandom
2. in the anticipated direction
3. sufficiently strong given the sample size that it was not a chance occurrence

Contingency tables indicate the direction of the relationship and, with measures of association, provide information so one can judge its strength. Tests of statistical significance indicate the probability that a relationship occurred by chance.

Table 13.1 shows how residents of Large County reacted to a proposal to consolidate the county government with the government of the county's largest city. Analysts wanted to display how residents inside the large city and residents outside of the city differed in their responses. They had hypothesized that support for consolidation was related to where the resident lived—in the city or outside of it.

Figure 13.1 is a simple model diagram of that hypothesized relationship.

By convention, the values of the independent variable head the columns and the values of the dependent variables head the rows of the table. If the variables are ordinal, the values are arranged along a continuum. In this table, "no opinion" was put in the middle or neutral position. The research model guides analysts in deciding which values to include and how to include them. The study's purpose should dominate their decisions, but other factors, such as the distribution of the values and the clarity of the table, may be considered as well. In Table 13.1, the analysts examined the values "inside city" and "outside city" and "for," "against," and "no opinion." Opinions on support for consolidation originally included "strongly for" and "strongly against." In the table, the analysts defended combining "strongly for" with "for" and "strongly against" with "against" on the grounds that their job was to estimate the proportion of residents for and against consolidation and that more specific information was not needed. If relatively few residents were strongly against consolidation, the analysts may appropriately decide to report four categories:

Table 13.1 Support for City-County Consolidation by Residence

	Residence			
	Inside City		Outside City	
Support for Consolidation	N	%	N	%
For	273	54	54	37
No opinion	134	27	34	23
Against	98	19	57	39
Total (% rounded)	505	100	145	100

Note: Percentages may not add up to 100% because of rounding.

Figure 13.1 Support for City–County Consolidation by Residence

"strongly for," "for," "no opinion," and "against." Depending on the purpose of a report, such an arrangement may provide fully adequate information without adding to a table's size and clutter.

The percentages are computed "in the direction of the independent variable." That is, if the independent variable heads the columns, as in Table 13.1, the column percents sum to 100 percent; if the independent variable heads the row, the row percents sum to 100 percent. Occasionally, a report will focus on the percentage of respondents in each category. For example, 54 percent of the respondents who live inside the city support consolidation. In this study, local government officials were most interested in the difference between city and non-city residents. Each group's total was treated as 100 percent, and the relative frequency of each group who responded "for," "no opinion," and "against" was computed.

Interpreting Contingency Tables

To interpret contingency tables, analysts focus on the percentages and the percentage difference between the values of the independent variable. In Table 13.1, the population of city residents is so much larger than that of non-city residents that directly comparing the frequencies (raw numbers) for any particular response would be inappropriate. Using percentages corrects this. The *percent difference*, the difference in percents when subtracting across the rows, can be tested to determine whether city residents' level of support is *statistically* different from that of county residents. The analyst begins by reading across a row and noting how the percentages change. The difference can range from 0, for no difference, to 100, indicating maximum difference.

In comparing the two groups in Table 13.1, we find that 54 percent of city residents supported consolidation, but only 37 percent of non-city residents did, a difference of 17 percentage points. Nineteen percent of city residents and 39 percent of non-city residents were against consolidation, a difference of 20 percentage points. We can conclude that city residents differ from non-city residents in the distribution of opinions toward consolidation. Since these results are based on a sample, to fully conclude that city residents differed from non-city residents would require a test of the hypothesis to determine if the difference is statistically significant.

Bivariate Distribution

The joint distribution of the values of two variables is a *bivariate distribution*. Table 13.1 shows the bivariate distribution between residence and support for consolidation. In the box, Example 13.1 and Table 3.2 show a second relationship, the bivariate distribution, between the evaluation of city services and support for consolidation. The table shows that those who gave a higher evaluation to city services were more likely to support consolidation. The example's discussion spells out how the table was constructed and interpreted. Figure 13.2 shows a model diagram of the relationship.

Example 13.1 Organizing and Interpreting Contingency Tables

Situation: City and county residents were surveyed about their evaluation of local government services and whether they favored city–county consolidation. The researchers' implied hypothesis was: Support for city–county consolidation is directly related to respondents' evaluation of their local services.

Analysis: The data are organized into a contingency table, Table 13.2. The values of the independent variable head the columns, and the values of the dependent variable head the rows. Either the highest or lowest value can go in the first column or first row. To avoid clutter, only the percents are shown in the body of the table.

To determine the direction of the relationship, an analyst looks across the rows: From left to right along the "For" row, the percentages drop; and from left to right along the "Against" row, the percentages increase. The data are consistent with the hypothesis. The higher that respondents rate local services, the more likely they are to favor consolidation; the lower they rate local services, the less likely respondents are to favor consolidation. Another way to view the table is to look at how the "center of gravity" changes along the values of the independent variable. The "Excellent or Good" column has the highest center of gravity and the "Poor" column the lowest. This strategy helps to quickly identify non-linear relationships, where a middle value will have the highest or lowest center of gravity.

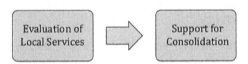

Figure 13.2 Support for City–County Consolidation by Evaluation of Local Services

Table 13.2 Support for Consolidation by Evaluation of Local Services

Evaluation of Local Services

Support for Consolidation	Excellent or Good (%)	Fair (%)	Poor (%)	Total (N)
For	55	47	36	331
No opinion	24	25	34	161
Against	21	28	31	155
Total (*N*)	387	224	36	647

Note: Percentages may not add up to 100% due to rounding.

A test of statistical significance establishes the probability that the distribution in the contingency table occurred by chance. The percentage distributions provide further support for a hypothesis, although whether one assesses the relationship as strong enough to support the hypothesis based on the percentage distributions is a matter of judgment.

Control Variables

In discussing model building and proposing two-variable relationships, investigators also consider control variables as well.

A *control variable* refers to a situation in which a third variable is included in the analysis to see if it influences the relationship between the other two variables.

For example, we may wish to see whether the relationship between public service rating and support for consolidation is the same for city residents as it is for non-city residents. To determine this, we would divide the original group of cases into two subgroups—city residents and non-city residents—and create two additional tables, one for each group. We would then analyze the relationship for each subgroup. In doing so, we would have controlled for place of residence. Table 13.3 shows the relationship among three variables:

- support for consolidation
- rating of local services
- place of residence

Inspection of the percentages in the table shows that those in the city were more likely than others to be in favor of consolidation if they rated service as "good" (65 percent), whereas non-city residents were very unlikely to be for consolidation if they rated service as "good" (1 percent). The number of variables displayed in a table can be expanded to three, four, and even more. Although occasionally we see a table with four variables, very few tables show the relationships among more than three.

These relationships can be illustrated by a schematic model. Figure 13.3 shows the three-variable model on which Table 13.3 is based. Note that place of residence and evaluation of local services are both related to support for consolidation. Place of residence is also related to

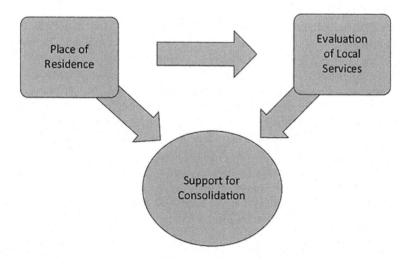

Figure 13.3 Support for Consolidation by Rating of Local Services and Place of Residence

Table 13.3 Support of Consolidation by Rating of Local Services by Place of Residence

	Residence					
	City Service Rating			*Non-City Service Rating*		
Support for Consolidation	*Good (%)*	*Fair (%)*	*Poor (%)*	*Good (%)*	*Fair (%)*	*Poor (%)*
For	65	46	30	1	49	35
No opinion	20	25	35	45	25	34
Against	15	29	35	54	25	31
Total (*N*)	329	161	15	59	63	21

evaluation of local services. The analyst was able to demonstrate this by reviewing the percents in the three-variable table. It is important to demonstrate that all three variables are related. For a more complete analysis, he could prepare a separate contingency table showing the relationship between place of residence and support for consolidation. The table would also be useful if the relationship were a weak one.

Tables also are described by their *size* and *dimensions*. A table's size is indicated by the numbers of rows and columns in the table. The number of rows is the number of categories of the dependent variable, and the number of columns is the number of categories of the independent variable. The dimension of a table refers to the number of variables whose joint distribution is being displayed. A table showing the joint distribution of two variables is a two-dimensional table, one with three variables is a three-dimensional table, and so on.

Selecting and Using Measures of Association

The percent distributions in a contingency table suggest the strength of the relationship between two variables. However, determining strength from percents alone can be difficult, and what constitutes a "strong" or a "weak" relationship partially depends on the observer. Measures of association are descriptive statistics that indicate the strength of a relationship between two variables. They provide a standard criterion to summarize the relationship in a table.

Perfect and Null Relationships

Each measure of association has a criterion for a perfect relationship. A perfect relationship is one in which a change in the independent variable is always associated with the same change in the dependent variable. An analyst may observe a positive relationship in which an increase in the value of the independent variable always coincides with the same increase in the dependent variable. Alternatively, she may observe an inverse relationship in which an increase in the value of the independent variable always coincides with the same decrease in the dependent variable. Typically, measures of association designate a perfect relationship with an absolute value of 1.00.

Similarly, each measure of association has a criterion for a *null relationship*. A null relationship occurs if a change in the independent variable is as likely to coincide with an increase as with a decrease, or with no change, in the dependent variable. Typically, measures of association designate a null relationship with a value of 0.00.

The closer the value of a measure of association is to 0.00, the weaker the association; the closer it is to +1.00 or −1.00, the stronger the association. The sign, a plus or a minus, indicates the direction, not the strength, of a relationship. A +1.00 represents a perfect positive or direct relationship, and a −1.00 represents a perfect negative or inverse relationship. Identifying direction is only possible for ordinal, interval, and ratio variables. Relationships involving nominal variables cannot properly be described as either positive or inverse (negative). If two variables are not related at all, the relationship is considered *null*.

Tables 13.4, 13.5, and 13.6 illustrate in turn these three types of relationships.

A Perfect Positive Relationship

A positive relationship in which an increase in the value of the independent variable always coincides with an increase in the dependent variable is a perfect positive relationship, and it would have a strength of +1.00. This is illustrated in Table 13.4.

A Perfect Inverse or Negative Relationship

Alternatively, an analyst may observe an inverse relationship, in which an increase in the value of the independent variable always coincides with a decrease in the dependent variable. The measure of association would designate this perfect relationship with a value of 1.00. Since the variables are ordinal, the measure would be −1.000. (See Table 13.5.)

Null Relationship

Table 13.6 illustrates a null relationship. A null relationship occurs if a change in the independent variable is as likely to coincide with an increase as with a decrease or with no change in the dependent variable. Typically, measures of association designate a null relationship with a value of 0.00.

Table 13.4 Length of Job-Training Program and Placement Success

Length of Job-Training Program		
Placement Rate (%)	*≤ 1 Month*	*2–3 Months*	*> 3 Months*
Low (<50)	100%		
Moderate (50–75)		100%	
High (>75)			100%

Table 13.5 Length of Job-Training Program and Placement Success

Length of Job-Training Program		
Placement Rate (%)	*≤ 1 Month*	*2–3 Months*	*> 3 Months*
Low (<50)			100%
Moderate (50–75)		100%	
High (>75)	100%		

Table 13.6 Length of Job-Training Program and Placement Success

Length of Job-Training Program		
Placement Rate (%)	*≤ 1 Month*	*2–3 Months*	*> 3 Months*
Low (<50)	20%	20%	20%
Moderate (50–75)	40%	40%	40%
High (>75)	40%	40%	40%

Table 13.7 Length of Job-Training Program and Placement Success

Length of Job-Training Program			
Placement Rate (%)	*≤ 1 Month*	*2–3 Months*	*> 3 Months*
Low (<50)	35%	30%	25%
Moderate (50–75)	45%	42%	40%
High (>75)	20%	28%	35%

Readers should note that the perfect relationships discussed and illustrated in Tables 13.4 and 13.5 are linear relationships. It is possible to have a perfect nonlinear relationship. It would look different than those in these tables. Nonlinear relationships will be discussed in later sections. The reader should be aware that perfect relationships in actual research are rare.

What about Table 13.7? This table illustrates why one needs to understand a statistic's model for a perfect or null relationship. Readers may decide that it shows a direct relationship. As the length of job training increases, the placement rate also increases. (Review the percents in the first and third rows of the table.) However, at least one common measure of association, lambda, would consider it null and produce a value of 0.00.[3] Table 13.4 also calls attention to the need for analysts to view actual tables and inspect the percentages in the table when practical rather than relying solely on measures of association to interpret results.

The Logic of a Measure of Association

Rather than surveying several nominal and ordinal measures of association, we present one nominal measure of association to give you a better idea of what a measure of association does. We also outline the major features of an ordinal measure of association to demonstrate how ordinal measures incorporate the additional information contained in ordinal variables and to alert you to some common problems encountered in working with ordinal measures of association. We then discuss a limited number of commonly used ordinal measures. With this understanding, you can refer to a statistics text to study and select a measure appropriate to your needs. In addition, you can find the calculations for the nominal and ordinal measures mentioned in the final section of the chapter. The "Recommended for Further Reading" section lists appropriate statistics texts to consult.

Calculating Lambda

The ratings of the performance of a city's human resources department by employees in other agencies in the city are reproduced in Table 13.8. Visual inspection of the table can help determine if a relationship exists. An analyst can get a sense of the relationship by observing if the mode of the dependent variable changes from one category of the independent variable to another and by comparing percentages for the dependent variable across categories of the independent variable. Knowing the strength of the relationship could be important for comparing it to the relationship between length of employment and rating of the human resources department.

To illustrate a measure of the strength of association for Table 13.8, we look at a nominal measure, lambda. Lambda is a *proportional reduction in error* (PRE) measure and is based on the mode. All PRE measures indicate how much knowledge of the distribution of the independent variable reduces the error in predicting the distribution of the dependent variable. How much error is in predicting the distribution of the dependent variable without knowing values of the independent variable? Lambda calls this error *original error*. Original error for lambda is equal to the number of nonmodal responses.

Table 13.8 Ratings of the Human Resources Department's Performance by Respondent Department

	Police	Fire	Public Works	Planning	Total
Poor	10	15	5	8	38
Satisfactory	5	10	15	2	32
Good	15	5	5	0	25
Total	30	30	25	10	95

Consider analyzing the ratings in Table 13.8 to assess the relationship. Imagine that you were to meet all 95 survey respondents but only knew the total number giving each rating. You know that 38 of the respondents rated the human resources department as "poor," 32 rated the department as "satisfactory," and 25 rated the department as "good." You guess how each individual rated the human resources department. To meet the assumptions of lambda, you guess that each person rated the department as "poor." In other words, you guess the most common response, or the mode, for each individual. With this guessing rule, you are wrong 57 times. Fifty-seven is the original error—the sum of the nonmodal responses—those that are not in the mode. If you guessed that everyone gave the department either a "satisfactory" or "good" rating, you would be wrong more than 57 times.

NEW ERROR VERSUS ORIGINAL ERROR

Does including data on an independent variable reduce the original error? Lambda considers the error for each value of the independent variable as *new error*. New error for lambda is equal to the nonmodal responses for each value of the independent variable. Let's return to our example. Note that the table gives the distribution of the independent variable, that is, the distribution of ratings given by each department's respondents. You use this additional information and ask each respondent in which department he or she works. You then guess that a respondent's rating was the same as the modal response for his or her department. You guess a police department employee rated the human resources department as "good," fire and planning employees rated it as "poor," and public works employees rated it as "satisfactory." In our example, you have incorrectly labeled 15 police employees, 15 fire employees, 10 public works employees, and 2 planning employees. With knowledge of a respondent's department, you made 42 errors. The 42 errors, based on the distribution of the independent variable, make up the new error.

To compute lambda, subtract the new error from the original error and divide the result by the original error. The solution for our example is:

$$\text{Lambda} = (\text{original error} - \text{new error}) / \text{original error} = (57 - 42) / 57$$
$$= .263$$

The .263 indicates that by knowing respondents' departments, we have reduced the error in guessing the ratings by 26.3 percent. This demonstrates that knowing department affiliation can improve our guess. There seems to be a relationship.

AN INSENSITIVE MEASURE

Lambda is an insensitive measure. That is, it does not detect small shifts or differences in the strength of a relationship. You can infer this by looking at its criterion for a null relationship. Lambda is calculated in such a way that many relationships with a discernible pattern between

Table 13.9 Ratings of the Human Resources Department's Performance by Respondent Department

	Police	Fire	Public Works	Planning	Total
Poor	12	15	9	8	44
Satisfactory	6	10	8	2	26
Good	12	5	8	0	25
Total	30	30	25	10	95

the independent and dependent variables have a lambda equal to 0.00. Consequently, even though lambda equals 0.00, an analyst or audience may see a relationship between the variables. We illustrate this by arranging our dataset on perceptions of the human resources department a little differently in Table 13.9. Do you see any departmental trends in the perception of the human resources department in that table?

The majority of respondents in the planning department rate the human resources department as "poor." Exactly half of the employees in the fire department and fewer than half the employees in the police and public works departments give a similar rating; in fact, 40 percent of the police employees rated human resources as "good." Lambda for these data equals 0.00. The lesson is not that your perception was wrong. Rather, it is to demonstrate that lambda is insensitive and ignores relationships if the mode of every value of the independent variable is the same category of the dependent variable.[4]

Salient Features of an Ordinal Measure of Association

Ordinal measures of association summarize relationships between ordinal independent and dependent variables. These measures take into account the additional information provided by ordinal measures. If an analyst has hypothesized a positive relationship, an increase in the value of the independent variable should be associated with an increase in the value of the dependent variable. If she has hypothesized an inverse relationship, an increase in the value of the independent variable should be associated with a decrease in the value of the dependent variable.

Ordinal measures of association vary in their statistical models of what constitutes a perfect relationship. Gamma is an ordinal measure with a generous criterion for detecting perfect relationships. Consequently, it may overestimate the strength of a relationship between two variables. An analyst who wishes to identify potentially interesting relationships may therefore prefer to work with gamma, which will screen out fewer relationships than other ordinal measures. The calculations for gamma will be discussed later in the chapter.

Ordinal measures cannot effectively estimate the strength of nonlinear relationships. (The reader may find it useful to review the discussion of types of relationships and their definitions.) If the analyst has hypothesized or suspects a nonlinear relationship, she may prefer to use a nominal measure of association. Nominal measures evaluate each category of a variable without assuming a rank ordering, and they will assign the same numerical value no matter how the categories are arranged. Ordinal measures assume a monotonically increasing or decreasing pattern; that is, two variables either steadily increase or steadily decrease.

Table 13.10 relates years of working experience to perceptions of the human resources department's effectiveness. Note the nonlinear pattern. Employees with the least working experience are most likely to rate the human resources department as "poor." Employees with between two and five years of working experience are most likely to rate the department as "good." Employees with more than five years of experience are most likely to rate the department as "satisfactory."

The data indicate a relationship that is neither distinctly positive nor inverse. The gamma value, suggesting a weak positive relationship, underestimates the strength of the relationship because

Table 13.10 Ratings of the Performance of the Human Resources Department by Years of Full-Time Working Experience

	< 2 years	2–5 years	> 5 years
Poor	20	0	5
Satisfactory	10	10	20
Good	5	20	5

Note: Gamma = .29.

it ignores the nonlinear pattern. A good analyst will examine tables analyzing hypothesized or anticipated relationships even if the measures of association suggest a weak relationship.

Measures of association can be characterized as symmetric or asymmetric. *Symmetric measures* have the same value no matter which variable is designated as the independent variable. *Asymmetric measures* may have different values depending on which variable is designated as the independent variable. Lambda is an asymmetric measure. Gamma is a symmetric measure of association. Cramer's V, which appears in several examples in this chapter, also is a symmetric measure.

Some Common Nominal and Ordinal Measures of Association

Nominal measures of association can be categorized as chi-square–based measures and proportional reduction in error measures. Chi-square, an inferential statistic, is the basis for several nominal, symmetric measures, including Cramer's V. Lambda, discussed earlier, is a PRE measure.

PRE measures tend to have lower values for a given relationship than do chi-square–based measures. Instead we might choose a chi-square–based measure to screen a set of contingency tables to select those that merit further analysis. Chi-square–based measures will normally result in more tables being reviewed, and the investigator is not likely to miss an interesting relationship.

PRE measures are less likely to overestimate the strength of an association. In fact, PRE measures may underestimate them. An investigator summarizing the effects of many independent variables might choose a PRE measure. Thus, the investigators reduce the risk of assuming that a trivial or nonexistent relationship is important.

Ordinal measures also are called rank order measures. Three common ordinal measures for contingency tables are:

* *gamma*
* *tau*
* *Somers's d*[5]

Gamma and tau are both symmetric measures, while Somers's d is asymmetric. If the variables in a contingency table are ordinal, researchers normally prefer ordinal measures of association over nominal measures. They are more likely to detect relationships, since they make fuller use of a table's information. Nominal statistics ignore the ranking inherent in an ordinal scale. However, in a table with one nominal and one ordinal variable, the analyst usually must rely on a nominal statistic.

Table 13.11 summarizes major characteristics of common measures of association for data measured at the nominal and ordinal levels. As you review this table, remember that the measures of association should add information to a contingency table. If the independent and dependent variables are ordinal, then gamma, tau, or Somers's d should be examined; however, if the relationship is nonlinear, these measures underestimate the strength of the relationship. In this

Table 13.11 Common Measures of Association for Data Measure at the Nominal or Ordinal Level

Measure	Type Data	Characteristics and Comments
Lambda (λ)	Nominal	Asymmetric, PRE measure; may report 0.00 even if variables are statistically related
Cramer's V	Nominal	Symmetric; chi-square–based; values of .20–.40 suggest a moderate relationship; values over .80 are rarely encountered[a]
Gamma (γ)	Ordinal	Symmetric; reaches ±1.00 for relationships that seem less than perfect; values of .30–.40 range suggest a moderate relationship[b]
Tau (τ)	Ordinal	Symmetric; tau[b] used for square tables, tau[c] for rectangular tables (tau[b] cannot reach ±1.00 for nonsquare tables); values of .30–.70 suggest a moderate relationship[c]
Somers's d	Ordinal	Asymmetric; values normally fall between gamma and tau[d]

[a]Interpretation of values is from T. H. Poister, *Public Program Analysis: Applied Research Methods* (Baltimore: University Park Press, 1978), 443.

[b]Interpretation of values is from Poister, *Public Program Analysis*, 456, citing J. A. Davis, *Elementary Survey Analysis* (Englewood Cliffs, NJ: Prentice-Hall, 1971), 49.

[c]Interpretation is from B. D. Bowen and H. F. Weisberg, *An Introduction to Data Analysis* (San Francisco: W. H. Freeman, 1977), 76.

[d]Poister, *Public Program Analysis*, 457.

case, the analyst may wish to use a nominal measure. Recall that what constitutes a "strong" or "moderate" relationship partially depends on the study topic. If an investigator has found few independent variables that are associated with the dependent variable, a weak relationship may merit attention. Furthermore, while ordinal measures of association assign negative values to inverse relationships, the sign indicates only the direction, not the strength, of the relationship.

Selecting a Measure of Association

The first criterion for choosing a measure of association is the consistency between the statistic and the level of measurement. If the variables are measured at the nominal level, the analyst must use a nominal statistic. If they are measured at the ordinal level, the analyst may use a nominal or an ordinal statistic. If the relationship includes one nominal and one ordinal variable, the analyst may use a nominal statistic or choose a measure of association designed to examine a relationship between nominal and ordinal variables.[6] It would not be appropriate to use a measure designed for ordinal variables. Note that dichotomous nominal variables, that is, variables with two categories, may be analyzed using ordinal measures of association and, in some cases, with interval measures.

The second criterion for selecting a measure is its criteria for a null and perfect relationship. The analyst should choose a statistic that is consistent with her theoretical purposes. She should select a measure that assigns values of "1" to relationships that she considers perfect and assigns a value of "0" to relationships that she considers random or null.

The third criterion for selecting a measure is its sensitivity. A *sensitive measure of association* is a statistic that detects small differences between the strengths of relationships. When we introduced measures of association, we argued that they helped the investigator quickly distinguish between stronger and weaker relationships. An insensitive measure assigns the same or nearly the same numerical value to relationships that an investigator would judge differed in their strength. A sensitive measure assigns different numerical values to relationships that may have slight, even subtle, differences in strength.

The fourth criterion is familiarity with the statistic. An analyst needs to be familiar with a statistic to know how sensitive it is and to decide if he agrees with its criteria for a null and perfect relationship. Furthermore, familiarity helps an analyst to judge the strength of a specific relationship.

The answer to what constitutes a strong relationship depends on the statistic and the research question. First, because they have different criteria for null and perfect relationships, some measures of association tend to have much lower values than others. Second, the factors affecting some dependent variables are well known. If a researcher were analyzing such a well-studied dependent variable, even moderate associations might be of limited value because stronger relationships may have been identified and explained. On the other hand, some dependent variables are not well studied or understood, so even relatively weak relationships may suggest research questions worth pursuing.

To compare one relationship with another, we should use the same measure of association for each. We cannot directly compare two different statistics measuring association since the different criteria for perfect and null relationships as well as the strength of the association affect the numerical value. Thus, we cannot assume that a relationship with a Cramer's V of .10 is weaker than a relationship with a gamma of .45. If, however, different measures of association all suggest a strong or weak relationship between two variables, we can be more confident in our conclusions.

Comparing relationships using a single statistic should be done cautiously. Some ordinal measures of association may underestimate or miss certain nonlinear relationships, so a measure that primarily identifies a strong direct or inverse linear relationship may completely miss a strong nonlinear relationship. Exact comparisons between tables of different sizes should be done with care because the number of rows and columns in a table can affect the value of a measure of association.

Assessing Relationships by Introducing Control Variables

Depending on the model, a researcher may add a control variable to eliminate a rival hypothesis or to explore further the original relationship. Researchers and administrators also often want to learn how a relationship changes if other variables are considered. This is called *specifying* or *elaborating* a relationship. They want to understand how three variables interact. In other words, they want to learn if the original relationship is differs under different conditions. The conditions are represented by the values of the control variables.

Including a control variable in a contingency table is relatively easy. The dataset is divided into subsets. Cases are assigned to subsets on the basis of their values of the control variable. For example, for the variable *gender*, the subsets are male and female. All the females in the dataset are assigned to one subset, and all the males are assigned to the other.

The investigator's model should guide the number of subsets created and studied. For some variables, such as gender, the content of subsets is easily determined. For other variables, the analyst may have to decide what categories are appropriate for the planned study. He needs to avoid having so many subsets that they contain too few cases for meaningful analysis. He may limit the subsets to the values of interest or combine similar values. The control variables he chooses and the way he arranges them depend on the purpose of the study. For example, if education were the control variable, with the values less than 12 years, high school graduate, bachelor's degree, and graduate degree, the analyst does not have to study each category. He may use all the existing values, combine some values, or examine select values. Table 13.12 lists three possible ways of grouping education levels for use as a control variable.

A third variable is often introduced into the analysis of a relationship to test for spurious relationships. A *spurious relationship* is one where the independent variable has been falsely assumed to cause the dependent variable. This occurs because the third variable is separately related to both the independent and dependent variables. If a relationship between two variables is spurious, it will disappear when a control variable is added to the analysis.

Table 13.12 Grouping Education Levels for Use as a Control Variable

Option 1	1	Less than 12 years
	2	High school graduate
	3	Associate's degree
	4	Bachelor's degree or graduate degree
Option 2	1	Less than 12 years
	2	High school graduate
	3	Postsecondary degree
Option 3	1	Less than 12 years
	2	High school graduate (includes those with a postsecondary degree)

Example 13.2 gives a vivid, "folksy," and classic example of a spurious relationship. Table 13.13 lends credence to the hypothesis based on the folk legend that storks deliver babies. Counties with large numbers of storks have high birth rates; conversely, counties with few storks have low birth rates.

Example 13.2 An Example of a Spurious Relationship

Hypothesis: **Storks Deliver Babies**

Table 13.13 Birth Rate of County by Number of Storks Sighted in County

Stork Count		
Birth Rate	High (n = 100)	Low (n = 100)
High	82%	18%
Low	18%	82%

Note: lambda = .36; Cramer's V = .64; gamma = .91; ^{2} = 82; df = 1, p 001.

Model Diagrams of a Spurious Relationship

Figure 13.4 shows a model diagram of this hypothesized relationship.

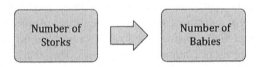

Figure 13.4 Model of a Spurious Relationship: Storks Deliver Babies

Undoubtedly, no reader of this text believes the hypothesis that storks deliver babies to be true. An alternative hypothesis that explains why the relationship was observed is that population density is related to both the independent and dependent variables. Specifically, rural areas may have both large numbers of storks and high birth rates, and urban areas may have few storks and lower birth rates. Table 13.14, introducing the control variable, county population density, shows that most rural counties have high birth rates no matter

Table 13.14 Birth Rate of County by Number of Storks Sighted by Population Density of County

Rural Counties Stork Count			*Urban Counties Stork Count*		
Birth Rate	*High (n = 90)*	*Low (n =10)*	*Birth Rate*	*High (n =10)*	*Low (n = 90)*
High	90%	90%	High	10%	10%
Low	10%	10%	Low	90%	90%

Note: For statistics involving rural and urban counties, lambda = 0.00; Cramer's V = 0.00; gamma = 0.00; $\chi^2 = 0$, $df = 1$, $p > .90$.

Source: Williamson, J. B., D.A. Karp, and J. R. Dalphin, *The Research Craft* (Boston: Little, Brown, 1977), 417–418.

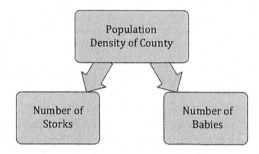

Figure 13.5 Three-Variable Model With Control Variable Separately Related to Both Independent and Dependent Variables

how many storks are present and most urban counties, with typically few storks, have low birth rates. You may be familiar with similar examples, such as ice cream sales being related to the number of violent crimes (the control variable that explains the association is outdoor temperature).

Figure 13.5 is a diagram to illustrate the model of the spurious relationship as described in Example 13.2. It shows that population density, which differs between rural and urban areas, is related separately to birth rates and to the number of storks. The original relationship between number of storks and the birth rate was shown to be spurious once the third variable of population density was included and therefore controlled.

Possible Outcomes When Controlling for a Third Variable

An analyst normally looks for one of four outcomes in elaborating a relationship. First, the original relationship may continue with little change in each category of the control variable. Second, the original relationship may become stronger or weaker in some, but not all, categories of the control variable. Third, the original relationship may disappear in some or all categories of the control variable. Fourth, the relationship may change direction in some or all categories of the control variable. If the relationship disappears in some, but not all, categories of the control variable, we say that the original relationship was conditional—that it existed in some conditions but not in others. For example, a survey showed a relationship between marital status and fear of walking alone in one's neighborhood at night. When gender was controlled, the relationship existed for men but not for women.

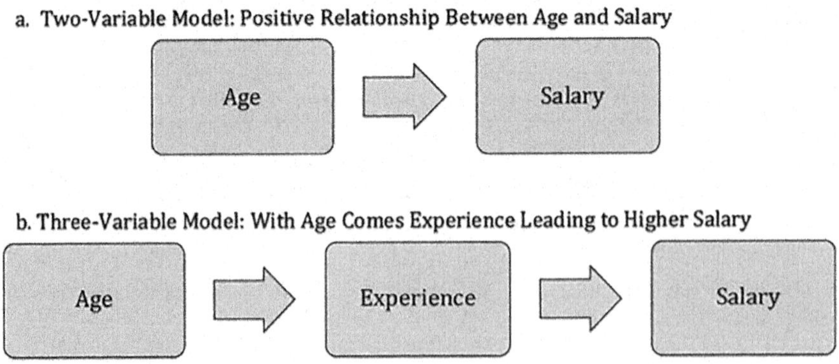

a. Two-Variable Model: Positive Relationship Between Age and Salary

b. Three-Variable Model: With Age Comes Experience Leading to Higher Salary

Figure 13.6 Two- and Three-Variable Models: a. Positive Relationship Between Age and Salary and b. Positive Relationship of Age to Experience Leading to Higher Salary

If an investigator finds that the relationship disappears in all categories of the control variable, the relationship is considered spurious. Occasionally, the control variable may reveal an association between two variables that was not observed when the relationship between only the independent and dependent variables was analyzed.[7] A common situation is when a third variable comes between the independent and dependent variables. In these cases, the control variable may be influenced by the original independent variable and in turn influence the original dependent variable. Without this link between the original variables, their association would not exist.[8]

Figure 13.7 is a model illustrating how a control variable can come between an independent and dependent variable. The original relationship was between employee age and annual salary. The control variable is employee length of experience within the industry. The analyst believed that as an employee gets older, he or she gains experience. With more experience, the employee earns a higher salary.

The analyst first found the relationship between age and annual salary. He then elaborated the original model by introducing the control variable and tested it by controlling for length of experience (see Figure 13.6b). If the second model is accurate, the original relationship should become weaker or perhaps disappear.[9] In this situation, the analyst would first evaluate the relationship between age and length of experience and the relationship between length of experience and salary separately. That would support her decision to control for length of experience.

Measures of Association and Control Variables

Measures of association can make it easier to determine the effect of a control variable. By looking at a measure of association, an analyst can quickly visualize how the control variable affects the original relationship. The analyst looks at the measure of association summarizing the relationship between independent and dependent variables for each value of the control variable to see how it varies among those values. To illustrate this role of control variables, we analyzed data on municipal employees to identify characteristics associated with the type of position employees held.

The dependent variable, position type, was assigned three values:

- administrator
- supervisor
- nonmanager

Table 13.15 Association Between Position Type and Selected Employee Characteristics Controlling for Gender

Measure of Association (Values of Cramer's V)

	Among All Employees	Among Males	Among Females
Department	.26	.30	.36
Years with city	.30	.30	.22
Sector of previous job	.09	.10	.10
Education	.17	.19	.08

Four independent variables that might account for variations in employee positions were examined:

- employee department
- number of years with the city
- sector of previous job (private or public)
- education

The analyst controlled for the variable gender to see if it affected the relationship between any of the independent variables and the dependent variable, position type.

Table 13.15 reports the statistical findings. (The contingency tables on which these measures are based are not included.) The column "Among All Employees" shows the relationship between each independent variable and whether a person worked as an administrator, supervisor, or nonmanager.

Cramer's V

To examine the table, we used Cramer's V, a statistic for nominal-level variables, which has values ranging from 0.00 to 1.00. Cramer's V is a chi-square–based measure which takes into account the sample size. It is more sensitive than a PRE measure. Consequently, it is less likely to report a value of 0.00 and more likely to pick up variations in relationships.

The Cramer's V statistic suggests that an employee's department ($V = .26$) and the years worked with the city ($V = .30$) are most closely related to the employee's position type. If we could examine the individual tables (not shown here), we would find that employees of the finance and planning departments were most likely to be administrators. Police and fire department employees were most likely to be supervisors. Employees with master's degrees were more likely than employees without master's degrees to hold either administrative or supervisory positions. Whether a person had previously worked in the private sector had little relationship to whether he or she currently held a managerial position.

The columns "Among Males" and "Among Females" report the relationship between the dependent variable (position type) and the three independent variables. The column "Among Males" shows the values of Cramer's V for the relationship between the dependent variable and each independent variable when only the data on men are analyzed. Similarly, the column "Among Females" shows the values when only the data on women are analyzed.

Using Nominal and Ordinal Measures

An analyst can quickly determine the influence of the control variable by reviewing a measure of association. Since Cramer's V is a statistic for nominal data, we had to check the tables relating each independent variable to the dependent variable to understand the exact nature of the

Table 13.16 Management Position by Years Employed With City by Gender

Males: Years Employed by City

	< 1 (72)	1–2 (153)	3–6 (208)	7–10 (243)	> 10 (830)
Non-manager	81.9%	79.7%	79.8%	68.5%	35.5%
Supervisor	12.5%	16.3%	13.5%	27.2%	57.0%
Administrator	5.6%	3.9%	6.5%	4.5%	7.5%

Note: Gamma = .56.

Females: Years Employed by City

	< 1 (19)	1–2 (45)	3–6 (41)	7–10 (51)	> 10 (72)
Non-manager	100.0%	91.1%	87.8%	84.3%	76.4%
Supervisor	0.0%	6.7%	12.2%	5.9%	22.2%
Administrator	0.0%	2.2%	0.07%	9.8%	1.4%

Note: Gamma = .39. Percentages may not add to 100% due to rounding.

relationship. Without the tables to look at, we could only guess which departments had relatively more administrators or supervisors and whether certain departments were more likely to employ women in managerial positions. Only finance and planning departments had women administrators. Women were most likely to work as supervisors in public safety and public works departments. The tables confirmed that the number of years spent working with the city seemed to benefit women less than men. Men were more likely to hold administrative positions the longer they worked for the city. This was less true for women. Education appeared to be more closely related to position type for men ($V = .19$) than for women ($V = .08$).

With an ordinal measure, an analyst can more clearly visualize the content of the table by just reviewing the measure of association. Look at the relationship between years with the city and management position in more detail. Table 13.16 shows the relationship separately for men and women.

The tables show that for both men and women, the longer they work with the city, the more likely they are to have a managerial position. Nevertheless, both the tables and gamma values imply that men are more likely than women to be promoted the longer they stay with the city. The tables and measures of association are said to show an interaction between gender and length of city employment. Being both male and employed by a city for more than 10 years increases the probability of working as either an administrator or supervisor. (The reader is encouraged to draw a three-variable model showing these relationships.)[10]

Exploring the Effect of Control Variables

Investigators are advised to develop an explicit research model when beginning a research project. This model should include the variables that the investigator considers important. Although the model is important, it should also be considered tentative. The tentative nature of a model becomes apparent as an analyst examines table after table of bivariate relationships. Then he or she explores the *effect of control variables*. For many analysts, the process is compelling. Each table raises more questions and possibilities. The analyst might wonder where to stop. The purpose of the study, the literature, and the hypotheses used in developing the model should suggest which control variables are worth examining. Having a theory to help generate the hypotheses and direct the research will be helpful. The purpose of the study may also help determine how exhaustive an investigation is warranted, keeping in mind that one purpose of a project is to help explain some phenomenon.

Tests of Statistical Significance and Measures of Association

Measures of association summarize the strength of the relationship between two variables; that is, a measure of association indicates the magnitude of the effect. It serves as a criterion for labeling an effect weak, moderate, or strong. A measure of association that suggests a weak relationship may motivate a researcher to discount a relationship found to be statistically significant. Indeed, a relationship may be nonrandom, but it also can be too small or trivial to merit any action. Recall that what constitutes a weak effect or a strong effect depends on the statistic, its criteria for a perfect relationship and a null relationship, and the degree of existing knowledge about the research question. Effect size is also related to the power of a statistical test. Power depends on the size of the effect that has been measured and on the number of cases in the dataset being studied. Other things being equal, the larger the effect and the larger the sample, the more powerful the test. When a test with low power fails to show a statistically significant relationship, one should not necessarily conclude that there is no effect.[11]

Tests of statistical significance are inferential statistics. They indicate the probability that an observed relationship occurred by chance. A researcher uses inferential statistics to infer from a sample to the population. Appropriate use of tests of statistical significance requires a probability sample. Measures of association are descriptive statistics, although analysts often test many to determine if they are statistically significant. A researcher uses descriptive statistics to summarize data whether or not they represent a random sample. Taken together, measures of association and tests of statistical significance produce evidence supporting or disconfirming a hypothesized relationship.

The more knowledge an investigator has about an entire research effort and the details contained in a contingency table, the more confident he can be in interpreting the study's statistics. The tests of statistical significance and measures of association may confirm his conclusions or lead him to look at some of the evidence more carefully. The statistics can elucidate the research findings, reinforce decisions about their implications, and help to summarize and communicate them effectively to an audience.

Using Analysis of Variance to Measure Strength of Relationships

As the material in Chapters 12 and 13 discussed, contingency tables are used to display and analyze data resulting from measuring variables at the nominal and ordinal level. The contingency table and accompanying statistics help the analyst determine the nature of a relationship—positive, negative, or non-linear—and its strength. Additional statistics are used to determine if the relationship in the table is statistically significant.

For many situations contingency-table analysis is inadequate. It ignores the statistical advantages of experimental designs and in general does not use the information available when variables are measured at the interval or ratio level. In doing so, it fails to utilize information provided by means and variances. For example, suppose school system administrators wanted to compare the number of days missed by children who attend schools outside their neighborhood with the number of days missed by children attending neighborhood schools. An analyst can create several categories of days missed, for example, none, 1–5, 6–10, and so on, and construct a contingency table with attending neighborhood or other schools. Alternatively, the analyst can average the number of days missed by children who attend schools outside their neighborhood and compare it with the mean number of days missed by neighborhood children. A *t*-test could be used to determine if the average days missed differed between the two groups. However, if three groups are compared, a different test is required. A measure of strength of association would also require another procedure.

Table 13.17 uses the data from the earnings for trainees in three different programs. This example was also used in Chapter 12. The analyst wanted to determine if the type of program was related to the level of wages after training ended. She also wanted information on the strength of any relationship. Table 13.17a shows the data from the analyst's study in a contingency table. Although the relationship could be tested for significance using chi-square and the strength of association measured with a statistic for nominal and ordinal data, much of the information from the original study data would not be used. In Chapter 12, the data from this study was presented in a posttest-only format and tested for significance using analysis of variance.

Table 13.17b shows the dataset up for ANOVA.

The primary statistical tool for analyzing experimental data and the differences between or among group means is either the *t*-test or analysis of variance. Although ANOVA is used to test the statistical significance of a relationship, it also provides information to measure the strength of relationship between a nominal or ordinal independent and an interval dependent variable. That information is used to calculate a measure of association, eta or E.[12] Eta can vary between 0.00 and 1.00. It compares the amount of variance in the dependent variable due to values being in different groups to the total amount of variance between the values of all cases. (see the Appendix to this chapter for more information on calculating eta). The value for eta from the data in Table 13.17b is .47. Analysts square eta to interpret it. In the example $(.47)^2$ equals .22, which indicates that 22 percent of the variation in monthly income after one year of work is associated with the independent variable, that is, the program in which the client was trained. Eta can also be obtained by requesting it with the contingency table procedure in SPSS called CROSSTABS[13]

This discussion focused on the basic analysis of variance model—one-way analysis of variance. In this model, the analysis involves one independent variable, and each case contributes one score to the analysis. In the school example mentioned at the beginning of this section, the independent variable was whether a student attended school in the neighborhood or elsewhere, and the piece of data contributed by each subject was the number of days missed from school. In the job-training

Table 13.17a Average Monthly Income by Training Program

	Basic Skills	Classroom Training	On-the-Job Training
Low ($300–500)	8	4	2
Medium ($501–700)	3	8	5
High ($701–1000)	0	9	1
N	11	21	8

Table 13.17b Average Monthly Earnings By Program

		Program		
		Basic Skills	Classroom Training	On-the-Job Training
Average monthly income ($)	$\bar{X}$	433.5	528.7	765.1
Sample size	N	11	21	8

example, the independent variable was type of job-training program, and the contributed score was each subject's monthly earnings.

The basic model can be adapted to include additional variables, which are essentially control variables. For example, the study could include another variable of whether the student attended a public or private school. Analysis of variance models including more than one independent variable are called multivariate analysis of variance.[14] Analysis of variance models may also include more than one measure on each subject. These are called repeated measures models. These topics are beyond the scope of this textbook. Interested readers are encouraged to consult a text on experimental design.

Summary

Administrators are normally comfortable with contingency tables and the information conveyed by percents. They also may be familiar with tests of statistical significance, especially chi-square, which indicate the probability that the relationship occurred by chance. Measures of association associated with contingency tables are less familiar to administrators and generally are of interest only to the analyst.

Contingency tables illustrate the relationship between variables. If the independent variable heads the table column, a user may compare the relative distribution of the dependent variable for each value of the independent variable. With the comparison, he sees how the percentage values of dependent variables differ for each category of the independent variable. The comparison, then, suggests if the data support a given hypothesis.

Measures of association reduce the table data to a single metric. All measures of association indicate a null relationship with a value of 0.00. Nominal measures of association indicate a perfect relationship with a value of 1.00. Ordinal measures of association assign +1.00 or −1.00 to a perfect linear relationship; they will miss or underestimate the strength of nonlinear relationships. The closer a measure is to 1.00, the stronger the relationship. The closer a measure is to 0.00, the weaker the relationship.

To compare contingency tables with measures of association, an analyst should use the same measure for all comparisons. In choosing a measure, she should select a measure she is familiar with and consider the level of measurement of the variables and the measure's criteria for a perfect and a null relationship. Comparisons must be made with caution, especially if the number of rows and columns varies from table to table.

The criteria for a strong or weak relationship depend on the measure of association and the nature of the research question. In general, ordinal measures produce higher values than nominal measures. Symmetric measures produce higher values than asymmetric measures. Similarly, some variables have been extensively studied, and moderate associations without a theoretical basis may not add to our knowledge. Other dependent variables are poorly understood, and weak relationships may generate some productive leads.

Measures of association help an analyst to determine the effect of control variables. The control variable may suggest that the original relationship is spurious or that the control variable has no effect. More often, the control variable elaborates the relationship by showing how a relationship changes for different values of a control variable—it may become stronger for some values, it may disappear for some values, or it may change directions in some or all values.

Analysis of variance, originally developed for experimental studies, may also be used to analyze hypotheses with an interval dependent variable and a nominal independent variable from nonexperimental studies. The subjects must constitute a random sample, however. Analysis of variance provides information on both the statistical significance and the strength of a relationship. The statistic eta is calculated to measure the strength of association. The values of eta range from 0.00 (a null relationship) to 1.00 (a perfect relationship). Eta squared indicates the amount of variation in the dependent variable that is explained by the independent variable.

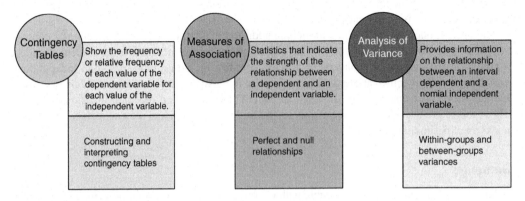

Figure 13.7 Examining Relationships Among Variables

One-way analysis of variance applies in designs with one independent variable in which each subject contributed one piece of data. Other models of analysis of variance are used to analyze designs with more than one independent variable or which collect data from the subjects more than once over the course of the study.

Other techniques, correlation and regression, are used for studying the linear relationship between variables measured at the interval level. Nominal and ordinal measures of association act primarily to identify relationships for closer study. Interval statistics are widely reported and provide investigators with valuable information to explain, describe, or predict relationships.

Notes

1. D. B. Wright, *Understanding Statistics: An Introduction for the Social Sciences* (London: Sage Publications, 1997), discusses the relationship between ANOVA and regression and cites seminal work in the social sciences on this topic. See page 132 especially.
2. See Shelly Cashman Series, Misty Vermaat, *Microsoft Excel, 2013. Introductory*, 14th ed. (Belmont, CA: Cengage Learning, 2014). The following web site gives instructions on creating pivot tables: www.stat.tamu.edu/stat30x/stat-tools/pivot.pdf. Also see www.stats.stackexchange.com/questions/44834/what-is-the-difference-between-the-pivot-table-and-contingency-table.
3. Lambda, demonstrated earlier, for Table 13.4 would be 0.00.
4. The reader is encouraged to calculate percents for Table 13.6 and also use them to interpret the relationship.
5. Other association measures for ordinal variables are used when there are many cases and the variables have a large number of categories. In these situations, data are usually not in contingency tables. We do not discuss these measures of association here. See Sidney Siegal, *Nonparametric Statistics for the Behavioral Sciences* (New York: McGraw-Hill, 1956), 202–239; Hubert M. Blalock, Jr., *Social Statistics*, 2nd ed. (New York: McGraw-Hill, 1972), 415–421.
6. See Siegal and Blalock, cited earlier for information on statistical tests and measures of association for nominal, ordinal, and interval combinations. Also see Larry Giventer, *Statistical Analysis for Public Administration*, 2nd ed. (Boston, MA: Jones and Bartlett Publishers, 2008).
7. For a complete discussion on how a control variable can affect the relationship between an independent and dependent variable, see T. Poister, *Public Program Analysis: Applied Research Methods* (Baltimore: University Park Press, 1978), 153–171. Also see M. Rosenberg, *The Logic of Survey Analysis* (New York: Basic Books, 1968).
8. For more information on types of control variables and how they can affect original two-variable relationships, see: B. W. Pellham and H. Blanton, *Conducting Research in Psychology*, 4th ed. (Wadsworth: Cengage Learning, 2014). Also see Rosenberg, *The Logic of Survey Analysis*.
9. Earl Babbie, *The Practice of Social Research*, 9th ed. (Belmont, CA: Thomson-Wadsworth, 2001), discusses the role of control variables on pages 421–430 and the names often used for them. Also see Russel Schutt, *Investigating the Social World: The Process and Practice of Research*, 8th ed. (Thousand Oaks, CA: Sage Publications, 2015). Schutt also discusses control variables and their various roles in three-variable models. He categorizes and defines several. Schutt defines some control variables as mediators or moderators. (See pages 206–210.)

10. First draw two separate two-variable models showing the relationship between the independent variable and the dependent variable. Then draw the model to illustrate the three-variable relationship.
11. R. S. Pindyck and D. L. Rubinfeld, *Econometric Models and Economic Forecasts*, 4th ed. (Boston, MA: Irwin and McGraw-Hill, 1998), 43–45.
12. We discuss using ANOVA as a test of statistical significance in detail in Chapter 12. G. W. Borhnsteadt and D. Knoke, *Statistics for Social Science Analysis*, 2nd ed. (Itasca, IL: F. E. Peacock, 1988), 234–236, 290–294 discuss eta.
13. www.ibm.com>support>sslvmb.23.0.0>spss>base.
14. Wright, *Understanding Statistics*, 116–145 is recommended for an introductory discussion of ANOVA and MANOVA models. Wright includes the SPSS printout for several of these.

Terms for Review

contingency table
measure of association
percentage difference
bivariate distribution
perfect relationship (between two variables)
null relationship
sensitive measure of association
proportional reduction in error (PRE) measure
original error
new error
symmetric measure
asymmetric measure
spurious relationship
effects of a control variable
eta

Questions for Review

The following questions should indicate whether you have a basic competency in this chapter's material.

1. Use the data array in Table 13.18 to:

   a. Create a contingency table showing the relationship between county and causes of injury. Write a sentence describing the relationship.
   b. Create a contingency table showing the relationship between cause of injury and its severity. Note that you will have to group together values of the severity index. Write a sentence describing the relationship.

2. Table 13.19 was constructed in an evaluation of a pilot project conducted by a city to encourage citizens to recycle more of their trash. Calculate percentages for this table. Write a short report to the head of the engineering department summarizing the results. Which of the neighborhoods were most active in recycling?

3. In a related activity, the city obtained information regarding the support of a sample of citizens for bond financing for a waste-to-energy facility. Table 13.20 shows support by income level. Table 13.21 shows the support for two areas of the city.

   a. Calculate the percentages for Table 13.18 and use the percentage difference to describe the variations in support.
   b. Calculate the percentages for Table 13.19 and describe what happens.
   c. What control variable was used in this analysis? How did it affect the original relationship?

Table 13.18 Data Array for Three Variables: County, Cause, and Severity of Injury

Case	County	Cause	Severity
01	Baker	Fall	3
02	Charlie	Car accident	4
03	Charlie	Violence	6
04	Able	Car accident	4
05	Charlie	Violence	5
06	Baker	Fall	9
07	Charlie	Car accident	10
08	Baker	Fall	1
09	Able	Violence	5
10	Charlie	Violence	5
11	Charlie	Fall	7
12	Able	Car accident	4
13	Charlie	Car accident	7
14	Baker	Fall	6
15	Able	Fall	3
16	Baker	Car accident	5
17	Charlie	Car accident	5
18	Baker	Fall	6
19	Able	Car accident	4
20	Charlie	Violence	7

Table 13.19 Recycling Participation by Neighborhood

Number of Times Participated in Recycling Program

Neighborhood	0–2	3–6	7–10	11–13
Steel Creek	13	16	8	5
Hidden Valley	6	15	17	12
East Side	2	5	14	19

Table 13.20 Support for Bonds to Finance Waste-to-Energy Facilities by Income Level

Income Level

Support for Bonds	Below $14,500	Over $14,500
Oppose	185	153
Favor	95	167

Table 13.21 Support for Bonds for Waste-to-Energy Facilities by Income and Place of Residence

Residence

Support for Bonds	West Side Income Level		East Side Income Level	
	Below $14,500	Over $14,500	Below $14,500	Over $14,500
Oppose	158	79	27	74
Favor	42	21	53	146

4. Write a one-sentence explanation for each of the criteria one should use to choose a measure of association. Which one do you think is most important? Why?

5. Consider the hypothesis: "Employees who are trained during their second quarter on the job make greater use of computer software than employees trained during the first, third, or fourth quarters of their first year on the job." The variables were categorized as: extent of use (limited, moderate, extensive) and time of training (first quarter of employment, second quarter, third quarter, or fourth quarter).

   a. If time of training and extent of use have a perfect relationship, what would the data look like? (Suggestion: Create a table with four columns and three rows and enter percents to illustrate the hypothesized relationship.)
   b. If time of training and extent of use have a null relationship, what would the data look like?
   c. Would you be more likely to consider using a nominal or an ordinal measure of association to study this hypothesis? Explain.

6. Explain why a rule of thumb such as the following is poor: "Any measure of association with a value greater than .45 depicts a strong relationship."
7. For Example 13.2, explain what Table 13.10 shows. What happens when a control variable is included in the analysis? What could explain this outcome?
8. Tables 13.22 through 13.24 present data collected on soup kitchen users. The tables related age to the number of meals eaten in a day.

   a. Write a sentence describing the relationship depicted in each table.
   b. What is the control variable? How does it affect the original relationship?
   c. Which measure of association did you use to compare the tables? Why did you select it?

Table 13.22 Number of Meals Eaten Daily by Respondent's Age

Age			
Number of Daily Meals	< 30 (66)	31–54 (213)	> 55 (144)
1.	38%	29%	27%
2.	47%	49%	40%
3.	15%	22%	33%

Table 13.23 Number of Meals Eaten Daily by Age for Respondent Males

Age			
Number of Daily Meals	< 30 (48)	31–54 (172)	> 55 (108)
1.	35%	32%	31%
2.	50%	48%	41%
3.	15%	20%	29%

Note: Cramer's V = .09; lambda = .000; gamma = .12. Percentages may not add up to 100% because of rounding.

Table 13.24 Number of Meals Eaten Daily by Age for Respondent Females

Age			
Number of Daily Meals	< 30 (18)	31–54 (38)	> 55 (35)
1.	44%	18%	17%
2.	39%	53%	40%
3.	17%	29%	43%

Note: Cramer's V = .21; lambda = .070; gamma = .34. Percentages may not add up to 100% because of rounding.

9. Discuss when an investigator would use analysis of variance and eta to evaluate a relationship instead of Cramer's *V* or gamma.
10. The average murder rate for western states is lower than the average murder rate for southern states. Is murder rate in states related to region? Explain.

Problems for Homework and Discussion

1. An investigator is curious about whether respondent characteristics were associated with completion of a lengthy questionnaire. Seven contingency tables were produced; their measures of association are given in Table 13.25.

Table 13.25 Associations Between Completion of Survey and Respondent Characteristics

	Lambda	Cramer's V	Gamma
Years worked for city	.00	.12	.14
Department	.00	.14	−.19
Sector of previous employment (public, private, first job)	.00	.00	−.01
Age	.00	.08	−.14
Gender (M/F)	.00	.01	−.04
Position (management, nonmanagement, unskilled, clerical)	.00	.14	−.375
Degree (MPA, BS/BA, Other)	.00	.09	.18

a. Which characteristics of respondents seem clearly to be randomly related to completion of the survey?
b. Which characteristics appear to be associated with whether a person completed the questionnaire?
c. What information did you use to answer a and b?
d. For what relationships (independent variables), if any, does lambda appear to be an inappropriate measure of association? Explain.
e. For what relationships (independent variables), if any, does gamma appear to be an inappropriate measure of association? Explain.
f. For this particular study, do you think that the investigator should use lambda or Cramer's *V* as a screen to identify tables for further examination? Explain.
g. Do you need to consult the tables to interpret the measures of association? Explain.

2. Tables 13.26 through 13.31 are from a printout examining data on people who volunteer for direct service activities.

a. Write a sentence describing the findings of each table.
b. Does education or age seem to play a greater role in whether a person decides to volunteer? Justify your answer.
c. Interpret the gammas for those tables. If you were reporting the findings, would you keep the signs for gamma the same as on the printout, or would you reverse them? Justify your answer.
d. Briefly discuss how you would use this information to recruit volunteers.

Table 13.26 Volunteering by Respondent Age

Age				
Volunteers	18–29 (331)	30–49 (529)	50–64 (301)	>65 (181)
Yes	25%	35%	20%	18%
No	75%	65%	80%	82%

Note: Lambda = 0.00; Cramer's V = .16; gamma = .14.

Table 13.27 Volunteering by Respondent Education

Years of Education					
Volunteers	>8 (203)	9–11 (238)	12 (504)	13–15 (206)	16 (178)
Yes	8%	19%	28%	34%	48%
No	92%	81%	72%	66%	52%

Note: Lambda = 0.00; Cramer's V = .26; gamma = −.41.

Table 13.28 Volunteering by Education for Respondents Aged 18–29

Years of Education					
Volunteers	< 8 (11)	9–11 (56)	12 (162)	13–15 (67)	16 (33)
Yes	18%	14%	25%	27%	42%
No	82%	86%	75%	73%	58%

Note: Lambda = 0.00; Cramer's V = .17; gamma = −.26.

Table 13.29 Volunteering by Education for Respondents Aged 30–49

Years of Education					
Volunteers	< 8 (31)	9–11 (98)	12 (212)	13–15 (95)	16 (91)
Yes	6%	18%	34%	41%	60%
No	94%	82%	66%	59%	40%

Note: Lambda = 0.04; Cramer's V = .31; gamma = −.46.

Table 13.30 Volunteering by Education for Respondents Aged 50–64

Years of Education					
Volunteers	< 8 (89)	9–11 (56)	12 (89)	13–15 (31)	16 (30)
Yes	6%	23%	23%	23%	37%
No	94%	77%	78%	77%	63%

Note: Lambda = 0.06; Cramer's V = .25; gamma = −.39.

Table 13.31 Volunteering by Education for Respondents Over Age 64

Years of Education

Volunteers	< 8 (72)	9–11 (28)	12 (41)	13–15 (13)	16 (24)
Yes	10%	25%	20%	38%	21%
No	90%	75%	80%	62%	79%

Note: Lambda = 0.08; Cramer's V = .21; gamma = −.29.

Working With Data

1. Review the General Social Survey variables. Then, using the GSS data, formulate and test a hypothesis relating the two variables: DEGREE and POLVIEWS. Make POLVIEWS the dependent variable. Use a software program that will provide you with a contingency table and the following measures of association: lambda, Cramer's V, and gamma.
   Describe the relationship between the two variables using the percents in the contingency table.

   a. Interpret the measures of association. Which one would be most useful? What does it tell you about the relationship?
   b. Obtain the chi-square information for this table from the software program. Does it allow you to reject a null hypothesis of no relationship between the variables? Explain your answer.

2. Using the data from the General Social Survey, formulate and test a hypothesis relating the two variables: MARITAL and FEAR. Make FEAR the dependent variable. Use a software program that will provide you with a contingency table and the following measures of association: lambda, Cramer's V, and gamma.

   Describe the relationship between the two variables using the percents in the contingency table.

   a. Interpret the measures of association. Which one would be most useful and appropriate? What does it tell you about the relationship?
   b. Obtain the chi-square information for this table from the software program. Does it allow you to reject a null hypothesis of no relationship between the variables? Explain your answer.

3. Using the information from number 2 in this section, analyze a three-variable model with SEX as the control variable.

   a. Draw a three-variable symbolic model that you believe would explain the relationships among the variables.
   b. Using a software program that allows you to introduce a control variable, obtain a three-variable contingency table. You should have two separate tables with the original two variables: one for Male, another for Female.
   c. Using percents and a measure of association, compare the two subtables with each other and with the original two variable tables. Has the relationship changed? If so, how? What is your interpretation of the effect of the control variable on the original relationship?

4. As the number of cells in a contingency table increases, the difficulty of interpreting the table also increases. This exercise is intended to improve your table-reading skills and to demonstrate the role of measures of association. Using the County Data File, create ordinal

variables representing low, medium, and high spending per capita on public safety and cultural/recreational expenditure for counties. Categorize each county according to low, medium, or high on these variables. Analyze these variables to see if region of the state and county funding levels per capita for these services are related.

a. Use computer software to create contingency tables for each type of expenditure, similar to Table 13.1 or Table 13.2, to examine the relationship between region and spending per capita in each area.
    If available on the statistical software, request lambda, Cramer's *V*, and gamma.

b. Write a very brief report (no more than one page, single spaced) describing the relationships you studied using information from the resulting contingency tables and the values of lambda, Cramer's *V*, and gamma.

Recommended for Further Reading

Elifson, K., R. P. Runyon, and A. Haber, *Fundamentals of Social Statistics*, 3rd ed. (Boston, MA: McGraw Hill, 1998), discuss the construction and interpretation of contingency tables and calculation of measures of association for contingency tables and analysis of variance.

Keppel, G., and T. D. Wickens, *Design and Analysis: A Researcher's Handbook*, 4th ed. (Englewood Cliffs, NJ: Prentice-Hall, 2004), provide a thorough discussion of experimental design, analysis of variance, and related topics, including details on statistical calculation and interpretation. The discussion is more accessible than that available in other excellent but more technical works.

Kerlinger, F. N., and H. B. Lee, *Foundations of Behavioral Research*, 4th ed. (Fort Worth: Harcourt College Publishers, 2000), have an excellent, nonstatistical discussion of analysis of variance. Wright, D. B., *Understanding Statistics: An Introduction for the Social Sciences* (London: Sage Publications, 1997), Chapter 6, contains an excellent introductory discussion of analysis of variance models and statistics.

Meier, K. J., J. L. Brudney, and J. Bohte, *Applied Statistics for Public and Nonprofit Administration*, 9th ed. (Belmont, CA: Thomson-Wadsworth, 2014), discuss the preparation and analysis of contingency tables, including several examples with control variables. Calculations for statistics are worked out in detail and easy to follow.

Rosenburg, M., *The Logic of Survey Analysis* (New York: Basic Books, 1968). Rosenburg thoroughly discusses the use of control variables in the analysis of survey data. He covers virtually all possible situations in a very readable fashion, using contingency tables and percentages as the only measure of association. A useful appendix on how to read contingency tables is included.

Appendix 13.1

Calculating Common Measures of Association for Nominal and Ordinal Data

This section presents the equations and calculations for some common measures of association for variables measured at the nominal and ordinal levels. We use Table 13.18, relating citizens' rating of the police department to their ages, to illustrate the calculations. Since both variables in this table are ordinal, either an ordinal or nominal measure can be used. If we were actually selecting a measure of association for this table, we would use an ordinal statistic because both variables are ordinal and the relationship is linear.

Nominal Measures

Nominal measures of association are required if the independent and dependent variables are nominal. If only one of these variables is nominal and not dichotomous, the analyst would need to use a nominal measure or a measure specifically intended for nominal-ordinal combinations.

This chapter discusses lambda, also known as Goodman's lambda, at length, and nothing needs to be added here. The value of lambda for Table 13.32 is 0.0.

Cramer's V, a chi-square–based measure, also is used with nominal measure of association. The equation for Cramer's V is

$$V = \sqrt{\frac{\chi^2}{n.\min(\text{row} - 1, \text{column} - 1)}}$$

For the data in Table 13.18: $\chi^2 = 13.9$, $n = 830$, min (row − 1, column − 1) = 2. The value of min (row − 1, column − 1) is either the row minus 1 or the column minus 1, whichever yields the lower figure. If Table 13.15 had four rows and three columns, the min (row − 1, column − 1) would still equal 2.

$$V = \sqrt{\frac{13.9}{830.2}} = 0.0915$$

Table 13.32 Ratings of Police Department by Respondent's Age

Age			
Rating of Police Department	*< 18*	*18–59*	*> 59*
Poor (D to F)	2	33	6
Satisfactory (C)	10	137	27
Good (A or B)	12	453	150

The measure of association, phi (ϕ), is used for 2 × 2 tables. It has essentially the same equation as Cramer's V. Since the min (row − 1, column 1) = 1, the equation reduces to

$$\phi = \sqrt{\frac{x^2}{n}}$$

Ordinal Measures

This section includes three common ordinal measures of association for data in contingency tables:

1. gamma
2. Somers's d
3. Kendall's tau$_b$ and tau$_c$

This chapter considers measures of association in two roles:

1. to filter out trivial relationships
2. to show the effects of control variables

All three measures have the same numerator, so we will start by explaining how to calculate its components. First, the values of the variables must be in rank order; that is, they must be ordered along a continuum. The analyst begins the calculations by identifying the number of cases that are in agreement (A) or disagreement (D) with the rank ordering of both variables. The rank order suggested by Table 13.18 is: "The older the respondent, the higher the rating." Note the placement of the values of the variables, particularly those of the dependent variable.

Calculating A and D

To calculate A, we start with 2 in the upper left-hand corner and count how many respondents are older than 17 who rate the police as better than "poor"; that is, 137 + 453 + 27 + 150 = 767. We then go to 10 and count the 603 (453 + 150) cases where people are older than 59 and rate the police as better than "poor." We repeat the procedures in the remaining columns. The value of A is based on the calculation:

$$A = 2 \times (137 + 453 + 27 + 150) + 10 \times (453 + 150) + 33 \times (27 + 150) + 137 \times 150$$
$$= 1,534 = 6,030 = 5,841 = 20,550$$
$$= 33,955$$

To calculate D, we go through a similar procedure, but we begin with the top right-hand cell. Starting with 6, we count the number of people who are younger than 60 and rate the police as better than "poor"; that is, 137 + 453 + 10 + 12 = 612. The value of D is based on the calculation:

$$D = 6 \times (137 + 453 + 10 + 12) + 27 \times (453 + 12) + 33 \times (10 + 12) + 137 \times 12$$
$$= 3,672 + 12,555 + 726 + 1,644$$
$$= 18,597$$

Calculating Gamma

Gamma is an ordinal measure of association. It is a PRE measure—proportional reduction in error—and symmetric; that is, it makes no difference which variable is identified as the independent variable and which is the dependent variable. To calculate gamma (γ), the equation is:

$$\gamma = \frac{A-D}{A+D}$$
$$= \frac{33,955-18,597}{33,955+18,597}$$
$$= .29$$

Somers's d (d_{yx})

Somers's d (d_{yx}) is an asymmetric measure that builds on the gamma calculation. To calculate Somers's d, the independent variable must be identified. The analyst calculates the number of ties, that is, the number of cases with the same value of the dependent variable as the cases being observed. Starting with 2, we count the number of people older than 17 who also rated the police department as "poor"; that is, $33 + 6 = 39$. The value of T_r is based on the calculation:

$$T_r = 2\times(33+6)+33\times(6)+10\times(137+27)+137\times27+12\times(453+150)+453\times150$$
$$= 78+198+1,640+3,9+699+7,236+67,950$$
$$= 80,801$$

To calculate Somers's d, the equation is:

$$d_{yx} = \frac{A+D}{A+D+T}$$
$$= \frac{33,955-18,597}{33,955+18,597+80,801}$$
$$= .12$$

Somers's $d_{yx} = .12$

The calculations for T_r assume that the independent variable heads the columns of Table 13.32. (The subscript r refers to rows. Note that we counted ties in the rows. In the next paragraph, we show how to count ties in the column.) A table can be organized so that the dependent variable heads the columns and T_c, column ties, can be substituted for T_r.

Tau

Tau$_b$ is another ordinal measure of association. It considers both ties on the dependent variable and the independent variable; hence, it is a symmetric measure. The discussion of gamma and Somers's d has covered all but one component of the equation to calculate tau$_b$. We have shown you how to calculate agreements, disagreements, and ties on the dependent variable, leaving only ties on the independent variable to be covered. To calculate the ties on the independent variable (T_c), we start again with 2 and note how many people younger than 17 rated police as better than "poor"; that is, $10 + 12$. The value of T_c was based on the calculation:

$$T_c = 2 \times (10 = 12) + 10 \times 12 + 33 \times (137 + 453) + 137 \times 453 + 6 \times (27 + 150) + 27 \times 150$$
$$= 44 + 120 + 19,470 + 62,061 + 1,062 + 4,050$$
$$= 86,807$$

To calculate tau_b, the equation is:

$$Tau_b - \frac{A - D}{\sqrt{A + D + T_c} \sqrt{A + D + T_r}}$$

$$= \frac{33,955 - 18,597}{\sqrt{33,955 + 18,597 + 86,807} \sqrt{33,955 + 18,597 + 80,801}}$$

$$\frac{15,358}{373.3.365.2}$$

$$= .113$$

$Tau_b = .113$

For nonsquare tables tau_c, a symmetric ordinal measure, should be used. Tau_c is not a PRE measure. The equation for tau_c is:

$$Tau_c \frac{2(\min(row, column))(A - D)}{N^2 (\min(row - 1, column - 1))}$$

Similar to Cramer's V, "min(row, column)" is replaced with the number of rows or the number of columns, whichever is lower. For the sake of simplicity, we will calculate tau_c for Table 13.32 *(displayed again here)*:

$$Tau_c = \frac{2.3.(33,955 - 18,597)}{830^2.(3 - 1)}$$

$$= \frac{6.15,358}{1,377,800}$$

$$= .067$$

$Tau_c = .067$

Eta

The formula for eta is:

$$E = \sqrt{ss_b / ss_t}$$

where:
SS_b = Between-groups sum of squares
SS_t = Total sum of squares

For the example's data:

$$E = \sqrt{528,965 / 2,398.195}$$
$$= .47$$

Analysts square eta to interpret it. In the example, $(.47)^2$ equals .22, which indicates that 22 percent of the variation in monthly income after one year of work is associated with the independent variable, that is, the program in which the client was trained.

Sum of squares information is taken from the printout for analysis of variance.

14 Regression Analysis and Correlation

In this chapter, you will learn:

1. How to interpret a linear regression equation with one dependent and one or more independent variables.
2. About tests to determine how well a linear regression model fits a set of data.
3. How to interpret the correlation coefficient, a measure of association for quantitative variables.
4. Common mistakes in interpreting a linear regression equation.
5. How to interpret logistic regression statistics.
6. Use of regression to analyze time-series data.

Once administrators go beyond studying a single variable, their interest lies in studying the relationships among variables. Examining bivariate relationships is typically only a first step in studying complex relationships. With a little practice, administrators can easily interpret contingency tables. Nevertheless, these tables quickly become cumbersome as more variables are added to the analysis. None of the contingency table examples in the chapter on nominal and ordinal variable relationships had more than three variables, and the control variables only had two values. Try to visualize what the tables would look like if you wanted to control for several variables at the same time or if you were studying a control variable that had eight or nine values. Interpretation would become elusive. Furthermore, contingency tables may not be the best way to analyze interval-level data.

Administrators who study multivariate relationships usually desire simplicity while maintaining statistical accuracy. Statistics for interval-level data offer an attractive alternative to contingency-table analysis. The major advantage of these statistics is their ability to make full use of the information contained in interval variables and to come up with more precise descriptions of the relationships in a model. These statistics greatly facilitate complex analysis and allow research findings to be reported in relatively direct, easily understood formats.

Furthermore, statistics to measure associations among interval variables are versatile. They can be used to:

- summarize or describe the distribution of interval variables
- infer population characteristics from a sample
- provide evidence that variables are causally linked
- forecast outcomes

In this chapter, we limit our presentation of interval statistics to linear regression and correlation. Both are widely used and appropriate for analysis of outcomes that are interval-level measures. Furthermore, they are fundamental to understanding analogous statistical models and more sophisticated linear models that have been developed to analyze complex relationships. Regression and correlation can also be easily adapted to analyze nonlinear relationships.

An administrator can use and interpret regression findings profitably and appropriately with an understanding of a few basic principles. The chapter's final section and endnotes present equations for deriving the basic regression statistics. We recommend first reading through the description of the principles and concepts of regression and correlation to understand them and review the examples before attempting to absorb the calculation details. The references and our recommendations for further reading direct you to some excellent texts appropriate for the administrative analyst.

Analyzing Relationships Between Two Interval Variables

Although administrators and analysts typically want to learn about the relationships among several interval variables, we will begin with the two-variable model. The principles of bivariate regression analysis and correlation apply to multiple regression models, that is, models containing more than one independent variable. The major benefit of starting with a two-variable example is that we can illustrate the logic of linear regression with graphs. We cannot easily produce similar graphs for more than two variables. Similarly, some of the statistics are easier to explain and understand in the two-variable case.

To examine the relationship between two-interval variables, an analyst produces a *regression equation* and a *correlation coefficient*. The regression equation describes the relationship between two-interval variables. The correlation coefficient, a measure of association, indicates the amount of variation in the dependent variable associated with the independent variable.

The Regression Equation

The regression equation is the equation for the straight line that best describes a set of data points. The equation is asymmetric; therefore, the analyst must designate which variable is independent and which is dependent before calculating the components of the equation.[1] Deriving a regression equation involves four steps:

1. Obtaining data on two-interval variables.
2. Plotting the data on a graph:

    a.   The dependent variable, Y, is graphed on the vertical axis
    b.   The independent variable, X, is graphed on the horizontal axis

3. Determining whether the relationship between the two variables represented by the plotted data can be appropriately summarized by a straight line.
4. Calculating the regression equation.

Step 1. Obtaining Data

To see how these steps are carried out, let's consider a simple study, reported in Table 14.1, to learn how cumulative car mileage is related to annual repair costs. We obtained data on 10 randomly selected cars from a motor pool and used data on each car's mileage on the first of January and its repair costs over the following 12 months.

Step 2. Plotting the Data on a Graph

Next, we plotted the data on the graph. The dependent variable, the Y of the regression equation, was placed on the vertical axis. The independent variable, the X of the regression equation, was placed along the horizontal axis. The graph of plotted points is called a *scatterplot* (Figure 14.1a).

Table 14.1 Data on Car Mileage and Repair Costs

	Total Miles on Odometer (X) January 1	Repair Costs (Y) from January–December
Car 1	80,000	1,200
Car 2	29,000	150
Car 3	42,000	600
Car 4	53,000	650
Car 5	73,000	1,000
Car 6	80,000	1,500
Car 7	60,000	600
Car 8	13,000	200
Car 9	14,500	0
Car 10	45,000	325

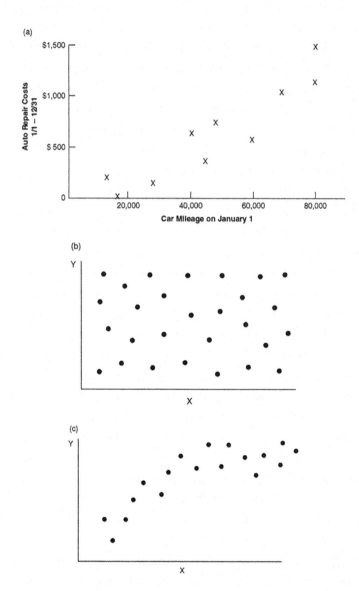

Figure 14.1 Scatterplots

Step 3. Reviewing the Scatterplot

Examining the scatterplot is necessary so that the analyst can decide if a linear model can be used to summarize the relationship between the two variables. A linear model assumes that the variables can be appropriately described by a straight line. The two scatterplots in Figures 14.1b and c illustrate cases where a straight line or linear model are not appropriate. Figure 14.1b shows no discernible pattern; the relationship between the two variables appears to be random. Figure 14.1c illustrates a relationship but one that is nonlinear. Using linear regression to analyze the data in scatterplots Figures 14.1b and c would be a mistake because the analyst would be forcing a linear model onto nonlinear data.

Step 4. Calculating the Regression Equation

If a linear model is deemed appropriate, then the regression equation is calculated. (See the final section in the body of this chapter for equations and calculations.) In practice, analysts rely on computer programs or calculators to carry out the actual calculations.[2] (The reader is encouraged to enter the data from Table 14.1 into a statistical program, such as SPSS, or a spreadsheet—Excel, for example—to obtain the statistics discussed here and to produce a scatterplot.)

The regression equation has the general format

$$Y = a + bX$$

where:
a = the constant or Y intercept
b = the regression coefficient or slope
Y = predicted value of Y, the dependent variable
X = the independent variable

For our example, the equation of the regression line is

$$Y = -267 + .018X$$

where:
Y = repair costs from January 1 through December 31
X = car mileage in thousands of miles on January 1

From the *regression coefficient*, we learn how a unit change in X, miles driven, affects Y, repair costs. For example, for every 100 miles driven, we expect the next year's repair costs to increase by $1.80. We obtained the figure $1.80 by multiplying .018 by 100. For every 1,000 miles driven, we expect repair costs to increase by $18.

Figure 14.2 shows the relationship between the regression line and the plotted points. On the graph, you can locate the point with an X value of 60,000 and a Y value of 600. The 600 is the cost of auto repairs for a car in the dataset that has been driven 60,000 miles. The Y equal to 600 is referred to as an *actual value of Y*.

Now, recall your algebra. For a given value of X, you can substitute its value in the equation and calculate the value of Y. For a car driven 60,000 miles, $813 is the predicted value of the cost of a year's car repairs, Y, for a car that has been driven 60,000 miles.

$$Y = -267 + .018X$$
$$Y = -267 + .018 (60,000) = \$813$$

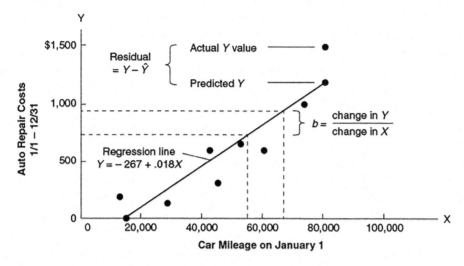

Figure 14.2 Scatterplot and Regression Line of Car Repair Costs and Miles

The $813 is the predicted value of the cost of a year's car repairs, Y, for a car that has been driven 60,000 miles. The *predicted value of Y* is the Y value on the regression line at the point where it intersects with X equal to 60,000.

The actual value of Y is often designated by Y_i and the predicted value of Y by $\bar{Y}$. Specifically, Y_i means the actual value of Y when X equals X_i, and $\bar{Y}$ the predicted value of Y *when X equals* X_i. The difference between Y_i and $\bar{Y}$ is referred to as the *residual*. In the example, the residual is –213, calculated from (600 – 813). The minus sign, "–", indicates that Y_i falls below the regression line; in other words, the regression equation overestimated the repair costs for the specific car.

A major use of the regression equation is to estimate population characteristics.[3] It is said to be the best estimator of the linear model because on the whole, it yields the smallest residuals. If any other straight line were placed on the scatterplot, the sum of the squared residuals, the distance between the Y_is and $\hat{Y}$s would be larger. As a practical matter, this means that with these data, the calculated regression equation is the best predictor of repair costs.

Values of the dependent variable are estimated from the regression line. The predicted $\hat{Y}$ represents the expected value of Y for a specific value of X. Values can be predicted for values of X that do not appear in the dataset; however, predictions should not be made for values that are much higher or lower than values of X in the dataset. To predict the repair costs for the entire motor pool, an investigator calculates the values of $\hat{Y}$ for each car in the pool and then adds all the $\hat{Y}$s together.

Tests of *goodness of fit* let an analyst decide how well the linear regression equation fits the data. A poor fit implies that the equation does a poor job of describing the data. One goodness of fit test is the visual evidence provided by a scatterplot. An analyst should routinely examine the scatterplot and satisfy themselves that a linear model is reasonable.[4]

The Correlation Coefficient

A second test of goodness of fit is the correlation coefficient. The correlation coefficient is a measure of association for interval-level data. Common terms used to identify the correlation coefficient include *Pearson's r, r,* and the *zero order correlation coefficient.* We shall refer to it as "r." An r of 1.00 or –1.00 indicates that every point on the scatterplot falls on a straight line, the regression line. An r equal to 1.00 indicates a direct, that is, positive, relationship; an r equal to

−1.00 indicates an inverse or negative relationship. Similar to other measures of association, an *r* equal to 0.00 designates that no linear relationship exists, similar to Figure 14.1b. The closer the value of *r* is to 1 or −1, the stronger the linear relationship. (The equation for *r* and a worked-out example are given in the appendix to this chapter.)

A valuable and intuitively attractive interpretation of *r* is given by r^2, formally called the coefficient of determination. The r^2 value indicates the proportion of the variance in the dependent variable associated with or explained by the independent variable. In the example, *r* equaled .93. The r^2 equaled .87, indicating that 87 percent of the variation in annual car repair costs was explained by knowing a car's cumulative mileage. The primary value of reporting the *r* instead of r^2 is that *r* gives direction.

The question of what size *r* is sufficiently large to merit attention does not have a definite answer. Instead, the value of a particular finding depends on the nature of the subject. A geneticist, for example, might consider an *r* of less than .95 to be of little value. Few, if any, social science studies approach that level of association. Values of *r* between .40 and .60 seem quite strong. In some studies, a modest *r* may be considerably larger than previous findings and therefore warrant attention. Nevertheless, remember that an *r* equal to .30 explains less than 10 percent of the variation in a dependent variable. And keep in mind that r^2 is also a measure of effect size.

Other Factors in Interpreting a Linear Regression Equation

Four other factors should be considered in the correct interpretation of a linear regression equation. They are:

1. the size of the regression coefficient
2. the standard error of the regression coefficient
3. the range of *X* values used to calculate the regression equation
4. the time interval between the measurement of *X* and *Y*

Regression Coefficient Size

It is important to understand that the size of the regression coefficient does not necessarily imply the strength of a relationship. The size of the regression coefficient is partially a function of the measurement scale.

In the motor pool example, if the *X* value were measured in thousands of miles, the regression equation would be $Y = -267 + 18X$. Remember, when *X* represented actual miles, the equation was $Y = -267 + 0.018X$. Eighteen is 1,000 times larger than .018; however, the two equations describe the same relationship and yield the same $\hat{Y}$ and the same *r*.

Not only does the measurement scale affect the size of the regression coefficient, but it also may cause estimation errors. Users who ignore the specific measurement of a variable may substitute the wrong value for *X* in predicting *Y*. Students commonly puzzle over how to handle *X*s that represent a percentage value. A percent should be treated as a whole number (10 percent, not 0.1). However, if the predicted values of *Y* do not make intuitive sense, you may investigate whether the calculations were based on proportions and not percents.

Standard Error (b) of the Regression Coefficient

We can estimate the relationship between two variables in a population by taking a sample. The standard error of *b* (se_b), also called the standard error of the slope, estimates the variability of the regression coefficient from sample to sample. The se_b has two uses. Analysts use it to estimate the true value of the regression coefficient. se_b also determines whether the relationship is statistically significant, that is, the probability that b in the population equals 0.00.

The specific values of both the constant and the regression coefficient normally vary from sample to sample. In the motor pool example, the se_b equaled .0025. At the 95 percent confidence level, an investigator estimates that the actual regression coefficient for the entire motor pool falls between .013 and .023. From the original equation, we stated that car repair costs would increase by $1.80 for every 100 miles driven. We can refine this statement. For the motor pool as a whole, we can estimate at the 95 percent confidence level that the repair costs range between $1.30 and $2.30 for every 100 miles driven. Recall that with a 95 percent confidence level, we assume that 5 out of 100 samples will fall outside this range.[5]

The se_b also is used to calculate the t-statistic and determine the statistical significance of the regression coefficient. The value of t, the t-ratio, equals the regression coefficient (b) divided by se_b. One can use a t-table to determine the associated probability. The null hypothesis is that the actual regression coefficient is 0.00; that is, there is no relationship between car repair costs and the number of miles driven.

In the motor pool example, the t-statistic equals 7.2 and the associated probability is less than .001. If the actual slope were 0.00, the probability of obtaining a slope as large as .018 would be quite small. A convenient rule of thumb to remember is that if the t-ratio is greater than 2, the regression coefficient is significant at the .05 level.[6]

Why the Range of X Values Is Important

The regression equation is computed for a specific dataset from the data in it. One should not estimate a value of Y for values of X that are well below the minimum or well above the maximum X values in the dataset. To illustrate the danger of estimating the value of Y for an X value outside the range of values used to compute the regression equation, let us return to the motor pool example. Consider a car that had been driven 160,000 miles by January 1. The predicted repair bill for the following year would be $2,613. In fact, the actual car repair bill was far less, only $100. Such "junkers" tend to stay in use as long as the costs of repairs do not get out of hand. In our dataset, the most miles a car had been driven was 80,000, which was half as much as the junker. Clearly, the dataset did not reflect the repair record of superannuated cars!

We also need to be aware of the influence of *outliers* on regression results. Points well outside the range of the other data are called *outliers*. In addition to checking the scatterplot for random or nonlinear patterns, you also must look for them. An outlier will bias the statistics. The scatterplot in Figure 14.3 illustrates an outlier that may yield a regression equation and correlation coefficient that make the relationship appear stronger than it actually is.

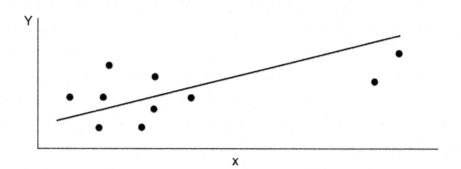

Figure 14.3 Scatterplot of a Relationship Where an Outlier Strengthens a Weak Relationship

Especially in a small dataset, the regression equation and correlation coefficient may show substantial changes with the addition or deletion of a single case that is an outlier. One should be especially careful in drawing implications from a small dataset.

For example, the addition of just one car can have a marked impact on a regression equation. If the car with 160,000 miles and repair costs of $100 had been included in the original dataset, the regression equation would have changed from $Y = -267 + .018X$ to $Y = 404 + .003X$. The r^2 would drop from .87 to .06. Such marked changes are most likely if the one case has an extreme value. Here, the outlier draws the line away from the other points and makes the relationship appear weaker than it actually is.[7] An outlier that is well below or well above the regression line, indicating that it is unique for its X value, will cause the correlation to be lower than it would be without the outlier. It will also affect the regression coefficient.

To handle outliers, one may:

1. Limit the analysis to the appropriate values of X and note that the model does not apply to other values of X.
2. Consider if a nonlinear model would better describe the data.[8]
3. Look for reasons the particular case is different from other cases and decide how to handle the case. For example, if the data for the case were collected at a different time, one might seek data that were contemporaneous with the other cases. Alternatively, the case might be deleted from the analysis. If so, the decision and the reasoning behind the deletion should be documented.[9] These unique cases often warrant additional investigation.

Time Relationship Between Variables

Another consideration is the time relationship between the independent and dependent variable. In many studies, time order may be of little concern, but sometimes it is important. *Lagging a variable* means that X is measured at a time preceding Y and that the time interval between X and Y is sufficient for X to have affected Y. In our example, we assumed that the cumulative mileage on a car affected its next year's repair costs. X was measured on January 1, and Y, representing repairs over the following 12 months, was measured on December 31, one year later. Similarly, one may assume a change in tax rate will affect spending three months later or that habits in one's youth will affect an aspect of one's old age. In the case of tax rate, Y may measure spending three months after the tax rate change. In the case of tracing the effect of youthful habits, an interval of 40 or more years may be created.

Multivariate Regression Analysis

Multiple regression has many advantages when compared to contingency tables. The analyst can present several variables in one equation. Furthermore, the regression equation gives the independent effect of each variable while controlling for the other variables in the equation.

The Multiple-Regression Equation

A common format for the regression equation is:

$$Y = a + b_1X_1 + b_2X_2 + b_3X_3 + b_4X_4$$

A more elaborate equation, which helps illustrate the information provided by multiple regression, is:

$$Y = a + by_{1.234}X_1 + by_{2.134}X_2 + by_{3.124}X_3 + by_{4.123}X_4$$

where:

Y = dependent variable

a = constant

$by_{1.234}$ = regression coefficient for X_1 associated with Y, while controlling for X_2 through X_4

X_1 = independent variable X_1

$by_{2.134}$ = regression coefficient for X_2 associated with Y, while controlling for X_1, X_3, and X_4

X_2 = independent variable X_2

$by_{3.124}$ = regression coefficient for X_3 associated with Y, while controlling for X_1, X_2, and X_4

X_3 = independent variable X_3

$by_{4.123}$ = regression coefficient for X_4 associated with Y, while controlling for X_1 through X_3 X_4 = independent variable X_4

The equation is normally written without the string of subscripts under each b, but viewing the subscripts may help you to understand the information in the equation. Each b, a partial regression coefficient, indicates the effect of the independent variable on the dependent variable while controlling for all other variables in the equation. The dot in the equation with subscripts separates the two variables being related from the variables being controlled.

$$Y = a + b_1X_1 + b_2X_2 + b_3X_3 + \ldots + b_nX_n$$

We arbitrarily chose to include four independent variables in the equation. We could have included fewer or more variables. The basic form of the equation remains the same and is indicated subsequently. A given independent variable would be related to the dependent variable while controlling for all of the other independent variables.

Applying the Multiple-Regression Equation

A good way to explain the multiple-regression equation is to look at an example. A study published in 1986, despite its age, clearly illustrates aspects of multiple regression.[10] The author gathered data on selected employee characteristics from a sample of federal government personnel records to derive a regression equation to explain a person's entry-grade level into the federal career service. The regression results are given in Table 14.2.

To interpret the results, first examine the variables. Several of the variables do not represent interval measures. Years of education, potential experience, and veteran's preference are the only variables measured at the interval level. The other variables represent dichotomies.

Dichotomous variables also are called *dummy variables*. A person either has the indicated characteristic or not. If the person has the characteristic, the variable is assigned value "1"; otherwise, the value "0" is assigned. In Table 14.2, major field of study was broken into a series of dummy variables, including business or management, law, and the social sciences. For each variable, a person was assigned the value 0 or 1. Although regression is intended for interval-level data, use of dichotomous independent variables in regression equations is a common and acceptable practice.

Potential experience represents the author's attempt to measure previous working experience of a newly hired federal employee. Potential experience was the difference between the person's age when he or she completed school and age upon starting federal employment. Veteran's preference is a bonus added to certain applicants' scores; some applicants may be entitled to a 5- or 10-point veteran's preference.

Table 14.2 Regression Equation to Predict Entry Grade into U.S. Civil Service ($N = 2,146$)

Variable	Regression Coefficient
White female	−.08
Minority female	−.29
Minority male	−.17
Years of education beyond high school	.50
Major field of study	
Biology, agriculture, or health profession	.97
Business or management	.82
Engineering, mathematics, or physical sciences	1.59
Law	4.37
Social sciences	.36
Other	.06
Potential experience	.17
Veteran's preference	.10
Constant	2.42
$R^2 = .69$	

Source: Lewis, G. B., "Equal Employment Opportunity and the Early Career in Federal Employment," *Review of Public Personnel Administration* (Summer 1986): 1–18.

The equation shows that female or minority status has a slight negative effect on a person's entry grade. Education has a positive effect, and degrees in either law or the physical and mathematical sciences markedly boost a person's entry grade. The coefficients for minority and female status control for years of education, veteran's preference, and specific degrees.[11]

Example 14.1 shows how the equation can be used to find a predicted entry grade for specific cases. By predicted, we mean the entry grade expected. The predicted value represents the estimated mean Y value for persons with the given characteristics.

Example 14.1 Predicting Individual Entry Level Grade From a Multiple-Regression Equation

Case 1: What entry level grade would you predict for a white male law school graduate, who has no veteran's preference, was 26 when he finished school, and is now 28? Recall that potential experience is the difference between school-finishing age and entry-level age.

The predicted grade level can be calculated by substituting the data from Case 1 into the equation that can be derived from the data in Table 14.2. The equation may be fully written with the zero values or without them. Keeping the same order for the variables as given in Table 14.2, the equation is written:

Predicted Entry Grade = 2.42 − .08(0) − .29(0) − .17(0) + .50(7) + .97(0) + .82(0) + 1.59(0) + 4.37(1) + .36(0) + .06(0) + .17(2) + .1(0)

The 0 values may be deleted in calculating the equation. The equation would then look like this:

Predicted Entry Grade = 2.42 + .50(7) + 4.37(1) + .17(2) = 10.63

Case 2: What entry grade would you predict for an African American female with an M.S. degree in biology who has no veteran's preference, was 24 when she finished school, and is now 25?

Predicted Entry Grade = 2.42 .29(1) + .50(6) + .97(1) + .17(1) = 6.27

Case 3: What entry grade would you predict for yourself?

Discussion: This example shows how to use the equation to calculate predicted values for the dependent variable. You actually predicted the average value of Y for people with a given set of characteristics. When you tried to apply the equation to yourself, you may have wanted to know more about how the variables were operationally defined. For example, if you studied engineering in school, graduated 10 years ago, and are now studying administration in an evening program, you may want more details on how the researcher defined potential experience and major field of study. The sample consisted of employees who had entered the federal civil service shortly after leaving school. Thus, you may fall outside the range of Xs included in calculating the equation.

The regression equation predicts or estimates the average entry grade for people in the population, that is, persons with records in the sampled personnel files. The dataset used to calculate the equation did not include persons who applied for the federal civil service but either were never selected or turned down a federal position. Also excluded from the dataset were persons who never applied for the federal civil service. Furthermore, the dataset represented federal personnel policies during a specific time period. Over time, policies or demand for specific skills may change; consequently, estimates based on an old dataset may become less accurate. If you are familiar with current civil service entry-level grades, you may find that the estimate for your civil service grade level is too low. Grade inflation isn't confined to universities.

We expect that the example helps you to appreciate the value of multiple regression. With the equation, you could identify the factors affecting entry grades for people in the civil service. The coefficients help identify the factors having the greater impact. The equation is easy for most people to work with, it generates useful data, and it requires little mathematical ability to understand or to estimate values of Y.

By far the most important reason for keeping these variables in the equation was the purpose of the study. The study was conducted to see what, if any, patterns of discrimination occurred in entry-level employment. Including the gender and race variables meant that the researcher could document how sex or race affected placement after eliminating the effects of education, previous experience, academic discipline, and veteran's preference. If race and gender were removed, a reader would not know if they had any impact. If the gender and race variables were deleted, the regression equation would need to be recalculated, since it no longer controls for them. With gender and race removed, the other regression coefficients in this equation should change only slightly, since gender and race had a weak statistical association with entry-level grade.

Measure of Association for Multiple Regression

How Accurate Were the Predictions From the Equation?

The R^2 = .69 means that the equation explains 69 percent of the variation in entry grade. R^2, called the *coefficient of multiple determination*, is a multivariate measure of association. R^2 indicates the degree of variation in the dependent variable explained by the model, that is, the independent variables included in the equation. By convention, r is used to indicate a bivariate relationship, while R indicates a multivariate one. Unlike r, which indicates direction, R can only be positive. Thus, reporting R as opposed to R^2 gives the user no additional information.

With the addition of variables, R^2 may either stay the same or become larger. The addition of a variable will never make R^2 smaller. This observation implies that, with enough variables, one can explain a large portion of the variation in a dependent variable. Trivial relationships may add .03, .02, or even smaller amounts to R^2.

With experience, you will find that few variables, often five or less, will explain much of the variation in the dataset. For example, in the civil service study, the author found that when years of education was the only independent variable, R^2 equaled .60. When the dummy variables measuring field of study were added, R^2 increased to .68. The remaining variables brought the value of R^2 up to .69. In other words, potential experience, gender, race, and veteran's preference explained an additional 1 percent of the variation in a person's entry-level grade.[12]

A serious mistake is to have a large number of independent variables to analyze a relatively small dataset. Another mistake is to have more independent variables than cases. The model then explains virtually all the variation in the dataset, but it is worth very little because the equation is literally custom designed to fit the dataset.

To avoid being misled and misleading others, researchers may report the adjusted R^2. The adjusted R^2 equation "shrinks" the value of R^2 by "penalizing" for each additional independent variable, k (k = number of independent variables). The adjusted R^2 cannot be larger than R^2. It may be noticeably smaller if the model applies a relatively large number of independent variables to a small number of cases.[13]

Adjusted $R^2 = 1 - (1 - R^2)(n - 1)/(n - k - 1)$

You might wonder when to stop adding variables to a model. The best strategy is to follow the example of the researcher in the civil service study and include the variables that are theoretically important to the model. As you may infer from this discussion, simply seeking the highest value for R^2 is not recommended. Some investigators eliminate variables that are not statistically significant; however, with a large enough sample, even the smallest regression coefficients may be "statistically significant."

Assessing the Importance of Each Independent Variable

Note that with the information in Table 14.2, you cannot rank the independent variables in order of their contribution to explaining the dependent variable. Remember that the regression coefficients are affected by how the independent variable is measured. Thus, the .5 for years of education has a far greater impact on a person's entry grade than the .97 for a degree in a life science such as biology. Why? The independent variable, life science, has the value of either 0 or 1. The largest impact it could have is .97 (1 times .97). Years of education after high school could easily range between 0 and 10, that is, from only high school to a Ph.D., and the impact of education could range between 0 and 5. So, in the equation, a degree in biology raises a person's predicted entry grade by roughly one grade. Two years of school have the same effect, and a college degree would raise a person's predicted entry grade by two grades.

Beta Weights

Beta weights, the standardized regression coefficients, allow one to rapidly and precisely compare the relative importance of each variable. Beta weights are calculated by normalizing the variables, that is, rescaling a variable so that its mean equals 0 and its standard deviation equals 1. In other words, all the variables are in the same standard units.

The larger the Beta weight, the stronger that variable's relationship to the dependent variable. Beta weights are usually calculated as part of the regression statistics produced by computer software. Since the Beta weight is on a standard scale, analyzing Beta weights allows the analyst to rank the relative strength of the association between the various independent variables on the dependent variable.

Example 14.2 illustrates the use of a regression equation in a study of student admissions. The researcher studied students admitted to a graduate business program to determine whether adding a scale quantifying personal achievements would improve admissions decisions.

Example 14.2 Developing a Regression Equation to Quantify Admissions Criteria

Background: The analysts developed a 12-point scale to summarize personal qualities, such as previous work experience and evidence of undergraduate leadership. They calculated scale scores for admitted students and then computed a regression equation to see whether the scale would improve the quality of admissions decisions. Graduate school performance was the dependent variable. Two sets of equations were calculated. One equation examined performance throughout the graduate degree program; the other examined performance during the first semester. The first-semester equation is reported in Table 14.3.

Table 14.3 Regression Model to Predict Student Grade Point Average

	Unstandardized Regress		
	Coefficient	*(s.e.)*	*Beta Weight*
Undergraduate average	.367	(.088)	.371
GMAT score	.00099	(.00045)	.188
Admissions scale	.036	(.016)	.188
Sex	−.019	(.080)	−.021
Years since college graduation	.014	(.012)	.104
Prior graduate work	−.055	(.054)	−.088
Constant	1.437	(.080)	
R^2	.188		

Note: GMAT = Graduate Management Admissions Test.

Discussion: To predict a student's first-semester grade point average, the equation with the unstandardized regression coefficient would be used. To determine which admissions factors are most closely associated with first-semester grades, the standardized regression coefficients, Beta weights, would be examined. Probably most striking is the importance of the Graduate Management Admissions Test (GMAT) score's tie with that of the admissions scale; the size of the unstandardized regression coefficient might lead one to conclude that GMATs had little to do with predicting a person's first-semester grades.

Source: Sobol, M. G., "GPA, GMAT, and Scale: A Method for Quantification of Admissions Criteria," *Research in Higher Education* 20 (1984): 77–88.

Note particularly the unstandardized regression coefficient for the GMAT score. The coefficient is so small that one might erroneously conclude that GMAT scores had no relationship to GPA. The coefficient of .00099 represents the impact of a 1-point increase in GMAT; at the time the data were collected, the GMAT had a minimum score of 200 and increased in 10-point increments to a maximum score of 800.

Some studies report both pieces of information. Beta weights alone tend to appear in studies where the meaning of the regression coefficient cannot be clearly interpreted. For example, neither the analyst nor the user will have a clear image of what a unit increase in an ordinal variable, such as satisfaction with a policy, means.

The t-Ratio and Statistical Significance

Two tests of statistical significance, the *t*-ratio and the *F*-ratio, are commonly reported along with regression findings. The *t*-ratio tests the hypothesis that the regression coefficient does not equal 0.00. A researcher may report the *t*-ratio, its associated probability, or se_b. If se_b is reported, the *t*-ratio value can be computed by dividing the regression coefficient by its standard error (b/se_b).

The standard errors of *b* also were reported in the example. Recall the rule of thumb that if the *t*-ratio is larger than 2, the relationship is statistically significant at the .05 level. Five times out of 100, a regression coefficient as large as the one obtained may occur if the value of the slope in the population is actually zero. The regression coefficients for undergraduate average, GMAT scores, and admissions scales were the only ones that had a *t*-ratio greater than 2; these also were the variables with the largest Beta weights.

To refine the scale further, the authors checked how R^2 improved as different variables were entered into the admissions criteria. The findings are shown in Table 14.4.

The Beta weights, the *t*-ratio, and the changes in R^2 with each additional variable all supported the researchers' conclusion that only undergraduate grade point average, GMAT scores, and the score on the admissions scales should be considered in predicting success in graduate school.

The regression equation shows that the score assigned on the basis of undergraduate grade point average, GMAT scores, and the admission's scale explains only 18 percent of the variation in first-semester grades. In other words, 82 percent of the variation in first-semester grades is associated with some other factor(s) not present in the model.

The F-Ratio and Statistical Significance

The *F*-ratio tests the model, as represented by the equation, as a whole. The *F*-ratio is the ratio between explained and unexplained variance. It indicates the probability that the regression equation could have occurred by chance. (See Chapter 13 to review the *F*-ratio and use of the table of *F*-values.) The null hypothesis is $b_1 = b_2 = b_3 = \ldots = b_k = 0.00$; *k* equals the number of independent variables. The hypothesis is that at least one of the independent variables is not equal to 0. The equation for the *F*-ratio is

$$F = R^2(n - k - 1)\,(1 - R^2)(k)$$

where:
R^2 = coefficient of multiple determination
n = number of subjects
k = number of independent variables

When several models are tested, the Beta weights, R^2, and significance data are examined to decide which model best describes the data.

Table 14.4 Relationship Between First-Semester Grades and Admissions Criteria

Regression Equations	R^2
Undergraduate average (UA)	.104
UA + GMAT	.148
UA + GMAT + admissions scale (AS)	.181
UA + GMAT + AS + Sex	.184
UA + GMAT + AS + Sex + Previous graduate study	.188

Note: GMAT = Graduate Management Admissions Test.

Source: Sobol, M. G., "GPA, GMAT, and Scale: A Method for Quantification of Admissions Criteria," *Research in Higher Education* 20 (1984): 77–88.

Three Example Models

Example 14.3 shows three models developed to identify factors associated with the frequency with which management reports are used. Note that the statistics reported for Equations 2 and 3 are similar. A researcher may decide that Equation 2 is the better model. Equation 2 explains the same amount of variation as Equation 3 but with less information, that is, with fewer variables; consequently, it is more efficient. Variable X_4, which is not included in Equation 2, is statistically significant, but it has the lowest level of significance, and it has the weakest relationship with the dependent variable.

Example 14.3 Statistical Significance and Multiple Regression

Situation: An investigator wants to identify management report characteristics associated with their actual use. He measured several variables:

Y = average frequency that a manager discussed report information with others
X_1 = report frequency (e.g., 1 = annual, 12 = monthly, 52 = weekly)
X_2 = medium (0 = distributed printed report, 1 = accessible by terminal display)
X_3 = technical quality of the report
X_4 = accessibility of report information
X_5 = perceived value of the report

Strategy: Evaluate alternative exploratory models using multiple regression analysis (see Figure 14.4).

Discussion:

The first model, Equation 1, includes just two independent variables, X_1 and X_2; the model explains 10 percent of the variation in the frequency of report usage. In the second model, Equation 2, two additional independent variables are added to the model, X_3 and X_5. The addition of the new variables changes the Beta values for X_1 and X_2 slightly. The model now explains 16 percent of the variation in the frequency of report usage.

The researcher found that Equation 2 was the best model. It produced the highest R^2 value. Equation 3 did not improve upon Equation 2. Equations 2 and 3 each explained 16

Independent Variables	Beta Weights		
	Equation 1	Equation 2	Equation 3
X_1	.41[c]	.39[c]	.39[c]
X_2	−.30[c]	−.27[c]	−.28[c]
X_3		−.26	−.29
X_4			.11[a]
X_5		.31[c]	.27[b]
R^2	.10	.16	.16
F-Ratio	10.94[c]	9.02[c]	7.53[c]

[a]$p < .05$.
[b]$p < .01$.
[c]$p < .001$.

Figure 14.4 Beta Weights for Three Multiple-Regression Equations

percent of the variation. Thus, the addition of variable X_4 to Equation 3 did not add to the explanation already provided by the other variables.

The Beta weights indicate that report frequency is the most strongly related to frequency of use and that the perceived value of the report also is strongly related. The footnotes imply that a *t*-ratio was computed, and the researcher concluded that the relationships between frequency of use and report frequency, value, and medium were probably nonrandom (Beta not equal to zero). The lack of statistical significance suggests that the technical quality of the report may not be associated with the frequency of use.

The *F*-ratio corroborates the other information. There is a low probability that any of three equations happened by chance. The *F*-ratio and the *t*-test findings strongly imply that the regression coefficients for frequency of use, medium, and perceived quality were not equal to 0.00. All three equations are statistically significant at the .001 level.

Source: Swanson, E. B., "Information Channel Distribution and Use," *Decision Sciences* 18 (1987): 131–145.

Multicollinearity

Sometimes in postulating a regression model, independent variables that are highly related will be incorporated into the model. The problem of having highly correlated independent variables in a regression model is termed *multicollinearity*. When two independent variables are highly correlated, the regression equation cannot accurately estimate their independent effects on the dependent variable. A variable that actually is related to a dependent variable may appear to be unimportant. Multicollinearity can affect the decisions of policy makers. In his data analysis text, E. R. Tufte gives an example of trying to study the relationship of family income and air quality on health.[14] If the family income and neighborhood air quality of subjects are closely related, an analyst cannot determine how much the variation in health was associated with air quality and how much was due to socioeconomic factors. Yet an administrator will want this information to decide whether investing in improving air quality is necessary.

The analyst will be particularly concerned with identifying multicollinearity. Symptoms of multicollinearity can be identified, and an investigator should review the analysis to make sure that it is free of these symptoms. Two symptoms of multicollinearity are that:

1. The multiple regression equation is statistically significant, but none of the *t*-ratios is statistically significant
2. The addition of an independent variable radically changes the values of the Beta weights or the regression coefficients.

If you review Figure 14.4 in Example 14.3, you may note that neither of these conditions exist. Although all three equations are statistically significant, none of the equations has nonsignificant Beta weights. Similarly, the addition of new independent variables did not markedly change the value of Beta weights in the equations. Researchers commonly examine the data matrix that contains the correlation coefficient (*r*) for each pair of independent variables. If any *r* is .80, the researcher may suspect multicollinearity. Nevertheless, this strategy is not perfect; multicollinearity may occur with lower correlation coefficients, and it is not necessarily present when *r* is as large as .80.[15]

Multicollinearity is considered a problem of sampling in that both multicollinearity and small sample size inflate the variance estimate. Essentially, the high degree of correlation between the independent variables does not allow for an individual effect to be detected with the available data. By adding more cases, the problem may disappear, but gathering more data is often impractical or impossible. Sometimes, the researcher can do no better than note the problem of multicollinearity and her inability to solve it.

Sometimes a researcher can modify the original model. She may decide two variables are measuring the same underlying concept and drop one of the variables. She may combine the variables into a single measure.[16] If a researcher includes city budget and number of full-time municipal employees as independent variables, she may find that the personnel costs constitute a sizable portion of the city budget and that the two variables are highly correlated. She may argue that the two variables represent the financial investment in city services and drop one of the variables. She may create a new variable, number of full-time employees for each $100,000 in the city budget.

Multicollinearity is not a problem if the purpose of the regression analysis is to forecast. When forecasting, the user is concerned with the accuracy of the prediction and not necessarily the accuracy of a specific regression coefficient. The accuracy of forecasting depends on past conditions interacting in the future the same way they did at the time the data were gathered.

Regression and Non-Interval Variables

One of the assumptions of regression is that the variables are measured at the interval level. In reading social science journals, you will often see ordinal variables in regression equations. If the values of an ordinal variable form a uniform scale, the assumption may be relaxed. There are different rules of thumb. A classic psychometric text suggests an analyst can treat ordinal variables with 11 or more values as interval. This is not a hard and fast rule. Level of education measures that may have only seven categories seem to work reasonably well.[17] If an ordinal variable has few values, the analyst should examine contingency tables or scatterplots to see that important dynamics of the relationship are not being missed.

As we pointed out earlier, interpreting dichotomous variables as interval measures is acceptable, as this only shifts the intercept of the model. This means that dichotomous variables may be included as independent variables in a regression equation. If the dependent variable is dichotomous, however, linear regression should generally not be used.[18] Rather, an analyst will often use something called logistic regression.[19]

Applying Logistic Regression

The logistic regression model allows us to identify a relationship between a dichotomous dependent variable and one or more independent or predictor variables. A dichotomous variable has two values that are mutually exclusive—a case can only be in one category or the other. The details of logistic regression are beyond the scope of this text.[20] Nevertheless, a short discussion of how to interpret the *logistic regression coefficients* and a strategy for determining the goodness of fit of the model is important. The review should enable you to identify situations where logistic regression could yield valuable information. The discussion should also enable you to understand logistic regression findings.

The Use of Cost Accounting Plans in Local Government

As an example of logistic regression, we will use a study of the use of cost accounting plans by local governments.[21] Cost accounting plans are a type of cost accounting used to account for both direct costs of a good or service, as can be found in the department budget, and additional resources called indirect costs or overhead costs that are provided in other areas of the organization like human resources, information technology, finance, and others. The use of cost accounting in government has been historically driven by the need to recover overhead for the purpose of grants but is increasingly being used to set rates and fees, to deal with fiscal stress, and for performance management.[22] In spite of its age and importance in public financial management, little has been known about its use in local governments in the United States until recently.

In the following study that looked at whether local governments were using a cost accounting plan, the researchers analyzed whether grant funds were still predictive of whether the government was creating a cost accounting plan. They also were interested in whether having more business funds, as measured by enterprise fund expense, was predictive of whether a government had a cost accounting plan. The researcher also wanted to know if performance measurement and fiscal stress were also associated with the use of a cost accounting plan. The model also controlled for revenue, median income, total expenses, population, the hierarchy of the organization, whether it was a city manager form of government, and right to work laws.

Odds Ratios

The logistic regression equation calculates the probability that an event—a value of the dependent variable—will occur. The calculations, however, produce the logarithm (log) of the odds ratio. This obviously complicates interpretation of the coefficients, especially since the log is the natural logarithm. Analysts can use this information to calculate the probability of the event occurring. The dependent variable (z) in this example was whether a city "has a cost accounting plan" (1 = has a cost accounting plan, and 0 = does not have a cost accounting plan).

The form of the logistic regression equation is similar to linear regression:

$$\text{log odds } (z) = \text{Constant} + b_1X_1 + b_2X_2 + \ldots b_iX_i$$

The log odds value can be converted to a probability, and the equation can be used to predict a value for each case. We show how to do this next.

The logistic regression coefficients provide information about how each independent variable in the equation relates to the dependent variable event occurring. Table 14.5 reports the logistic regression coefficients for our example. These coefficients, in the column headed by "B," have the same function as linear regression coefficients in that they show the direction and amount of change in the dependent variable for a unit change in the independent variable. For example in

Table 14.5 Determinants of Cost Accounting in Large U.S. Cities

Variables	Model 1			Model 2		
	B	Odds Ratio		B	Odds Ratio	
Grant Funding Percentage	−0.018	0.982		–	–	
Enterprise Expense Percentage	0.044	1.045	*	0.026	2.842	*
Performance Measurement	−0.268	0.765		–	–	
Fund Balance	−0.125	0.883	**	–	–	
Total Revenue/Population	0.001	1.001		–	–	
Median Income (in thousands)	−0.037	0.964		–	–	
Total Expense (in thousands)	−0.001	0.999		–	–	
Population (in thousands)	0.005	1.005		–	–	
Hierarchy	1.240	3.456	**	–	–	
Manager	−1.294	0.274		–	–	
Right to Work	0.808	2.244		–	–	
Intercept	−2.494	0.083		−0.117	0.890	
N	81			81		
Model Deviance	76.14			100.77		
Pseudo R^2	0.27			0.035		

$*p \le .1$, $**p \le .05$ One-tailed significance

Model 1 adapted from Z. Mohr "An Analysis of the Purposes of Cost Accounting in Large U.S. Cities" *Public Budgeting and Finance* (Spring 2015): 95–115

Model 2, the coefficient for Enterprise Expense is .026, which indicates that as enterprise expense goes up, the likelihood of the city having a cost accounting plan increases.[23]

The regression coefficients are difficult to interpret, however, as they are measured on a log scale and report the log of an odds ratio, more commonly referred to as log-odds. They represent the difference in the log-odds of the dependent variable for a one-unit change in the independent variable. A common way to interpret the coefficient is to convert from the log-odds to an *odds ratio*. An odds ratio is the ratio of the probability of an event occurring divided by the probability of the event not occurring.[24] Converting regression coefficients to odds ratios does not affect their level of statistical significance. If the coefficient is statistically significant, then the corresponding odds ratio will be statistically significant. The odds ratio is simply a way to interpret the coefficient in a more meaningful way.

To obtain the odds ratio, the natural log is removed from the logged coefficient by exponentiation. The columns in Table 14.5 headed "Odds Ratio" contains the converted coefficients. Computer programs such as SPSS that do logistic regression can provide these. They are also easy to obtain with a spreadsheet program such as Excel. You simply exponentiate the coefficient term (i.e., =exp(coefficient)).

To further interpret the coefficients and the predictions made by the model, we can convert the odds into probability. We can use the following formula to convert odds to a probability:

Probability = Odds/(1 + Odds)

Due to the mathematical properties of logs, the odds ratio measures how many times higher the odds of occurrence are for each one-unit increase in the independent variable. For example, when the coefficient for Enterprise Expense (the indicator variable for the amount of business-like activity that a city does) in Model 2 is converted to an odds ratio, we see that for each percentage point increase in the Enterprise Expense that the odds of the city having a cost accounting plan increase approximately by a factor of 2.8.[25]

A negative logistic regression coefficient will translate into an odds ratio between 0 and 1, while a positive coefficient will translate into an odds ratio exceeding 1. You can see this in Table 14.5.

Applying the Equations to Determine the Probability of a City Using Cost Accounting

Let's apply the equation derived from the results from Model 2 in Table 14.5 to a city to determine the probability that it will have a cost accounting plan. Suppose that we want to figure out the probability of a city with an average amount of enterprise expense using a cost accounting plan. In this dataset, the average enterprise expense is 31 percent. To calculate the probability, we first have to figure the regression formula for z, which is measured on the log-odds scale. Then we have to remove the logarithmic term from the log-odds estimate, and then we can convert the odds to probability. To do this, we will follow a three-step process:

1. First, we use the logistic regression coefficients and obtain the log odds.

   $Z = -.117 + .026(31) = .689$

2. This value is then used to calculate the probability of the city using a cost accounting plan as follows:

   Convert log odds to odds: Exp(.689) = 1.992

3. Finally, the odds are converted to a probability.

   Probability = 1.992/(1 + 1.992) = .667

The estimated probability from Model 2 that the city will have a cost accounting plan is 67 percent (i.e., .667 × 100 = 67%).[26] Because the probability can be easily calculated and many statistical programs can easily create them, a newer way of interpreting the effect of predictors on the probability of the outcome of a logistic regression is to use a predicted probability plot. A predicted probability from the model can be plotted along the entire range of values of the independent variable of interest to show how the probability changes over the range to give a better idea of the probability for all values of a predictor.[27]

While Model 2 does a decent job of estimating the probability of the average city having a cost accounting plan given that there is only one predictor in this model, we can see that there are other significant factors that also influence whether a city uses a cost accounting plan. We can see from Model 1 that other significant predictors of the use of cost accounting include fund balance and the levels of hierarchy in the organization. A high fund balance indicates lack of fiscal stress, which would lead a city to possibly not needing to be as concerned about its costs. The number of levels of hierarchy, on the other hand, seems to be positively related to a higher likelihood of cost accounting plans in cities, as the cities may have more overhead expenses to allocate down to service departments.

Interpreting the Fit of Logistic Regression Models

A safe way to describe the overall fit of a logistic regression model is to simply describe the deviance of the model (also known as the -2log-likelihood).[28] The *deviance* is how well a logistic regression model fits relative to an ideal model. The deviance is similar to the coefficient of determination in that it shows how well the predictors match the data, but it does not have a direct interpretation like R squared. With deviance, the closer to zero, the better the fit of the model.

As expected with more significant predictors, you can see that the deviance in Model 1 is lower than that in Model 2 (76.14 to 100.77). Deviance is well suited for comparing nested models (a nested model is a second model that is the same as the first model, but it does not contain all of the variables in the first model), but it does not have a direct interpretation other than to show that models with a lower deviance tend to fit the data better than models with higher levels of deviance.[29]

A more contentious but still often used measure is a pseudo-R^2 measure. Like deviance, pseudo-R^2 measures do not have a direct interpretation like R^2 from linear regression. However, they are designed to be interpreted in similar fashion to the R^2 from a linear regression in that a higher value of a pseudo-R^2 shows a better-fitting model. It is *not* appropriate to say that the value of a pseudo-R^2 measure accounts for 27% of the variation in the data as is done with linear regression. Rather, various pseudo-R^2 measures show that models with a higher pseudo-R^2 are better-fitting models than models with lower pseudo-R^2. For example, Model 1 has a pseudo-R^2 of .27 and Model 2 has a pseudo-R^2 of only .035. Clearly, Model 1 is a better-fitting model in terms of both lower deviance and higher pseudo-R^2.

While there is no general guidance on what constitutes a good fit for a logistic regression model, there are some rules of thumb that are generally employed in the literature. For example, some researchers note that a pseudo-R^2 value greater than .2 is a generally acceptable model fit. This may be even a bit statistically conservative, as useful models may have pseudo-R^2 values below .2; however, it is a useful guide found in the literature.[30]

Regression Models to Analyze Time-Series Data

A researcher can use regression to analyze longitudinal data. Time-series data are particularly valuable for making forecasts. Administrators commonly forecast future demands to make hiring, scheduling, and purchasing decisions. Economic decisions depend on forecasts; governments forecast tax revenues, while businesses forecast demand for their products.

The linear regression model is appropriate for forecasting if over time the variable's values can be reasonably described by a straight line. Nevertheless, predictions based on linear regression may be inaccurate. First, the basic *linear regression model* fails to react to changes in direction or in the rate of change. Second, short-term estimates, such as estimates of demand for tax forms in any given month, may be highly inaccurate if a model ignores seasonal and cyclical data.

Example 14.4 presents a model to illustrate this use of the linear regression; the model was built to forecast the size of the local government workforce in the United States. Our example is rather elementary. It involves few years and ignores seasonal and cyclical fluctuations.

Example 14.4 Forecasting the Value of a Variable With Linear Regression

Situation: A policy analyst wants to estimate the size of the local government workforce in the United States two years from now.

Strategy: He examines employment data for the previous nine years.

To estimate the size of the local government workforce in two years, that is, Year 11:

$Y = 2,357 + 29(11) = 2,676$ or 2,676,000 employees

Table 14.6 Employment Over Time

Number of Local Government Employees in 1,000s (Y)	Year
2,357	1
2,423	2
2,434	3
2,467	4
2,494	5
2,541	6
2,570	7
2,569	8
2,642	9

Regression model:

$Y = 2,357 + 29$ (Year)

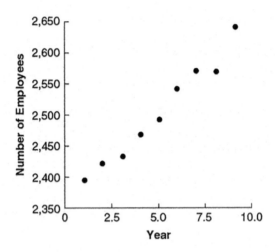

Figure 14.5 Employment Over Time

Source: U.S. Bureau of the Census, *City Employment, 1990*, Series GE-90–2. (Washington, DC: U.S. Government Printing Office), vi.

Before analyzing time-series data to determine the relationships among variables measured over time, a researcher needs to determine if autocorrelation is present. *Autocorrelation* refers to a nonrandom relationship among a variable's values at different time periods. A monograph on time series uses the variable "U.S. defense budget" to illustrate autocorrelation. The author posits two scenarios. In one scenario, the defense budget follows a consistent pattern of feast or famine. If the budget is quite large at Time 1, one may expect a decrease at Time 2. At Time 3, the defense budget again increases to offset the period of relative deprivation. In another scenario, if the defense budget is quite high, the forces in favor of keeping the spending high may thwart efforts to decrease spending; conversely, if spending is low, those favoring less defense spending may be able to successfully forestall efforts to increase spending. Either scenario is symptomatic of autocorrelation; that is, the defense spending at one time is associated with defense spending at a later time.[31]

Autocorrelation violates an assumption of the regression model that the residuals are independent of one another. If autocorrelation is present, the model will produce biased *t*-ratios, confidence limits, and tests of statistical significance.[32] To check for autocorrelation, researchers begin by plotting the residuals.[33] If autocorrelation is present, the plot of residuals will have a distinctive pattern. For example, if the first scenario accurately describes the pattern of defense spending, every other residual will lie beneath the regression line. If the second scenario accurately describes the pattern of defense spending, the size of the residuals may increase for a while, then decrease for a time, and then begin to increase again.

Autocorrelation may be eliminated if the researcher can identify and include an independent variable that explains some of the unexplained variance. One author notes that an apparently strong relationship between the number of suicides and the number of persons unemployed between 1900 and 1970 may have been due to population growth during the time period. If population size were included as a variable, the problem of autocorrelation might disappear.[34] If autocorrelation cannot be eliminated, a researcher should use a statistical model other than the ordinary least squares (OLS) regression model, the model described in this chapter.[35] Analyzing time-series data presents other challenges. One rule of thumb suggests at least 30 cases for a stable regression model. Working with 30 years' worth of data may introduce other problems. A complete set of data may not be available, or the operational definitions or the measurement intervals may have changed. To add cases, an analyst may divide the data into months or quarters, but this strategy may introduce seasonal variations. The OLS model is inappropriate if seasonal variations or autocorrelation occur. Statistical methods exist that take into account autocorrelation and seasonal variations. In addition, in working with social data, one may question how consistent the relationship between two or more variables is over a long time period because conditions in the distant future may not reflect the conditions that created the data in the previous years.[36] An investigator with time and patience can work through practical examples, such as those found in the Woodridge text, and develop the skills and understanding necessary to analyze time-series or panel data.[37]

Regression and Causality

Many topics of interest to administrators and policy analysts cannot be studied experimentally. Multiple regression techniques, for example, path analysis and structural equation modeling, have been employed to produce statistical evidence of theoretic causality.[38] A major value of these techniques is in their ability to eliminate alternative hypotheses by statistically controlling for possible causal variables. The shortcoming in all of these techniques, however, is that the quality of the theoretical model affects the accuracy of the statistical model. For example, control variables, which can demonstrate the spuriousness of the model, may not have been included; thus, a causal relationship may be erroneously assumed. Control variables, which interact with other variables, may have been left out, also affecting the accuracy of the estimated relationships.

In general, administrators are well advised to remember the shibboleth "Correlation does not equal causality." An assumption of causality seems defensible only after extensive model building, testing, and refinement.[39] Some would argue that claims about causality are impossible where the key explanatory variable of interest was not produced through random assignment.[40]

Summary

Regression analysis is a statistical technique that efficiently describes complex relationships. A regression equation allows the user to describe a dataset, to estimate population parameters, to obtain information supporting a claim of causality, and to forecast (see Figure 14.6).

The linear regression equation for the two-variable case is the equation for the straight line that is the best linear description of a dataset. An estimated value of the dependent variable for specific values of the independent variables can be calculated from the equation. The regression coefficients estimate the amount of impact of the independent variable.

To assess the adequacy of the linear-equation model, the analyst should examine the scatterplot to make sure that the relationship is reasonably well described by a straight line. Furthermore, the analyst should look for outliers that may bias the statistics.

The correlation coefficient, r, measures the strength and direction of the association between variables. The r^2 value represents the percentage of variation in the dependent variable that is explained by the independent variable.

Multiple regression extends the two-variable regression model. Partial regression coefficients estimate the impact of an independent variable while controlling for the other independent variables in the equation. Beta weights, the standardized regression coefficients, indicate the relative strength of the association between each independent variable and the dependent variable. The user who assumes that the size of the unstandardized regression coefficient indicates an independent variable's importance in the model may be seriously misled.

The R^2, similar to r^2, reports the variation in the dependent variable that is explained by the variables included in the model. The R^2 may increase slightly with the addition of even trivial variables and should be replaced with the adjusted R^2. Consequently, the analyst should be skeptical of models with large values of R^2 and many variables. To determine what variables to include in a model, the analyst should first consider the purpose of the model and why each variable should be included. Second, the analysis should examine the statistical evidence supporting each variable's relationship to the dependent variable. The analyst also needs to check for multicollinearity, that is, when two variables vary together. If unresolved, the effects of the related variables will not be correctly estimated. Whether a particular value of r^2 or R^2 merits attention depends on the nature of the study. Physical science researchers generally expect far higher values for these statistics than do social scientists. The differences reflect the disciplines' research questions and models. High R^2 values may occur if a dataset contains more variables than cases; the model is essentially hand tailored. When examining a regression model, one should not assume that a low R^2 value is offset by statistical significance. If the model is based on a large sample, a non-random, statistical relationship may be very weak.

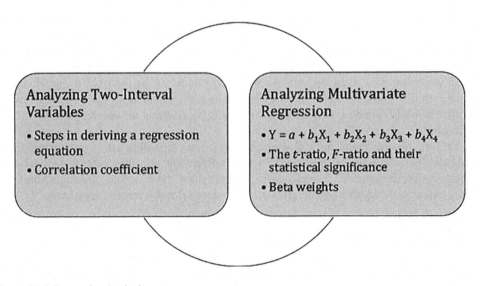

Figure 14.6 Regression Analysis

Although regression techniques have been used in forecasting and to infer causality, the administrator should be skeptical of such inferences. In forecasting, the length of the series, changes in the longer-term trends, and other fluctuations in the series may distort the estimate. A model that suggests a causal relationship based on time-series data may not be credible if the investigators did not test for autocorrelation and make appropriate adjustments. Additionally, forecasting very far into the future is unwise, as many conditions can change well into the future.

Regression statistics assume that the variables are measured at the interval level. If ordinal variables can be assumed to form a uniform scale, the assumption is relaxed. Regression appears to work reasonably well if an ordinal variable has a sufficient number of values and the distribution of outcomes approximates the normal distribution. Dichotomous independent variables may be used in multiple regression. Ordinary least squares regression should not be used to analyze dichotomous dependent variables; rather, logistic regression may be used. Logistic regression predicts the probability that an event will occur and indicates the statistical model's ability to correctly categorize the values of the dependent variable.

Surveying the major approaches that administrators use to collect, examine, and analyze data is important for an understanding of research methods. Another major stage of the research process remains, however. That is reporting the findings and communicating effectively to relevant audiences. Two issues in addition to proper and effective reporting remain: ethical considerations in reporting findings and requirements for keeping research records. A remaining chapter addresses each of these.

Notes

1. For an asymmetric measure, the results will be different depending on which variable is identified as independent and which is dependent. For a symmetric measure, the results will be the same regardless of which variable is independent and which is dependent.
2. SPSS, Stata, and other statistical programs such as R and even Excel will do regression analysis. It is easy to generate graphs called scatterplots with these programs.
3. For a further discussion on using regression to estimate parameters, see L. L. Giventer, *Statistical Analysis for Public Administration*, 2nd ed. (Boston, MA: Jones and Bartlett, 2008), Chapter 13, which is organized around estimation. The Giventer chapter and K. J. Meier, J. L. Brudney, and J. Bohte, *Applied Statistics for Public and Nonprofit Administration*, 9th ed. (Belmont, CA: Cengage Learning>, 2014), Chapter 18, cover the standard error of the estimate (used to calculate confidence limits around *Y* for a specific value of *X*), which we do not include in this chapter.
4. In addition to problems of randomness or a nonlinear relationship, the analyst should look for patterns suggesting that statistical assumptions have been violated. For a discussion on the statistical assumptions, illustrated with scatterplots, see Meier, Brudney, and Bohte, *Applied Statistics for Public and Nonprofit Administration*, Chapter 19. Giventer has a similar discussion on pages 284–286. T. H. Poister, *Public Program Analysis* (Baltimore: University Park Press, 1978), 523–525, has a clear discussion on how plots of the residuals can be evaluated to determine goodness of fit.
5. Methodologists increasingly emphasize the importance of reporting confidence intervals when reporting statistical results. The example here illustrates this for regression coefficients. See Geoff Cummings, *Understanding the New Statistics: Effect Sizes, Confidence Intervals, and Meta-Analysis* (New York: Taylor & Francis, 2012). He discusses confidence intervals in Chapters 3 and 4.
6. M. S. Lewis-Beck, *Applied Regression: An Introduction*, University Paper Quantitative Applications in the Social Sciences, no. 22 (Newbury Park, CA: Sage Publications, 1980), 32–33.
7. E. R. Tufte, *Data Analysis for Politics and Policy* (Englewood Cliffs, NJ: Prentice-Hall, 1974), 108–131. Tufte has an interesting real-world policy example of how an outlier can affect the regression and correlation statistics, especially with a small number of cases.
8. See Meier, Brudney, and Bohte, *Applied Statistics for Public and Nonprofit Administration*, Chapter 19, for illustrations of nonlinear scatterplots. A common strategy to deal with these is to use logarithmic transformations. For a detailed discussion of logarithmic transformations, see Tufte, *Data Analysis for Politics and Policy*.
9. See Giventer, *Statistical Analysis*, 284, for an elaboration of this point. Edward Tufte has an example of an outlier, discusses the rationale for removing it, and shows the effect of removing it. His example also

underlines the benefit of examining outliers as unusual cases. They can often be identified by examining the residuals in a normal probability plot or using the Cook's *d* statistic. See Regression with Stata, Chapter 2—Regression Diagnostics. UCLA: Statistical Consulting Group. From https://stats.idre.ucla.edu/other/mult-pkg/faq/general/faq-how-do-i-cite-web-pages-and-programs-from-the-ucla-statistical-consulting-group/ (accessed April 8, 2020).

10. G. B. Lewis, "Equal Employment Opportunity and the Early Career in Federal Employment," *Review of Public Personnel Administration* (Summer 1986): 1–18.

11. Ibid., 8. The regression coefficient for minority female status was statistically significant at the .01 level. In other words, if the actual coefficient was 0, there was a less than 1 percent chance of obtaining a regression coefficient as high as 2.29. The regression coefficients for white females and minority males were not significant at the .05 level.

12. Technically, the percentage of contribution of potential experience, gender, race, and veteran's preference depends on how the analyst built the model. If these had been entered before the education variables, the strength of their relationships might have been greater. This aspect of model-building strategies is beyond the scope of this text, but you should recognize that the analyst makes decisions that determine the specific values in a multiple regression equation.

13. This equation for adjusted R^2 is from J. Cohen and P. Cohen, *Applied Multiple Regression/Correlation Analysis for the Behavioral Sciences*, 2nd ed. (Hillsdale, NJ: Erlbaum, 1983), 105–106.

14. Tufte, *Data Analysis for Politics and Policy*, 150.

15. W. D. Berry and S. Feldman, *Multiple Regression in Practice* (Thousand Oaks, CA: Sage Publications, 1985), 43. Two common symptoms are high intercorrelations between independent variables or very large standard errors of *b*. For further discussion on symptoms of multicollinearity and possible solutions, see Tufte, *Data Analysis for Politics and Policy*, 148–155; Lewis-Beck, *Applied Regression*, 58–63; Berry and Feldman, *Multiple Regression*, 42–50.

16. If multicollinearity is suspected, the analyst may increase the size of the sample, combine the involved variables to form a single variable or keep only one of the variables, or consider an experimental design. For further discussion, see Tufte, *Data Analysis for Politics and Policy*, 148–155.

17. Jum C. Nunnally and Ira H. Bernstein, *Psychometric Theory*, 3rd ed. (New York: McGraw-Hill, 1994), 115–116. Observation about uniform scales is from Giventer, *Statistical Analysis*, 162.

18. At times, a researcher will use a "linear probability model" with a dichotomous dependent variable because the results are easier to interpret than are those from logistic regression. That researcher will, however, virtually always also provide the results from logistic regression when doing so.

19. Probit regression, not discussed here, is another approach sometimes used with dichotomous dependent variables.

20. The SPSS manual, *SPSS Advanced Statistics User's Guide* (Chicago: SPSS Inc., 1990), by Marija J. Norusis, has an accessible explanation of logistic regression. Also see Stephen Sweet and Karin Grace-Martin, *Data Analysis with SPSS: A First Course in Applied Statistics*, 4th ed. (Boston, MA: Longman-Pearson, 2010), Chapter 8 is on logistic regression. D. W. Hosmer and S. Lemeshow, *Applied Logistic Regression* (New York: John Wiley & Sons, Inc., 1989), is recommended as an introduction to the method.

21. The example is based on Zachary Mohr, "An Analysis of the Purposes of Cost Accounting in Large U.S. Cities," *Public Budgeting and Finance* (Spring 2015): 95–115.

22. Ibid. Also see Zachary Mohr, *Cost Accounting in Government: Theory and Applications* (New York: Routledge, 2017).

23. Model 1 shows the entire logistic regression model from the original. Model 2 shows a reduced model estimated with only enterprise expense to show how to interpret the coefficients on a simpler model.

24. To understand an odds ratio, it is helpful to understand odds, and this note provides an example of interpreting odds. If the probability of an event occurring is 75 percent, then the probability of the event not occurring is 100 − 75 percent, or 25 percent. The odds ratio is 75 to 25 or 3. We usually then say that the odds of the event occurring are 3 to 1.

25. Sweet and Grace-Martin, *Data Analysis with SPSS*.

26. This estimate corresponds closely with the observed percentage of cities with cost accounting plans. In the study, 65% of cities had cost accounting plans. While empirically it is good that the estimation is close to the observed percentage, this is actually a lower percentage than what theory and public finance had previously expected to see. However, more recent research shows that there are a significant percentage of cities that do not have a cost accounting plan.

27. To see a predicted probability plot for Model 1, see Figure 2 of Mohr, "An Analysis of the Purposes of Cost Accounting in Large U.S. Cities," 106.

28. For an interesting discussion of why the model deviance is multiplied by −2 and interpretation, see "Why Is the Deviance Expressed as the Log-Likelihood Multiplied by −2?", ReStore National Center

for Research Methods. From www.restore.ac.uk/srme/www/fac/soc/wie/research-new/srme/modules/mod4/6/extensionf/index.html (accessed April 8, 2020).

29. A special case of analyses of the deviance is using an F statistic to look at the fitted model relative to a model with no predictors. If the F is significant, it rejects the null hypothesis that the model has significantly lower deviance than the null model. Statistical programs like Stata calculate and report this F statistic automatically. However, this is a relatively simplistic measure of model fit and should only be reported in conjunction with other measures of model fit.

30. One additional word of caution about pseudo-R^2 measures—there are many different ones that are all slightly different. Some are like linear regression R^2 and range only from 0 to 1. Others can be greater than 1. This difficulty of interpretation and the fact that there are multiple types of pseudo-R^2 have earned it a bit of a bad reputation. We have reported the McFadden's pseudo-R^2 in our example because it is generally considered a fairly conservative pseudo-R^2; however, like most things in statistics, there may be value in other pseudo-R^2 measures.

31. C. W. Ostrom, Jr., *Time Series Analysis: Regression Techniques*, University Paper, Quantitative Applications in the Social Sciences, no. 9, 2nd ed. (Newbury Park, CA: Sage Publications, 1990), 11.

32. For discussions on detecting autocorrelation and its effects, see Ostrom, *Time Series Analysis*, 12–17, 25–35; S. Makridakis and S. G. Wheelwright, *Forecasting Methods for Management*, 5th ed. (New York: John Wiley & Sons, Inc., 1989), have a relatively clear discussion on autocorrelation, 126–130, 192–194. Also see Meier, Brudney and Bohte, *Applied Statistics for Public and Nonprofit Administration*, Chapter 22, for using regression to analyze interrupted time series and a discussion of autocorrelation.

33. Statistical tests, such as the Durbin-Watson, will test for autocorrelation.

34. M. M. Mark, "The Causal Analysis of Concomitances in Time Series," in *Quasi-Experimentation: Designs and Analysis Issues for Field Settings*, eds. T. D. Cook and D. T. Campbell (Boston, MA: Houghton Mifflin, 1979), 323.

35. For discussion of alternative models, see Mark, "The Causal Analysis . . .," 321–339, or Ostrom, *Time Series Analysis*. The works of S. Makridakis and S. G. Wheelwright, Armstrong, and McCleary and Hays (cited in the "Recommended for Further Reading" section) provide more detailed information.

36. The 30 rule of thumb is from Makridakis and Wheelwright, *Forecasting Methods*, 208. For more elaboration on this discussion of the drawbacks of using linear regression, see Mark, "The Causal Analysis . . .," 335–339.

37. J. M. Wooldridge, *Econometric Analysis of Cross Section and Panel Data*, eds. T. D. Cook and D. T. Campbell (Cambridge: MIT Press).

38. See Cohen and Cohen, *Applied Multiple Regression* (Mahwah, NJ: Lawrence Erlbaum Associates, 2003), Chapter 9; R. D. Cook and D. T. Campbell, *Quasi-Experimentation* (Boston, MA: Houghton Mifflin, 1979), Chapter 7.

39. This argument is effectively made by Tufte, *Data Analysis for Politics and Policy*, 146–147.

40. Substantial work has been done on making causal inferences from observational data. See Christopher Winship and Stephen Morgan, "The Estimation of Causal Effects from Observational Data," *Annual Review of Sociology* 25 (1999): 659–707, available at www.scholar.harvard.edu/files/cwinship_causal_observational_99.pdf. Also see: Herbert Asher, *Causal Modeling*, Sage series Quantitative Applications in Social Sciences (Thousand Oaks, CA: Sage Publications, 1983).

Terms for Review

regression equation
correlation coefficient
scatterplot
residual
goodness of fit
regression coefficient
outliers
lagging a variable
partial regression coefficient
dummy variables
coefficient of determination
Beta weights
multicollinearity

linear regression model
logistic regression
logistic regression coefficient
odds ratio
deviance
autocorrelation

Questions for Review

The following questions should indicate whether you have a basic competency in this chapter's material.

1. Data are gathered on 40 countries to study variations in birth rate. Consider an equation derived in the study:

   $Y = 32 - .0018X$
   $r = -.78$
   $se_b = .00024$

   where:
   Y = birth rate per 1,000 population
   X = per capita income

   a. Identify the following:

   - the independent and dependent variables
   - the regression coefficient
   - the constant
   - the correlation coefficient
   - the standard error of the slope
   - the linear regression equation

   b. What percent variation in birth rate is associated with per capita income?
   c. What is the direction of the relationship between per capita income and birth rate? What evidence supports your answer?
   d. What goodness of fit information is provided? How should you interpret this information?
   e. Calculate the *t*-ratio. What can you infer about the relationship between per capita income and a nation's birth rate?
   f. A country has a per capita income of $2,000. Estimate its birth rate.
   g. Per capita income in the dataset ranges from $400 per year to $12,000. Can you estimate the birth rate for a country with a per capita income of $20,000 per year? Justify your answer.
   h. If per capita income is measured as "per capita income in thousands of dollars," the regression equation is $Y = 32\ 1.8X$. Is this a better model of the relationship? Justify your answer.
   i. Should you examine the scatterplot for this data? Why or why not? What pattern would you need to decide that the linear regression model is appropriate? What would you do if you found a nonlinear pattern or no pattern at all?
   j. Should you examine a scatterplot of residuals for this data? Why or why not? What would you hope to find?

2. The equation is recalculated to include information on the percentage of the population living on farms:

$Y = 36 - .0018X_1 - .186X_2; R^2 = .67$
(.066)

where:
Y = birth rate per 1,000 population
X_1 = per capita income
X_2 = percentage of population living on farms
se_b = reported in parentheses

   a. Interpret the information provided by ".0018" and ".186" in the equation.
   b. What does the se_b indicate about the relationships between birth rate and per capita income and percentage of the population living on farms?
   c. What variation in birth rate is explained by the multiple regression model? Do you consider this high, low, or in between? Explain.
   d. What is the direction of the relationships between birth rate and per capita income and percentage of the population living on farms? What evidence supports your answer?
   e. Predict the birth rate for countries with the following characteristics:
      i. Per capita income of $10,100 and a farm population of 13%
      ii. Per capita income of $1,230 and a farm population of 35%
      iii. Per capita income of $4,100 and a farm population of 4%
   f. What additional information would Beta weights provide that you do not currently have?

3. Use the equation shown to predict values of Y for the following cases:

   a. a department with an average grade point of 2.5 and no placement assistance
   b. a department with an average grade point of 2.5 and formal placement assistance
   c. a department with an average grade point of 2.7 and no formal placement assistance
      $Y = 5 + 20X_1 + 10X_2; R = .5$

where:
Y = percentage of department graduates placed within a month of graduation
X_1 = mean grade point average of department graduates
X_2 = formal placement assistance (X_2 = 1 if formal assistance; X_2 = 0 if no formal assistance)

   d. What variation in placement success is explained by the multiple regression model?
      What is meant when we note that the linear regression line is asymmetric?
      Why should you learn the range of values of X that are used to calculate the regression equation?

Problems for Homework and Discussion

1. The following two problems refer to the multiple regression equation given in Review Question 3.

   a. If the data had been collected last year, how would they affect the ability of the university to forecast placement outcomes for this year's graduates?

   b.   The original model included X_3, average aptitude test scores, for example, Scholastic Aptitude Tests and Graduate Record Exams. The variable was dropped because of a problem with multicollinearity. Explain what is meant by multicollinearity in this context. How could it have been detected?

2.   Table 14.7 appeared in a study of the income of planners. Note that planners' income is the dependent variable.

   a.   Compare how well the model explains the variation of income among male and female planners.
   b.   What factors have the greatest effect on the income of male planners? On female planners? What information did you use to reach this conclusion?
   c.   Does the evidence suggest that job turnovers were associated with higher incomes? Briefly explain how you reached this conclusion.
   d.   Does the evidence suggest planners' salaries were affected by racial considerations or marital status? Briefly explain how you reached this conclusion.
   e.   What does the evidence suggest about the value of taking on many different planner roles?
   f.   What is the difference between the starred ([a]) regression coefficients and the other coefficients?

Table 14.7 Planner's Income Beta Coefficients by Gender

	Men	*Women*
Age	.02	.08
Education[b]	.25[a]	.30[a]
Marital (married = 1)	.05[a]	.01
Race (white = 1)	.01	.02
Agency size	.19[a]	.17[a]
Centralization	.04	−.03
Job turnover	−.14	−.10
No. planner roles	.03	−.01
Yrs. professional experience	.53[a]	.37[a]
R^2	.600	.426

[a]Significant at the .05 level.
[b]Education rescaled: (1) high school; (2) 1–2 years college; (3) 3 or more years college; (4) bachelor's degree; (5) 1 or more years graduate work; (6) master's degree; (7) 1 or more years graduate work after master's; (8) doctorate.

Source: Mayo, J. M. Jr., "Job Attainment in Planning: Women Versus Men," *Work and Occupation* (May 1985): 152. (Copyright 1985 by Sage Publications, Inc. Reprinted with permission: Sage Publications, Inc.).

3.   To study factors associated with frequency of managers' use of report information, a researcher examined five linear models, reported in Table 14.8. The independent variables consisted of:

   Frequency of report (annual = 1; monthly =12; weekly = 52; daily = 260)
   Medium (printed = 1; terminal screen = 0)
   Technique (factor score representing objective information quality, for example, data reliability, precision, and timeliness)
   Accessibility (factor score representing ability to obtain and interpret information)
   Value (factor score representing value of information to specific user)

   a.   Based on this information, which model best explains variations in managers' information use? Why did you choose this model?

b. Based on this information, should an agency go to the extra expense of printing reports? Justify.

c. Based on this information, should an agency concentrate on continually improving the quality of its information? Explain.

d. Which variable has the greatest association with the frequency of report use?

e. "Technique" has a Beta weight of .07 in Equation 2 and a markedly higher value in Equations 3 through 5 (.17, .26, and .29). Briefly explain.

Table 14.8 Standardized (Beta) Regression Equations

	1	2	3	4	5
Independent Variables					
Frequency of report	.41	.42	.41	.39	.39
Medium	−.30	−.2	−.31	−.27	−.28
Technique		−.07	−.17	−.26	−.29
Accessibility			.19		.11
Value				.31	.27
R^2	.10	.10	.12	.16	.16

Source: Swanson, E. B., "Information Channel Disposition and Use," *Decision Sciences* 18 (1987): 131–145.

4. The following model was used to study variations in water consumption (measured as total sales) in communities:

$$Y = 600 + .019X_1 - 15X_2 - .02X_3 - 6X_4 - 682D_1 - 736D_2 - 626D_3 - 356D_4 + 40D_6 + 244D_7 + 226D_8 + 2D_9 - 183D_{10} - 532D_{11} - 720D_{12}$$

where:
Y = total water sales (measured by acre feet)
X_1 = community population
X_2 = average yearly rainfall (in inches)
X_3 = public relations costs
X_4 = real price/acre
D_1–D_{12} = January–December
  Adapted from L. J. Mercer and W. D. Morgan, "Impact of a Water Conservation Campaign," Evaluation Review 4 (1980): 112.

a. What would be the predicted water consumption in January for a community of 50,000 with an annual rainfall of 10 inches, $1,000 in costs for public relations, and real price of $80/acre?

b. What is the effect on water consumption of every $100 spent on public relations?

c. What month is water consumption the highest? The lowest?

5. Table 14.9 reports data on verbal SAT scores and average daily attendance for 15 high schools.

The equation for the regression line is:
$Y = -46 + 4.4X$; $r = .66$

where:
Y = verbal SAT score
X = average daily attendance

Note: Percents were entered as whole numbers; for example, the value entered for X was "20" for a school with 20 percent daily attendance.

a. Create a scatterplot for the data in Table 14.9.
b. Plot the regression line.
c. Calculate the residuals for Adams, Day, and Lee.
d. Do the data indicate that students who attend school regularly do better on their verbal SATs than students who do not? Justify your answer.
e. Interpret the relationship between school attendance and SAT verbal scores.

Table 14.9 Average Verbal Scholastic Aptitude Test Scores and Average Daily Attendance for County High School

School	Average Verbal SATs	Average Daily Attendance
Adams	281	77%
Bell	312	75%
Clark	439	96%
Day	282	91%
Edwards	294	81%
French	316	83%
Grant	365	84%
Hill	294	80%
James	308	79%
Key	257	85%
Lee	308	76%
Mann	387	95%
Nash	287	71%
Parr	394	91%
Rhodes	319	88%

6. Researchers created an accounting test to replace a standardized test. They wanted to validate the test and to establish that the test did not discriminate against minority or female job applicants. To validate the test, the researchers tested agency accountants and gathered data on two criteria:

Criterion 1 (C_1): Behavioral scale demonstrating job knowledge
Criterion 2 (C_2): Rating of overall job performance

Table 14.10 reports the correlation coefficients (r) between each test and validation criteria for various types of subjects. For instance, the .56 at the bottom of column 1 represents $r = .56$ for the relationship between the score on the behavioral scale demonstrating job knowledge (C_1, a dependent variable) and the score on the created test for all nonminority subjects.

Table 14.10 Correlations Between Performance on Two Job Tests and Two Performance Measures by Subject Characteristics

Test	Created Test		Standardized	
	C_1	C_2	C_1	C_2
All subjects	.44	.28	.12	.18
Male subjects	.41	.20	−.03	.14
Female subjects	.48	.56	.25	.11
Minorities	.37	.26	.18	−.21
Nonminorities	.56	.29	.20	−.10

Source: Kesselman, G. A., and F. E. Lopez, "The Impact of Job Analysis on Employment Test Validation for Minority and Nonminority Accounting Personnel," *Personnel Psychology* (1979): 99.

a. What type of validation did the researchers conduct?
b. Which test does the data suggest the agency should use to hire accountants? Why? Do you see any weaknesses in this test that should be rectified? What are they?
c. Based on the data, would you recommend using both criteria to validate future tests? Justify your answer.

7. Tables 14.11 and 14.12 were reproduced from a General Accounting Office report, "Infant Formula in the WIC Program." The study estimated the effects of various factors on the cost of infant formula in the federal Women, Infant and Children's program.

a. Estimate the cost of infant formula in a program in the West that uses sole-source bidding, has an average WIC population of 5,000, serves approximately 60 percent of the eligible population, and does not have an infant formula manufacturing facility in the state.
b. What percent of the variance in the cost of infant formula is explained by the model?
c. Based on these data, would you recommend that a WIC program use sole-source competitive bidding or multisource? Justify your answer.
d. Do states that have an infant formula manufacturer in their state benefit? Justify your answer.

Table 14.11 Definition of Variable Used in Regression Analysis

Variable	Description
SOLE BID	Equals 1 if WIC agency uses sole-source competitive bidding, 0 otherwise
MULTI BID	Equals 1 if WIC agency uses multisource competitive bidding, 0 otherwise
WICPOP	Average monthly WIC infant population (June 1988 to May 1989) in thousands
MFGSTATE	Equals 1 if infant formula manufactured in the state, 0 otherwise
ELIG89	Proportion of the eligible population served by the WIC agency in fiscal year 1989
INDIAN	Equals 1 if the WIC agency represents an Indian tribe, 0 otherwise
MIDWEST	Equals 1 if the WIC agency is located in the Midwest, 0 otherwise
WEST	Equals 1 if the WIC agency is located in the West, 0 otherwise
SOUTH	Equals 1 if the WIC agency is located in the South, 0 otherwise
PRICE	Wholesale price minus rebate amount per 13-ounce can of formula as of June 1, 1989

Table 14.12 Regression Results

Variable	Parameter Estimate (B)	Standard Error	T-stat
INTERCEPT	0.53	0.10	5.39
SOLE BID	−0.36	0.05	−7.91
MULTI BID	−0.24	0.09	−2.85
WICPOP	0.0002	0.00082	−0.25
MFGSTATE	−0.01	0.07	−0.11
ELIG89	0.40	0.15	2.64
INDIAN	0.27	0.06	4.45
MIDWEST	−0.04	0.06	−0.66
WEST	0.03	0.06	0.58
SOUTH	0.01	0.06	0.12

Notes: Number of observations, 56.
Adjusted *R*-square, 0.824.
F-statistic, 29.67.
Probability of *F*-statistic, 0.0001.
WIC = Women, Infants, and Children.

Working With Data

1. Researchers continuously try to understand why people do or do not vote in elections. Using the County Data file, examine the relationship between the dependent variable "percent of the population voting" and three independent variables of your own choice representing education, poverty, unemployment, crime, or income. Use only interval data for this exercise.

    a. Write a hypothesis about the relationship between the dependent variable and each of three independent variables you chose from the list; remember to indicate an anticipated direction. You will have three hypotheses.
    b. Create a scatterplot for each hypothesized relationship. Do the data fit a linear model? Explain why or why not.
    c. Use a statistical software program and obtain the following statistics for each relationship: constant, slope or regression coefficient, r, standard error of the slope.

        i.   Write the regression equation for each relationship.
        ii. How well is each relationship described by the linear regression model? Cite evidence to support your observation.
        iii. For each relationship, decide whether it supports the hypotheses you postulated in 1.a.

2. Using the same variables and data, use a statistical software program to test a multiple regression model that tests these relationships simultaneously.

    a. Obtain the overall R^2, the b (unstandardized regression coefficient), B (standardized regression coefficient), and standard error of b or value of t for each independent variable.
    b. Report your findings in a table similar to Table 14.3.
    c. Report your findings as an equation (for an example, see problem 4 for homework and discussion).
    d. How much variation in political activity across counties, as measured by the percent of the population voting, is explained by the model?
    e. Would your model be better if unemployment was dropped from the model? Justify your answer.

3. Add the independent variable region to your model using dummy variables, one dummy for each region (remember to use only two dummies in the model, leaving the third region as the benchmark). Report your findings in a table similar to Table 14.3.

    a. Which variables have the strongest relationship to percent of the population that voted? What criteria did you use to select these variables?
    b. Which variables have the weakest relationship to percent of the population which voted? What criteria did you use to select these variables?

4. Use the regression model in Exercise 2 and report separate equations for each region. Compare the models for the three regions.
5. Use the variables in the County Data file to create and test a multiple regression model of your own design.

    a. Select a dependent variable and write a verbal model that includes three independent variables. Your write-up should suggest the value of studying each relationship, hypothesize the direction of the relationships, and provide a justification or evidence supporting each.

b.   Test your model and present your findings in a table or equation.

c.   Write a verbal summary of your findings. If your hypotheses were not supported, suggest possible reasons for the lack of support.

6.   (Note: This problem is somewhat advanced.) Test a model that predicts that a county will have a Republican majority county council. Select three variables as independent variables to include. Justify the selection of each. Using a logistic regression routine in a statistical package or spreadsheet program, test your model. Interpret the coefficients.

Recommended for Further Reading

Tufte, E. R., *Data Analysis for Politics and Policy* (Englewood Cliffs, NJ: Prentice-Hall, 1974). This is a classic text with excellent examples and valuable observations about the interpretation of regression findings. Our discussion owes much to Tufte's presentation. He provides more detail and considers a variety of statistical models, such as logarithmic transformations. The book is short and quite readable.

For accessible explanations of regression statistics along with worked-out examples, see Giventer, L. L., *Statistical Analysis for Public Administration*, 2nd ed. (Boston, MA: Jones and Bartlett Publishers, 2006), Chapters 12–14, and Meier, K. J., J. L. Brudney, and J. Bohte, *Applied Statistics for Public and Nonprofit Administration*, 9th ed. (Belmont, CA: Thomson-Wadsworth, 2014), Chapters 18–19, 21, 23. Both texts include computer printouts. Meier, Brudney, and Bohte also cover cubic and quadric transformations. The Sage series Quantitative Applications in the Social Sciences has several monographs on regression.

Information on forecasting and analyzing interrupted time-series designs can be found in Meier, Brudney, and Bohte, Chapters 20 and 22; Ostrom, C. W. Jr., *Time Series Analysis: Regression Techniques*, University Paper Quantitative Applications in the Social Sciences, no. 9, 2nd ed. (Newbury Park, CA: Sage Publications, 1990); R. McCleary and R. A. Hay, Jr., *Applied Time Series Analysis for the Social Sciences* (Beverly Hills, CA: Sage Publications, 1980), provide more sophisticated discussions of how to analyze time-series data.

Weiberg, Sanford, *Applied Linear Regression* (New York: Wiley Interscience, 3rd ed., 2005, and 4th ed., 2013) are also highly recommended. Accompanying material including datasets are available at www.stat.umn.edu/alr.

Appendix 14.1

Calculating Bivariate Regression and Correlation Statistics

This section reviews the equations for deriving a simple linear regression equation, the correlation coefficient (r), standard error of the estimate of the regression coefficient (se_b), and the t-ratio (see Table 14.12).

The calculations to obtain the regression equation involve two major steps. First, the value of the slope, or the regression coefficient, symbolized by b, is calculated.

The equation to calculate the regression coefficient is:

$$b = \frac{\sum(X_i - \bar{X})(Y_i - \bar{Y})}{\sum(X_i - \bar{X})^2}$$

where:
X_i = the value of each X
$\bar{X}$ = the mean value of all the Xs
Y_i = the value of each Y
$\bar{Y}$ = the mean value of all the Ys

Table 14.13 Calculations for b, a, and the linear regression

Car	Miles (1000s) (X)	Repair Costs (Y)	$X_i - \bar{X}$	$Y_i - \bar{Y}$	$(X_i - \bar{X})(Y_i - \bar{Y})$	$(X_i - \bar{X})^2$	$(Y_i - \bar{Y})^2$
1	80	1,200	31.05	577.5	17,931.38	964.10	333,506.3
2	29	150	−19.95	−472.5	9,426.38	398.00	223,256.3
3	42	600	−6.95	−22.5	156.38	48.30	506.25
4	53	650	4.05	27.5	111.37	16.40	756.25
5	73	1,000	24.05	377.5	9,078.88	578.40	142,506.3
6	80	1,500	31.05	877.5	27,246.38	964.10	770,006.3
7	60	600	11.05	−22.5	−248.62	122.10	506.25
8	13	200	−35.95	−422.5	15,188.88	1,292.40	178,506.3
9	14.5	0	−34.45	−622.5	21,445.13	1,186.80	387,506.3
10	45	325	−3.95	−297.5	1,175.13	15.60	88,506.25
Sum	489.5	6,225			101,511.29	5,586.20	2,125,562.8
Mean	48.95	622.5					

$b = (101511.3)/5586.225 = 18.17$
$a = 622.5 - 18.17(48.95) = -266.92$
$Y = -267 - 18X$

Second, the value of the Y intercept, or constant, symbolized by a, is calculated. The equation to calculate the intercept is:

$$a = \bar{Y} - b\bar{X}$$

Using the same figures, we can calculate the correlation coefficient r with the equation:

$$r = \frac{\Sigma(X_i - \bar{X})(Y_i - \bar{Y})}{\Sigma(X_i - \bar{X})^2 \, \Sigma(Y_i - \bar{Y})^2}$$

$$r = \frac{(101,511.3)}{\sqrt{(5,586.225)(2,125,563)}}$$

$$r = .93$$

The Standard Error and t-Ratio

To calculate the se_b, the equation is

$$\sqrt{\frac{\Sigma(Y_i - \bar{Y})^2 / (n-2)}{\Sigma(X_i - \bar{X})^2}}$$

Table 14.14 uses data and regression equation to generate the information needed to show the calculations for obtaining the value se_b for the car repair dataset.

$$se_b = \sqrt{(281,800 / 8) / 5,586.225} = 6.31$$

The t-ratio $= b/se_b$; in our example, the t-ratio is $18/2.5 = 7.2$. A table of the t distribution shows that the associated probability for $t = 7.2$, $df = n - 2 = 8$, is less than .01. In other words, a t-ratio as great as 7.2 would occur less than 1 time out of 100 if the regression coefficient in the population were 0.

Table 14.14 Data and Calculations for *se*

Car	Miles (1000s) (X)	Repair Costs (Y)	$\bar{Y}$	Residual	Residual2	$(X_1 - \bar{X})^2$
1.	80	1,200	1,173	27	729	964.1025
2.	29	150	255	−105	11,025	398.0025
3.	42	600	489	111	12,321	48.3025
4.	53	650	687	−37	1,369	16.4025
5.	73	1,000	1,047	−47	2,209	578.4025
6.	80	1,500	1,173	327	106,929	964.1025
7.	60	600	813	−213	45,369	122.1025
8.	13	200	−33	233	54,289	1,292.403
9.	14.5	0	−6	6	36	1,186.803
10.	45	325	543	−218	47,524	15.6025
Mean	48.95	622.5				
Sum					281,800	5,586.225

15 Completing the Project and Communicating Findings

In this chapter, you will learn:

1. The information to include when presenting research findings.
2. How to include tables and graphs in a presentation.
3. Elements of effective oral briefings and written reports.
4. Ethical considerations in reporting research findings.
5. About various venues for presenting research findings.
6. Guidelines for storing data.

Most investigators find conducting research engaging and satisfying. From building the initial model through to the analysis, they get caught up in the excitement of testing and refining their ideas. After they have obtained and refined the data and completed the analysis, the researchers prepare to report the findings. They do this in informal meetings, formal briefings, project reports, and academic papers. Their audiences rarely share the investigators' level of enthusiasm or interest in the research. If researchers want their efforts to influence policy decisions or even be remembered, they must tailor their presentations to engage audience interest.

This chapter examines the last stages of a research project. It focuses on the components of quantitative reports and how to present them. The chapter argues for clear, focused presentations tailored to the needs of a particular audience. The desire to be relevant and to enhance one's reputation can conflict with ethical practices. The last section of the chapter, therefore, discusses ethical considerations in reporting research findings and in storing research data. Administrators should review the data storage section; failure to establish ownership of the data and determine where they will be kept and for how long may waste resources and prevent efforts to validate or refine findings. Administrators who lose track of information also risk serious problems if promised confidentiality is compromised.

Variations in Audiences and Their Needs

In presenting a report and its findings, investigators should focus on the presentation's purpose and the characteristics of its intended audience. Whether preparing a briefing, giving a presentation, or writing a report, researchers should bear in mind their audience, its interests, and its preference for how it receives research information. Professionals are bombarded by requests for their time and attention. Researchers have to contend with administrators and policy makers who must protect their time and ignore information that they do not want or feel like they need. Beyond getting an audience's attention, investigators want listeners and readers to correctly identify important findings and implied actions.

Academic Versus Practitioner Audiences

Technical communications textbooks consider audience analysis preparatory to drafting oral or written presentations. Academic paper writing may not prepare students very well for assessing practitioner or work environments or "real-world audiences." The student writer can assume, whether true or not, that instructors are knowledgeable, that they read each paper in its entirety, and that they try to understand unclear sections. "Real-world" readers may not understand the paper, may not care about the topic, may ignore unclear discussions, or may browse through a few sections. Listeners in the work environment may let their minds wander, fall asleep, or slip out of the room. Of course, these behaviors may be witnessed in the classroom as well.

What Information Should Be Included When Presenting Research Findings?

Undergraduate and graduate students are often taught to include more information and detail in a paper than the general public and busy administrators are willing to absorb. Many recent graduates are surprised when a manager gives back to them the student's lengthy, thoroughly researched and well-written report with the instructions to: "Condense it to two—no more than three—pages"! Not all policy research reports are treated in this fashion. However, recent graduates often learn that their reports must be more concise and yet convey the important and necessary information.

To counteract the freedom of audiences to ignore or misunderstand information, researchers should identify their audiences and those audiences' characteristics and anticipated needs. They should also consider the various alternatives now available, along with traditional methods, for reporting research findings.

Identifying and Analyzing the Audience

The following five-step procedure for audience analysis seems well suited to researchers conducting research to be used by administrators (see Figure 15.1).[1]

Step 1, identifying likely audiences, identifies probable audiences and suggests how they will use the information.

Step 2, identifying uses of the communication, applies primarily to written communication. Once a report is written and released, its routing is uncertain. Administrators may give a report or handouts to their supervisors, staff, agency analysts, interest-group members, professional acquaintances, legislators, students, or the agency library. Reports are posted on web sites and to Listservs. To satisfy diverse audiences, researchers need to write clear reports that fully document procedures and distinguish objective findings from opinions.

Step 3, identifying the primary target, is recommended so that researchers can learn if their perspective differs from that of an audience. To appreciate differences between themselves and their audiences, the researchers may wish to identify an audience's educational background and experience, knowledge of the subject, and expectations about the report presentation. Then the researchers should identify the same information about themselves.[2] This information will help them critique the clarity of a planned presentation and avoid making unwarranted assumptions about an audience's interests, knowledge, and values.

Step 4, communicating with clarity, recognizes the competition for an audience's attention. Providing a comprehensive discussion of the study information can diminish a presentation's clarity or accessibility. Researchers resolve the dilemma by putting important information first and placing lengthy complicated or technical details in footnotes or appendices.

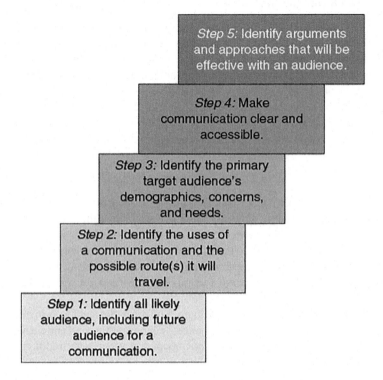

Figure 15.1 Identifying and Preparing for the Audience

Step 5, identifying an approach, builds on the information developed with the preceding steps and requires the investigators to consider how to tailor their presentation so that they communicate effectively to an audience. As part of this step, they may look for models for organizing a presentation. To learn agency preferences for organizing and presenting information, researchers may attend oral briefings or ask agency contacts to identify especially effective reports. They should also determine if a top-level administrator in the organization is likely to be interested in the research reported. Is this a busy administrator who reads dozens of similar items each day? The researchers use this information to infer what features of reports or briefings generate audience interest and involvement. As you read professional reports, save ones that seem particularly well done. You also may want to save an example of a poor report to remind yourself of features to avoid. When you attend oral presentations, note what seems to work and what does not.

Components of the Quantitative Report

A quantitative research report includes a summary of the report, a discussion of the study's background, a review of the literature, a description of the methodology, and the findings and implications. Recommendations, references, and a bibliography are included as appropriate. We assume that readers are familiar with standard documentation requirements, so we will not discuss them in this section.

Example 15.1 reproduces a table of contents for a report on citizen attitudes toward a local police department. It is typical of quantitative reports. As is true of all reports, the specific sections and their content depend on the nature of the study.

Example 15.1 Sample Table of Contents for a Quantitative Report

Comments: The report does not have some sections that commonly appear in quantitative reports. Specifically, it does not include a background discussion, recommendations, bibliography, or methodological appendices. The purpose of the study was to conduct and report on a citizen survey to meet accreditation requirements. This tended to focus the study and enable investigators to skip some of the detail that is found in other papers. The actual paper had more than three tables, but we wanted to give you an idea of what information was included.

Source: Barlow, J. et al., "An Analysis of Survey of Citizens' Opinions of City Police," (Raleigh: North Carolina State University, Unpublished paper, 1992).

This study was designed to meet accreditation requirements. Its background section was brief and included as part of the discussion of purpose. The methodology was straightforward and could be communicated clearly in the body of the report. If the methodology is complex or a detailed discussion is needed, the investigators may place the technical details in an appendix. This study did not make recommendations to the police department; if it had, the recommendations would normally appear in the executive summary and at the end of the body of the report.

Executive Summaries and Abstracts

The executive summary highlights a report's content. The intended audience is the executive who has little time to read reports. Busy administrators and policy makers scan an executive summary to decide if and when to read the entire report or to refer it to an associate. Administrators with a limited interest in the topic may skim a summary to keep themselves current. Policy actors may distribute summaries to communicate and endorse the report's findings. Investigators doing literature reviews can use the executive summary to infer if a report is appropriate to their needs.

The executive summary is the last part of the report to be written. It includes only information in the report, but it can be read and understood independently of the report. Researchers decide what they want a reader—who will spend just a few minutes learning about the study—to know. They may visualize the impatient administrator who asks: "What's the headline?" The researchers go through the report and find sentences that concisely describe why the study was done, who the subjects were, how the data were collected, limitations in the methodology or its implementation, and what the major findings were. An executive summary may include recommendations.

Example 15.2 reproduces the executive summary for the report on citizen attitudes toward the police department. The summary relies on clear, direct sentences and visual cues to allow an individual to read it quickly. Its length and degree of detail are consistent with the length and complexity of the report, agency expectations, and the importance of the findings. In preparing an executive summary, the writer should avoid trying to compress as many details as possible into the summary; otherwise, its benefits are defeated.

Example 15.2 An Executive Summary

Purpose: To meet accreditation standards, the police department must annually survey citizen attitudes and opinions. The department may combine survey findings with other information to aid departmental decision making. This report analyzes data gathered during the 1992 citizen survey.

Methodology: A telephone survey was conducted during October 1992 to gather opinion data from a sample of 898 city residents. The respondents were randomly selected from the city directory so that the same percentage of people came from each police beat.

Police department personnel took the lead in writing the survey questionnaire. The questions focused on how citizens felt about the department in the following areas: overall agency performance, competence of agency employees, officers' attitudes and behaviors toward citizens, concerns over safety and security.

Major findings:
In all districts among respondents who have contact with officers:

- More than 90 percent agree or strongly agree that police officers are competent.
- More than 80 percent agree or strongly agree that police officers have good attitudes.
- More than 85 percent agree or strongly agree that the overall performance of the officers present was good.
- City residents (91 percent) feel safe moving around the city, but 69 percent do not feel safe in downtown at night.
- Twenty-eight percent of respondents feel unsafe in the Southern District.
- Over 50 percent of Southern District respondents expressed safety and security concerns. The most frequently mentioned concerns were related to drug problems.
- The most frequently mentioned concern in the rest of the city was break-ins.

Source: Barlow, J. et al., "An Analysis of Survey of Citizens' Opinions of City Police," (Raleigh: North Carolina State University, Unpublished paper, 1992).

Journal articles have abstracts instead of executive summaries. Abstracts and executive summaries have a similar purpose and content. Abstracts are shorter, about 200 words, with highly condensed writing. They may liberally use technical jargon, more so than most executive summaries. In addition to prefacing articles, abstracts are reprinted in various compilations, such as Sage Public Administration Abstracts.

Background Information

Background information, which puts the study in context, may be included in the introduction as a separate section or in a review of the literature. The report should include sufficient information for the audience to understand why the study was designed the way it was. The specific information needed depends on the report's audience and the study's purpose. A knowledgeable audience may need little background information.

Management and administrative research studies typically start with a problem or a question to be answered. The background information includes a definitive statement of the study's purpose and a definition and history of the problem or the source of the question. If the study involves a program or policy, the report may discuss its origins, implementation history, goals, relevant constituents or stakeholders, resources, and processes. If a formal research proposal was written, the investigators may include its background material in the final report instead of writing a new version.

To develop the background information, investigators rely on documents, summaries of interviews, and previous research on the topic. In applied research reports, information from the literature may be woven into the background presentation, assigned to an appendix, or abbreviated in an annotated bibliography. In academic research, the literature review may appear as a separate section.

Literature Review

The literature review establishes the value of the study and how it fits in with previous relevant investigations of the topic. It brings the reader up to date on previous research that applies to the given study.

In academic publications, the literature review demonstrates that the research does the following:

1. addresses a question not investigated in previous studies,
2. fills in a gap in previous research,
3. tests a model under different conditions,
4. corrects for errors in previous research, or
5. resolves conflicting research findings.

In research conducted for administrators to address a specific issue, the literature review may be much shorter and less comprehensive than that prepared for an academic audience. However, an administrator considering an innovative practice or program may assign an employee to research and summarize the literature on the topic. This literature review may be more extensive and is likely to include case studies of organizations having adopted the innovation. The report based on this literature search would be important to those who will decide about adopting the innovation.[3]

Literature Review Formats

Three literature review formats are common. The review may discuss previous research in chronological order. This strategy works best if there is a discrete body of studies that build upon each other. The first study done on a topic is generally the most basic, and the most recent are

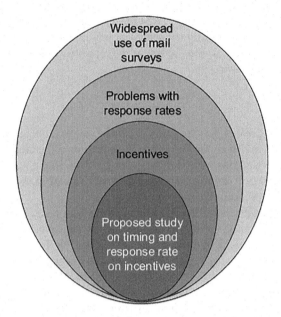

Figure 15.2 Concentric Circles As a Way to Organize Research Literature

generally the most sophisticated or complex. If the research is a further refinement of the models and research previously done, this format may make sense.

Another format for the literature review is to organize the discussion around key variables or concepts. An article on citizen satisfaction with the police organized its literature review around three sets of independent variables. The authors first discussed studies that linked demographic variables (age, race, gender, and education) to attitudes about the police. Then they considered studies that investigated the effect of individual political variables, such as political efficacy. Finally, they wrote about studies that examined the effect of various modes of service delivery.[4] An analogous approach organizes the discussion around theoretical or methodological approaches. Organizing the discussion around variables, theories, or methodologies works best if there are diverse studies examining the same dependent variable.

The third format may be visualized as a set of concentric circles. The discussion begins by identifying studies relating to the general research topic; this is the outermost circle. In each subsequent section, the researcher moves closer and closer to the literature that generated the specific research question. In a study on incentives and response rates to mail surveys, the author first identified literature that confirmed the widespread use of mail surveys. He next cited a source documenting problems with response rates. He worked his way through research on the effectiveness of incentives in increasing response rates, comparisons between the effectiveness of monetary versus nonmonetary incentives, and whether incentives should be mailed with a survey or sent after the response is received. At this point, the stage was set for his study on the combined effects of timing and type of incentive on response rate.[5] This format seems to work best for research that is narrowly focused and fills in a research gap or applies a model under different conditions (see Figure 15.2).

Space Devoted to an Article

The amount of space devoted to any particular article depends on its relevance to the given project. At a minimum, a researcher cites an article, establishes its relevance, and summarizes its major finding. For articles that directly apply to the study, the researcher gives more details. These details allow a reader to compare and contrast the research findings with other research. Such details may include information on the model (what variables were included), the methodology (characteristics of the measures, sample, or research design), the method of analysis, and the findings.

Methodology Section

The methodology section describes a study's measures, sample, and research design. The amount of detail it contains varies widely depending on the audience. At a minimum, it should have enough information so that audience members can decide if the findings are credible. The final project report should be comprehensive enough for others to verify the findings or replicate the research.

The measurement section implicitly or explicitly discusses the operational definitions, how numerical values were assigned and combined, how indices and scales were created, and evidence supporting the reliability and operational validity of the measures. Customarily, the only reliability evidence researchers report are the findings from mathematical tests of reliability.

If the research used survey data or any other kind of data from a sample designed to represent some target population, the researcher should present information about that sample. The discussion of the sample identifies the target population, sampling frame, sampling design, response rate, and when and how the data were collected. There are different ways to compute the response rate. To avoid ambiguity, the investigator should report the sample size, how many members of the sample were contacted, how many of those contacted belonged to the target population, how many refused to provide data, and how many supplied incomplete data. Statistics comparing respondents and nonrespondents should be reported. Any other sources of nonsampling error should be mentioned. For example, if researchers suspect that their sampling frame was inadequate, they should explicitly note their suspicion and explain the basis for it.

Minimal discussion is needed to describe cross-sectional or time-series designs. For a cross-sectional design, the measurement and sampling discussions are usually enough. For a time-series design, the researcher notes the beginning and ending points of the time series and the frequency of intervals. Apparent sources of nonrandom fluctuations and problems with data comparability should be mentioned.

Laboratory or field experiments require more extensive discussion, distinct from the measurement and sampling sections. The narrative should imply or state how the design and its implementation controlled for various threats to internal validity. Researchers should describe the experimental environment so that subsequent investigators can determine what, if any, features of the research conditions constituted a threat to external validity. The final project report should contain enough detail so that another researcher could repeat the experiment.

An audience should remember that quantitative investigations are not flawless. A study's purpose and limited resources require researchers to settle for imperfect measures, samples, or research designs. Similarly, researchers should not hesitate to report the flaws and limitations in their methodology or other factors that may significantly alter or limit the results. They should be sure to note shortcomings that may lead audience members to misinterpret the findings.

The Findings

The findings section reports the research results. The authors include an explicit statement of what was found and the statistical or qualitative evidence to support it. This section may include interpretation of the findings and tentative explanations for unexpected findings.

To communicate the findings effectively, researchers must do the following:

1. organize the findings into a coherent presentation
2. focus on the important findings and avoid overwhelming the audience with unnecessary detail
3. decide on how to present the findings, including the use of tables or graphs

Traditionally, a study's background, model, and methodology are presented before the research results. The audience is told what questions were asked, why they were asked, and how they were asked. This prepares the audience to listen to and remember the findings. The findings are organized around related themes, variables, or questions.

The logical structure of the presentation may help focus the audience's attention. For example, when we discussed how to present the literature review, we cited a study of citizen satisfaction with the police. The authors discussed how each variable was measured, using the same order presented in the review of the literature, that is, (1) demographic variables, (2) political variables, and (3) objective indicators of police performance. They continued this strategy in presenting their findings; they first looked at the effects of the demographic variables, then the political variables, and finally the objective indicators.

Graphs and Tables

Just as the executive summary acknowledges that administrators may only give a report a passing glance, a report presentation should facilitate a review by hurried and distracted audience members. Presented with an uninteresting analysis or an overwhelming amount of detail, audience members may stop listening or reading. Graphs and tables, which may complement the verbal presentation, underscore important points and exhibit data efficiently. The location and the amount of space devoted to words or graphics should signal the importance of the findings. Unimportant and trivial findings do not deserve major emphasis, and researchers should not waste space on graphics that illustrate unimportant points.

A detailed discussion of a large amount of data or many numbers is not very effective. An audience may find the presentation difficult to follow and not take advantage of the opportunity to question the findings and explore their implications. Attractive graphics and clear explanations allow readers to access the richness of the data. Tables show exact numerical values and work well when the data presentation requires many specific comparisons. Examples of conventional types of tables include data arrays, summary tables of regression statistics (see Table 15.1), and

Table 15.1 Regression Model to Predict Student Grade Point Average

| | Unstandardized Regression | | |
	Coefficient	(s.e.)	Beta Weight
Undergraduate average	.367	(.088)	.371
GMAT score	.00099	(.00045)	.188
Admissions scale	.036	(.016)	.188
Sex	−.019	(.080)	−.021
Years since college graduation	.014	(.012)	.104
Prior graduate work	−.055	(.054)	−.088
Constant	1.437	(.080)	
R^2	*.188*		

Table 15.2 Support for Consolidation by Evaluation of Local Services

Evaluation of Local Services			
Support for Consolidation	*Excellent or Good (%)*	*Fair (%)*	*Poor (%)*	*Total (N)*
---	---	---	---	---
For	55	47	36	331
No opinion	24	25	34	161
Against	21	28	31	155
Total (N)	387	224	36	647

contingency tables (Table 15.2). Graphs are conventional for presenting time-series data and scatterplots. Graphs permit an audience to pick out movements over time including long-term trends, cycles, and seasonal fluctuations, as well as differences before and after an intervention.

Careful labeling of tables and graphs is important. Spell words out, avoid abbreviations, and use clear variable names. Try to place labels on the graphic rather than in a legend. It is often helpful to include a caption below graphs stating the point that the graph conveys. Words on and around graphics are effective in telling viewers how to focus attention on the various parts of the display.[6] However, be careful to use only those that are useful or necessary so as to keep the graphic or chart uncluttered.

Additional Advice on Presenting Data

The following summarizes several points and offers additional advice on presenting data:

1. Tables or graphs should have a precise, descriptive title. A title may list the dependent variable by independent variable by control variable (if any). Alternatively, a title may summarize a major finding supported by the graphic, for example, "City homicide rates have dropped over the past 20 years."
2. A good table supplements, not duplicates, the text. Each table and its data should be referred to in the text, but only the highlights should be discussed.
3. All variables and reported values or categories should be clearly labeled, and appropriate units (e.g., years) should be indicated.
4. By convention, the independent variable is listed along the column, or top, of the chart, and the dependent variable is listed along the row. (Presenting regression coefficients is usually an exception to this. See Table 15.1.)
5. If percents are used, a percent sign (%) should be entered at the top of the columns. Also, avoid unnecessary decimals—they are easily overlooked.
6. The number of cases on which the 100 percent is based should be indicated. The total number of cases used in analysis should also be indicated.
7. Statistical measures, if any, should be listed at the bottom of the table.
8. All terms open to interpretation should be defined in the footnote section below the chart.
9. The source of the data should be indicated in the footnote section.

Recommendations

Program evaluations, policy analyses, and other studies done for a legislative or administrative client may include recommendations. Analysts may be reluctant to make recommendations. Analysts are trained to accurately describe a program or policy and what it is accomplishing.

Recommendations often are normative statements about changes that should be made in a program or policy. An analyst may be ill equipped or uncomfortable in making normative statements or telling clients what they should do.

Research clients, on the other hand, may expect recommendations. Investigators may find that making recommendations improves the probability that their studies will be used. Recommendations should be carefully formulated; they may be developed and evaluated throughout the course of a study. They should naturally follow from the research findings, for example, a reader should be able to understand why the recommendations were made. In making recommendations, the researcher should address changes that the client agency can make; for example, recommending a change in federal program requirements will not be of any value to a local social service agency. In some cases, the costs and benefits of adopting a recommendation may be identified and included. Alternatively, the researcher may suggest several options for agencies to consider.[7]

Oral Presentations of Research Findings

Researchers who do studies for administrators need good oral presentation skills. Employees often do research projects as work assignments or as consultants for an organization. The presentation of the results is then to a supervisor or as a representative of the contracting organization. Sometimes the researcher is required to report to an advisory group or representative of a funding organization. Administrators and policy makers, depending on their learning styles and time demands, may prefer listening to an oral presentation and asking questions rather than reading a report. They may find that an oral presentation alerts them to what to look for in the report or, conversely, the oral presentation gives them an opportunity to ask follow-up questions. Oral presentations also have distinct advantages for researchers. They can better anticipate their audience and tailor their presentation to meet its needs and interests. Audience members may feel compelled to pay attention; analogous pressure to concentrate on a written report is virtually nonexistent. The social interactions among audience members may increase understanding of the results and may motivate the group or individuals to discuss and follow up on the findings.

The training of researchers stresses careful attention to detail and a willingness to examine findings from various perspectives. These skills can translate into tedious, unfocused presentations. In contrast, good oral presenters avoid covering too much; they concentrate on a few important points. A presenter may fill in the details or elaborate on alternative interpretations only if an opportunity opens up in a conversation or during a question-and-answer session. Putting the detailed information about alternative models and assumption checks in appendix slides may help nervous researchers to feel that they have the visual information available if they get questioned on a specific point, but concentrating on the most important points during the presentation makes for a much more interesting and impactful presentation.

One-on-One Discussions

Researchers should not overlook the importance of one-on-one informal discussions of findings. Their informality can be deceptive. They offer an important opportunity to develop an administrator's interest in a project. A researcher should think about what she expects from informal communication. Does she want to alert an administrator to a problem? Does she want his insights about an unexpected finding? Does she want him to think about the value of the study and how its findings can be implemented? Her ability to communicate information clearly and to understand why she wants to communicate it to a particular administrator may improve her effectiveness.

Explain Clearly and Concisely

Whether speaking to one person; a small group; or a large, formal audience, the ability to explain your work clearly will serve you well. Professionals feel continually pressed for time. Many people are "oral learners." Others value the chance to debate information and discuss it with researchers and other administrators. As teachers, we have observed that many talented students avoid making oral presentations. These students lose opportunities to present their ideas clearly, to understand listeners' questions, and to provide effective answers (as well as to hone an important skill). If you are still in school as you read this chapter, consider the value of getting practice and feedback on your speaking and presentation skills. Employers often look for presentation skills in those considered for promotion. Many organizations offer workshops on effective presentations for employees. Investigate if your organization does and take advantage of them if available. Also seek out experienced colleagues known to be excellent presenters for their advice. For additional materials on oral communications, see a technical communications text.[8]

Planning a Presentation

An effective presentation requires planning and practice. To plan a presentation, researchers select the points they want to emphasize, the evidence they will use to support these points, the order in which material will be presented, and visual aids. The researchers should have determined something about what the audience knows about their topic, what the audience wants to get from the presentation, and how long members will listen attentively. This will depend somewhat on the circumstances. For instance, the conference panel presentation is different than that of the single researcher giving a longer presentation on a single topic to an audience. The style of the reports should reflect that.

The traditional order for a research presentation (background, methodology, and findings) usually works well. It develops the material logically. People with training in the sciences, including the social and behavioral sciences, have come to expect it. If audience members are informed about the program or policy, identifying the study's purpose and summarizing the methodology may be sufficient. Otherwise, a description of the program or policy is necessary to put the information in context and to help audience members to follow the presentation. Except for specialized audiences, technical details are not presented in the discussion of methodology. Presenters should encourage audience members to ask for the details which interest them, especially those details that affect their willingness to accept the findings.

Presenting at Conferences

The major value of many research projects comes when the results are disseminated publicly. The majority of academic and professional research is reported at conferences and in journals, although many other options are available.

Most conference organizers provide instructions for presenters in advance of the conference. You should read and follow those. If you are scheduled to present a paper at a conference and have not received panel instructions, contact the program or conference chair and request them. Traditionally in the conference setting, presenters would make copies of their papers available to those interested. Today the papers can be sent or made available electronically. A researcher should always offer to do this unless restricted from doing so.

In the typical format for oral presentations at most academic and professional conferences, researchers with papers on the same or closely related topics are scheduled on panels.[9] Three or four researchers present their research to an audience of those interested in the topic. The sessions may be scheduled for one and a half hours, during which time each presenter summarizes the research and audience members can ask questions and discuss the reports. In these situations, presentations must be condensed and brief, usually 12 to 15 minutes each.

The following are some recommendations for accomplishing this.[10]

- Concentrate on only one or two points. Omit most details of the research design and process.
- State what you studied, why you studied it, and how you studied it. In explaining how, provide a general description of the research design.
- Report what you found.
- Explain why the results are important.

Depending on the nature of the research question or objective, you might spend more time on one or more of these points. Do not read the presentation. (You will not have time to read to the audience a printed version of your paper in any case.) Use the list of points as an outline to speak from. Just talk to the audience. And be sure to rehearse ahead of time.

Visual Supplements

Presenters usually find it helpful to have visual aids to engage the audience. PowerPoint slides and Prezis[11] are common. Visual aids may be used throughout a presentation. PowerPoint slides, tables, or graphs focus the presenter and the audience. To select a visual aid, consider whether it distracts attention, slows down the presentation, requires special equipment, or communicates information effectively and clearly. Too many visuals can bore or confuse an audience. Technically sophisticated presentations can be fun to put together, but they are often distracting and draw attention away from the content of the presentation. Fumbling around with unfamiliar equipment creates a serious distraction. Poor lighting can also cause problems. Detailed images leave people in the back rows squinting or feeling left out; however, the problem can be solved if a handout contains the same information. Also, be prepared to work without planned visual aids. Equipment failures happen, and you should make sure that they don't undermine a presentation.

PowerPoint slides can be a great aid in the presentation and, with a little instruction and practice, easy to prepare. However, they should be used sparingly. More is not necessarily better. And the information on each slide should be kept within strict bounds. Publications offer tips on using slides in presentations.[12] Some of those tips are:

1. Limit the number of slides to approximately one per three minutes of discussion. If you need more, consider putting the information in a handout.
2. No more than five lines per slide. A general rule here is: do not put too much on the slide—keep it simple and clear.
3. Use topic phrases only, not complete sentences.
4. Make sure the type is large enough so everyone in the room can read slides easily.[13]

Except for definitions or important quotes, avoid reading what is on the slide to the audience. A few simple, short phrases give the presenter an outline to speak from, will get the audience's attention, and provide a key term to remember. Remember to talk to the audience. A distracting presentation error is for the researcher to turn away from the audience toward the screen and read the slide to the audience. Another is to stand between the audience and the screen and block the view. Take care to avoid these simple presentation errors.

Handouts

In addition to slides, if there are tables with statistics, consider having handouts. Handouts, however, should be kept to a few pages. Complicated tables shown on PowerPoints and in handouts should have the pertinent figures circled or highlighted. Some presenters also provide the audience with a one- or two-page handout with a few important details: title of presentation, major research objective, statement of design, and a statement of the findings. This handout would also include the name, title, and contact information for the presenter.

Selecting a visual aid without thinking about the audience can result in ludicrous situations. One-size-fits-all presentation methods and using only one type of visual aid may not be effective. For instance, employees of an agency for the blind were astonished that professionals, including ophthalmologists, consistently supplemented their presentations to agency staff with slides. Most of the staff had severe visual impairments and derived no benefit from the slide shows.

The Importance of Practice

Inexperienced presenters may overlook the importance of practice. A researcher who has pored over a study may feel confident in her ability to deliver the report spontaneously. Unfortunately, this person may end up bogging down on some of the study's less important details or moving erratically from point to point. One should practice with an audience of colleagues, team members, or friends. Practice session observers should make sure that the major points are clearly presented, the reiteration of key points does not become repetitious or condescending, the transitions are smooth, and the equipment operates correctly. The presenter should ask the observers to ask questions about the methodology or the interpretation of the findings. Preparing answers to "hard" questions avoids the embarrassment of stumbling around. If questions challenging the quality of the research go unanswered, the effectiveness of the entire report may be undermined.

Poster Presentations

Poster presentations can be an effective way to report research. They have become common at professional and academic conferences. Good posters have unique features not important to papers. These presentations allow the researcher to interact with audience members individually and in groups. Interested people can view the poster to judge how much interest they have in the topic and spend more or less time reviewing and discussing it. Because the audience can "pass by" any poster, the poster must quickly portray its subject and catch the eye of that passerby. A poster combines text and graphics to present the research project in a way that is visually interesting and informative. The poster makes the information accessible to a large group and provides opportunities for an audience to discuss details of the research and for the researcher to gain feedback on the project.

The following are some basic tips for preparing a poster.

- Define the purpose of the poster.
- Decide: What do you want the audience to learn from the poster?
- Design the poster to "sell" your work in 10 seconds.
- The title is important: design a good one.
- Take advantage of the unique advantages of posters. You can provide more information than in most oral presentations.

- Layout and format are critical; do not crowd the information; design the poster to lead the viewer in a sequence explaining the research.
- The impact of a poster happens during and after the poster session.[14]
- Posters are usually 36" high by 48" wide, although some are 48" by 56".[15] Some conferences allow more than one poster per project presentation.

Online and Social Media

Even just a few years ago, the digital version of research provided online was not a substitute for the content that was provided in the print journals, and the digital versions were at best providing only complementary content to the print version. Top publication outlets were all print journals, and they simply published their articles online. Fast forward to today and open access journals that are only published online, such as the *Journal of Statistical Software* that publishes research on open statistical platforms like R, which is the top journal in statistics according to the Social Science Citation Index. Also, publishing digital and more accessible versions of research is now often more important for achieving impact and reaching more readers than publishing in the top journals. This section outlines a few of the primary ways that researchers and public agencies use online and social media to disseminate their research broadly online.

The oldest form of mass online communication was the webpage and blog, where researchers and their organizations wrote descriptions of research to be viewed by anyone that came to the page. Blogs have been used extensively by academic researchers, but they are also used by organizations to communicate regularly with a wider audience. Agencies like the National Aeronautics and Space Administration (NASA) and the Government Accountability Office (GAO) use blogs to communicate technical work in a more accessible manner.[16] Even small cities often have a blog that is written by the city manager or a member of the communication staff. These blogs can be used to highlight the important research that is done within the agency or organization.

The informal style of a blog makes it accessible and interesting to a broader audience than a typical research report. Blog postings are likely to be shorter than a journal article or a conference paper, but they are typically longer than the very short content that would be found in most social media posts. The benefit of blogs and individual webpages is that they can be widely accessed and have more discretion in terms of content length and formatting, and readers can provide feedback to the blog or the organization.

While many programs are useful for marketing or communicating smaller amounts of information and messages between and among researchers, some are used as depositories for sharing larger documents. The latter include cloud-based services such as Dropbox and Google Docs. Many researchers use these to collaborate with colleagues in different locations. They can post full-length research reports on these sites to share. However, these are not usually shared beyond the individuals involved unless others are given permission to access the site. These programs also allow researchers to easily send full research reports to those requesting them. Social media has benefitted researchers by enabling them to work together more effectively but also greatly expanded their ability to communicate results to each other and to a wider audience.

Shorter in length than a blog but with more potential for sharing, researchers increasingly use social media[17] to disseminate the results of research. The content on social media is often much shorter than a blog. For example, Twitter posts are capped at 280 characters. While researchers get around this by posting strings of posts or threads (see Figure 15.3), research on social media finds that one of the main benefits is that it forces researchers to distil their findings into much shorter, more specific, and more accessible points. For example, the local economic development researcher in this post talks about the growth of her city by comparing it to a weed, and then follows it up with another post that shows the cost for other peer cities. The accessible and short

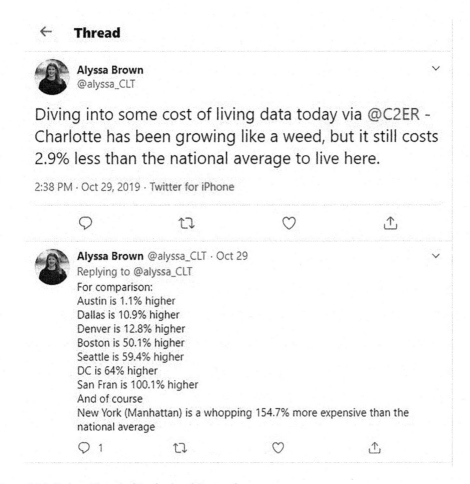

Figure 15.3 Twitter Thread of Institutional Research

nature of posts and threads allows even those that are casually interested in a topic to follow it on social media and like or share the post on their networks.

Social media is increasingly important in both research practice and in disseminating government research. Facebook alone has 2.4 billion active users, and YouTube, WhatsApp, and Instagram all have over a billion active users.[18] Having a strong social media presence may be critical for bridging generational gaps where both Millennials and Gen Z are digital natives that use social media to find out information about products and do preliminary research.[19] Savvy researchers and organizations have caught on to this and have expanded their social media presence. At a minimum, researchers and organizations link their social media accounts to the blogs so that people can follow the blog on social media. More sophisticated organizations make an intentional effort to create and curate content that will appeal to and engage their audience on social media. Again, organizations like NASA and the GAO have extremely interesting content for social media. Recently, state and local governments have become active on social media. The State of New Jersey and the Lawrence, Kansas, police department have received a lot of popular attention on Twitter for funny Tweets that show their unique culture and brand. However, with hundreds of thousands of followers (NJ) and more followers than residents (Lawrence PD), these organizations can inform citizens of extreme conditions, get help with identifying criminals, or provide information about public health hazards like COVID-19. Similarly, institutional researchers use their own social media and their organizations' social media accounts to distribute the main findings of their research broadly to people in their own social and professional networks.

Finally, podcasts and video blogs are increasingly becoming popular. Public Administration Review, a well-known journal in the field of public administration, routinely asks researchers to turn their research into a podcast or video blog. Individual researchers may also have their own podcasts (i.e. Academics of PA). Organizations like the Pew Research Center have their own YouTube channels that produce very high-quality videos to complement their reports, blogs, and social media. Like social media generally, it takes a lot of work to create a short and interesting podcasts and videos, but the benefit is that these media may reach an audience that does not have the time or is not willing to take the time to read a journal article.

As younger generations become researchers and managers in public organizations, there is no doubt that more content will be distributed online in ever more accessible ways. This is just a short list of ways that researchers have found to distribute their research more broadly and to increase visibility of their research.

Ethical Issues

The following sections discuss some issues involving clearly inappropriate and, in some cases, dishonest, behavior on the part of researchers. In the 1980s, professional research ethics received increased systematic attention. Respected scientists were charged with scientific misconduct, including plagiarism and making up data.[20] In order to more effectively maintain the integrity of scientific research, the National Academy of Sciences, the National Academy of Engineering, and the Institute of Medicine convened a joint study committee in 1989. They drew committee members from the biological and physical sciences, and the committee's report focused on practices of these sciences. The social sciences and administrative researchers did not receive a similar degree of scrutiny. They still operate on an honor system, which assumes that researchers behave properly. Social scientists are probably no more ethical than biological and physical scientists. The stiff competition for basic research monies and the potential financial windfalls associated with patents may make the latter professions more susceptible to misconduct and therefore to having their misconduct discovered and disclosed.

The following discussion relies heavily on the committee's final report. The report categorized inappropriate scientific research behavior into three categories: research misconduct, questionable research practices, and other misconduct. Research misconduct consists of acts of fabrication, falsification, or plagiarism. Questionable practices concern data retention and sharing, record quality, authorship, supervision of research assistants, statistical analysis, and release of information. Other misconduct refers to acts that are unacceptable but not unique to researchers—for example, misuse of funds, vandalism, violations of government research regulations, and conflicts of interest.[21] Our discussion focuses on research misconduct, handling research errors, and record-keeping—issues that affect all researchers. For a more complete picture of issues, readers should consult the committee's report and background papers.[22] We also discuss some general standards that further increase the integrity of reported research. Researchers should feel obligated to adhere to procedures and standards that go beyond the merely acceptable.

Research Misconduct

The study committee defined fabrication as making up data or results, and falsification as changing data or results. We assume that every reader knows that fabrication is wrong and recognizes if he or she has made up data or findings. Falsification can be a bit more ambiguous.

Falsification

An easy way to falsify results is to drop cases from a dataset. Dropping selected cases can rescue a weak statistical model. The researcher may rationalize his decision. For example, he may argue that the dropped cases were tainted by measurement error and therefore do not belong in the dataset. If he thinks that measurement error occurred, he should try to confirm the measurement error. If he does not have time to track down the source of error or he cannot confirm the error, he can remove the cases. If he removes cases, however, he must indicate in his reports what he did, why, and its effect on the results. The more marked the effect, the more diligent he must be in alerting an audience of his decision. A decision that markedly affects the finding should not get buried in the small print.

Plagiarism

Plagiarism is falsely presenting another's ideas or words as one's own. Quoted material should be placed in quotation marks and references cited. Closely following another author's diction or paraphrasing his words is wrong. The writer should either use her own words and sentence structure or quote directly from her sources.

Relying on the works of others is inevitable in research. No one knows this better than a textbook writer. We have referenced sources that we relied on to write segments of this manuscript that provided a unique or valuable perspective on the material or that contained additional information. We have not referenced sources for ideas and perspectives that we know are part of the common knowledge of social science researchers.

Younger researchers seem to cite sources for their ideas more frequently than experienced researchers. This probably does not occur because of different ethical standards. Younger researchers more often work on cutting-edge topics, and they read more original research. In our experience, younger researchers are more likely to listen to panels at professional meetings and to read a wider range of journals. The experienced researcher is more familiar with the subject area. He can better distinguish between unique contributions to his thinking and ideas from the general body of knowledge about the subject.

Sometimes newer researchers may feel overwhelmed by the need to avoid charges of falsification or plagiarism. This need not be the case. To avoid charges of falsifying, a researcher documents her decisions and her reasons for them. The documentation makes her decisions accessible for peer review.

Avoiding charges of plagiarism should not be difficult. Diligent referencing and avoiding using another's wording should be adequate. If a report is to be published, the writer needs to pay attention to copyright laws. Researchers must get permission from the copyright holder to reproduce graphs; tables; long quotes; and other materials, including song lyrics, poetry, and cartoons. Government documents are not covered by copyright, and their contents can be reproduced without obtaining permission. Nevertheless, the researcher should use standard referencing procedures to cite a government document.

Handling Research Errors

Error is inevitable in research. Errors arise from constraints that require investigators to compromise the quality of their efforts. Errors also arise from any one person's or group's point of view, type of knowledge, and degree of ability. The potential for error occurs throughout the research process. The study committee mentioned earlier identified four potential sources of error: the accuracy and precision of measurements, the generalizability of experiments, the quality of the experimental design, and the interpretation of the practical significance of the findings.[23]

Researchers must ascertain that data entry and analysis procedures were conducted correctly. Errors that appear to be innocuous, such as miscoding data, can seriously distort the findings. If research findings differ from what is expected, researchers should pore over the data to discover what happened. Good procedure requires checking the raw data and computer programs. Unexpected results should be reported. Expected results rarely receive similar attention with regard to checking for errors. Richard Feynman, a winner of the Nobel Prize for physics, reminded researchers that "the easiest person to fool is yourself." He suggests that researchers analyze the data to see if other typical relationships between variables exist. If possible, researchers should also analyze the data to test alternative explanations.[24] In publishing a final report, the researchers indicate what alternative explanations they examined, what they found, and what explanations they were unable to examine and their possible effects.

Correcting Errors

To provide opportunities to identify and correct errors, researchers fully disclose their research procedures and subject their work to peer review. Researchers are expected to acknowledge and correct errors that are discovered after research is reported and disseminated. Full disclosure allows others to scrutinize the research. Errors may be found by examining the research documents or attempting to replicate the research. Concealing limitations is deceptive. Research reports should clearly identify and evaluate the limitations. The more troublesome a limitation, the more emphasis it should receive.

Including complete information about research procedures can overwhelm readers with details and seriously diminish a report's effectiveness. The professional standards for program evaluation recognize the competing demands of providing useful information and full disclosure. To provide useful information, the standards advise evaluators to write clearly, present information that their audiences can understand, and indicate the relative importance of their findings and recommendations. To achieve full disclosure, the standards advise evaluators to state their assumptions, their constraints, and how readers may obtain full information on research procedures, including data analysis.[25] The standards relieve evaluators of the burden of providing complete research information in every report, but they must take reasonable actions to ensure the accessibility of the database and documentation.

Peer Review

Peer review helps detect errors prior to a report's publication. Peer reviewers are specialists who read and evaluate manuscripts and proposals submitted to academic journals or conferences. The specialists should be conversant with the research topic. The reviewers may recommend the research for publication or presentation. They may identify weaknesses that need to be addressed—for example, additional sources for the researcher to consult, sections of the papers requiring amplification, or alternative analytical procedures. A blind review means the reviewer does not know who conducted the research. Blind reviews diminish personal factors that may color a reviewer's judgment.

The peer-review process primarily interests academic researchers. Nevertheless, the public relies on peer review to "protect" it from flawed research. Whether we eat red meat, drink wine, or exercise regularly may be influenced by reports of research findings. Researchers may be tempted to go directly to the mass media with interesting results, but such behavior is viewed as inappropriate. Rather, researchers are expected to release their study first to their professional colleagues, who are in a better position to detect its flaws. Russell Schutt, however, discusses how the media can help to engage the public in policy issues addressed by research findings and gives examples.[26]

"Hot" Topics

In 1993, an error in a study on a treatment for the human immunodeficiency virus (HIV) made headlines. An American research team reported an apparently successful drug therapy for treating HIV. The article appeared in the prestigious scientific journal *Nature*. Six months later, the researchers announced their research was in error.[27] They had, in fact, fooled themselves. They had misread lab results. At the 10th cycle of two experiments, the virus was undetectable, but by the 30th cycle it was flourishing. A columnist in the *New York Times* cited the error as a failure of the peer-review system. He suggested that the original research was flawed.

After the paper was published, researchers in the United Kingdom tried to replicate the experiment; their findings were contrary to the original findings. The Americans repeated their experiment and identified an error. The researchers sent a letter to *Nature*, correcting their findings. The researchers also publicly announced the flaw in the original research; their announcement was reported widely in national news media.

The example implies the dynamics of research on "hot" topics. The potential rewards for finding a successful drug therapy to treat HIV and AIDS are tantalizing. They may lead researchers to be less careful in their work and to push for acceptance of their theories. On the other hand, the research is more likely to receive scrutiny and criticism. Ultimately, researchers must be willing to listen and respond to challengers. Researchers are expected to acknowledge and take responsibility for significant errors they subsequently detect. Such errors change the interpretation of the findings.[28] The *New York Times* columnist noted, "Many other scientists, who have made mistakes in far less visible and less important research, have not been as quick to publish corrections or retractions, if they reported them at all."[29]

Guidelines for Ethical Reporting

Russell Schutt discusses a set of guidelines for full ethical reporting. He presents as a general standard that "the researcher has a duty to provide an honest accounting of how the research was done."[30] Many of these guidelines have been discussed earlier in this text; some have not. All should be emphasized. Some of Schutt's guidelines are:

- Inform the reader of major changes in hypotheses or research design made during the research.
- Report if the major hypothesis is not supported.
- Give an honest evaluation of strengths and weaknesses of the research design.
- Set the research within the literature of prior research and interpret results within this body of research.
- Maintain a full record of the project so questions can be addressed.
- Record tests conducted but not included in the report. (Schutt does not go into detail on this guideline or the previous one. However, many researchers keep a daily log of research activities and all personnel who worked on the project each day. We recommend doing so as well.)
- Be careful to not mislead the reader with statistics or graphics.
- Acknowledge sponsors and criticisms.
- Decide on authorship in advance.[31]

Saving the Data

Data must be saved and be accessible to allow research audits, replication of results, refinement of the analysis, additional analyses, or incorporation of data into contemporary research designs. Research data include completed data collection instruments, procedures for collecting and entering data, experimental procedures, data files, computer printouts, field notes, videotapes, or audiotapes.[32] With this information, an investigator can reconstruct or replicate the research. Audits may be a component of ensuring scientific integrity; they can substantiate or repudiate charges of falsification or fabrication. Replication may be a part of a formal audit, or it may arise from an investigator's desire to confirm the findings. Researchers may work with "old" data to see if including different variables or changing the statistical analysis affects the original results. Existing data may be incorporated into a time-series design or a cross-sectional design.

Funders may have specific policies covering data ownership and data retention.[33] In other cases, an investigator may have to make her own decisions. Misunderstandings can be avoided if investigators and administrators agree on who will retain data, how long they will be kept, and the conditions governing data sharing. If such agreements are not made explicit, administrators may find that they cannot access the data later for further analysis. The investigators may have discarded or misplaced data, information to retrieve data may not exist, or research documents may be scattered.

Increasingly, academic journals are requiring data to be archived in data repositories such as the Harvard Dataverse, the ICPSR at the University of Michigan, or the Qualitative Data Repository at Syracuse University. Initiatives such as the Data Access and Research Transparency (DA-RT)[34] initiative in political science now require that data be made available through an online data repository for future analysis. Increasingly, even public administration journals such as *Public Administration Review* (PAR) are implementing the Transparency and Openness Promotion (TOP) guidelines, which strongly encourage researchers to deposit data in an online data repository.[35] Of course, these initiatives are aimed at encouraging public administration, public policy, political science, and related disciplines to make data available and increase the ability to reproduce analysis.

Implicit in the question of storing data is deciding who can have access to the data. Issues affecting the confidentiality of respondents must be resolved prior to sharing data. Research that uses data that is limited by contracts may not be able to be posted in online digital repositories. Researchers should consult the journal guidelines on what data and transparency standards are required before they submit an article to an academic journal.

Administrative research ranges from exploratory studies to research designed to answer specific one-time questions to longitudinal studies conducted to evaluate programs and to guide policy making. A set of specific guidelines on storage may not adequately cover all types of studies. During any study, researchers should describe research processes and decisions and label all research documents. At the end of the study, researchers should organize and store the data and documents for a reasonable amount of time. Since more documents and data are now recorded digitally, the storage length of time may not be a major problem. However, organizing, archiving, and providing convenient access may require additional planning and record keeping.

The exact amount of time may depend on the assumed "shelf life" of the data, agency custom, research agreements, or contracts. Completed questionnaires and records, needed primarily for research audits or to correct data-entry errors, may be kept the shortest time. Data files, data dictionaries, data collection instruments, and other research protocols may be saved indefinitely. Materials relating to a published work may be kept longer than those relating to an unpublished or in-house study. Alternatively, if an agency centralizes data collected on its behalf, it should develop a mechanism to exert quality control so that it does not get overwhelmed with data that are of little or no use.

Summary

Before presenting their findings, researchers identify the audiences for their information. The researchers identify each audience's characteristics, how to get its attention, and how to motivate its members to follow through on the report's findings and recommendations. Perusal of agency

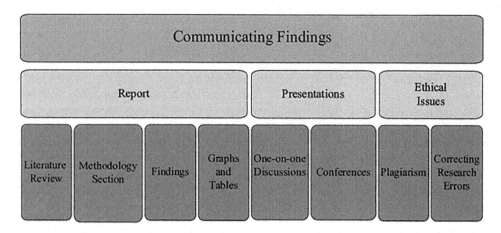

Figure 15.4 Communicating Findings

reports or listening to oral presentations may suggest how to organize information effectively for an agency (see Figure 15.4).

Presentations Versus Reports

Presentations tend to include the same material, although the amount of time or detail will vary according to agency needs. These sections are an executive summary, a statement of the study's purpose, background information, the description of the methodology, and results, along with supporting analyses. The methodology section describes the measures, sample, and research design. It explicitly identifies any limitations that may change the findings. The results section includes, as appropriate, the supporting statistical models, tables, or graphs. Researchers may include their own observations in a report; however, they should clearly separate observations based on data and observations based on judgment. Recommendations should be based on findings.

Reports are tailored for different audiences and different needs. Reports are organized so that important information is clearly distinguishable and receives major emphasis. Nevertheless, a researcher should remember that written reports may be passed around. He should not assume that every reader will know about a study's background or design. Consequently, he needs to include necessary elaboration and documentation in the report. The methodology and findings sections in particular should be sufficiently documented so that the careful reader can make her own judgment on the adequacy of the report.

Oral presentations involve a captive audience. Still, the presenters are not guaranteed its attention. They should practice the presentation so that it flows logically and listeners can easily follow it. Visual aids should be easily seen and interpreted. If visual aids require special equipment, the presenters should know how to operate the equipment with minimal effort and be prepared in case it fails.

Written reports include research proposals, final project reports, and academic papers. A research proposal's sections on the problem, related background information, the literature review, and the proposed methodology are later incorporated in the final project report. Final project reports and academic papers include similar material, but the project report may be organized to facilitate the needs of administrators, who vary in their degree of interest. Final project reports rely on clear chapter headings, subheadings, and highlighting to direct a reader's attention. Technical details are placed in an appendix.

Ethical Practice

Empirical researchers are expected to adhere to ethical practice. Plagiarism, fabrication, and falsification clearly violate research ethics. Although human error is inevitable, researchers should take reasonable steps to minimize errors in their work. They should check entered data, computer programs, and data analyses to make sure that their findings are reported accurately. The descriptions of the study's methodology should disclose problems encountered, design limitations, and other information needed to facilitate critical review. Where possible, researchers should subject their work to peer review. After a research report has been reviewed by peers and published, researchers are expected to acknowledge any errors they subsequently discover.

Sharing and Keeping Data

When a study is initiated, investigators and administrators need to agree on who will retain data, how long data will be kept, and any conditions governing data sharing. During the study, investigators should write up descriptions of research processes and decisions and label all research documents. Many researchers keep a research calendar or diary as well as dating when major parts of the study were conducted and by whom. At the end of the study, data and documents must be organized and saved. These steps ensure that data are available for audit, replication, or further analysis.

Notes

1. L. A. Olsen and T. N. Huckin, *Technical Writing and Professional Communication*, 2nd ed. (New York: McGraw-Hill, 1991), 66–69. For an alternative approach, see D. E. Zimmerman and D. G. Clark, *The Random House Guide to Technical and Scientific Communication* (New York: Random House, 1987), 90–100.
2. See Zimmerman and Clark, *Random House Guide*, 87–90.
3. Efforts of this nature are common among local and state governments. The interested reader can find case examples of organizations adopting innovative practices in budgeting, human resources management, pension investment, and so on.
4. K. Brown and P. B. Coulter, "Subjective and Objective Measures of Police Service Delivery," *Public Administration Review* (January–February 1963): 50–51.
5. A. H. Church, "Incentives in Mail Surveys: A Meta-Analysis," *Public Opinion Quarterly* 57 (1993): 62–63.
6. Edward R. Tufte, *The Visual Display of Quantitative Information* (Cheshire, CT: Graphics Press, 1983), 182. Also see the following for information on preparing graphs and charts. Steven Few, *Now You See It: Simple Visualization Techniques for Quantitative Analysis* (Burlington, CA: Analytics Press, 2009); Steven Few, *Show Me the Numbers: Designing Tables and Graphs to Enlighten* (Burlington, CA: Analytics Press, 2012).
7. For further discussions of recommendations see R. C. Sonnichsen, "Evaluators As Change Agents," in *Handbook of Practical Program Evaluation*, eds. J. S. Wholey, H. P. Hatry, and K. E. Newcomer (San Francisco: Jossey-Bass, 1994), 534–548; M. Q. Patton, *Utilization-Focused Evaluation: The New Century Text* (Thousand Oaks, CA: Sage Publications, 1997), 324–329.
8. For example, see Mike Markel, *Technical Communications*, 11th ed. (Boston, MA: Bedford/St. Martin's Press, 2014).
9. Although we refer to academic conferences here, the format for professional organizations is similar. Virtually all professions hold conferences to present current research and innovative practices. Those employed in government organizations will be familiar with conferences for budget, human resources, finance, engineering, and solid waste management specialists. Private organizations have specialists in many of these areas as well.

10. See Larry Christensen, R. B. Johnson, and Lisa Turner, *Research Methods: Design and Analysis*, 12th ed. (Boston, MA: Longman-Pearson, 2014), 474.

11. See www.prezi.com.

12. George Grob, "Writing for Impact," in *Handbook of Practical Program Evaluation*, 3rd ed., eds. J. Wholey, H. Hatry, and K. Newcomer (San Francisco: Jossey-Bass, 2010), Chapter 25, 595–619.

13. Ibid.

14. Thomas Erren and Phillip Bourne, "Ten Simple Rules for a Good Poster Presentation," available at www.ncbi.nim.nih.gov/pmc/articles/PME1876493; George Hess, Kathryn Tosney, and Leon Liegel, "Creating Effective Poster Presentations," available at www.ncsu.edu/project/posters/NewSite/index.html. A link on this web site will also take the reader to a video library where Professor Hess discusses posters.

15. Scott Plunkett, "Poster Presentations at Professional Conferences," available at www.csun.edu/plunk/documents/poster_presentation.pdf.

16. A curated site for all U.S. federal government blogs can be found at blog.usa.gov.

17. Social media is a web site or app that allows users to create, view, and share content in a virtual community or social network. Examples include Facebook, Twitter, Instagram, and Foursquare.

18. Ashley Viens, "This Graph Tells Us Who's Using Social Media the Most," *World Economic Forum, Blog Post*, October 2, 2019, available at www.weforum.org/agenda/2019/10/social-media-use-by-generation/.

19. Ibid.

20. For statistical information on the history of scientific misconduct, see National Academy of Sciences, "Misconduct in Science—Incidence and Significance," in *Responsible Science: Ensuring the Integrity of the Research Process, Vol. 1, Report by U.S. Committee on Science, Engineering, and Public Policy, Panel on Scientific Responsibility and the Conduct of Research* (Washington, DC: National Academy Press, 1992), 80–97. The chapter's bibliography identifies sources for information on specific cases.

21. *Responsible Science: Ensuring the Integrity of the Research Process*, vol. 1, 5–7. The report referred to research misconduct as "misconduct in science." The same behaviors are labeled as "research misconduct" in the Massachusetts Institute of Technology's. Panel on Scientific Responsibility and the Conduct of Research, "Report of the Committee on Academic Responsibility," in *Responsible Science: Ensuring the Integrity of the Research Process*, vol. 2 (Washington, DC: National Academy Press, 1992), 171.

22. Panel on Scientific Responsibility and the Conduct of Research, "Report of the Committee on Academic Responsibility," vols. 1, 2.

23. Ibid., vol. 1, 56–57.

24. R. P. Feynman, "Cargo Cult Science," *Surely You're Joking Mr. Feynman!* (New York: Bantam Books, 1989), 308–317.

25. The Joint Committee on Standards for Educational Evaluation, *The Program Evaluation Standards: How to Assess Evaluations of Educational Programs*, 2nd ed. (Thousand Oaks, CA: Sage Publications, 1994). Seventeen standards address reporting issues. The standards' application is not limited to educational studies.

26. Russel Schutt, *Investigating the Social World: The Process and Practice of Research*, 8th ed. (Thousand Oaks, CA: Sage Publications, 2015), 583. Also see Peter Rossi, "Half Truths with Real Consequences: Journalism, Research, and Public Policy. Three Encounters," *Contemporary Sociology* 28 (1999): 1–5.

27. Information on this case is drawn from D. Brown, "Scientists Acknowledge Flaw in 3-Drug Attack on AIDS Virus," *Washington Post*, July 23, 1993, A–3m, and L. K. Altman, "The Doctor's World: Faith in Multiple-Drug AIDS Trial Shaken by Report of Error in Lab," *New York Times*, July 27, 1993, B6. The research paper and the follow-up letter were published in *Nature* by M. S. Hirsch and his colleagues, researchers at Massachusetts General Hospital.

28. Reporting of significant errors is part of the American Psychological Association's Ethical Principles and Code of Conduct (2002), Section 8.10(b). M. McGue, "Authorship and Intellectual Property," in *Ethics in Research with Human Participants*, eds. B. D. Sales and S. Folkman (Washington, DC: American Psychological Association, 2000), discusses the author's responsibility in reporting errors.

29. Altman, "The Doctor's World," B6.

30. Schutt, *Investigating the Social World*, 582.

31. Ibid., 582–583.

32. Panel on Scientific Responsibility and the Conduct of Research, "Report of the Committee on Academic Responsibility," vol. 1, 47.

33. F. L. Macrina, *Scientific Integrity*, 3rd ed. (Washington, DC: American Society for Microbiology Press, 2005), Chapters 9 and 11 cover data ownership and data records.

34. B. A. Nosek et al., "Promoting an Open Research Culture," *Science* 348, no. 6242 (2015): 1422–1425, doi:10.1126/science.aab2374.

35. "Data Access and Research Transparency (DA-RT): A Joint Statement by Political Science Journal Editors," *Political Science Research and Methods* 3, no. 3: 421, https://doi.org/10.1017/psrm.2015.44.

Terms for Review

executive summary
literature review
fabrication
falsification
plagiarism
peer review
blind review
poster session
blogs
social network
thread
podcast

Questions for Review

The following questions should indicate whether you have a basic competency in this chapter's material.

1. A study of citizen satisfaction with an urban police force found that residents living in the downtown area are markedly less satisfied with police performance than other residents. Based on the data, the researchers recommend that the city increase police visibility in downtown neighborhoods.

    a.   Identify possible audiences for the report.
    b.   How might an oral briefing to police administrators differ from an oral briefing to the city council?
    c.   Contrast the information the researcher would include in the following three presentations: (i) an oral briefing to the city council, (ii) the information in the final project report, and (iii) the information in an article submitted to *Public Administration Review*.

2. Create a checklist for researchers to use to make sure that they have covered the necessary topics in a presentation.
3. Examine three different quantitative research articles. Compare and contrast (a) their content and organization and (b) the literature review as presented.
4. To study the benefits of educational programs in prisons, researchers surveyed inmates at five institutions. The surveys were administered by prison staff. The surveys from one institution show little variation, and the researchers feel certain that the data were fabricated. Should the researchers include the data in their analysis? What should they say about the data in their report?
5. Write a paragraph on plagiarism. Plagiarize this text without using a direct quotation.
6. A state agency plans to contract with a victims' advocacy group to collect and analyze data on victims' experiences with criminal-justice agencies. Make recommendations on who should keep the data, where, and for how long.

Problems for Homework and Discussion

1. Find an article of interest in a public administration journal such as the *Public Administration Review*.

    a.   Write an executive summary for the article.
    b.   Prepare a 10-minute oral briefing describing the research.

c. Participate in a debriefing session in which one person or a team presents the research and others ask questions. The questioners should take on appropriate roles—for example, policy analysts, political representatives, community activists.

2. Read an assigned article and see if and how the authors included the information on your checklist (see Review Question 2 in the previous section).
3. Write two memos, "Guidelines for Effective Oral Presentations of Research Results" and "Guidelines for Effective Written Presentations of Research Results." Work with your classmates to develop a class guide.
4. If you have worked on a quantitative study this semester, prepare an oral presentation that includes visual aids.
5. Find out the data storage and retention policy or practices in your academic department or place of employment.
6. Find out the data storage and retention policies for an academic journal.
7. Some organizations seem to keep data and data collection instruments for a long time, much longer than would be reasonably useful. Other organizations seem to discard data and records at the earliest possible moment. Why? In what ways are these organizations likely to differ? How and why are their goals and motives likely to differ?
8. Find a Twitter, Instagram, or Facebook post (or thread) from a government or government researcher. What is the point that they are trying to make? What data are used in the post? Is the post accessible? Does the post link to a longer version of the research?

Working With Data

Using the information in the "Focus Group Report" accompanying this text, complete the following steps. The course instructor will assign the class members to teams of three. Half of the teams will prepare "conference presentations" and deliver them to the class. The other half of the teams will prepare poster presentations. Then members of the class, as a group, will discuss which method of reporting they prefer and which they think is more effective.

Recommended for Further Reading

For information on effective oral and written research presentations, refer to technical communications texts. Most university libraries have a large number of suitable texts.

Elliott, Deni, and Judy E. Stern, eds., *Research Ethics: A Reader* (Hanover, NH: University Press of New England, 1997); this book was designed for use by engineers and scientists in a graduate course on research ethics. It includes articles on scientific misconduct and reporting research issues. A similar text, Macrina, Francis L., *Scientific Integrity: An Introductory Text with Cases*, 3rd ed. (Washington, DC: American Society for Microbiology, 2005), has interesting, readable chapters on record keeping, data ownership, peer review, and relationships with mentors.

Oliver, Paul, *The Student's Guide to Research Ethics* (Philadelphia: Open University Press, 2003), Chapter 8, has an accessible discussion of academic publishing and issues associated with plagiarism.

Strunk, W., Jr., and E. B. White, *The Elements of Style*, 3rd ed. (New York: Palgrave Macmillan, 1979), is a book often recommended to writers. Chapters 2 and 5 are especially valuable. Strunk and White's book also is available at www.bartleby.com/141/.

Students interested in ethical practices in research should consult Responsible Science: Ensuring the Integrity of the Research Process, vols. 1 and 2, *Report by U.S. Committee on Science, Engineering, and Public Policy, Panel on Scientific Responsibility and the Conduct of Research* (Washington, DC: National Academy Press, 1992).

Torres, R. T., H. S. Preskill, and M. E. Piontek, *Evaluation Strategies for Communicating and Reporting: Enhancing Learning in Organizations*, 2nd ed. (Thousand Oaks, CA: Sage Publications, 2005), cover strategies for effectively communicating findings and common reporting formats.

For more information on how researchers are using social media to promote their research, see Jaring, P., and A. Bäck, "How Researchers Use Social Media to Promote Their Research and Network with Industry," *Technology Innovation Management Review* 7, no. 8 (2017). Available at https://timreview.ca/article/.

Glossary

Accuracy. A measure of the size of the sampling error, indicating how close a sample's finding is to the parameter.

Aggregate data. Data on groups or jurisdictions such as cities, counties, or organizations that are the result of a measure of smaller units within the larger group.

Alpha level. The probability that the investigator will reject as false a true null hypothesis. The probability of committing a Type I error.

American Community Survey. An annual survey initiated in January 2005 by the U.S. Census Bureau to collect social, economic, housing, and demographic data.

Analytic mapping. Plotting the values of one or more variables on a map to analyze the relationships between variables and geographic location.

Anonymity. Collecting information so that researchers cannot link any piece of data to a specific, named individual.

ANOVA (analysis of variance). The primary statistical tool for analyzing experimental data and the differences between group means. Provides information on the statistical significance of a relationship.

Applications program. Software designed for a specific activity or set of activities other than operating the computer. Statistical packages, spreadsheets, and database managers are examples of application programs.

Arithmetic mean. A measure of central tendency determined by adding together the values of a variable for all cases in the distribution and dividing this sum by the total number of cases. The use of the arithmetic mean requires that data be measured at the interval or ratio level.

Array. A listing of the values for each variable for all cases.

Associated probability. The probability that a specific value of a test statistic will occur if the null hypothesis is true; usually combined with other evidence to make inferences about hypotheses.

Asymmetric measure. A measure of association between two variables which may have different values depending on the variable designated as the independent variable.

Autocorrelation. A nonrandom relationship among the values of a variable at different time periods violating the assumption that residuals are independent of one another; may be eliminated if the researcher can identify and include an appropriate independent variable.

Average deviation. Calculation obtained by adding the absolute value of the deviation of each case from the mean of the distribution and dividing by the number of cases.

Bar graph. A graph showing the variable and its values or categories along one side and a scale for the frequency or percentage of cases along the other. The length or height of a bar indicates the number or percentage of cases with each value of the variable. The width of a bar does not have meaning.

Behavioral public administration. An emerging way to study public administration that utilizes concepts and methods from psychology to improve our understanding of individual behavior and attitudes.

Beta weights. Standardized regression coefficients, calculated by first standardizing the measure of all variables so that they are in standard units. Beta weights indicate the relative influence of each of the independent variables on the dependent variable.

Between-group variances. Indications of how the mean and variances of each group differ from the other groups.

Bias. A systematic difference between a sample statistic and the population parameter it is to estimate. Because of a flaw in the sampling design, its implementation, or the data collection procedure, the sample either systematically under- or overestimates the population parameter. In survey research, it includes inaccurate information resulting from unreliable or invalid questions.

Biased question. A survey question that elicits inaccurate information because it is worded in such a way that respondents are encouraged to give one answer rather than another. (See Loaded question)

Big data. Massive amounts of data so large that they are difficult to process using traditional software.

Bivariate distribution. The joint distribution of the values of two variables.

Bivariate statistics. Statistics that summarize the association between two variables.

Blind review. Review of a report or study when the reviewer does not know who conducted the research.

Block numbering area. Smallest statistical area defined by the U.S. Census Bureau; a well-defined piece of land bounded by a street, railroad, or similar physical feature.

Blog. A word condensed from the term "web log." A regularly updated web site maintained by an individual or organization. Usually written in an informal style.

Box plot. A graphic technique that illustrates the median of a variable, its minimum and maximum values, quartile locations, inter-quartile range, and, in some cases, outliers. It gives a quick view of both central tendency and spread.

Case. One unit of analysis.

Case study. A type of study in which a single person, program, agency, or some other unit of analysis is examined in detail.

Categorical variables. Nominal and ordinal variables.

Cell phone. A small, portable telephone that people can carry with them.

Census block. The smallest geographic unit for which the Census Bureau tabulates 100-percent data. Many blocks correspond to individual city blocks bounded by streets.

Census of Population and Housing. The census conducted every 10 years by the U.S. Census Bureau to count the population. Required by the Constitution of the United States.

Census tract. Statistical area encompassing a large neighborhood, averaging 4,000 in population and generally with a population between 1,500 and 6,000 within county lines; every metropolitan area is mapped with census tracts.

Census undercount. The underestimating of the total size of the U.S. population. Certain groups, particularly urban minorities, are more likely to be undercounted.

Chi-square. A test of statistical significance intended for use with nominal measures. Often applied to data in contingency tables.

Circle graph. A graph or pie chart showing the proportion of a total made up by each component. The size of each wedge of the circle indicates what proportion of the whole it encompasses.

Class interval. The grouping of values of a variable.

Classical experimental design.   A design with at least one experimental group and one control group, with subjects randomly assigned to each by the investigator, with the independent variable under the control of the experimenter, and with a pretest and a posttest.

Closed-ended question.   A type of survey question in which the respondent is given a list of possible answers and is requested to select an answer or answers from that list.

Cluster sampling.   Probability sampling in which groups or jurisdictions comprising groups of units are randomly selected for a sample.

Coding.   The process of converting information collected by surveys or other means into symbols, usually numbers, for storage, management, and analysis.

Coefficient of determination.   The proportion of variation in the dependent variable that is accounted for, statistically, by one or more independent variables. Symbolized by r^2 or R^2.

Cohort.   Cases experiencing the same significant event in a specific time period.

Comparison group design.   A quasi-experimental design similar to the classical experimental design. However, the subjects in each group are not randomly assigned by the researcher, and the occurrence of the independent variable may not be controlled by the researcher.

Computer-assisted telephone interviewing (CATI).   Administration of surveys in which the interviewer reads items from a computer terminal and keys in the responses. The computer paces the interview by branching for contingency questions and keeping track of the number of calls made and which numbers were called.

Concept.   An abstract or general characteristic.

Conceptual definition.   A definition of a concept or variable in terms of other concepts. Essentially a dictionary definition of a concept or variable.

Confidence interval.   An interval placed around the value of a sample statistic. The investigator expects the value of the corresponding parameter to be located within the confidence interval.

Confidence level.   The confidence that an investigator has that a sample estimate is within a specified range of the parameter.

Confidentiality.   Protection of information, so that researchers cannot or will not disclose records with individual identifiers.

Constants.   Elements in a research model that do not vary.

Content-based evidence of validity.   Evidence that the items included in the operational definition are relevant to the concept being measured and adequately represent it.

Contingency table.   A table showing how the distribution of one variable is related to the values of one or more other variables.

Control group.   The group that is not exposed to the independent variable but is similar to the experimental group in all other respects.

Control variable.   A variable included in an analysis to determine whether it affects the relationship between two other variables. The values of the control variable are "held constant," while the relationship between the other two variables is analyzed.

Convenience sampling.   Nonprobability sampling in which units are chosen primarily on the basis of their availability.

Correlation coefficient.   Usually meant to refer to the Pearson product-moment correlation coefficient, r, it is a measure of the strength and direction of association between quantitative variables. The correlation coefficient is also a test of the goodness of fit for a regression equation.

Covariation.   The patterned relationship between an independent and dependent variable.

Cramer's *V*.   A measure of the strength of association between two nominal variables. Typically used for contingency tables, it is based on the chi-square statistic.

Criterion-based evidence of validity.   Statistical evidence that compares the data produced by a measure with the data produced by an alternative measure.

Cross-sectional design. A type of research design with all relevant variables measured at the same time.

Cross-walk. A linking of different files in a relational data base with a common element in each file.

Current Population Survey. An extensive survey conducted monthly in person or by telephone by the U.S. Census Bureau to gather current population labor-force data.

Cyclical variations. Changes in a variable recurring over time at regular intervals, usually of one to five years.

Database. A set of related data records storing information shared by several users for multiple purposes.

Database management system. An application program to enter, store, organize, and retrieve data. Some database programs can relate data from separate files.

Data control log. A record listing users' names and when they entered, used, or accessed data.

Data dictionary. A component of a database manager that names and describes each element and its location in the database, how it can be obtained, and the code for each value.

Data snooping. Using a computer program to relate many variables in a set of data to every other one with no model or hypothesis to guide the effort.

Data visualization. Presentation of data in pictorial or graphic format such as charts and graphs.

Data warehouse. A relational database that links files and provides managers and analysts the ability to query records and conduct statistical analysis.

Decennial Census of Population and Housing. The U.S. Census Bureau's major effort of counting the population every 10 years.

Deductive disclosure. When information in a research project can be used to sort through the data to identify a specific person or otherwise compromise the privacy of the subjects.

Degrees of freedom. A value needed for many statistical significance tests. This is the number of parameters or values that can vary independently of others. For many tests, it is based on the number of cases.

Demographic questions. Survey questions that ask about a respondent's age, race, education, occupation, religion, and so forth.

Dependent variable. One of the variables in a hypothesis that represents or measures the characteristic or event being explained. It is sometimes referred to as an "outcome" or an "effect."

Descriptive statistics. Statistics used to describe and summarize a dataset.

Dichotomous variable. A variable that has only two possible values.

Direct relationship. A relationship between variables in which an increase in the value of one variable is associated with an increase in the value of the other variable.

Discriminant analysis. Also known as discriminant function analysis, this is a type of statistical analysis similar to regression in which the dependent variable is a dichotomous variable.

Disproportionate stratified sampling. A probability sampling procedure in which the sample comprises a larger percentage of the units in some strata than of others.

Dummy variable. A dichotomous variable, usually called a dummy variable when included in a regression equation.

Effect size. A measure quantifying the impact of an independent variable on a dependent variable.

Effects of a control variable. The addition of a control variable may show that the relationship between two variables (1) stays the same, (2) is stronger for some values of the control variable than for others, (3) changes direction, or (4) disappears.

Elements. The variables that are identified for inclusion in a model.

Empirical validity. Empirical demonstration that an instrument measures what it has been designed to measure. Predictive validity is a type of empirical validity.

Enumeration.   The process of ordering data by variable value and grouping and counting the number of cases with similar values.

Equivalence (of a measure).   Different investigators using the same measurement procedure would produce the same results. Different versions of the same measuring procedure should produce the same results.

Eta.   A measure of association between a nominal or ordinal and interval variable; used with analysis of variance.

Executive summary.   A section located at the beginning of a report that highlights its content and may include recommendations. The intended audience is the executive who has little time to read reports; thus, this section is designed to be understood independently of the report.

Expected frequency.   The number of cases expected to have a particular value or set of values if the null hypothesis is true.

Experiment.   A type of study in which randomly assigned subjects are exposed to a deliberately manipulated treatment and compared to other subjects, also randomly assigned, who are not exposed to the same treatment.

Experimental design.   A type of design in which the researcher can assign subjects to different research groups and control who is exposed to the independent variable, when they are exposed to it, and the conditions under which the experiment takes place.

Experimental group.   The group that is exposed to the independent variable.

External validity.   The extent to which a study produces evidence that the findings of a study apply to cases not in the study.

_F_-test.   A statistical test of significance that evaluates the ratio between the total amount of variance in a variable and that explained by an independent variable or set of independent variables. Commonly used for analysis of variance (AVOVA) procedures. It is named after its developer, Sir Ronald Fisher.

Fabrication.   Making up, or inventing, the data or results of a research effort.

Factor analysis.   A statistical procedure for identifying a small number of concepts underlying a large number of related measures and for reducing the larger number of measures to a few indices. Useful in identifying measures to include in indices and in creating indices.

Factor loading.   A component of factor analysis that indicates how closely each individual item is associated with the underlying concept or factor.

Factor score.   The value of a case or unit of analysis for an index developed by factor analysis.

Factor score coefficient.   Weights for each item in factor analysis that are multiplied by a case's value on that item to arrive at a value on the index for each case.

Factorial design.   A factorial design is an experimental design used to efficiently test the impact of more than one intervention—independent variable—on one dependent variable. (See Two by two design.)

Falsification.   Changing the data or results of a report or research by a researcher; may be more ambiguous than fabrication, because altering information may involve dropping a case or group to tailor the report to the researcher's needs.

Federal Register.   A publication of the federal government listing regulations and proposed regulations of federal agencies. Used in this text in referring to regulations of the Department of Health and Human Services regarding research on human subjects.

Field.   The columns occupied by the data for a variable.

File.   A set of records.

Filter question.   A type of survey question used to identify respondents who should answer specific follow-up questions.

Fixed format.   Data entered in such a way that the values for each variable are in the same location in the data record for each case.

Focus group. A research tool utilizing small-group interviews to obtain qualitative data as well as items for questionnaires or surveys. Group interaction is an important component of a focus group.

Forced-choice question. A closed-ended question that requires the respondent to choose among available options with no provision for an "other" or "none of the above" response.

Forms manager. A computer program that takes data entered from a terminal and sets it up in a preset format in a record.

Frequency distribution. A table listing the variable values along with the number of cases with each value.

Frequency polygon. A line graph showing the frequency distribution of quantitative variables. A horizontal axis shows the values of the variable, and the vertical axis is a scale showing the frequency. The height of the polygon at any specific place indicates the number of cases with a particular value.

Gamma. A measure of the strength and direction of association between two ordinal variables.

Geographic information system (GIS). A system of hardware and software that integrates computer graphics with a relational database to manage data about geographic locations. A GIS can display information on maps it generates to show the relationship between variable values and location.

Geometric mean. A measure used to calculate an average for rates of change measured over several successive time periods.

Goodness of fit. Evidence of how well the statistical model describes a dataset. Evidence of goodness of fit for linear regression includes a linear scatterplot, the size of the correlation coefficient, and the standard error of the regression coefficient.

Hardware. Computer hardware is the computer itself and peripheral equipment such as disk drives, printers, and other storage and input devices.

Histogram. A graph similar to a bar chart except that both the length and width of the bar have meaning. The widths represent the range of values included in a category; histograms can be used for quantitative (interval and ratio) variables.

Human subject. A living individual about whom an investigator obtains data through intervention or interaction with the individual or by obtaining identifiable private information.

Hypothesis. A tentative explanation for an observation, phenomenon, or problem that can be tested empirically. Typically a hypothesis specifies the relationship between two variables.

Hypothesis testing. Also called significance testing, it is a procedure used to determine the probability that a statement, called a null hypothesis, about variables in a population is false given data from a sample of cases selected from that population.

In-person interviewing. Also known as face-to-face interviewing. A form of survey research in which the interviewer meets personally with the respondent and records the answers as the respondent provides them.

Incidence. Common rate measure referring to the number of people who get a disease over a specified period of time, usually a year. Also applicable to other situations, for example, being a victim of a crime or having an accident.

Independent variable. One of the variables in a hypothesis that is used to explain the variation in the characteristics or event of interest. It is sometimes referred to as an "input" or "cause."

Index. A set of variables combined to measure a more abstract concept.

Index number. A number expressing the relationship between two figures, one of which is the base, that is used to describe changes over time in such things as prices, production, wages, and unemployment. The Consumer Price Index (CPI) is a well-known index number.

Inferential statistics. Statistics used to estimate the value of population characteristics from data obtained from a probability sample.

Informed consent. The principle that (1) prospective subjects of research must be informed of the purpose of the research and of any risks and benefits that may be incurred by participating and (2) the subjects must give their explicit consent prior to participating in a study.

Institutional review board (IRB). An internal body created by an institution that receives federal money for research involving human subjects; the IRB reviews all institutional research involving human subjects to determine whether it conforms to ethical practices.

Intensive interviewing. An in-depth, lengthy, and extensive interview of respondents. It usually requires a skilled interviewer using an unstructured format. Typically a study using this technique involves relatively few respondents.

Internal consistency. The extent to which all items in a measuring procedure relate to the concept measured.

Internal validity. The extent to which a design provides evidence that a specific independent variable caused a change in a dependent variable.

Internet surveys. A type of research design where surveys are posted on the Web or sent as part of an e-mail message.

Inter-quartile range. The range of values encompassing the middle one-half of the observations in an ordered distribution.

Inter-rater reliability. The extent to which two different observers using the same instrument to measure a concept obtain the same results.

Interrupted time-series design. A quasi-experimental design incorporating an independent variable other than time into the time-series design. Several measures of the dependent variable are taken before and after the occurrence of the independent variable.

Interrupted time-series design with comparison group. An interrupted time series is compared to another time series that is not interrupted or is interrupted at a different time.

Interval scales. A measurement scale that measures quantitative differences between values of a variable. Each unit of the variable has the same quantitative value as every other unit. Equal-distance intervals are measured between values of the variable.

Inverse relationship. A linear relationship between variables in which an increase in the value of one variable is associated with a decrease in the value of the other variable.

Irregular fluctuations. Changes over time in a variable that cannot be attributed to long-term trends or cyclical or seasonal variations.

Lagging (of a variable). A process in which the value of the independent variable is measured at an earlier time than the value of the dependent variable.

Lambda. A measure of association between two nominal-level variables.

Level of analysis. Either individual, in which case the data are from individuals, or aggregate, in which case the measures are of group characteristics.

Level of measurement. The level of measurement categorizes the relationships among a variable's values; nominal measures categorize variables; ordinal measures, or ordinal scales, rank them along a continuum; with interval measures the values can be added and subtracted; ratio measures have an absolute zero, and ratios of different values of ratio measures can be computed.

Likert scaling. A method of index construction, also known as summated rating. A numerical value is given to the response to each of a number of items, and the values are added or averaged to obtain a value for each case.

Linear model. A model that assumes that the relationship between the variables can be appropriately described by a straight line; if the linear model is appropriate, then a linear regression equation may be calculated.

Linear regression model. The relationship between variables when the regression equation defines a straight line; $y = a + bx; \ldots bx_n$.

Literature review. The process of reviewing studies relevant to the research question. The literature review section of a report establishes the value of a research project and how it fits in with other research. The review may document that the research addresses a question not investigated in previous studies, fills a gap in previous research, tests a model under different conditions, corrects for errors in previous research, or resolves conflicting research findings.

Loaded question. A biased question worded in such a way that the respondent perceives that only one way of answering is acceptable. (See Biased question.)

Logistic regression. Regression model in which the dependent variable is dichotomous. The regression equation calculates the probability that the dependent variable will have one of two values for given values of the independent variable or variables.

Logistic regression coefficient. Measures the predictive capability of the independent variables in a logistic regression.

Longitudinal design. A type of study utilized to collect information on each variable for two or more distinct time periods.

Long-term trend. General upward or downward movement in the values of a variable over a number of years.

Macrodata. Data aggregated by political or statistical area.

Mailed questionnaires. A type of survey in which respondents are contacted by mail and are asked to fill out a questionnaire and return it by mail.

Mainframe computer. A very large computer capable of handling large amounts of data and performing extensive analysis. Most large agencies and universities have at least one mainframe computer.

Matrix. A listing of data in columns and rows. The rows constitute cases, and the columns are fields containing the information for variables. The vertical dimension is related to the number of cases and the horizontal dimension to the number of fields.

Measurement. The process of assigning numerals to the values of variables according to a set of rules.

Measurement levels. Categorize the relationships among a variable's values; nominal measures categorize variables, and ordinal levels rank them along a continuum; with interval levels, values can be added and subtracted; ratio levels have an absolute zero, and ratios comparing different cases can be computed.

Measures of association. Statistics that measure the strength of the relationship between variables. Common ones include lambda, gamma, and the Pearson product-moment correlation coefficient (Pearson's *r*).

Measures of central tendency. Measures indicating the value of a distribution that is representative, most typical, or central. These measures include the mode, median, and arithmetic mean.

Measures of dispersion. Measures that indicate the extent to which values in a distribution are different from each other. These include the range, inter-quartile range, percentiles, variance, and standard deviation.

Median. A measure of central tendency; the value of the middle case of an ordered distribution of values. The median requires ordinal or quantitative-level measurement.

Median absolute deviation. A measure of dispersion calculated by determining the average deviation of a set of cases from the median of the distribution.

Mediating variable. A variable that transmits or link the indirect effects of an independent variable to a dependent variable.

Meta-analysis. A systematic technique utilized by researchers to analyze a set of existing studies. It is conducted to draw general conclusions from several empirical studies and to identify hypotheses that merit further testing.

Metadata.   A term for data that provides information about other data. Metadata is often listed in categories of types of data including descriptive metadata, structural metadata, and administrative metadata, among others.

Microdata.   With reference to U.S. Census data, a sample of individual records with all identifying information removed.

MIS.   Management information system.

Mode.   A measure of central tendency; the value of a variable that occurs most frequently.

Model.   A representation of reality, it delineates certain aspects of the real world as being relevant to the problem under investigation and makes explicit the relationships among these aspects; it enables the formulation of empirically testable propositions regarding the nature of these relationships.

Model building.   The process of constructing a research model to answer a research question, including selecting elements and postulating the nature of their relationship.

Moderating variable.   A variable that affects the direction and/or strength of a relationship between an independent and a dependent variable.

Multicollinearity.   A condition existing when independent variables in a regression equation are closely related to each other and their independent effects on the dependent variable cannot be estimated.

Multiple regression.   Regression analysis in which more than one independent variable is included in the regression equation and analyzed.

Multistage sampling.   Probability sampling that proceeds in at least two stages. In each stage, a grouping of units is selected as a sample, and then, from the previous group, a smaller grouping of units is selected as a sample.

Multivariate statistics.   Statistics that apply to relationships among more than two variables.

New error.   Used in the calculation of lambda, a measure of association. Equal to the nonmodal responses for each category of the independent variable.

Nominal scales.   Scales that categorize and label the values of a variable. The investigator can group cases by the variable categories but cannot order them.

Nonexperimental design.   Designs that do not control for the threats to internal validity.

Nonlinear relationship.   A distinct pattern, other than a straight line, describing the relationship between variables.

Nonprobability sampling.   Sampling done in such a way that the probability that any unit or set of units will be selected for the sample is unknown.

Nonrandom variations.   Variations in a time series that are not associated with long-term trends, cyclical variations, or seasonal variations but that can be explained by an event or a specific set of conditions.

Nonresponse rate.   The proportion of people who do not respond to a survey.

Nonsampling error (bias).   Error resulting from a flaw in the sampling design, from unreliable or invalid measures or from faulty data collection.

Normal curve.   A theoretical distribution with the following characteristics: it is bell-shaped and symmetrical; the mode, mean, and median have the same value; and a fixed proportion of the observations lies between the mean and any other value.

Null hypothesis.   A hypothesis stating that two variables are not related in the population. It is the null hypothesis that is actually tested by tests of statistical significance.

Null relationship.   Relationship that occurs if a change in the independent variable is as likely to coincide with an increase as with a decrease or no change in the dependent variable.

Numerical variables.   Variables measured on an interval or ratio scale.

Odds ratio.   The ratio of the probability of something occurring to the probability of it not occurring; ratio of p to $(p - 1)$.

Open-ended question. A type of survey question in which respondents are required to provide their own answers without a listing of possibilities from the researcher.

Operational definition. Detail of the procedures for measuring a concept or variable and for assigning values to a variable for each case.

Operational validity. The extent to which measuring procedures or instruments actually measure what they have been devised to measure.

Ordinal scale. A measuring scale that orders the values of a variable and allows an investigator to order cases based on their variable value. Ordinal scales do not, however, measure the quantitative difference between cases.

Original error. Indicates how much error there is in predicting the distribution of the dependent variable without knowing the values of an independent variable.

Outliers. In regression analysis, these are data points well outside the range of the remainder of the data.

Panel design. A type of longitudinal design examining the same cases individually at each successive period; may reveal which individual cases change.

Parameter. The value of a characteristic of a population. Analysts draw samples in order to estimate parameters.

Partial regression coefficient. Indicates the effect of the independent variable on the dependent variable while controlling for all other variables in the equation.

Pearson's *r*. A measure of the strength and direction of the association between two interval-level variables.

Peer review. A process in which individuals well versed in a particular topic read and evaluate manuscripts and proposals submitted to academic journals, conferences, or funding agencies to detect errors or weaknesses prior to funding research or publication.

Percentage. A relative frequency calculated by dividing the frequency of cases with a value of a variable by the total number of cases and then multiplying this result by 100.

Percentage change. A relative frequency that converts to a percentage the amount of change in the value of a variable from one time to another for the same case.

Percentage difference. The difference between the columns in the percent of cases in a row (category of the dependent variable). Calculated by subtracting across the columns of a contingency table; can be used to analyze a relationship between two variables.

Percentage distribution. A table showing the values of a variable and the percentage of all cases having each value.

Percentile. A value below which a certain percent of the ordered observations in a distribution are located.

Perfect relationship between two variables. Relationship in which a change in the independent variable is always associated with the same change in the dependent variable.

Pie chart. A visual chart represented by a complete circle, indicating a quantity that is sliced into a number of wedges. The graph conveys what proportion of the whole is accounted for by each component and facilitates visual comparisons among parts of the whole.

Pilot study. A small study designed to test the adequacy of a proposed data collection strategy. A pilot study should test planned measures and analysis on a sample that represents the target population.

Plagiarism. The presentation of another's ideas as one's own.

Population. A total set of units sharing at least one characteristic of interest. It is from this set that the sample is selected.

Population variability. The extent to which members of a population differ from each other on the variables in which an investigator is interested.

Power. Refers to the probability that a test of significance results in the rejection of a false null hypothesis; it is related to the sample size and the strength of a relationship.

Practical significance. The extent to which a statistical finding is important or can be applied to solve a problem.

proportional reduction in error (PRE). A measure that indicates how much knowing the distribution of the independent variable reduces the error in predicting the distribution of the dependent variable.

Predicted value. In regression analysis, it is the value of the dependent variable for a case computed by using the regression equation.

Pretest. An initial test of a proposed questionnaire given to a small group of subjects who represent common variations found in the target population.

Prevalence. Common rate measure referring to the total number of people who have a disease at a given time. Also applicable to other situations, for example, total number of people who have been a victim of a crime or have had an accident.

Privacy. An individual's ability to control the access of other people to information about himself.

Probability sampling. A type of sampling in which each unit of the population has a known, nonzero chance of being in the sample.

Program file. A computer file containing instructions to the computer for running a program.

Proportion. A relative frequency determined by dividing the number of cases with a value of a variable by the total number of cases. A proportion is a decimal figure that is less than 1.

Proportionate stratified sampling. Probability sampling in which the same percentage of the units in each strata is selected for the sample.

Purposive sampling. Nonprobability sampling in which units are selected because the investigator judges that the units somehow are representative of the population. Also known as judgmental or expert-choice sampling.

Qualitative research. Refers to research involving detailed, verbal descriptions of characteristics, cases, and settings. Qualitative research usually involves relatively few cases investigated in depth.

Qualitative case analysis (QCA) A method that integrates case studies with variable approaches usually in samples ranging from 8 to 200. It relies on Boolean algebra that requires each case to be reduced to a series of variables where context of the case and mechanisms produce an outcome.

Quality control. A set of procedures used by organizations to monitor the quality of goods produced or services delivered and to take corrective action if quality drops below a set standard. Probability sampling is important in quality-control programs.

Quantitative research. Refers to research in which values of variables are characterized by numbers. Data are summarized and analyzed with statistical techniques.

Quantities of interest. The value of a variable that researchers are interested in estimating in a population. Some concern surrounds which quantities to evaluate and how best to report them.

Quantitative scales. A more inclusive term referring to both interval and ratio scales.

Quartile. Specification of a range of values in four parts; the first quartile is that value below which 25 percent of the cases are found. The second quartile is the median, the third quartile is the value below which 75 percent of the cases are found, and the fourth quartile is the maximum value in the distribution.

Quasi-experimental design. A design having some but not all of the characteristics of the experimental design. Often used in applied-research situations when the control necessary for a true experiment is not possible or practical.

Question sequencing. Arranging the questions in a survey in such a way as to obtain the most information and to encourage the maximum number of respondents to answer.

Questionnaire design. Question wording, sequencing, placement, and length of the physical layout of the questionnaire or survey.

Quota sampling. Nonprobability sampling in which the units are chosen for the sample in the same proportion as selected characteristics are believed to exist in the population.

Random assignment. Subjects are assigned in such a way that there is no systematic difference between the groups. Each subject has the same chance as any other subject of being in either an experimental group or control group.

Random digit dialing. A method of selecting a sample of respondents for telephone interviewing. A set of telephone numbers is selected in such a way as to result in a representative sample.

Random relationship. Two variables are independent of each other; that is, a change in one variable is not associated with a change in the other variable.

Random variations. Unexplained variations that are not associated with long-term trends, cyclical variations, or seasonal variations.

Randomized posttest design. An experimental design with at least two groups and a posttest but no pretest.

Range. A measure of dispersion giving the quantitative distance between the lowest and the highest values in an ordered distribution.

Rate. A relative frequency used to standardize the occurrence of some event. Rates are calculated by dividing the frequency of occurrence of an event by a total frequency count, such as the population of a jurisdiction in which the event takes place. The result is multiplied by a base number which is a power of 10, for example, 10, 100, or 1000.

Ratio. A relative frequency found by comparing the number of cases with one value of a variable with the number of cases with another value of the variable.

Ratio scale. A measurement scale that allows researchers to rank objects on a scale and determine the exact difference between them and also to use ratios to describe relationships between the values of the scaled objects. Ratio scales have an absolute zero point. In administrative work, almost all scales that are interval are also ratio.

Reactive effects of experimental arrangements. The fact that study situations often are necessarily artificial, and the study setting itself may affect the outcome.

Record. The data from a case or unit of analysis. A line of data in a computer file.

Rectangular array. A method of organizing data in which each case or unit of observation is represented by a row and each variable by one or more columns, and each row is the same length.

Regression coefficient. The value in the regression equation (b or slope) that indicates how much the dependent variable changes with a unit increase in the independent variable.

Regression constant. Often symbolized by the letter a. It is a component of the regression equation and is added to the value of the slope times the value of the independent variable to obtain a predicted value for the dependent variable. It is also the Y intercept and is the value produced by the regression equation when the value of X is zero.

Regression equation. An equation computed from the data in a specific dataset to find the line that best describes the set of data points. The equation relates the values of the independent to a dependent variable.

Relational database. A set of files in which two or more files have some variables in common, allowing information from separate files to be linked for analysis and reports.

Relationships in model. The links between elements in a research model.

Relative frequency. The number of times that a value of a variable occurs divided by another frequency count such as the total number of cases, the frequency of another value of the same variable, or the frequency of a second variable. Percents, ratios, and rates are all relative frequencies.

Reliability. The degree of random error associated with a measurement. Reliable measures produce consistent or dependable data.

Replication.   Repeating a study to see if the original findings hold. Sometimes replication is done with new subjects, as is the case with experiments. It may also refer to statistical replication where the statistical analysis is performed again on the original data.

Replication data.   Are the data used to conduct a statistical replication of research findings. These data are often posted to an external replication data repository or database.

Research design.   Two meanings: (1) general plans that guide decisions as to what data to gather, from whom, when and how to collect the data, and how to analyze the information; (2) specific plans for when to collect data on dependent, independent, and control variables.

Research methodology.   The steps needed to test a research model, including plans for selecting study subjects, measuring each variable, collecting the data, and analyzing the data.

Research records.   Individual or aggregated data for the observations in a research project.

Residual.   The difference between a value of the dependent variable for a case predicted by the regression equation and the actual value of the variable for that case.

Respondent.   Someone from whom we collect information, usually through survey research procedures.

Response rate.   The percentage of survey respondents from a sample who respond to a questionnaire or who are interviewed.

Response set.   The tendency of respondents to a questionnaire or those interviewed to answer all items in the same way.

Responsible conduct of research.   Practice of scientific investigation with integrity. It involves the awareness and application of established professional norms and ethical principles in the performance of all activities.

Sample.   A subset of units selected from a larger set of the same units. The subset is used to infer something about the larger set, the population.

Sampling bias.   A systematic misrepresentation of the population by the sample. It usually comes about because of a flaw in the sample design or its implementation.

Sampling design.   The set of procedures for selecting the units from the population to be in the sample.

Sampling error.   Refers to the difference between a sample estimate and the parameter, which is attributed to the fact that any one probability sample cannot precisely estimate a parameter.

Sampling fraction.   The percentage of the population units selected for the sample.

Sampling frame.   A list of the members of the population from which the sample is actually drawn.

Sampling unit.   A unit or set of units considered for the sample.

Scale.   A measurement procedure that assigns a value to a case based on how that case fits into a pattern of indicators.

Scatterplot.   A graph of plotted points where the dependent variable (Y) is placed on the vertical axis and the independent variable (X) is placed along the horizontal axis. Examining the scatterplot is useful in deciding whether a linear model should be used to summarize the relationship between the two variables.

Schematic model.   A model that illustrates the important links and relationships between elements with lines and arrows or similar graphical representations.

Seasonal variations.   Changes in a variable over time that occur regularly during certain times of the year.

Secondary data.   Existing data that were collected for a purpose other than the given study.

Selection bias.   The introduction of error due to systematic differences in the characteristics between those selected and those not selected for a given study.

Sensitivity.   The ability of a measure to detect meaningful differences among cases.

Sensitivity (of measure of association).   A criterion for selecting a measure of association. The ability of a statistic to detect small differences in the strength of a relationship by assigning different numerical values to relationships that may have slight, even subtle, differences in strength.

Simple random sampling. Probability sampling done in such a way that the following are true: (1) each unit of the population has the same, nonzero probability of being selected for the sample as every other unit; (2) the selection of one unit of the population does not affect the probability that some other unit will be selected.

Skew. A characteristic of a frequency distribution. If the mode or median have values different from the arithmetic mean, the distribution is skewed. A negatively skewed distribution is one in which the median is larger than the mean; in a positively skewed distribution, the median is smaller than the mean.

Skewed distribution. Occurs when a distribution has a few extreme values, either high or low, affecting the arithmetic mean. A distribution will be skewed to the left, or negatively, if the extreme values are low, and to the right, or positively, if the extreme values are high.

Skip interval. The number of units in a list to be skipped when drawing a systematic sample.

Snowball sampling (referral sampling). A type of nonprobability sampling where members of a population know or are aware of each other and recommend subjects for the sample; snowball sampling is used when members of the population cannot be easily accessed by other methods.

Social media. Forms of electronic communication through which users create online communities to share information, ideas, personal messages, and other content.

Software. Computer programs and instructions necessary for computers to operate and carry out various tasks.

Somers's d (d_{yx}). An asymmetric measure of association for ordinal variables.

Sponsored research. Research financially supported by government agencies, foundations, or other organizations.

Spreadsheet. A rectangular matrix of rows and columns.

Spreadsheet program. An application program that displays a matrix on the computer screen and allows the user to enter labels, data, and mathematical formulas into the cells of the matrix. The user can manipulate data in the matrix and perform statistical and mathematical analysis.

Spurious relationship. An apparent, but untrue, relationship between two variables that occurs because of the association of a third variable to each of the original two.

Stability. The ability of a measure to yield the same results time after time, if and only if what is being measured has not changed.

Standard deviation. A measure of dispersion. It is calculated by squaring the difference between the arithmetic mean of a distribution and the value of each case, summing these values, dividing by the number of cases, and taking the square root of the result. It is used in a number of statistical measures and tests.

Standard error. A statistical measure of the size of the sampling error based on theoretical distribution of a variable for a large number of probability samples of the same size. The standard error is normally distributed.

Standard error of the slope. An estimate of how much the regression coefficient, b, is likely to vary from sample to sample.

Standard score. Also called a z-score, this expresses the values of a distribution in terms of units of the standard deviation of the distribution. It is calculated by subtracting the arithmetic mean from the value of a particular case and dividing the result by the standard deviation.

Statistic. The value of a characteristic of a sample. Investigators use statistics to estimate parameters.

Statistical significance. The probability that the value of a statistic could have been found from a population in which the value of the corresponding parameter was much different.

Statistical software packages. Application programs to perform statistical analysis of data.

Statistically significant relationship. A relationship between variables in a sample that a statistical test suggests would also be found if the entire population from which the sample was selected could be studied; the relationship in the population may differ in strength from that in the sample.

Strata. Subgroups within a larger population.

Stratified random sampling. Probability sampling in which the population is divided into subgroups (strata). A probability sample is selected from each strata. Stratified sampling can be either proportionate or disproportionate.

Structured interviewing. Surveys in which respondents are asked the same questions, in the same order, and in the same way.

Subjects at risk. A research design in which participants could experience negative consequences. Such consequences could range from permanently harmful effects to less significant reactions such as anger or humiliation.

Survey research. A type of data collection procedure in which investigators question individual subjects in face-to-face interviews, over the telephone, by sending a questionnaire to them through the mail or by posting it on the Internet.

Symbolic model. A model using words, mathematical equations, or computer programs to illustrate the links and relationships among elements.

Symmetric measure. A measure of association between two variables that has the same value no matter which variable is designated as the independent variable.

Systematic sampling. Probability sampling which has an investigator select units from a list of members of the populations using a skip interval (population size divided by the sample size) and randomly determining a starting place in the list.

Tablet (computer). A small computer that can be hand held and which provides many functions of larger desktop and laptop computers. Tablets have touchscreen panels.

Target population. The set of units to which investigators wish to apply their results.

Test-retest. A technique that requires that an instrument or test be administered to a subject at two points in time. This establishes the stability of a measure.

Text editor. A computer program allowing the user to enter information, store, edit, and retrieve it for use.

Threats to external validity. Factors that limit the ability to apply the findings of a study to cases not involved in the study.

Threats to internal validity. Factors that could be the cause of a change in a dependent variable and represent alternatives to the tested independent variable.

Time series. A set of data obtained at regular intervals for a single case; a type of line graph with the units of time displayed along the horizontal axis and the values of a variable or frequency of an occurrence on the vertical axis.

Time-series design. A design that collects data on a quantitative variable at regular intervals.

True experiment. A study in which the experimenter can assign subjects to either an experimental or control group and can manipulate the value and occurrence of the independent variable.

***t*-Test.** A test of statistical significance requiring an interval-dependent variable. It has wide application but is often used to test whether the difference between the arithmetic averages of two groups is significant.

Type I error. Rejecting a null hypothesis as false when, in fact, it is true.

Type II error. Accepting a null hypothesis as true when, in fact, it is false.

Two By Two Design [also 2 × 2]. A 2 × 2 factorial design is an experimental design used to efficiently test the impact of two interventions—two independent variables—on one dependent variable.

Unit of analysis. The object whose characteristics are measured and in which the analyst is interested.

Univariate analysis. The statistical analysis of one variable at a time.

Univariate distribution. The distribution of the values of a single variable.

Variable. A measurable characteristic that has more than one value.

Variance. A measure of dispersion and variation. It is calculated by squaring the difference between the arithmetic mean of distribution and the value of each case, summing these values, and dividing the result by the number of cases. The variance is the square of the standard deviation. It is used in other statistical measures and tests.

Vital records. Records that include primarily vital statistics.

Vital statistics. Information collected by federal, state, and local governments on births, deaths, marriages, divorces, abortions, communicable diseases, and hospitalizations.

Voluntary research participation. When potential research subjects have clear and realistic information on the benefits and risks of participation and have the ability to withdraw from a study at any time.

Weighting. A process by which different items are weighted to reflect their importance in an index.

Index